NOVEL APPLICATIONS OF THE UWB TECHNOLOGIES

Edited by **Boris I. Lembrikov**

Novel Applications of the UWB Technologies
Edited by Boris I. Lembrikov

Published by InTech
Janeza Trdine 9, 51000 Rijeka, Croatia

Publishing Process Manager Viktorija Zgela

Technical Editor Teodora Smiljanic

Cover Designer InTech Design Team

Additional hard copies can be obtained from orders@intechopen.com

Novel Applications of the UWB Technologies, Edited by Boris I. Lembrikov
p. cm.
978-953-307-324-8

For EU product safety concerns:
IN TECH d.o.o., Prolaz Marije Krucifikse Kozulić 3, 51000 Rijeka, Croatia,
info@intechopen.com or visit our website at intechopen.com.

Meet the editor

Boris I. Lembrikov is a senior lecturer at the Faculty of Electronics, Electrical and Communication Engineering of the Holon Institute of Technology (HIT), Holon, Israel. B. I. Lembrikov received his Ph.D. in Nonlinear Optics at the Technion – Israel Institute of Technology in 1996. Since then he was an invited researcher at the Haifa University, at the Max Planck Institute High Magnetic Field Laboratory at Grenoble, France, at the Technion, Haifa, Israel. Dr. B. I. Lembrikov is an author of the book "Electrodynamics of Magnetoactive Media", a number of chapters in scientific books, a large number of papers in international peer reviewed journals and reports delivered at the international scientific conferences. He actively participated in a number of research projects concerning optics of nanoparticles, optical communications, UWB communications. The main research fields of interest of Dr. B. I. Lembrikov are nonlinear optics, optical and UWB communications, nanostructures, quantum dot lasers.

Contents

Preface

Ultra wideband (UWB) communication systems are characterized by high data rates, low cost, multipath immunity, and low power transmission. They are widely used in wireless communications, networking, radar, imaging, and positioning systems. In 2002, the Federal Communication Commission (FCC) legalized low power UWB emission between 3.1 GHz and 10.6 GHz for indoor communication devices stimulating many novel UWB applications such as personal area networks (PANs), accurate tracking and location, safety and homeland security.

Since then, UWB technologies attracted a great research and practical interest. In recent years, UWB technologies have been rapidly developing. The number of scientific articles concerning different aspects of UWB technologies is enormous and hardly observable. For this reason, it is important to present to the UWB community an adequate review of novel UWB technology applications. We have tried our best in order to provide such a review in the proposed book **Novel Applications of the UWB Technologies.** The book is divided into five parts concerning the UWB communication systems, and UWB applications in PANs, medicine, radars and localization systems.

Part 1, **General Aspects of UWB Communication Systems,** includes chapters 1 - 6 describing the general problems of UWB communication systems such as modulation formats, transmitter and receiver architecture, UWB communication system performance.

Part 2, **Novel UWB Applications in Networks,** includes chapters 7 - 10 related to novel UWB applications in wireless personal area networks (WPANs), wireless sensor networks (WSNs), femtocells, and vehicles.

Part 3, **Novel UWB Applications in Cognitive Radio Systems,** consists of chapters 11 - 13 where the problems of UWB cognitive radio are considered.

Part 4, **Novel UWB Applications in Medicine**, includes chapters 14 - 16 where the novel UWB technology applications in medicine are presented.

Finally, part 5, **Novel UWB Application in Radars and Localization Systems**, consists of chapters 17 - 20 and describes the UWB radar and localization problems.

Consider briefly the outline of the chapters.

In Chapter 1, the multiband orthogonal frequency division multiplexing (MB-OFDM) modulation and demodulation are considered. In order to optimize the UWB system performance the authors proposed a cost-effective and high performance modulation scheme based on quadrature phase shift keying (QPSK) and dual carrier modulation (DCM).

In Chapter 2, the time hopping (TH)-UWB system model based on orthogonal pulse waveform is presented. The chapter contains a detailed analysis of the pulse based modulation schemes over multipath channel and power spectral density (PSD) for pulse shape modulation (PSM), bi-orthogonal PSM (BPSM), and combined modulation scheme of orthogonal pulse position modulation (OPPM)-BPSM.

In Chapter 3, a fully integrated CMOS 6-9 GHz 9 bands transceiver for double carrier (DC) OFDM UWB system realized on the chip is described in detail. It is shown in particular that 9 carrier frequencies from 6336 MHz up to 8712 MHz with a frequency gap of 264 MHz are available.

In Chapter 4, a novel simulation methodology of impulse radio (IR) UWB energy detection receiver has been developed.

In Chapter 5, the correlation properties of the time-hopping (TH) sequences for TH spread spectrum (THSS) UWB systems are analyzed. The theory of TH sequences is developed and a method to improve TH sequences correlation properties is proposed.

In Chapter 6, the problem of UWB system performance in single-user and multi-user environments is discussed. A fine synchronization algorithm for pulse amplitude modulation (PAM) and pulse position modulation (PPM) UWB signals with a spread spectrum involving TH is proposed.

In Chapter 7, a comprehensive review of the latest developments in the field of high-speed WPANs is provided. UWB technology is especially promising for WPAN applications due to high bandwidth and small communication ranges.

In Chapter 8, the UWB technology applications in WSNs are analyzed. UWB technology manifests high robustness to interference and provides low complexity and low energy consumption receivers and transmitters for WSNs.

In Chapter 9, a novel concept of the fourth generation femtocell is proposed. The system is based on a recently developed UWB-radio-over-optical-fiber (UROOF) technology. It is characterized as a green femtocell since its energy consumption is significantly lower as compared to wireless systems.

In Chapter 10, a novel interesting application of UWB technology is described that combines the advantages of the low cost radio-over-fiber (RoF) links and wireless propagation of UWB radio signals inside a vehicle. UWB is able to provide high data rates while RoF technology extends the UWB radio transmission over comparatively long distances in trains, trams and airplanes.

In Chapter 11, the UWB communication is presented as a promising candidate for cognitive radio (CR) technology. CRs are the intelligent radios adopting itself by sensing and learning the radio environment in order to optimize the transmission strategies. Several possible scenarios and applications for the UWB based CR are discussed.

In Chapter 12, the detection-and-avoidance (DAA) cognitive UWB scheme is proposed. It is shown that DAA as a CR scheme is effective for MB UWB group.

In Chapter 13, the design and performance of the UWB cognitive radio system with DAA technique are investigated.

In Chapter 14, the UWB applications in medicine are reviewed in detail. The UWB systems for the 3D localization, the research systems, the aspects of the microwave radiation interaction with biological tissues, and state-of-the-art in wireless medical systems are discussed. Novel experimental results for electromagnetic interference in the operating room are presented.

In Chapter 15, the experimental aspects of the breast cancer detection based on the UWB imaging are discussed.

In Chapter 16, a novel efficient method for the UWB cancer detection is proposed. A corresponding mathematical model is developed and successfully applied for the signal reconstruction.

In Chapter 17, the ground penetrating radar (GPR) operating at frequencies of about 0.5-1.6GHz is considered theoretically. The obtained numerical results describe adequately the mechanisms of subsurface electromagnetic wave scattering and antenna mutual coupling processes.

In Chapter 18, the effect of the UWB interference on the next generation weather radar has been investigated. Different scenarios have been discussed concerning the UWB antenna configurations and the environment influence on the UWB system performance.

In Chapter 19, the UWB system applications in indoor ranging/localization are reviewed. A novel algorithm for the time-of-arrival (ToA) estimation is developed, and the simulation results are checked experimentally.

In Chapter 20, the current state-of-the-art localization techniques for large-scale and single-loop networks are discussed. UWB technology attracted a significant interest due to its accurate ranging capabilities and energy efficiency. Useful models for the localization problem are presented, and numerical simulation results for UWB ranging model are obtained.

We believe that this book will be helpful to engineers and researchers occupied in the field of UWB technology.

July 2011

Dr. B. I. Lembrikov,
Faculty of Electronics, Electrical and Communication Engineering,
Holon Institute of Technology (HIT)
Israel

Part 1

General Aspects of UWB Communication Systems

1

Multiband OFDM Modulation and Demodulation for Ultra Wideband Communications

Runfeng Yang[1] and R. Simon Sherratt[2]
[1]*Dongguan Polytechnic*
[2]*University of Reading*
[1]*China,*
[2]*United Kingdom*

1. Introduction

This chapter considers the Multiband Orthogonal Frequency Division Multiplexing (MB-OFDM) modulation and demodulation with the intention to optimize the Ultra-Wideband (UWB) system performance. OFDM is a type of multicarrier modulation and becomes the most important aspect for the MB-OFDM system performance. It is also a low cost digital signal component efficiently using Fast Fourier Transform (FFT) algorithm to implement the multicarrier orthogonality. Within the MB-OFDM approach, the OFDM modulation is employed in each 528 MHz wide band to transmit the data across the different bands while also using the frequency hopping technique across different bands. Each parallel bit stream can be mapped onto one of the OFDM subcarriers.

Quadrature Phase Shift Keying (QPSK) and Dual Carrier Modulation (DCM) are currently used as the modulation schemes for MB-OFDM in the ECMA-368 defined UWB radio platform. A dual QPSK soft-demapper is suitable for ECMA-368 that exploits the inherent Time-Domain Spreading (TDS) and guard symbol subcarrier diversity to improve the receiver performance, yet merges decoding operations together to minimize hardware and power requirements. There are several methods to demap the DCM, which are soft bit demapping, Maximum Likelihood (ML) soft bit demapping, and Log Likelihood Ratio (LLR) demapping. The Channel State Information (CSI) aided scheme coupled with the band hopping information is used as a further technique to improve the DCM demapping performance.

ECMA-368 offers up to 480 Mb/s instantaneous bit rate to the Medium Access Control (MAC) layer, but depending on radio channel conditions dropped packets unfortunately result in a lower throughput. An alternative high data rate modulation scheme termed Dual Circular 32-QAM that fits within the configuration of the current standard increasing system throughput thus maintaining the high rate throughput even with a moderate level of dropped packets.

2. MB-OFDM in ECMA-368

2.1 UWB standardization

The fundamental issue of UWB is that the transmitted signal can be spread over an extremely large bandwidth with very low Power Spectral Density (PSD). In early 2002, the

USA Federal Communications Commission (FCC) agreed to allocate 7500 MHz RF spectrum in 3.1-10.6 GHz band for unlicensed use for the UWB devices (Federal Communications Commission [FCC], 2002a), and limit the UWB Effective Isotropic Radiated Power (EIRP) to -41.3 dBm/MHz (FCC, 2002b). In later 2002, Ellis *et al* (Ellis et al., 2002) published the initial requirements specification for UWB systems.

Many UWB proposals were made to converge on an agreed solution. Two clear candidates quickly emerged under the Institute of Electrical and Electronic Engineering (IEEE) 802.15.3a working party for Wireless Personal Area Network (WPAN), which were Direct-Sequence (DS) UWB (Fisher et al., 2005) and MB-OFDM (Batra, et al., 2004a). In parallel with the IEEE standardization attempted, the Multiband OFDM Alliance Special Interest Group (MBOA-SIG) forged ahead to standardize their UWB system based on MB-OFDM. A key activity in MBOA-SIG development was MB-OFDM being selected by the USB implementers forum for the new Wireless-USB Physical layer (PHY) standard (USB Implementers forum, 2005). In 2005, the WiMedia Alliance working with European Computer Manufacturers Association (ECMA) announced the establishment of the WiMedia MB-OFDM UWB radio platform as their global UWB PHY and Media Access Control (MAC) standard, ECMA-368, based on the previous MBOA-SIG proposal (Multiband OFDM Alliance, 2004) with only minor changes. A third updated version of ECMA-368 was published in December 2008 with additions for regulatory flexibility and maintained as ISO/IEC 26907 (ECMA, 2008).

ECMA-368 specifies an MB-OFDM system occupying 14 bands with a bandwidth of 528 MHz for each band. This technique has the capability to efficiently capture multipath energy with a single RF chain. The first 12 bands are grouped into 4 band groups (BG1-BG4), and the last two bands are grouped into a fifth band group (BG5). A sixth band group (BG6) containing band 9, 10 and 11 is also defined within the spectrum of BG3 and BG4, in agreement to usage within worldwide spectrum regulations. The advantage of the grouping is that the transmitter and receiver can process a smaller bandwidth signal while taking advantages from frequency hopping. Figure 1 depicts the band group allocation.

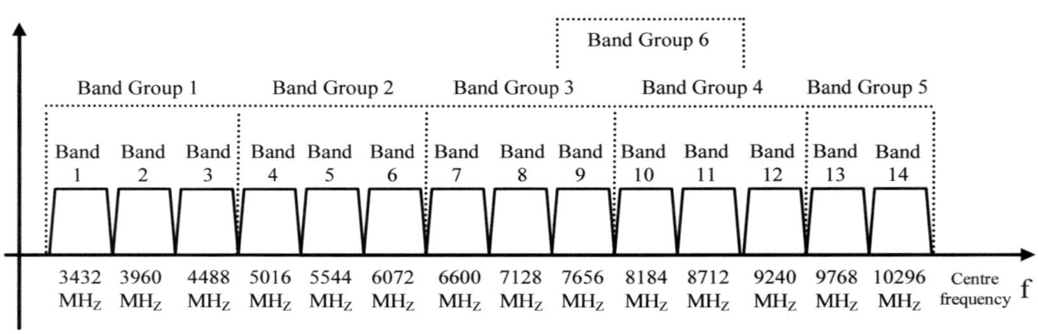

Fig. 1. Band group allocation (ECMA-International, 2008)

2.2 PHY operation in MB-OFDM

To operate the PHY service interface to the MAC, a Physical Layer Convergence Protocol (PLCP) sublayer is defined to provide a method for converting a PSDU (PHY Service Data Unit) into a PPDU (PLCP Packet Data Unit) composed from three components: the PLCP preamble (containing the Packet/Frame Synchronization and the Channel Estimation sequence), the PLCP header, and the PSDU, as illustrated in Figure 2. They are added with appropriated error detection and correction schemes to robust a communication channel as practically possible. When transmitting the packets, the PLCP preamble is sent first, followed by the PLCP header, and finally the PSDU. At the receiver, the PLCP preamble and PLCP header are processed to aid in the demodulation, decoding and delivery of the PSDU.

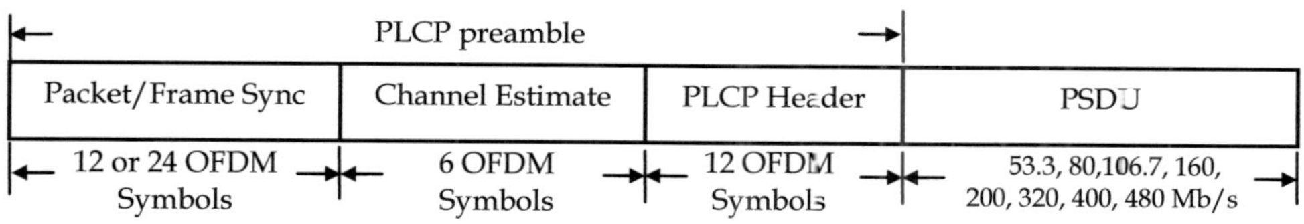

Fig. 2. PPDU structure

To transmit a PSDU that contains the information bits, ECMA-368 has eight transmission modes by applying various levels of coding and diversity to offer 53.3, 80, 106.7, 160, 200, 320, 400 or 480 Mb/s. Firstly the PSDU needs to be scrambled by the scrambler, then it is encoded by puncturing the convolutional code to achieve the appropriate coding rate, 1/3, 1/2, 5/8 or 3/4, interleaved and modulated onto a QPSK complex constellation for data rates 200 Mb/s and lower, or DCM for data rates 320 Mb/s and higher. After bit interleaving, the coded and interleaved binary data sequence is mapped onto a QPSK or DCM complex constellation. The resulting complex numbers are loaded onto the data subcarriers of the OFDM symbol implemented using an IFFT to create real or complex baseband signal. Figure 3 depicts the encoding process for the PSDU at the transmitter.

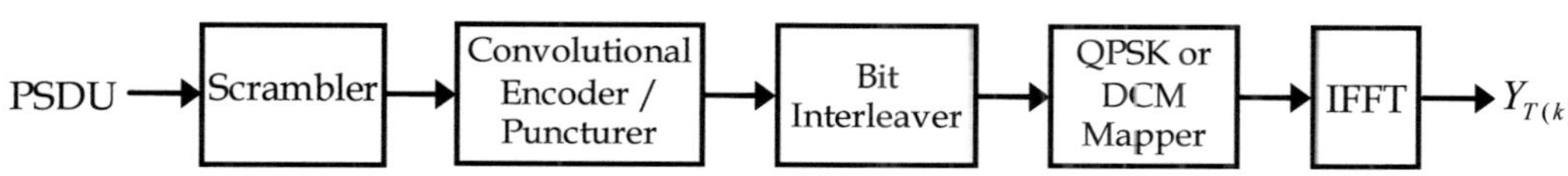

Fig. 3. Encoding process for the PSDU at Transmitter

At the heart of ECMA-368 lies a 128-pt IFFT with a 242.42ns IFFT period resulting in each IFFT subcarrier being clocked at 528MHz. The subcarriers in each OFDM symbol include 100 data subcarriers, 12 pilot subcarriers, 6 NULL valued subcarriers and 10 guard subcarriers. The 10 guard subcarriers used for mitigating Inter Symbol Interference (ISI) are located on either edge of the OFDM symbol and have same value as the 5 outermost data subcarriers. Each OFDM symbol is separated with a Zero Padded Suffix (ZPS) of 70.08ns (37 zeros as the FFT rate) to aid multipath interference mitigation and settling times of the transmitter and receiver. The IFFT/FFT operation ensures that subcarriers do not interfere with one other. Moreover, Frequency-Domain Spreading (FDS) and Time-Domain Spreading (TDS) can be used to obtain further bandwidth expansion within the OFDM modulation process depending on the coding scheme in ECMA-368.

2.3 Frequency hopping

ECMA-368 employs a frequency hopping technique in which the band is hopped by using a Time-Frequency code (TFC) known to both transmitter and receiver. The transceiver transmits a single OFDM symbol in one band, and then the next transmitted OFDM symbol is hopped to the next band. Figure 4 depicts OFDM symbols transmitted in RF signal utilizing a TFC within Band group 1 (BG1). There are two types of TFCs: Time-Frequency Interleaving (TFI), where the coded information is interleaved over three bands; and Fixed-Frequency Interleaving (FFI), where the coded information is transmitted on a single band. BG1 is a mandatory mode targeted for the first generation UWB devices.

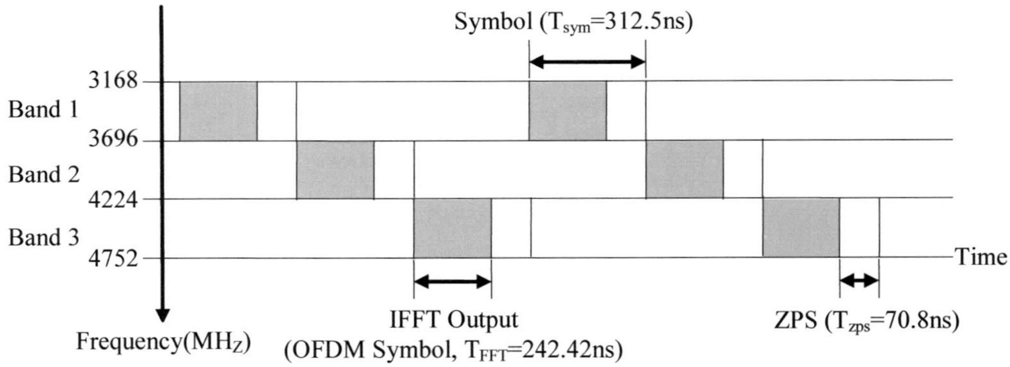

Fig. 4. OFDM symbols transmitted in RF signal utilizing a TFC within BG1

2.4 System performance measurements

Simulating system performance is an important criterion in order to compare to current literature. However, the literature on MB-OFDM system performance for measuring propagation with respect to distance is surprisingly sparse. We followed the original MBOA-SIG MB-OFDM proposal settings and adopted the assumptions (described in the following section 2.4.2) to simulate the MB-OFDM system with standard UWB channels.

2.4.1 Propagation distance measurement

The received signal power is calculated the difference between the total transmit power and path loss. Since the FCC defines the average power as 1mW per Megahertz, the total transmitted power P_{TX} can be obtained from the PSD and the operating bandwidth, as in (1) assuming no power loss at the transmitter and 0 dBi transmit antenna gain.

$$P_{TX} = -41.25 + 10\log_{10}(f_U - f_L) \text{ dBm} \tag{1}$$

where -41.25 dBm/MHz is the UWB EIRP/MHz, f_L = 3168 MHz is the lower frequency of the operating bandwidth, f_U is upper frequency varying from BG1 to BG6. However, BG1 is used for first generation of UWB devices, thus f_U = 4752 MHz is assigned. The free-space propagation model is defined under IEEE 802.15.3a, which specifies the path loss attenuating the transmitted signal as a function of the lower and upper frequencies of the operating bandwidth. The path loss P_L can be expressed as in (2).

$$P_L = 20\log_{10}(\frac{4\pi f_g d}{c}) \text{ dB} \tag{2}$$

where f_g = 3882 MHz is the geometric mean of the lower and upper frequencies in BG1. The geometric mean offers a more reasonable value for the expected path loss in the system (Batra et al., 2004b). d is the distance measured in meters between the transmitter and receiver. c = 3x108 m/s is the speed of light. As a result, the function of received signal power, as described in (3), can be derived from (1) and (2) with transmit and receive antenna gain (G_T, G_R).

$$P_{RX} = P_{TX} + G_T + G_R - P_L \text{ dBm} \tag{3}$$

2.4.2 Receiver sensitivity

Receiver sensitivity is the lowest power level at which the receiver can detect an RF signal and demodulate data. Sensitivity is purely a receiver specification and is independent of the transmitter. In ECMA-368, the minimum receiver sensitivity values in Additive White Gaussian Noise (AWGN) for different data rates are listed in Table 1 for BG1 to achieve a Packet Error Rate (PER) of less than 8% with a payload of 1024 octets each in the PSDU, where a noise figure of 6.6 dB (referenced at the antenna), an implementation loss of 2.5 dB, and a margin of 3 dB have been assumed (ECMA, 2008).

Data Rate (Mb/s)	Minimum Receiver Sensitivity (dBm)
53.3	-80.8
80	-78.9
106.7	-77.8
160	-75.9
200	-74.5
320	-72.8
400	-71.5
480	-70.4

Table 1. Minimum receiver sensitivities for BG1 (ECMA, 2008)

2.4.3 System configuration

The proposed UWB system is simulated in a realistic multipath channel environment of 100 channel realizations in the four UWB channel models CM1-CM4 (Foerster, 2003). The simulation results are averaged over at least 500 packets with a payload of 1024 octets each in the PSDU and 90th-percentile channel realization (the worst 10% channels are discarded). The link success probability is defined as the 90th-percentile of channel realizations for which system can successfully acquire and demodulate a packet with a PER (a packet is in error if at least one bit is in error) of less than 8% (Multiband OFDM Alliance, 2004).

The original MBOA-SIG proposal specifies implementation loss affecting the practical system, which includes front-end filtering, clipping at the Digital-to-Analogue Converter (DAC), Analogue-to-Digital Converter (ADC) degradation, channel estimation, clock frequency mismatch (±20 ppm at the transmitter and receiver), carrier offset recovery, carrier tracking, etc. Similarly, ECMA-368 specifies the total implementation loss of 2.5 dB and a margin of 3 dB as an assumption. It should be noted that ECMA-368 only defines the

performance for reference sensitivity, not multipath. This research will revert back to multipath test performed in the original MBOA-SIG tests when appropriate.
This research will maintain strict adherence to timing (no frequency offset and perfect OFDM symbol timing) and use a hopping characteristic of TFC = 1, and incorporate 6.6 dB noise figure referenced at the antenna and 2.5 dB implementation loss in the floating point system model. PER being a function of distance will be used as a performance indicator for the system performance measurement.

3. QPSK modulation

3.1 QPSK mapping

QPSK constellation mapping is used when data rate is 200 Mb/s or lower combing with FDS and (or) TDS techniques. The FDS and TDS modes are not only used to create the varied data rates, but also maximize frequency diversity and improve the performance. ECMA-368 offers eight data rates with various levels of coding and diversity to achieve the required trade-off between speed and reliability. For the slowest 5 out of the 8 PSDU coding schemes (summarized in Table 2), or for the PLCP header, the binary coded and interleaved input data are divided into groups of two bits, and then mapped in one of the four Gray-coded QPSK constellation points, as in Figure 5. By mapping two bits per symbol, the output symbol values *Y[k]*, as in (4), are normalized by a normalization factor of $K_{MOD} = 1/\sqrt{2}$, to have constant average symbol power.

$$Y[k] = K_{MOD} \times \left[(2 \times b[2k] - 1) + j(2 \times b[2k+1] - 1)\right] \tag{4}$$

where $k = 0, 1, 2, \ldots n$; $n = 49$ is used when FDS is enabled, otherwise $n = 99$.

Data Rate (Mb/s)	Modulation	Coding Rate (R)	FDS	TDS	Coded Bits / 6 OFDM symbol	Info Bits / 6 OFDM symbol
53.3	QPSK	1/3	Yes	Yes	300	100
80	QPSK	1/2	Yes	Yes	300	150
106.7	QPSK	1/3	No	Yes	600	200
160	QPSK	1/2	No	Yes	600	300
200	QPSK	5/8	No	Yes	600	375
320	DCM	1/2	No	No	1200	600
400	DCM	5/8	No	No	1200	750
480	DCM	3/4	No	No	1200	900

Table 2. PSDU rate-dependent parameters (ECMA, 2008)

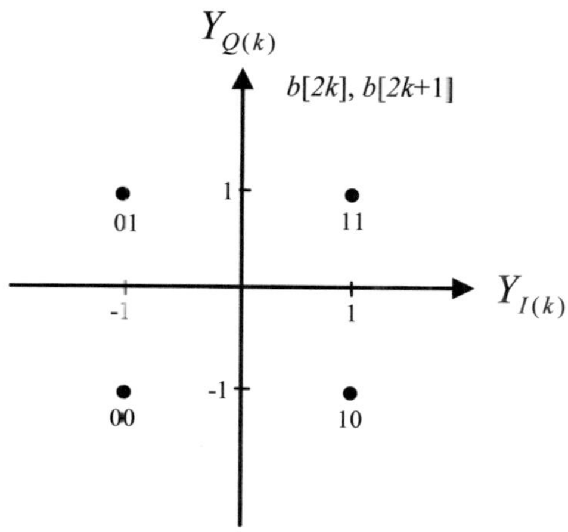

Fig. 5. QPSK constellation mapping

ECMA-368 supports TDS and (or) FDS to provide repetition of the same information (complex number) mapping over the OFDM symbols for the slowest 5 out of the 8 coding schemes. In conjunction with the FDS approach, 50 QPSK symbols and 50 of their respective conjugate values are mapped onto 100 OFDM data subcarriers as conjugate symmetric IFFT inputs. As a result, the transmitter only needs to implement the real portion of the IFFT output. As for the TDS approach, when 100 QPSK symbols are mapped onto an OFDM symbol, the copy of these 100 QPSK symbols is mapped onto the next OFDM symbol.

3.2 A dual QPSK soft-demapper exploiting TDS and guard interval diversity

Soft bit decision is used to demap the QPSK to maximize the post-QPSK baseband processing. However a soft bit based receiver will consume more memory in the deinterleaver and increase the computational complexity of the Viterbi decoder, and it will be required to store the soft bit trackback memory to search the 64 trellis states for each information bit. The soft-QPSK demapping process can improve WPAN receiver performance where each input complex number for the demapper outputs two soft bits being the symbol likelihood. The demapping process can be simply the process of outputting real part (I value) and then the imaginary part (Q value) of the input symbol for the first and second soft bit respectively. For a given code bit, a positive soft bit with large possible magnitude may indicate more confidence in '1' being sent for the code bit, while a negative soft bit with large possible magnitude may indicate more confidence in '0' being sent for the code bit, and a soft bit of zero may indicate equal likelihood of '0' or '1' being sent for the code bit. The soft bits from the QPSK demodulator are then input to the bit deinterleaver, then Viterbi decoder and descrambler to recover the original bit stream, as shown in Figure 6.

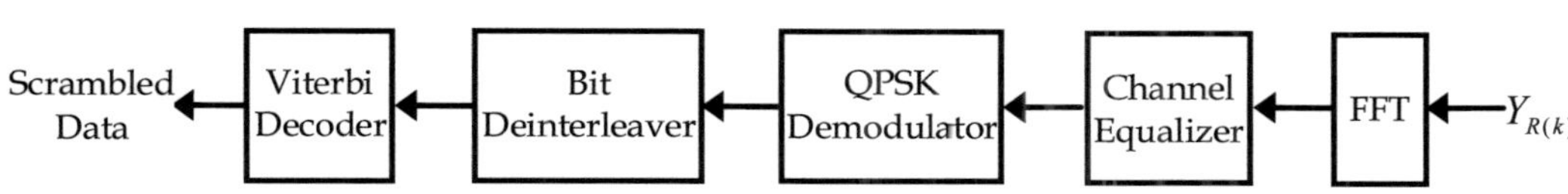

Fig. 6. Decoding process for the PSDU for low data rates or PLCP header at Receiver

3.2.1 Time-domain de-spreading and equal gain combing

As ECMA-368 implements frequency hopping, each time diversity pair in the TDS is transmitted over a different frequency band and therefore has independent channel fading characteristics. The possibility of both OFDM symbols having deep fades on the same subcarriers from different frequency bands is small. The receiver may decide to select and decode one received OFDM symbol or combine the two to maximize performance. Since the duration of an OFDM symbol is fixed, the receiver may implement a single serial decoding path for each symbol pair, or have two parallel decoding paths clocked at half the serial rate. ne main OFDM symbol and its spread OFDM symbol are received at the receiver. After the equalization, the data from those two OFDM symbols are demodulated by QPSK. Since QPSK soft demapping is used for a pair of main and spread QPSK symbols, the receiver must implement soft decision for the soft bit decoding. For two one-dimensional points (x_1, y_1) and (x_2, y_2), the Euclidean distance is computed in (5). The spread decision can be selected from calculating the Euclidean symbol distance between the corresponding QPSK constellation point to a main QPSK symbol, Ym_k, or a spread QPSK symbol, Ys_k, and then output the symbol that has shorter distance. d_m and d_s are the Euclidean distances for Ym_k and Ys_k respectively, as described in (6)-(7). Ym_k and Ys_k are the signals after the equalisation. Figure 7 depicts possible QPSK symbol pairs from the main and spread OFDM symbols.

$$d = \sqrt{(x_2 - x_1)^2 + (y_2 - y_1)^2} \tag{5}$$

$$d_m = \sqrt{\{\mathrm{Re}(Ym_k) - \mathrm{Re}(S_n)\}^2 + \{\mathrm{Im}(Ym_k) - \mathrm{Im}(S_n)\}^2} \tag{6}$$

$$d_s = \sqrt{\{\mathrm{Re}(Ys_k) - \mathrm{Re}(S_n)\}^2 + \{\mathrm{Im}(Ys_k) - \mathrm{Im}(S_n)\}^2} \tag{7}$$

where $k = 0, 1\ldots99$; S_n is the reference signal for one of the four constellation points. Then the soft value of Ym_k and Ys_k is used after deciding the Euclidean distance, as the following:

$$Soft(b_{2k}) = \mathrm{Re}(Ym_k) \qquad Soft(b_{2k+1}) = \mathrm{Im}(Ym_k), \quad if(d_m < d_s) \tag{8}$$

$$Soft(b_{2k}) = \mathrm{Re}(Ys_k) \qquad Soft(b_{2k+1}) = \mathrm{Im}(Ys_k), \quad if(d_s < d_m) \tag{9}$$

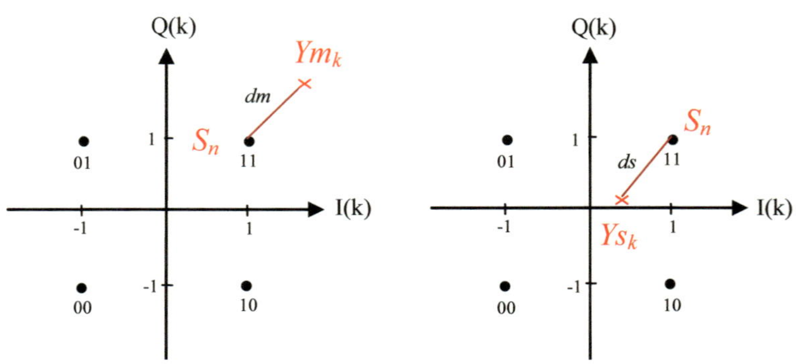

Fig. 7. A possible equalized QPSK symbol pairs from the main and spread OFDM symbols

Equal gain combing scheme can be employed to combine the spread sequence whether before the QPSK demapping process or after the QPSK demapping process. The demapping process can be integrated into the symbol decoding scheme and the combined QPSK soft-demapping and TDS symbol simply reduces to (10) and (11). The overall likelihood of a larger signal to noise ratio value is increased, thereby increasing more reliability for Viterbi decoder input. This equal gain combing scheme not only has better soft decoding performance than the spread decision, but also has much lower complexity.

$$Soft(b_{2k}) = \mathrm{Re}(Ym_k) + \mathrm{Re}(Ys_k) \quad k = 0, 1, \ldots 99 \qquad (10)$$

$$Soft(b_{2k+1}) = \mathrm{Im}(Ym_k) + \mathrm{Im}(Ys_k) \qquad (11)$$

3.2.2 Merged TDS and QPSK soft demapping architecture

Merging the TDS and QPSK soft-demapper architecture can be used to optimize the receiver performance. At the rates where TDS is implemented, the receiver architecture can process the main and spread OFDM symbols in parallel (to reduce hardware, but clock rates are often high). In this proposed dual QPSK soft-demapper, the parallel approach is assumed, as illustrated in Figure 8, but the serial approach can also be used where additional memory for holding one equalizer output buffer is required before the two equalized symbols are merged. The main and spread symbols have their own FFT and equalizer in the parallel implementation. This may be considered as a consumptive resource solution, because too much dynamic range is expensive in terms of power and floor size, particularly for the FFT. To implement the demapping and combining process, the basic operations include fetching I and Q values, add, shift, and store. The proposed dual QPSK demapper requires no memory to hold the QPSK demapping output before the add operation. Thus a reduction of 40% required memory has been achieved compared to having separate demapping and combining functions. The dual soft-QPSK demapping process and the management of guard subcarriers are implemented with zero overhead.

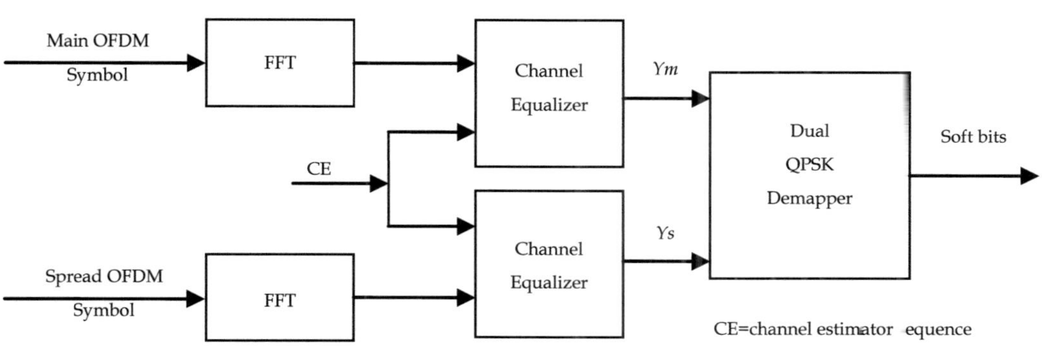

Fig. 8. Architecture of soft-output QPSK demapper with TDS symbol combining

3.2.3 Use of guard subcarrier diversity

In ECMA-368, an OFDM symbol contains 100 data subcarriers ($C_d[0..99]$), 10 guard subcarriers ($C_g[0..9]$), 12 pilot subcarriers ($C_p[0..11]$) and 6 Null subcarriers. The 10 guard subcarriers are aimed at minimizing ISI and are split into 2 sets of 5 carriers each set at either end of the IFFT input data set, and they are the same values as the 5 neighbouring data subcarriers. The power can be reduced on the guard subcarriers with the objective to relax the specifications of the analogue transmit and receive filters.

The guard subcarriers can be considered as another form of time and frequency diversity, at least to the 2 sets of 5 neighbouring data subcarriers, to improve the performance of the receiver. In this case, each of the 5 outer data subcarriers ($C_d[0..4]$ and $C_d[95..99]$) are computed as the sum of the data subcarriers and its associated guard subcarriers ($C_g[0..4]$ and $C_g[5..9]$), as described in (12). There are many Maximum Ratio Combing (MRC) (Proakis, 2001) schemes to utilize the guard subcarriers as another form of the diversity, but they are difficult to implement while considering the high data rate in ECMA-368, hence implementing as the sum of the data can simplify the computation and decoding process at the receiver, as described in the following:

$$C_d[k] = C_d[k] + C_g[n]\text{, k = 0, 1, 2, 3, 4; n = 0, 1, 2, 3, 4} \tag{12}$$

$$C_d[k] = C_d[k] + C_g[n]\text{, k = 95, 96, 97, 98, 99; n = 5, 6, 7, 8, 9}$$

3.2.4 Performance gain by exploiting guard interval diversity

By following system configuration stated in section 2.4.3, the system was simulated with exploiting guard interval diversity in conjunction with merged TDS and QPSK soft demapping architecture in the 200 Mb/s mode. Two models are presented here, an AWGN channel and channel realisation 1 in CM1. The simulation was performed both with the use of guard subcarriers and without the use of the guard subcarriers. Figure 9 depicts the performance gains in the order of 0.3 dB for utilizing the received guard symbols. At UWB frequencies of BG 1, FCC compliant transmitter power and expected distance of 10 meters then the 0.3 dB improvement equates to a propagation distance improvement to 10.4 meters.

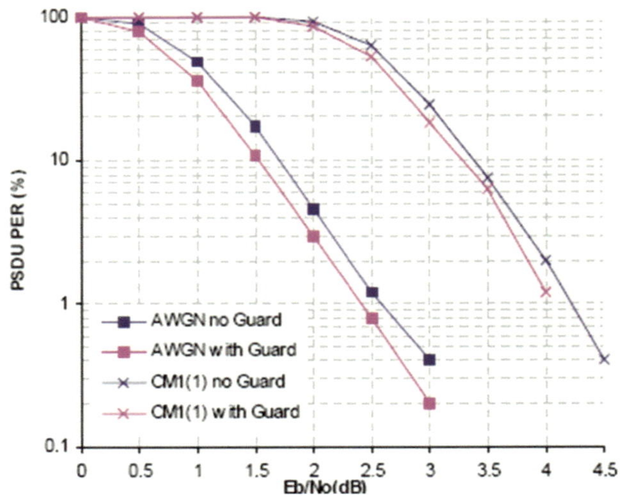

Fig. 9. PSDU PER Performance improvement from utilizing guard interval diversity in the AWGN and Channel realization 1 in CM1

3.2.5 System performance for 200 Mb/s mode

The system performance at a data rate of 200 Mb/s is simulated with using the UWB channel models to measure propagation distance. The proposed dual QPSK soft-demapper with TDS mode and guard interval diversity is used in the simulation. To obtain more accurate simulation results, 2000 packets per simulation with a payload of 1024 octets were used in each PSDU and 90th-percentile channel realization, including a noise figure of 6.6 dB and an implementation loss of 2.5 dB. The other settings remained the same as before. The system using the 200 Mb/s mode can achieve a successful link of 9.7 meters in CM1, and 9.8 meters in CM3, as illustrated in Figure 10. This performance has supported the WiMedia requirements for the extended distances of about 10 meters at the lower data rates. It is imperative to compare the system performance with current literature. As illustrated in Figure 11, the proposed system for the 200 Mb/s mode in CM1 has a better performance, while MBOA-SIG quoted 7.4 meters (Multiband OFDM Alliance, 2004) and 6.8 meters from aRenarti semiconductor (aRenarti Semiconductor, 2007).

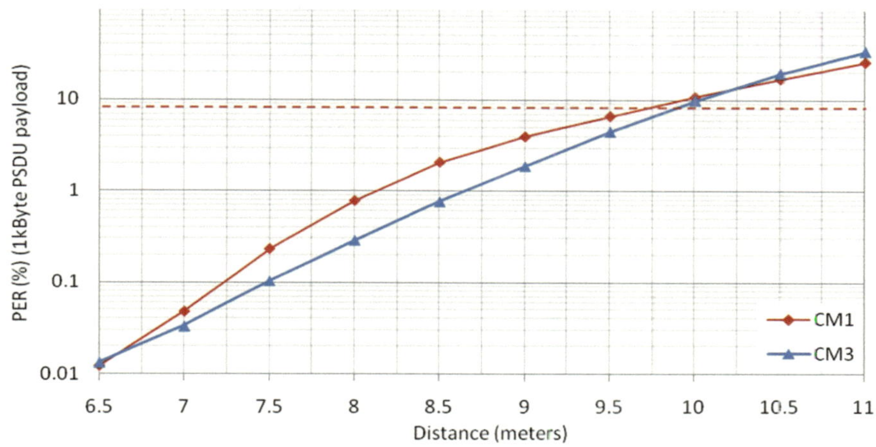

Fig. 10. System performance for 200 Mb/s mode in CM1 and CM3

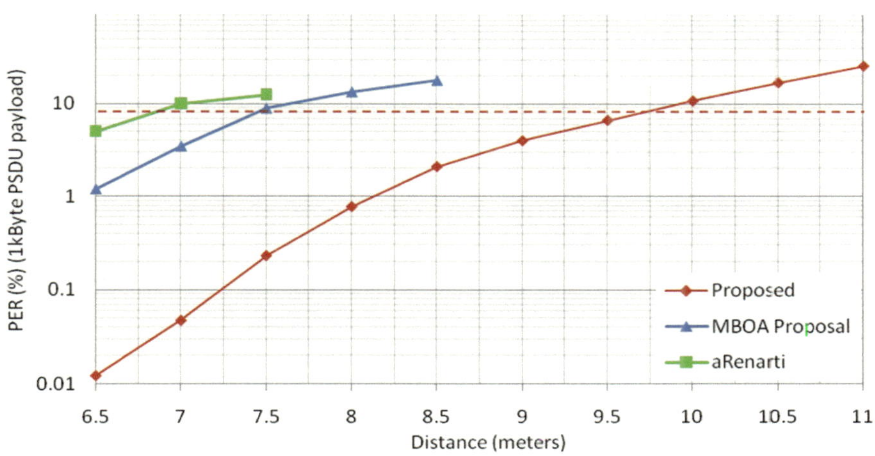

Fig. 11. Performance comparison for 200 Mb/s mode in CM1

4. Dual Carrier Modulation (DCM)

4.1 DCM mapping

DCM as a four-dimensional constellation is used for the highest 3 out of the 8 PSDU coding schemes. After bit interleaving, 1200 interleaved and coded bits are divided into groups of 200 bits, and further grouped into 50 groups of 4 reordering bits. Each group of 4 bits is represented as ($b_{g(k)}$, $b_{g(k)+1}$, $b_{g(k)+50}$, $b_{g(k)+51}$), where $k \in [0\ldots49]$ and

$$g(k) = \begin{cases} 2k & k \in [0\ldots24] \\ 2k+50 & k \in [25\ldots49] \end{cases} \tag{13}$$

These four binary bits are mapped to two QPSK symbols ($x_{g(k)}+jx_{g(k)+50}$), ($x_{g(k)+1}+jx_{g(k)+51}$) as in (14). Then the DCM mapper uses a DCM matrix H as in (15) to execute mapping of the two QPSK symbols into two DCM symbols ($y_{(k)}$, $y_{(k+50)}$) as in (16), where $1/\sqrt{10}$ is a normalization factor for normalizing the average symbol power to be a constant unit. The resulting DCM symbols are formed into two 16-QAM-like constellations, as in Figure 12.

$$\begin{bmatrix} x_{g(k)} + jx_{g(k)+50} \\ x_{g(k)+1} + jx_{g(k)+51} \end{bmatrix} = \begin{bmatrix} (2b_{g(k)} - 1) + j(2b_{g(k)+50} - 1) \\ (2b_{g(k)+1} - 1) + j(2b_{g(k)+51} - 1) \end{bmatrix} \tag{14}$$

$$H = \begin{bmatrix} 2 & 1 \\ 1 & -2 \end{bmatrix} \tag{15}$$

$$\begin{bmatrix} y_{(k)} \\ y_{(k+50)} \end{bmatrix} = \frac{1}{\sqrt{10}} \begin{bmatrix} 2 & 1 \\ 1 & -2 \end{bmatrix} \begin{bmatrix} x_{g(k)} + jx_{g(k)+50} \\ x_{g(k)+1} + jx_{g(k)+51} \end{bmatrix} \tag{16}$$

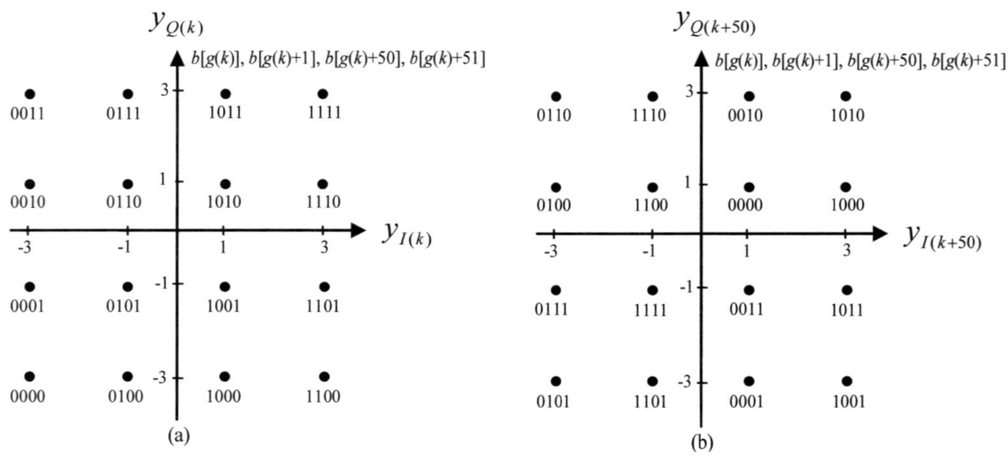

Fig. 12. DCM constellation mapping (ECMA, 2008): (a) mapping for y_k; (b) mapping for y_{k+50}

Information on a single tone is unreliable if a channel has deep fade. If the tone is completely attenuated by the multipath channel, the information will be completely lost. The probability of two tones experiencing a channel deep fade is extremely small if the two tones with the same information are separated by a large bandwidth. Frequency diversity is used in the DCM by mapping the same information but with different forms onto two different tones at different channel frequencies with a large bandwidth separation. The two resulting DCM symbols are allocated into two individual OFDM data subcarriers with 50 subcarriers separation to achieve frequency diversity. 100 DCM symbols (complex numbers) are given to the 128 points IFFT block for building an OFDM symbol. Each OFDM subcarrier occupies a bandwidth of 4.125 MHz (528 MHz/128). Therefore the bandwidth between the two individual OFDM data subcarriers related to the two complex numbers ($I_{(k)}$, $Q_{(k)}$) and ($I_{(k+50)}$, $Q_{(k+50)}$) is at least 200 MHz, which offers good frequency diversity gain against channel deep fading. Figure 13 depicts the DCM mapping process.

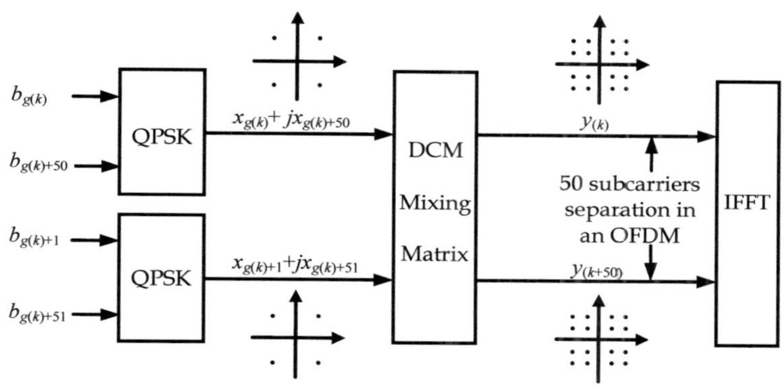

Fig. 13. DCM mapping process

4.2 DCM demapping

The receiver converts each time-domain OFDM symbol into the frequency-domain via the Fast Fourier Transform (FFT). Then, channel estimation and symbol equalization follow. The DCM demapper uses two separate subcarriers concurrently to decode the symbol pair. If one symbol within one subcarrier is lost or degraded, it can be detected, even recovered by the DCM demapper. It is required to repeatedly execute demapping of the two received DCM symbols to output groups of 200 soft bits. The soft bits from the DCM demapper are then input to the bit deinterleaver, the soft bit Viterbi decoder and then descrambled to recover the PSDU, as same as in Figure 6.

4.2.1 Channel State Information

In ECMA-368, the channel estimate section in PLCP preamble is used to form an estimate of the transmission channel, commonly termed the Channel State Information (CSI). It can be used as a dynamic estimation in the frequency-domain for data reliability in each subcarrier position (Li et al., 2005). Each OFDM data subcarrier has a potentially different CSI. The more CSI measurement that can be taken, the more reliable the CSI estimation is in the presence of thermal noise offering better decoding result. The LS algorithm having low complexity to implement is one of the popular methods for the OFDM based system channel estimation without using any knowledge of the statistics for the channels, as in (17). Furthermore, it has been shown that LS estimation has similar performance close to MMSE (Li et al., 2005).

$$\hat{H}_{LS} = \left(X_k^{-1} Y_k \right) \; (k = 0, 1, \ldots) \tag{17}$$

ECMA-368 defines 6 stored CE sequences (CE0…CE5) in blocks of 122 subcarriers contained in the 6 OFDM symbols of the PLCP preamble in order to simplify the required receiver processing. The proposed LS channel estimator uses the 6 received CE sequences with the priori CE sequences and TFCs to create 6 received merged and inversed CE sequences (channel response) by taking the average of each CE sequence sent on the same frequency. The channel estimator calculates the CE sequence with priori transmitted CE sequence. To maintain the polarity of each received CE value ($CEr = I_{rx\_CE} + jQ_{rx\_CE}$), each received CE value is divided by its expected value that are a priori CE value, CEs. The division is conducted with the stored values at 1+*j*1, 1-*j*1, -1+*j*1, -1-*j*1, hence a simple look-up table allows each division to easily sum the real part and imaginary part of the received CE value and permuting the polarity of the received CE value, as the following:

$$\begin{aligned} I_{estimated\_CE} + jQ_{estimated\_CE} &= \frac{I_{rx\_CE} + jQ_{rx\_CE}}{1+j} = \frac{(I_{rx\_CE} + Q_{rx\_CE}) + j(Q_{rx\_CE} - I_{rx\_CE})}{2} \\ or \quad &= \frac{I_{rx\_CE} + jQ_{rx\_CE}}{1-j} = \frac{(I_{rx\_CE} - Q_{rx\_CE}) + j(I_{rx\_CE} + Q_{rx\_CE})}{2} \\ or \quad &= \frac{I_{rx\_CE} + jQ_{rx\_CE}}{-1+j} = \frac{(Q_{rx\_CE} - I_{rx\_CE}) - j(I_{rx\_CE} + Q_{rx\_CE})}{2} \\ or \quad &= \frac{I_{rx\_CE} + jQ_{rx\_CE}}{-1-j} = \frac{-(I_{rx\_CE} + Q_{rx\_CE}) + j(I_{rx\_CE} - Q_{rx\_CE})}{2} \end{aligned} \tag{18}$$

Due to the assumption in MB-OFDM that a channel is linear and time invariant over each PPDU (Foerster, 2003), and also that at least 2 of the 6 channel estimation sequences will be on the same carrier frequency, the receiver may average channel estimates from the same carrier frequency to reduce the effect of Gaussian noise. Figure 14 depicts the averaged and inversed CE sequences. For the non-hopping schemes, all the 6 channel estimation bursts $\hat{H}_{CE1} \cdots \hat{H}_{CE6}$ can be averaged together and then inversed for one of the non-hopping schemes in mandatory mode (TFC=1, BG1). Finally, the average received CE sequences are inversed.

$\frac{\hat{H}_{CE1}+\hat{H}_{CE4}}{2}$	$\frac{\hat{H}_{CE2}+\hat{H}_{CE5}}{2}$	$\frac{\hat{H}_{CE3}+\hat{H}_{CE6}}{2}$	$\frac{\hat{H}_{CE1}+\hat{H}_{CE4}}{2}$	$\frac{\hat{H}_{CE2}+\hat{H}_{CE5}}{2}$	$\frac{\hat{H}_{CE3}+\hat{H}_{CE6}}{2}$
⇩					
$\hat{H}_{\overline{CE1}}$	$\hat{H}_{\overline{CE2}}$	$\hat{H}_{\overline{CE3}}$	$\hat{H}_{\overline{CE4}}$	$\hat{H}_{\overline{CE5}}$	$\hat{H}_{\overline{CE6}}$
⇩					
$(\hat{H}_{\overline{CE1}})^{-1}$	$(\hat{H}_{\overline{CE2}})^{-1}$	$(\hat{H}_{\overline{CE3}})^{-1}$	$(\hat{H}_{\overline{CE4}})^{-1}$	$(\hat{H}_{\overline{CE5}})^{-1}$	$(\hat{H}_{\overline{CE6}})^{-1}$
⇩					
H_{CE1}	H_{CE2}	H_{CE3}	H_{CE4}	H_{CE5}	H_{CE6}

Fig. 14. Averaged and inversed channel estimation sequences for TFC=1, BG=1

The CSI is estimated from each of the CE sequences transmitted on that band. The LS CSI for each equalized data is calculated from the received and stored CE sequences and given by (19). It should be noted that CEr/CEs includes both phase and amplitude information, i.e. the I and Q components of each frequency component of the sequences, whereas CSI is the modulus of CEr/CEs and therefore is a scalar term. Moreover, no division is required in the CSI calculation according to (18), where CE_r is the received CE sequence, CE_s is the priori stored CE sequence, which means the divider can be avoided in the hardware implementation, thus lowering the complexity of system implementation.

$$CSI \approx \left|\frac{CE_r}{CE_s}\right| \tag{19}$$

With the similarity of computing the channel estimation, taking the 6 CE sequences can create the 6 averaging blocks of CSI for the non-hopping schemes. Hence, averaging those different blocks of CSI can produce a more accurate CSI in the time invariant or slowly changing channel with respect to the frame time. Again, subject to the mandatory mode, TFC=1 and BG=1 is selected for the band hopping. The first block of CSI is averaged with the fourth block of CSI while the second one is averaged with the fifth one, and the third one is averaged with the sixth one. Then the new averaged CSI blocks are illustrated in Figure 15.

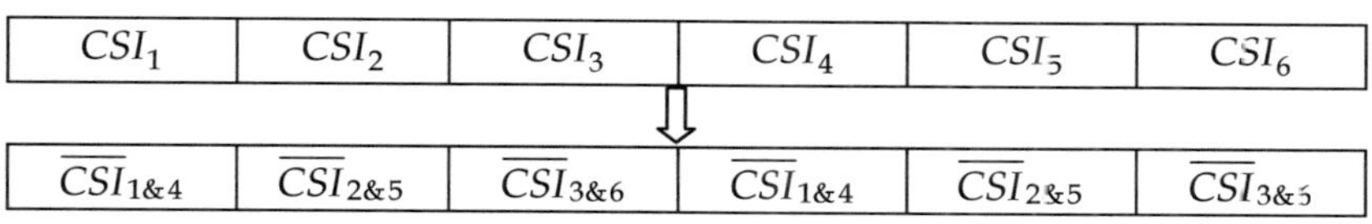

Fig. 15. Averaged CSI blocks allocation for TFC=1, BG=1

To avoid the cost of this CSI aided Viterbi decoder, the soft input of the decoding chain is obtained from the multiplication of the demodulation soft output R_m and its corresponding CSI_k, as described in (20). The receiver is arranged to modify the soft bits using the CSI, as illustrated in Figure 16. The overall data reliability is obtained from directly scaling the soft bit value by the corresponding CSI value. Therefore the reliability of received data is maximized. What is of upmost interest is to apply the CSI as a demapping technique for the MB-OFDM system at the higher data rates, where the DCM modulation scheme is used.

$$Softbit_m = R_m \times CSI_k \tag{20}$$

where m is the index of the numbers of soft bit value depending on the modulation scheme; k is the index into the 100 data subcarriers in an OFDM symbol.

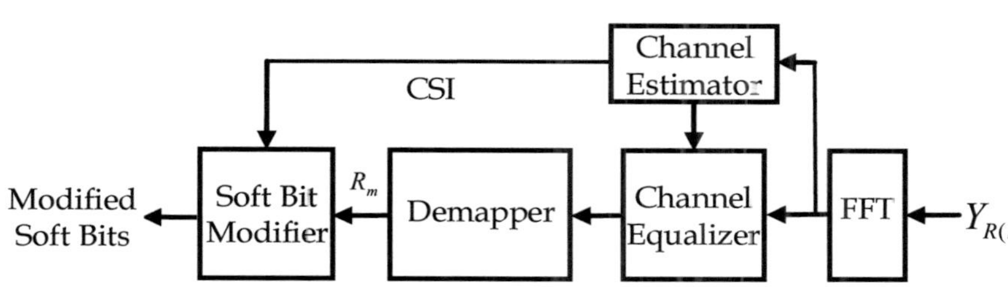

Fig. 16. Demodulation exploiting CSI

4.2.2 Soft bit demapping

The DCM demapper shall demap two equalized complex numbers ($I_{R(k)}$, $Q_{R(k)}$) and ($I_{R(k+50)}$, $Q_{R(k+50)}$) that previously transmitted on two different subcarriers back to two related DCM symbols by using the DCM mixing matrix. Then the DCM demapper outputs the corresponding real part and imaginary part as a group of 4 soft bits, as described in (21)-(24). However, demapping performance can remain the same without using the factor of $\sqrt{10}\,/5$. The group of 4 soft bits applying two CSI values are from two corresponding data subcarriers in an OFDM symbol, as described in (25)-(28).

$$Soft(b_{g(k)}) = \sqrt{10}\left(2I_{R(k)} - I_{R(k+50)}\right)/\,5 \tag{21}$$

$$Soft(b_{g(k)+1}) = \sqrt{10}\left(I_{R(k)} - 2I_{R(k+50)}\right)/\,5 \tag{22}$$

$$Soft(b_{g(k)+50}) = \sqrt{10}\left(2Q_{R(k)} + Q_{R(k+50)}\right)/\,5 \tag{23}$$

$$Soft(b_{g(k)+51}) = \sqrt{10}\left(Q_{R(k)} - 2Q_{R(k+50)}\right)/\,5 \tag{24}$$

$$Soft(b_{g(k)}) = \left(2I_{R(k)} + I_{R(k+50)}\right) \times \min\{CSI_k, CSI_{k+50}\} \tag{25}$$

$$Soft(b_{g(k)+1}) = \left(I_{R(k)} - 2I_{R(k+50)}\right) \times \min\{CSI_k, CSI_{k+50}\} \tag{26}$$

$$Soft(b_{g(k)+50}) = \left(2Q_{R(k)} + Q_{R(k+50)}\right) \times \min\{CSI_k, CSI_{k+50}\} \tag{27}$$

$$Soft(b_{g(k)+51}) = \left(Q_{R(k)} - 2Q_{R(k+50)}\right) \times \min\{CSI_k, CSI_{k+50}\} \tag{28}$$

4.2.3 Maximum likelihood soft bit demapping

The more reliable soft bit values that are given to Viterbi decoder, the more accurately the binary bits can be decoded. Maximum Likelihood (ML) offers finding parameters to obtain the most probable emitted symbols (Oberg, 2001). The DCM symbols are transmitted at different amplitudes and phases (I and Q values). The real part or the imaginary part in the two DCM symbols (signal amplitude) is always fixed with data pairs being -3 and +1, -1 and -3, +1 and +3, +3 and -1. In our case, the large probable soft bit value can be obtained from the two received DCM symbols with an appropriate parameter θ, as described in (29)-(32). The DCM symbol pair, $y_{R(k)}$ and $y_{R(k+50)}$, is received from the channel equalization.

$$Soft(b_{g(k)}) = 2I_{R(k)}\theta + I_{R(k+50)} \tag{29}$$

$$Soft(b_{g(k)+1}) = I_{R(k)} - 2I_{R(k+50)}\theta \tag{30}$$

$$Soft(b_{g(k)+50}) = 2Q_{R(k)}\theta + Q_{R(k+50)} \tag{31}$$

$$Soft(b_{g(k)+51}) = Q_{R(k)} - 2Q_{R(k+50)}\theta \tag{32}$$

To find the appropriate parameter θ, two conditions need to be satisfied.

a. If perfect, I and Q values received are input to the DCM demapper, applying θ to equations (29)-(32) to make the soft magnitude sufficiently large;

b. A symbol in the DCM symbol pair is transmitted with a large magnitude I (or Q), while another symbol in the DCM symbol pair is transmitted with a small magnitude I (or Q). The signal with smaller power can be more easily corrupted. Suppose the small magnitude I (or Q) in a DCM symbol is received as inverted, while the large magnitude I (or Q) in another DCM symbol is received as uncorrupted. In this case, a maximum θ is required to retain the sign of the soft bit value; otherwise using a larger θ can make the sign of the soft bit value inverted, which causes errors for the soft bit decoding.

θ is set to 1.5 as a threshold value according to the two conditions above. The ML soft bit is generated with the appropriate factor and CSI aided technique as described in the following:

$$Soft(b_{g(k)}) = \left(3I_{R(k)} + I_{R(k+50)}\right) \times \min\{CSI_k, CSI_{k+50}\} \tag{33}$$

$$Soft(b_{g(k)+1}) = \left(I_{R(k)} - 3I_{R(k+50)}\right) \times \min\{CSI_k, CSI_{k+50}\} \tag{34}$$

$$Soft(b_{g(k)+50}) = \left(3Q_{R(k)} + Q_{R(k+50)}\right) \times \min\{CSI_k, CSI_{k+50}\} \tag{35}$$

$$Soft(b_{g(k)+51}) = \left(Q_{R(k)} - 3Q_{R(k+50)}\right) \times \min\{CSI_k, CSI_{k+50}\} \tag{36}$$

4.2.4 Log likelihood ratio demapping

As well as improving the symbol reliability at the input of the Viterbi decoder, Log Likelihood Ratio (LLR) is another alternative demapping approach for the DCM. The generic format of LLR equation can be expressed in (37). In our case, a LLR is calculated from the received DCM symbols $y_{R(k)}$ and $y_{R(k+50)}$. In addition, the LLR functions related to the two 16-QAM like constellations are independent. Hence the LLR for a group of 4 bits ($b_{g(k)}$, $b_{g(k)+1}$, $b_{g(k)+50}$, $b_{g(k)+51}$) is formed from combining the two independent LLR, as in (38)-(41). σ_2 is noise variance associated with the channel.

$$LLR = \log\left(\exp(A) + \exp(B)\right) - \log\left(\exp(X) + \exp(Y)\right) \tag{37}$$

$$\begin{aligned} LLR(b_{g(k)}) = \log&\left\{ \sum_{b_{g(k)}=1} \exp\left[\frac{\left(I_{T(k)} - I_{R(k)}\right)^2}{-\sigma_k^2}\right] + \sum_{b_{g(k)}=1} \exp\left[\frac{\left(I_{T(k+50)} - I_{R(k+50)}\right)^2}{-\sigma_{k+50}^2}\right]\right\} \\ -\log&\left\{ \sum_{b_{g(k)}=0} \exp\left[\frac{\left(I_{T(k)} - I_{R(k)}\right)^2}{-\sigma_k^2}\right] + \sum_{b_{g(k)}=0} \exp\left[\frac{\left(I_{T(k+50)} - I_{R(k+50)}\right)^2}{-\sigma_{k+50}^2}\right]\right\} \end{aligned} \tag{38}$$

$$LLR(b_{g(k)+1}) = \log\left\{\sum_{b_{g(k)+1}=1} \exp\left[\frac{\left(I_{T(k)} - I_{R(k)}\right)^2}{-\sigma_k^2}\right] + \sum_{b_{g(k)+1}=1} \exp\left[\frac{\left(I_{T(k+50)} - I_{R(k+50)}\right)^2}{-\sigma_{k+50}^2}\right]\right\} - \log\left\{\sum_{b_{g(k)+1}=0} \exp\left[\frac{\left(I_{T(k)} - I_{R(k)}\right)^2}{-\sigma_k^2}\right] + \sum_{b_{g(k)+1}=0} \exp\left[\frac{\left(I_{T(k+50)} - I_{R(k+50)}\right)^2}{-\sigma_{k+50}^2}\right]\right\} \quad (39)$$

$$LLR(b_{g(k)+50}) = \log\left\{\sum_{b_{g(k)+50}=1} \exp\left[\frac{\left(Q_{T(k)} - Q_{R(k)}\right)^2}{-\sigma_k^2}\right] + \sum_{b_{g(k)+50}=1} \exp\left[\frac{\left(Q_{T(k+50)} - Q_{R(k+50)}\right)^2}{-\sigma_{k+50}^2}\right]\right\} - \log\left\{\sum_{b_{g(k)+50}=0} \exp\left[\frac{\left(Q_{T(k)} - Q_{R(k)}\right)^2}{-\sigma_k^2}\right] + \sum_{b_{g(k)+50}=0} \exp\left[\frac{\left(Q_{T(k+50)} - Q_{R(k+50)}\right)^2}{-\sigma_{k+50}^2}\right]\right\} \quad (40)$$

$$LLR(b_{g(k)+51}) = \log\left\{\sum_{b_{g(k)+51}=1} \exp\left[\frac{\left(Q_{T(k)} - Q_{R(k)}\right)^2}{-\sigma_k^2}\right] + \sum_{b_{g(k)+51}=1} \exp\left[\frac{\left(Q_{T(k+50)} - Q_{R(k+50)}\right)^2}{-\sigma_{k+50}^2}\right]\right\} - \log\left\{\sum_{b_{g(k)+51}=0} \exp\left[\frac{\left(Q_{T(k)} - Q_{R(k)}\right)^2}{-\sigma_k^2}\right] + \sum_{b_{g(k)+51}=0} \exp\left[\frac{\left(Q_{T(k+50)} - Q_{R(k+50)}\right)^2}{-\sigma_{k+50}^2}\right]\right\} \quad (41)$$

For a Gaussian channel, the LLR can be approximated as two piecewise-linear functions which depend on the amplitude of I/Q signals (Seguin, 2004). Furthermore, the maximum LLR value can be approximated to be soft magnitude with the associated bit completely depending on the amplitude of the I/Q signals. In our case, there are two bits associated with each of the two 16-QAM like constellations completely relying on their soft magnitude of the I/Q. The LLR functions related to these two bits from each constellation are considered to be partially linear. Therefore some terms of these LLR functions are approximated by soft magnitude, as in (42)-(45). The CSI is also used for LLR soft bit values scaling. The noise variance is obtained from mapping the ratio of received symbol and its average energy estimate has been taken into account to approximate the LLR value.

$$LLR(b_{g(k)}) = 3I_{R(k)} + \log\left\{\sum_{b_{g(k)}=1} \exp\left[\frac{\left(I_{T(k+50)} - I_{R(k+50)}\right)^2}{-\sigma_{k+50}^2}\right]\right\} - \log\left\{\sum_{b_{g(k)}=0} \exp\left[\frac{\left(I_{T(k+50)} - I_{R(k+50)}\right)^2}{-\sigma_{k+50}^2}\right]\right\} \quad (42)$$

$$LLR(b_{g(k)+1}) = \log\left\{\sum_{b_{g(k)+1}=1} \exp\left[\frac{\left(I_{T(k)} - I_{R(k)}\right)^2}{-\sigma_k^2}\right]\right\} - \log\left\{\sum_{b_{g(k)+1}=0} \exp\left[\frac{\left(I_{T(k)} - I_{R(k)}\right)^2}{-\sigma_k^2}\right]\right\} - 3I_{R(k+50)} \tag{43}$$

$$LLR(b_{g(k)+50}) = 3Q_{R(k)} + \log\left\{\sum_{b_{g(k)+50}=1} \exp\left[\frac{\left(Q_{T(k+50)} - Q_{R(k+50)}\right)^2}{-\sigma_{k+50}^2}\right]\right\} - \log\left\{\sum_{b_{g(k)+50}=0} \exp\left[\frac{\left(Q_{T(k+50)} - Q_{R(k+50)}\right)^2}{-\sigma_{k+50}^2}\right]\right\} \tag{44}$$

$$LLR(b_{g(k)+51}) = \log\left\{\sum_{b_{g(k)+51}=1} \exp\left[\frac{\left(Q_{T(k)} - Q_{R(k)}\right)^2}{-\sigma_k^2}\right]\right\} - \log\left\{\sum_{b_{g(k)+51}=0} \exp\left[\frac{\left(Q_{T(k)} - Q_{R(k)}\right)^2}{-\sigma_k^2}\right]\right\} - 3Q_{R(k+50)} \tag{45}$$

Now the LLR functions have been simplified by approximating with a linear part, to solve the non-linear part for the LLR function, the noise variance σ^2 needs to be estimated, which generally requires the mean of the absolute value of the received symbol components (m, as in (46)) and also estimates the average energy of the received symbol components (E, as in (47)). The ratio of m^2/E can be mapped to ratio a/m (a is signal amplitude, I or Q) and ratio σ^2/m. σ^2 can be determined from this mapping, but requiring large calculation in hardware and computation simulation.

$$m = \frac{1}{2k}\sum_{k=1}^{K}\left\{\left(\left|I_{R(k)}\right| + \left|Q_{R(k)}\right|\right)\right\} \tag{46}$$

$$E = \frac{1}{2k}\sum_{k=1}^{K}\left\{\left(\left|I^2{}_{R(k)}\right| + \left|Q^2{}_{R(k)}\right|\right)\right\} \tag{47}$$

4.2.5 System performance for 480 Mb/s mode

The system is simulated at the data rate of 480 Mb/s in UWB channel model 1 (CM1). The original MB-OFDM proposal settings of 2000 packets per simulation with a payload of 1024 octets each in the PSDU and 90th-percentile channel realization were followed. Strict adherence to timing was used. A hopping characteristic of TFC=1 was used. A 6.6 dB noise figure and a 2.5 dB implementation loss in the floating point system model were incorporated. The guard interval diversity is also used in the simulation.

The system performance exploiting soft bit, ML soft bit, and LLR DCM demapping methods with CSI as demapping enhancements were examined. From the simulation results shown in Figure 17, LLR with CSI is better demapping method and can achieve 3.9 meters in CM1. On closer examination for the performance at 8% PER, ML soft bit demapping method can achieve 3.9 meters in CM1 as well. In this case it is reasonable to conclude that ML soft bit demapping has same performance as LLR, but with slightly worse performance in shorter distance transmission. Soft bit demapping with CSI can only achieve 3.4 meters at 8% PER level in CM1. However soft bit or ML soft bit demapping method has lower computation complexity and reduces hardware implementation cost. Therefore ML soft bit demapping with CSI will be the best demapping method to implement hardware for ECMA-368.
The system performance in the 480 Mb/s mode was compared with current literature. It is difficult to compare the system performance with all the literature because most of them did not follow the conformance testing from WiMedia. This research used the simulation result from MBOA-SIG proposal (Multiband OFDM Alliance, 2004) for comparison. By implementing Kim's LLR DCM demapping method (Kim, 2007) with this proposed CSI further demapping technique, then the research will have the system performance using Kim's method for comparison. Figure 18 depicts the comparision for system performance for 480 Mb/s mode in CM1, wherein a performance gain can be achieved by the proposed LLR CSI method, while the system performance is 3.8 meters in MBOA-SIG proposal and the sytem using Kim's method. As can be seen, the proposed LLR CSI scheme performs the best at 8% PER.

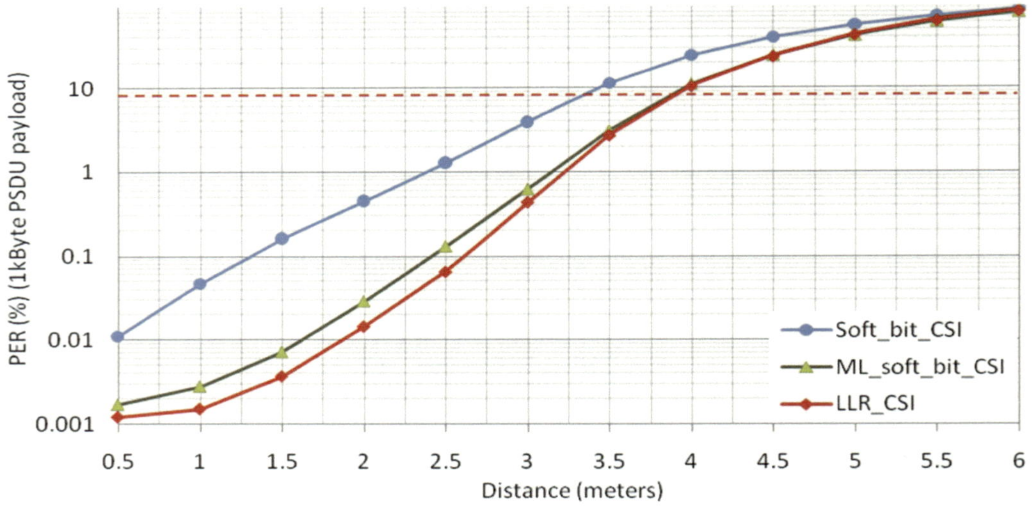

Fig. 17. Performance comparison for the proposed DCM demapping methods

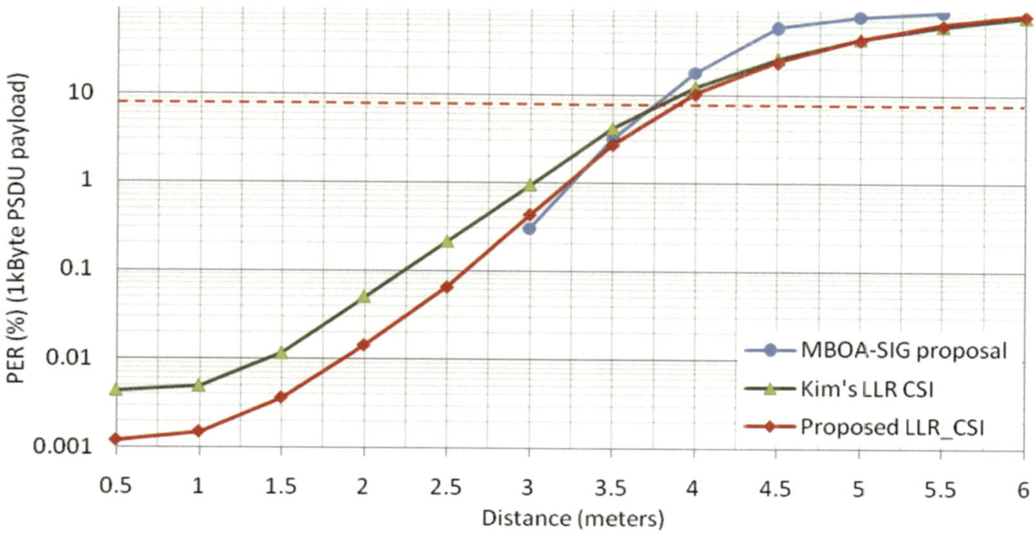

Fig. 18. Performance comparison for 480 Mb/s mode in CM1

4.3 Dual Circular 32-QAM

To enable the transport of high data rate UWB communications, ECMA-368 offers up to 480 Mb/s instantaneous bit rate to the MAC layer. However the maximum data rate of 480 Mb/s in a practical radio environment can not be achieved due to poor radio channel conditions causing dropped packets unfortunately resulting in a lower throughput hence need to retransmit the dropped packets. An alternative high data rate modulation scheme is needed to allow effective 480 Mb/s performance.

Two 16-QAM-like constellation mappings are used in the DCM. Obviously if only one 16-QAM-like constellation mapping is used for the modulation, this would result in less reliability but twice the number of bits can be transmitted per subcarrier offering faster throughput, which is from 640 Mb/s to 960 Mb/s comparing to DCM 320 Mb/s to 480 Mb/s mode (see Table 3). However there is no successful link under multipath environments (CM1...CM4) transmitting at 960 Mb/s or the system has poor performance only achieving 1.2 meters at 640 Mb/s. The simulation result will be shown in section 4.3.3. Hence 16-QAM is not the ideal modulation scheme for the high data rate MB-OFDM system.

4.3.1 Dual Circular 32-QAM mapping

Since 16-QAM is not a suitable modulation scheme for the high data rate MB-OFDM system, there is no need to consider higher order modulations, for instance 32-QAM, 64-QAM etc. Therefore if a new modulation scheme is proposed to fit into the existing system, the new modulation scheme comprising for an OFDM symbol shall not map the number of bits over 400 bits. Moreover, the new modulation scheme needs to be robust mapping 400 bits or less with successful transmission in a multipath environment.

A Dual Circular (DC) 32-QAM modulator is proposed to use two 8-ary PSK-like constellations mapping 5 bits into two symbols, which is basically derived from two QPSK symbols mapping 4 bits and taken the bipolarity of the fifth bit to drive the two QPSK

constellations to two 8-ary PSK-like constellations. Within a group of 5 bits, the first and second bit are mapped into one DC 32-QAM symbol, while the third and forth bit are mapped into another DC 32-QAM symbol, and then the fifth bit is mapped into both DC 32-QAM symbols offering diversity. 250 interleaved and coded bits are required to map by the DC 32-QAM mapper onto 100 data subcarriers in an OFDM symbol, hence increasing the system throughput to 600 Mb/s comparing to the DCM 480 Mb/s mode (see Table 3).
Figure 19 depicts the proposed DC 32-QAM modulator as an alternative modulation scheme that fits into the existing PSDU encoding process with the objective to map more information bits onto an OFDM symbol. After the bit interleaving, 1500 coded and interleaved bits are required to divide into groups of 250 bits and then further grouped into 50 groups of 5 reordering bits. Each group of 5 bits is represented as ($b_{g(k)}$, $b_{g(k)+50}$, $b_{g(k)+51}$, $b_{g(k)+100}$, $b_{g(k)+101}$), where $k \in [0\ldots49]$ and

$$g(k)=\begin{cases} 2k & k\in[\,0\ldots24\,] \\ 2k+50 & k\in[\,25\ldots49\,] \end{cases} \tag{48}$$

Four bits ($b_{g(k)+50}$, $b_{g(k)+51}$, $b_{g(k)+100}$, $b_{g(k)+101}$) are mapped across two QPSK symbols ($x_{g(k)}+jx_{g(k)+50}$), ($x_{g(k)+1}+jx_{g(k)+51}$) as in (49). Those two bits pairs are not in consecutive order within the bit streams. $b_{g(k)+50}$ and $b_{g(k)+100}$ are mapped to one QPSK symbol while $b_{g(k)+51}$ and $b_{g(k)+101}$ are mapped to another, which aids to achieve further bit interleaving against burst errors.

Data Rate (Mb/s)	**Modulation**	**Coding Rate (R)**	**Frequency Domain Spreading**	**Time Domain Spreading**	**Coded Bits / 6 OFDM symbol(N_{CBP6S})**
53.3	QPSK	1/3	YES	YES	300
80	QPSK	1/2	YES	YES	300
106.7	QPSK	1/3	NO	YES	600
160	QPSK	1/2	NO	YES	600
200	QPSK	5/8	NO	YES	600
320	DCM	1/2	NO	NO	1200
400	DCM	5/8	NO	NO	1200
480	DCM	3/4	NO	NO	1200
600	DC 32-QAM	3/4	NO	NO	1500
640	16-QAM	1/2	NO	NO	2400
960	16-QAM	3/4	NO	NO	2400

Table 3. PSDU rate-dependent parameters

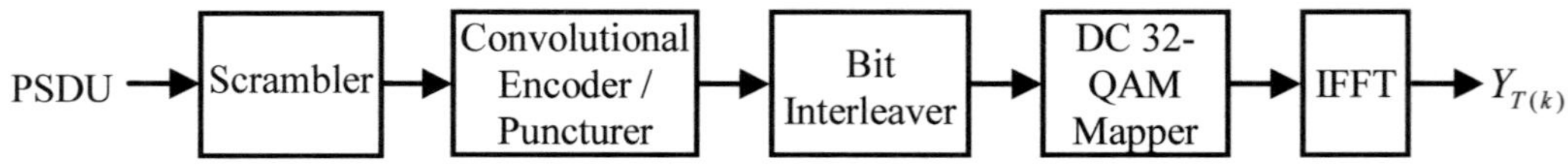

Fig. 19. PSDU Encoding process with DC 32-QAM

$$\begin{bmatrix} x_{g(k)} + jx_{g(k)+50} \\ x_{g(k)+1} + jx_{g(k)+51} \end{bmatrix} = \begin{bmatrix} (2b_{g(k)+50} - 1) + j(2b_{g(k)+100} - 1) \\ (2b_{g(k)+51} - 1) + j(2b_{g(k)+101} - 1) \end{bmatrix} \tag{49}$$

Then these two QPSK symbols are mapped into two DC 32-QAM symbols ($y_{T(k)}$, $y_{T(k+50)}$) depending on value of bit $b_{g(k)}$ as in (50)-(52), where $K_{MOD} = 1/\sqrt{6.175}$ as the normalization factor. Each DC 32-QAM symbol in the constellation mapping has equal decision region for each bit, as illustrated in Figure 20. The DCM symbols having two 16-QAM-like constellations do not have fixed amplitude. Thus the DCM will worsen the Peak to Average Power Ratio (PAPR) of the OFDM signals, resulting in more impact to the Automatic Gain Control (AGC) and ADC. In contract, the constellation points are positioned in circular loci to offer constant power for each DC 32-QAM symbol, which is of great benefit to the AGC and ADC.

$$\begin{bmatrix} y_{T(k)} \\ y_{T(k+50)} \end{bmatrix} = K_{MOD} \begin{bmatrix} \alpha x_{g(k)} + j\beta x_{g(k)+50} \\ \beta x_{g(k)+1} + j\alpha x_{g(k)+51} \end{bmatrix} \tag{50}$$

where

$$\alpha = \begin{cases} 1 & ,b_{g(k)} = 0 \\ 2.275 & ,b_{g(k)} = 1 \end{cases} \tag{51}$$

$$\beta = \begin{cases} 2.275 & ,b_{g(k)} = 0 \\ 1 & ,b_{g(k)} = 1 \end{cases} \tag{52}$$

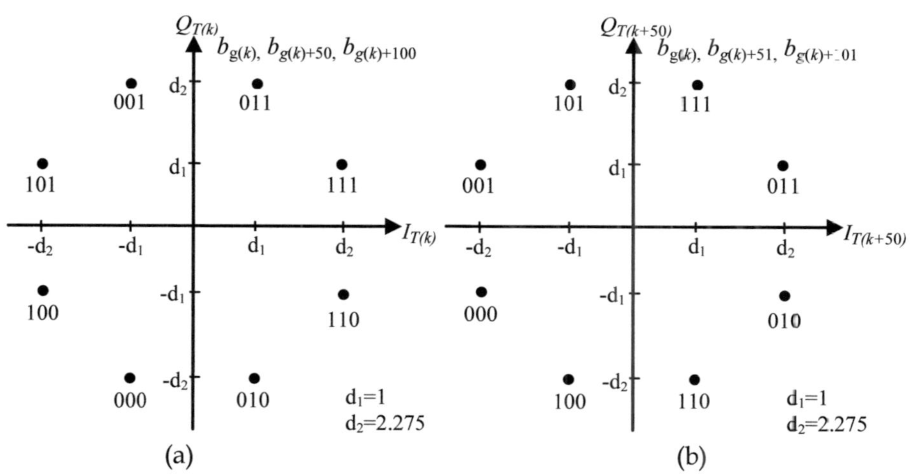

Fig. 20. DC 32-QAM constellation mapping: (a) mapping for $y_{T(k)}$; (b) mapping for $y_{T(k+50)}$

The two resulting DC 32-QAM symbols ($y_{(k)}$, $y_{(k+50)}$) are allocated into two individual OFDM data subcarriers with 50 subcarriers separation to achieve frequency diversity. An OFDM symbol is formed from the 128 point IFFT block requiring 100 DC 32-QAM symbols. Each OFDM subcarrier occupies a bandwidth of about 4 MHz, therefore the bandwidth between the two individual OFDM data subcarriers related to the two complex numbers ($I_{(k)}$, $Q_{(k)}$) and ($I_{(k+50)}$, $Q_{(k+50)}$) is at least 200 MHz, which offers a frequency diversity gain against channel deep fading. This will benefit for recovering the five information bits mapped across the two DC 32-QAM symbols. Figure 21 depicts the DC 32-QAM mapping process.

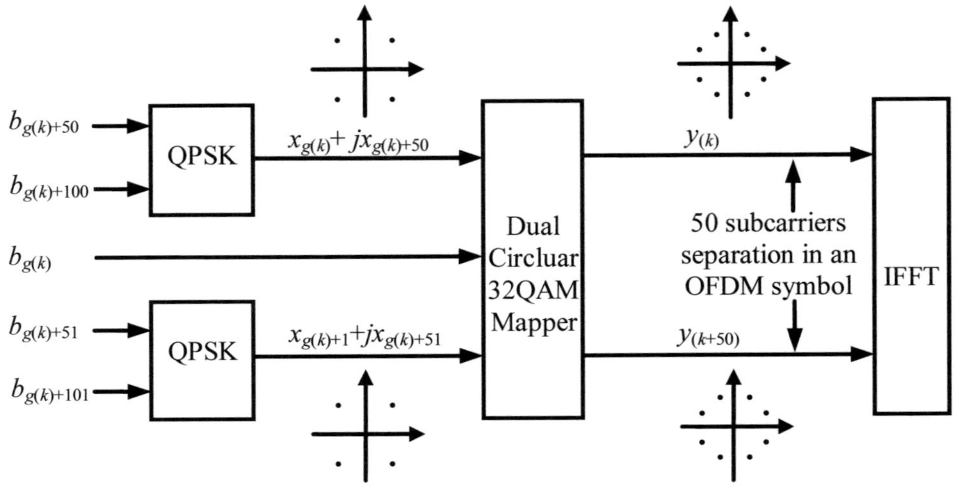

Fig. 21. DC 32-QAM mapping process

4.3.2 DC 32-QAM demapping

The proposed DC 32-QAM utilizes soft bit demapping to demap two equalized complex numbers previously transmitted on different data subcarriers into a subgroup of 5 soft bits, and then outputs groups of 250 soft bits in sequential order. The demapper is proposed to use the DC 32-QAM demapper, and other functional blocks are remained. The demapped and deinterleaved soft bits are input to Viterbi decoder to recover the original bit streams. Each soft bit value of $b_{g(k)+50}$, $b_{g(k)+51}$, $b_{g(k)+100}$ and $b_{g(k)+101}$ depend on the soft bit magnitude of the I/Q completely. In addition, each soft bit can be demapped from its associated ($I_{R(k)}$, $Q_{(k)}$) and ($I_{R(k+50)}$, $Q_{R(k+50)}$) independently. Furthermore, the demapping performance can remain without using the factor $1/K_{MOD}$. Hence the soft bit values for $b_{g(k)+50}$, $b_{g(k)+51}$, $b_{g(k)+100}$ and $b_{g(k)+101}$ are given by the following.

$$Soft(b_{g(k)+50}) = I_{R(k)} \tag{53}$$

$$Soft(b_{g(k)+51}) = I_{R(k+50)} \tag{54}$$

$$Soft(b_{g(k)+100}) = Q_{R(k)} \tag{55}$$

$$Soft(b_{g(k)+101}) = Q_{R(k+50)} \tag{56}$$

To demap $b_{g(k)}$ in the constellation for $y_{R(k)}$, the demapped information bit is considered to be '1' if the received symbol is close to the constellation point along with I axis, otherwise it is '0' if close to the constellation point along with Q axis. However, to demap $b_{g(k)}$ in the constellation for $y_{R(k+50)}$, the demapped information bit is considered to be '0' if the received symbol is close to the constellation point along with I axis, otherwise it is '1' if close to the constellation point along with Q axis. Figure 22 depicts Euclidean distances for a possible received DC 32-QAM symbol pair with region for $b_{g(k)}$. Since the bit regions of $b_{g(k)}$ in the two constellation mapping are different, the associated I and Q value from $y_{R(k)}$ and $y_{R(k+50)}$ cannot be simply combined. Hence the Euclidean symbol distance for each received symbol in the associated constellation mapping is calculated first, as in (57)-(60). Then the two Euclidean symbol distances are summed together as a soft bit value for $b_{g(k)}$, as in (61).

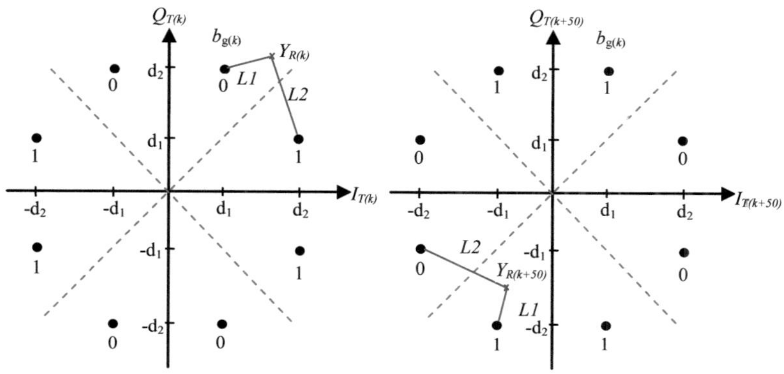

Fig. 22. Symbol distances for a possible received symbol pair $y_{R(k)}$ and $y_{R(k+50)}$ with decision region for $b_{g(k)}$

$$L1 = \sqrt{\left(\left|I_{R(k)}\right| - d1\right)^2 + \left(\left|Q_{R(k)}\right| - d2\right)^2} \tag{57}$$

$$L2 = \sqrt{\left(\left|I_{R(k)}\right| - d2\right)^2 + \left(\left|Q_{R(k)}\right| - d1\right)^2} \tag{58}$$

$$L3 = \sqrt{\left(\left|I_{R(k+50)}\right| - d1\right)^2 + \left(\left|Q_{R(k+50)}\right| - d2\right)^2} \tag{59}$$

$$L4 = \sqrt{\left(\left|I_{R(k+50)}\right| - d2\right)^2 + \left(\left|Q_{R(k+50)}\right| - d1\right)^2} \tag{60}$$

$$Soft(b_{g(k)}) = \frac{1}{2} \times (L1 - L2 + L3 - L4) \tag{61}$$

The proposed CSI aided scheme coupled with the band hopping information maximizes the DCM soft demapping performance. $b_{g(k)}$ mapped to two DC 32-QAM symbols are mapped onto two OFDM data subcarriers resulting in two CSI from the two associated data subcarriers. If a smaller or larger CSI value is chosen as a reliable scale term, it causes inequality of signal power for the different OFDM data subcarriers. The averaging CSI is adopted for $b_{g(k)}$. Therefore the soft bits incorporated with CSI for the DC 32-QAM demapping are given by the following:

$$Soft(b_{g(k)}) = \left(\frac{L1 - L2 + L3 - L4}{2}\right) \times \left(\frac{CSI_k + CSI_{k+50}}{2}\right) \tag{62}$$

$$Soft(b_{g(k)+50}) = I_{R(k)} \times CSI_k \tag{63}$$

$$Soft(b_{g(k)+51}) = I_{R(k+50)} \times CSI_{k+50} \tag{64}$$

$$Soft(b_{g(k)+100}) = Q_{R(k)} \times CSI_k \tag{65}$$

$$Soft(b_{g(k)+101}) = Q_{R(k+50)} \times CSI_{k+50} \tag{66}$$

4.3.3 System performance comparison for 16-QAM, DC 32-QAM and DCM

The system simulation setting is same as in section 4.2.4. To compare 16-QAM, DC 32-QAM and DCM performance, the system is set to the same configuration with the same coding rate. All the modulation schemes for the comparison use the best demapping solutions with CSI aided demapping scheme as presented in this thesis. While changing the modulation scheme and the associated bit interleaver, the system throughput can be increased to 600 Mb/s and 960 Mb/s by DC 32-QAM and 16-QAM respectively, while the DCM performs 480 Mb/s. As shown in Figure 23, there is no successful link if the system is operated with 16-QAM at the data rate of 960 Mb/s. Alternatively lowering the data rate to 640 Mb/s by changing the coding scheme (Table 3), the system performance is only 1.2 meters. However, implementing the DC 32-QAM scheme offers 3.2 meters at 600 Mb/s while the existing system using DCM can be achieved 3.9 meters at 480 Mb/s. The effective 600 Mb/s performance in practical multipath environment with moderate packet loss can offer an effective data rate at 480 Mb/s.

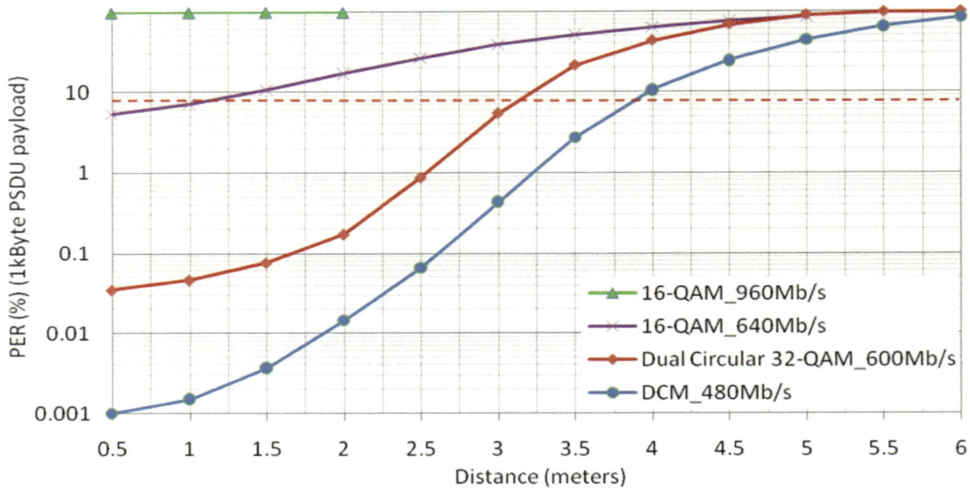

Fig. 23. System performance comparison for 16-QAM, DC 32-QAM and DCM

5. Conclusions

WiMedia Alliance working with ECMA established MB-OFDM UWB radio platform as the global UWB standard, ECMA-368. It is an important part to consumer electronics and the users' experience of these products. Since the standard has been set for the transmitter, optimization of the receiver becomes paramount to maximize the MB-OFDM system performance. Furthermore, the solutions of improving the MB-OFDM need to be cost-effective for implementing the low power and high performance device.

OFDM modulation is the important part for the multicarrier system. The proposed dual QPSK soft demapper exploiting TDS and guard interval diversity improved the system performance with requiring no overhead for ECMA-368. Three DCM demapping methods have been described and developed, which are soft bit demapping, ML soft bit demapping and LLR demapping methods. A CSI aided scheme coupled with the band hopping information maximized the DCM demapping performance, thus improving the system performance. Based on the QPSK and DCM, a cost-effective and high performance modulation scheme (termed DC 32-QAM) that fits within the configuration of current standard offering high rate USB throughput (480 Mb/s) with a moderate level of dropped packets, and can even offer a faster throughput for comparable propagation conditions. The contribution of this research can enable the UWB technology and help to ensure its success.

Hardware implementation at FPGA need solutions for ever increasing demands on system clock rates, silicon performance and long verification times etc. Not only logic and design size minimization, but also architecture solutions will be the challenge for the further research to handle large amounts of data through a fast UWB wireless connection.

6. References

aRenarti Semiconductor (2007). MB-OFDM UWB PHY: Baseband Processor (BBP), August 2007, Available from http://www.arenarti.com/docs/tb1000rB.pdf

Batra, A.; et al. (2004). Multi-band OFDM physical layer proposal for IEEE 802.15 task group 3a, *IEEE standard proposal P802.15-03*, March 2004

Batra, A.; Balakrishnan, J.; Aiello, G.; Foerster, J. & Dabak, A. (2004). Design of a multiband OFDM system for realistic UWB channel environments, *IEEE Transactions on Microwave Theory and Techniques*, Vol.52, No.9, (September 2004), pp 2123-2138, ISSN: 0018-9480

ECMA-368 (2008). High rate ultra wideband PHY and MAC standard (3rd Edition), ECMA International, December 2008

Ellis, J.; Siwiak, K. & Roberts, R. (2002). TG3a Technical Requirements, *IEEE P802.15-03/030-SG3a*, December 2002

FCC (February 2002). New public safety applications and broadband internet access among uses envisaged by FCC authorization of ultra-wideband technology, press released February 14, 2002

FCC (April 2002). Revision of Part 15 of the Commissions Rules Regarding Ultra-Wideband Transmission Systems. *ET Docket 98-153, FCC 02-48*, Released: April 22, 2002

Fisher, R.; et al (2005). DS-UWB Physical Layer Submission to 802.15 Task Group 3a, *IEEE standard proposal IEEE P802.15-04/0137r4*, January 2005

Foerster, J. (2003). Channel Modeling Sub-committee Report Final, *IEEE P802.15-02/490-SG3a*. February 7, 2003

Kim, Y. (2007). Dual Carrier Modulation (DCM) demapping method and demapper, *European Patent Application*, EP1858215A1, November 21, 2007

Li, W.; Wang, Z.; Yan, Y. & Tomisawa, M. (2005). An efficient low-cost LS equalization in COFDM based UWB systems by utilizing channel-state-information (CSI), *IEEE 62nd Vehicular Technology Conference*, Vol. 4, pp. 67-71, ISSN 1090-3038, Dallas, Texas, USA, September 2005,

Multiband OFDM Alliance (2004). MultiBand OFDM Physical Layer Proposal for IEEE 802.15.3a, *IEEE P802.15 Working Group for Wireless Personal Area Networks (WPANs)*. September 2004

Oberg, T. (2001). *Modulation, Detection and Coding*, John Wiley & Sons, ISBN 0471497665, Chichester, England

Proakis, J. G. (2001). *Digital Communications (Fourth edition)*, McGraw-Hill, ISBN 0072321113, New York, USA

Seguin, F.; Lahuec, Lebert, C. J.; Arzel, M. & Jezequel, M. (2004). Analogue 16-QAM demodulator, *IEE Electronics Letters*, Vol.40, No.18, (September 2004), pp.1138-1140, ISSN: 0013-5194

USB Implementers forum (2005). Wireless Universal Serial Bus Specification, Revision 1.0, May 12, 2005

2

Orthogonal Pulse-Based Modulation Schemes for Time Hopping Ultra Wideband Radio Systems

Sudhan Majhi[1] and Youssef Nasser[2]

[1]*Electrical and Electronic Engineering, Nanyang Technological University*
[2]*Faculty of Engineering and Architecture, American University of Beirut*
[1]*Singapore*
[2]*Lebanon*

1. Introduction

Ultra wideband (UWB) radio is a promising technology for short range wireless communications. It can be used for both high rate and low rate transmissions. High data rate can be achieved by using multiband (MB)-UWB approach whereas low data rate with robust system performance can be obtained by employing time hopping (TH)-UWB radio systems Majhi et al. (2006); Win & Scholtz (1998a). Nowadays, applications of UWB are spreading to various fields such as vehicle communications, wireless sensor networks, ad hoc wireless networks, and controller area networks. The most of the systems require low to moderate (1 kbs-100 mbs) data rates with an acceptable implementation cost. However, due to the presence of fast Fourier transform (FFT) and inverse FFT (IFFT), MB-UWB may not be a cost effective procedure for low data rate systems. Therefore, one needs an efficient system which adaptively changes the data rate from low to moderate with robust system performance. TH-UWB with OOK-PSM modulation provides low data rate with robust system performance Majhi, Madhukumar, Premkumar & Richardson (2008). However, it is possible to scale the TH-UWB radio system for low to moderate data rates by incorporating higher level modulation schemes with an adaptive method.

For TH-UWB systems, various M-ary modulation schemes such as pulse position modulation (PPM), pulse amplitude modulation (PAM), pulse shape modulation (PSM), and their combined forms have been proposed to improve data rates and system performance with low complexities Bin et al. (2003); Durisi & Benedetto (2003); Ghavami et al. (2002); Michell et al. (2003); Usuda et al. (2004). However, due to the increase of inter symbol interference (ISI) in the presence of multipath channel, M-ary PPM or M-ary orthogonal PPM (OPPM) are not effective for TH-UWB systems with RAKE reception when M is high Foerster (2003); Win & Scholtz (1998b). High-level M-ary PAM is rarely used in short range and low power consumption communications systems Guvenc & Arslan (2003). This is because that the Euclidian distances between constellations become small with increase in M. Due to its robustness against ISI and multiple access interference (MAI), pulse-based modulation such as PSM has become an interesting research topic in TH-UWB, direct sequence UWB (DS-UWB) and transmitted reference UWB (TR-UWB) radio systems Chu & Murch (2005); de Abrue et al. (2003); Gezici et al. (2006); Hwang et al. (2007); Kim & Womack (2007); Parr et al. (2003). However, high-level M-ary PSM cannot be used due to the limited auto correlation properties

of higher order orthogonal pulses. Moreover, its system complexity increases linearly with order of pulse waveform Gezici & Kobayashi (2005); Harada et al. (2004).
To deal with these challenges, combined modulation schemes such as M-ary biorthogonal PSM (BPSM), PPM-PSM, BPSK-PSM and OOK-PSM have been provided for M-ary TH-UWB systems Hu & Zheng (2005); Majhi, Madhukumar & Premkumar (2007); Majhi, Madhukumar, Premkumar & Chin (2007b); Michell et al. (2003); Usuda et al. (2004). However, M-ary BPSM still requires $M/2$ orthogonal pulses to transmit the signal. To improve system performance, PPM-PSM requires orthogonal coded modulation and memory at the receiver to maintain orthogonalities among the constellation vectors Mitchell & Kohno (2004). In the presence of multipath channel, BPSK-PSM and OOK-PSM cannot be used for higher level modulation scheme for higher data rates Majhi, Madhukumar, Premkumar & Chin (2007b); Usuda et al. (2004).
In order to address these problems, a combined modulation scheme (OPPM-BPSM) for TH-UWB systems was provided by the first author of this chapter to increase system data rate with good system performance Majhi et al. (2011). The proposed scheme was a combination of orthogonal PPM (OPPM) and BPSM modulation. In this chapter, we provide TH-UWB system design based on orthogonal pulse waveform. To show the robustness of orthogonal pulse waveform for TH-UWB systems we have provided performance analysis, capacity analysis and power spectral analysis of various orthogonal pulse based modulation schemes.
The rest of the chapter is organized as follows: section 2 describes used orthogonal pulses and its various modulation forms. Section 3 discusses system performance of OPPM-BPSM modulation schemes and its various interference issues. Section 4 provides the system capacity of TH-UWB systems for several orthogonal pulse based modulation schemes. Section 5 provides power spectral analysis of orthogonal pulse based modulation scheme of TH-UWB systems. Section 6 is provided for the simulation results. Section 7 provides the summary of chapter.

2. System model for TH-UWB

One of the essential functions in TH-UWB systems is the representation of a message symbol by a short duration pulse waveform for signal transmission through air de Abrue & Kohno (2003); Hu & Beaulieu (2004). The pulse waveform is an important design consideration which can affect UWB system performance considerably. The successful deployment of high data rate indoor TH-UWB systems strongly depends on the development of pulse waveforms and modulation schemes. Because of the short pulse waveforms, UWB is capable of providing high data rates for short range wireless communication. The chapter describes orthogonal pulse based TH-UWB system.

2.1 Orthogonal pulses

The commonly used orthogonal pulses for PSM modulation scheme are modified Hermite pulses (MHPs)Ghavami et al. (2002), Prolate spheroidal wave functions (PSWFs) Usuda et al. (2004), Battle-Lemarie wavelet orthogonal function Kim et al. (2005), and Haar wavelet orthogonal function Zhang & Zhou (2005). In this chapter all the analysis has been done based on MHPs and PSWFs. The system performance depends on autocorrelation and crosscorrelation properties. In addition, MAI is also reduced considerably by using crosscorrelation properties of orthogonal pulses. The time and frequency domain representation of MHPs are given in Fig. 1.

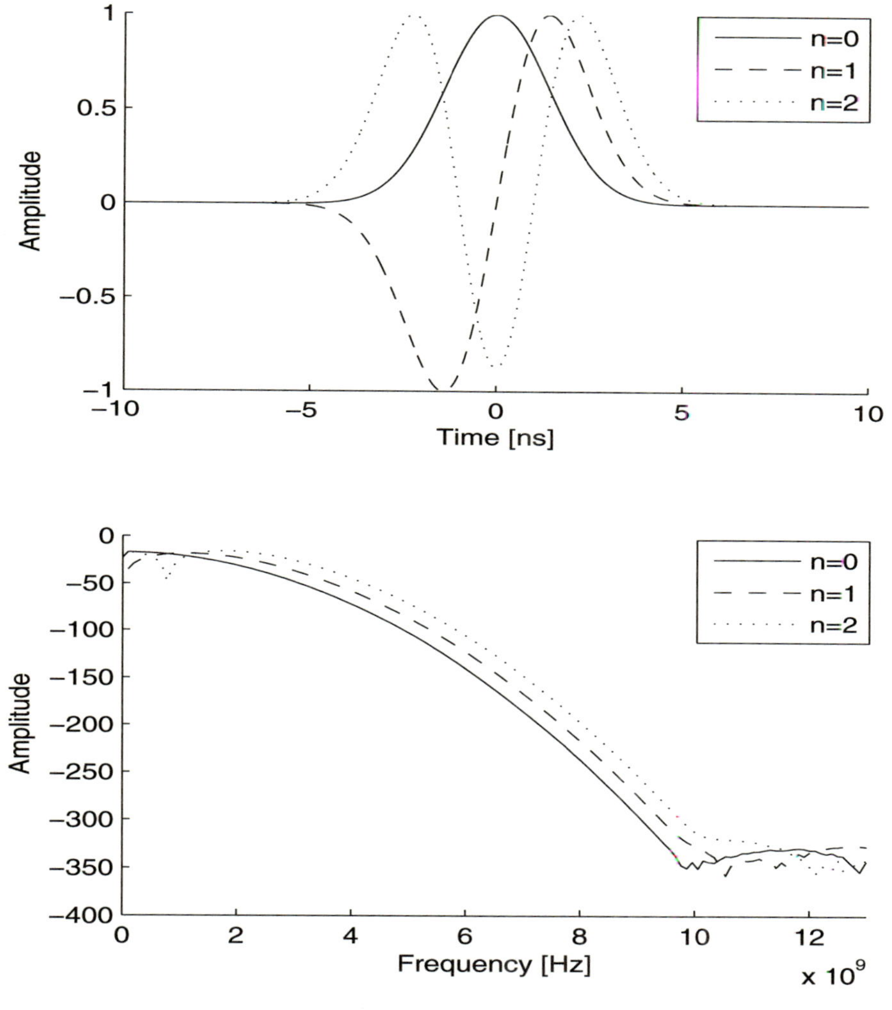

Fig. 1. Time and frequency (logarithmic plot) domains representation of modified Hermite pulses (MHPs).

2.2 M-ary Pulse Shape Modulation (PSM)

In pulse shape modulation, a set of symbols is assigned by a set of orthogonal pulses which are orthonormal. The M-ary signal set for PSM can be written as

$$s^{(k)}(t) = \sum_{j=-\infty}^{\infty} \sqrt{E_{tx}^{(k)}} w^{(k)}_{(\lfloor j/N_s \rfloor \% M)}(t - jT_f - c_j^{(k)} T_c) \tag{1}$$

where $\{w_0(t), w_1(t), \cdots, w_{M-1}(t)\}$ is a set of orthogonal pulses and in general $w_i(t) \perp w_l(t) = 0$ for $i \neq l$ and 1 for $i = l$ in normalized form. M correlators and corresponding orthogonal pulses are used as a reference signal to detect M-ary signal at the receiver Usuda et al. (2004)Kim et al. (2005). Any mutually orthonormal pulses can be used for this modulation scheme.

2.3 M-ary Biorthogonal PSM

Biorthogonal PSM (BPSM) modulation is similar to BPM scheme, the only difference is that it is used for M-ary signaling . M-ary BPSM uses $M/2$ orthogonal pulses in the transmitter and $M/2$ correlators in the receiver to transmit all M possible symbols. The number of correlators or matched filters used in this scheme drops to half of those used in M-ary PSM scheme, thus reducing complexity of TH-UWB systems. An M-ary BPSM modulation has been proposed in Wen & Guoxin (2005). The output of an M-ary BPSM can be a signal with $M/2$ possible pulse shapes which are biorthogonal. Orthogonal pulse shapes are represented as follows: $w_0(t), w_1(t), ..., w_{M/2-1}(t)$. The negative ones are defined as $w_{i+M/2}(t) = -w_i(t)$, where $i = 0, 1, ..., M/2 - 1$.

BPSM gives high data rate and makes it easier to map symbols into pulse waveforms. It has high power efficiency due to pulse polarity. However, similar to BPM scheme, it also requires two transmitters to generate BPSM signal. Maintaining bi-phase of orthogonal pulses is a challenging task. On the other hand, due to limitation of the possible number of orthogonal pulses it cannot be used for higher level modulation schemes.

2.4 M-ary OPPM-BPSM

OPPM-BPSM scheme is a combination of orthogonal PPM and biorthogonal PSM (OPPM-BPSM). In order to transmit M symbols, one has to use L orthogonal pulse positions and N biorthogonal pulses where $M = 2^k$, $L = 2^l$, $N = 2^{k-l-1}$, $k > 1$ and $0 \leq l \leq k-1$. Antipodal pulses are chosen to smooth the PSD of TH-UWB signal and to improve its coexistence ability with NB systems without any degradation in system performance Majhi, Madhukumar & Ye (2007). These biorthogonal pulses reduce the number of correlators in the receiver. Further, the system complexity is reduced by half when compared with a scheme that uses a combination of L pulse positions and N orthogonal pulses Kim et al. (2005). By changing the number of pulse positions and orthogonal pulses, one can construct a wide variety of symbols. For example, M-ary BPPM scheme can be constructed by using $M/2$ pulse positions and one biorthogonal pulse waveform, and M-ary BPSM scheme can be constructed by using one pulse position and $M/2$ biorthogonal pulses Wen & Guoxin (2005); Zhang & Gulliver (2005a). OPPM-BPSM scheme ensures relatively constant power envelope for transmitted symbol irrespective of the number of pulse positions and biorthogonal pulses. This multidimensional scheme increases the Euclidean distance of the transmitted signal, so power efficiency increases without affecting signal bandwidth Ramseier & Schlegel (1993). Since each position is able to transmit multiple orthogonal pulses, it does not require longer chip duration as well as longer time frame compared to M-ary OPPM or M-ary BPPM schemes. Theoretically, $N \times L$ correlators are required in the receiver for M-ary OPPM-BPSM scheme, however, only N correlators with L delay units are sufficient to receive the signal. These delays can be implemented through software which reduces hardware complexity of M-ary OPPM-BPSM scheme for $N < M$.

3. Performance of M-ary OPPM-BPSM

The system performance of the orthogonal pulse based modulation scheme decreases in the presence of multipath channel. The performance of OPPM-BPSM scheme is evaluated using the UWB multipath channel model based on the indoor channel measurement in the 2-8 GHz frequency band accepted by the IEEE802.15.3a study group Foerster (2003); Saleh & Valenzuela (1987). For simplicity, it is also assumed that signal is transmitted by using i^{th} $(0 \leq i \leq N-1)$ order pulse in the qth $(0 \leq q \leq L-1)$ pulse position. Therefore the signal in Majhi, Madhukumar, Premkumar & Chin (2007a) can be rewritten as

$$s_{iq}^{(k)}(t) = \sum_{j} \sqrt{E_{tx}^{(k)}} d_m^{(k)} w_i^{(k)}(t - jT_f - c_j^{(k)} T_c - \delta_q^{(k)}) \tag{2}$$

where $d_m \in \{-1, 1\}$. If there are N_u users and each experiences a different channel model, the simplified received signal can be expressed as

$$r(t) = \sum_{k=1}^{N_u} \sum_{l=1}^{L_p} \alpha_l^{(k)} s_{iq}^{(k)}(t - \tau_l^{(k)}) + n(t) \tag{3}$$

where $\tau_l^{(k)}$ is the delay of path of kth user which takes values in the continuous time-invariant model, $\alpha_l^{(k)}$ is the lth path gain of kth user, and L_p is the maximum number of paths among the users. It is assumed that the reference RAKE receiver is synchronized i.e. $\tau_l^{(1)} = 0$ for l^{th} RAKE finger of user 1. The receiver structure with RAKE fingers is shown in Fig.2. To receive the symbols, receiver requires L bank of correlators based on the L positions and each bank of correlators contains N correlators based on the order of orthogonal pulse. Further, each correlator contains RAKE fingers based on the number of estimated paths. The delay of the paths and fading are done by channel estimation.
The reference signal in correlator of ith order pulse and q^{th} pulse position of user 1 can be expressed as

$$\phi_{iq}^{(1)}(t) = \sum_{j=0}^{N_s-1} v_i^{(1)}(t - jT_f - c_j^{(1)} T_c - \delta_q^{(1)}) \tag{4}$$

where N_s is the number of pulse repetition interval for a symbol and

$$v_i^{(1)}(t) = \sum_{p=1}^{L_p} \alpha_p^{(1)} w_i^{(1)}(t - \tau_p^{(1)}) \tag{5}$$

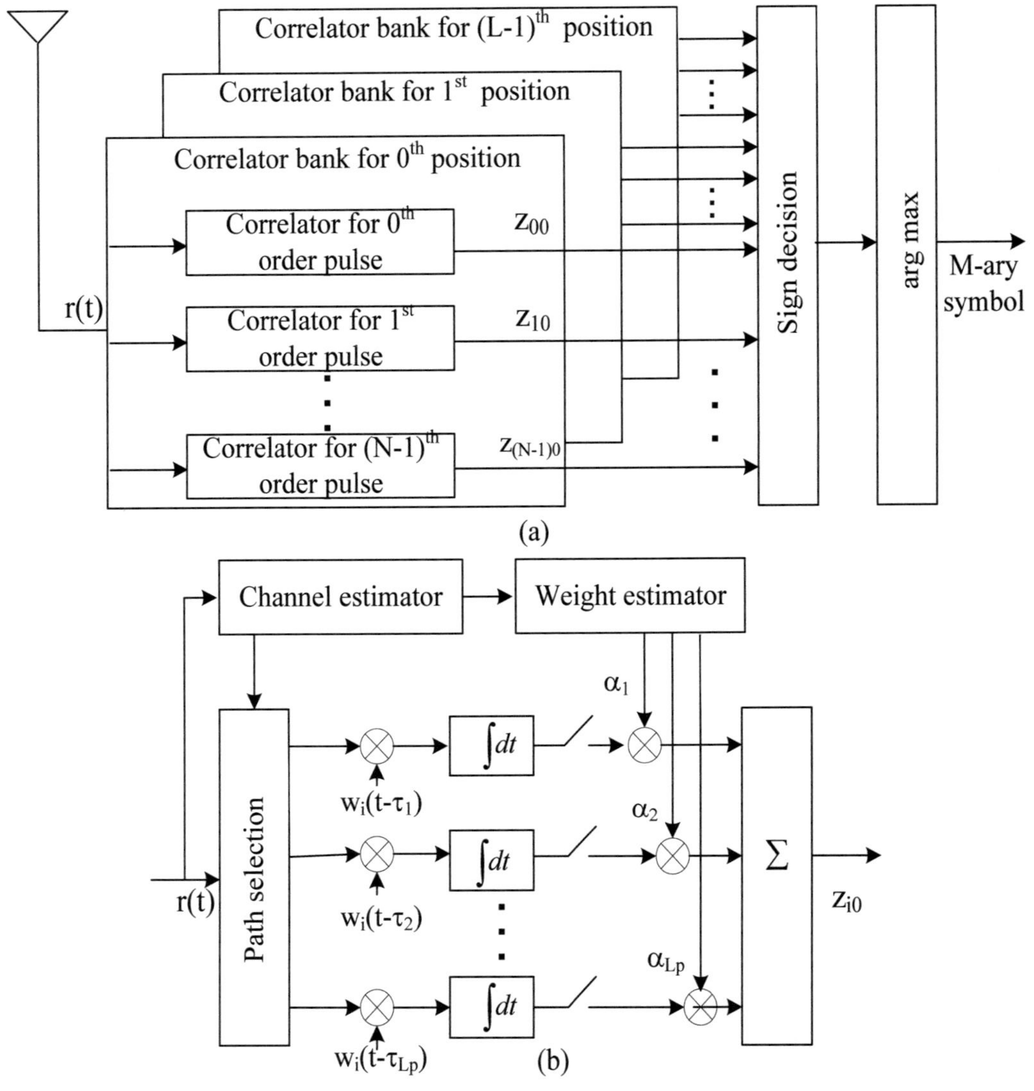

Fig. 2. Receiver structure for the OPPM-BPSM scheme (a) Block diagram for bank of correlators for the different pulse position and different order of orthogonal pulses. (b) Block diagram of correlators for 0th pulse position and $i^{th}(i = 0,\ 1,\ \ldots,\ N-1)$ order pulse which contains L_p RAKE fingers for different delays and different weights

3.1 Decision statistics

At the receiver, the received signal, indicated as a useful signal, is corrupted by mainly three additive noise components: ISI noise generated due to multipath components, MAI due to presence of multiple users, and thermal noise generated in receiver antenna and receiver circuitry. The problem of receiver design can thus be stated as follows: Finding a good, when possible optimal, way for extracting a useful signal from the received signal. Solving the general problem is a complicated task that leads to complex receiver structures and requires good modeling of noise components. It is assumed that the receiver has perfect channel estimation and uses partial RAKE combining. At sample time $t = (j+1)T_f$, the combined output of first L_p paths of the correlator for i^{th} order pulse and for q^{th} pulse position can be written as Jia & Kim (2005); Jiang et al. (2005)

$$\begin{aligned} z_{iq}^{(1)} &= \int_{jT_f}^{(j+1)T_f} r(t)\phi_{iq}^{(1)}(t)dt \\ &= S_{iq}^{(1)} + ISI_{iq}^{(1)} + MAI_{iq}^{(1)} + N_{iq}^{(1)} \end{aligned} \tag{6}$$

where $S_{iq}^{(1)}$ is the desired signal, $ISI_{iq}^{(1)}$ is the ISI term, $MAI_{iq}^{(1)}$ is the MAI term and $N_{iq}^{(1)}$ is the AWGN component of user 1 in the correlator of ith order pulse and q^{th} pulse position. In order to improve detector performance of the system, effects of ISI and MAI have to be canceled before a symbol decision is made. In communication systems, decision feedback equalizers are commonly employed for this purpose. The main idea behind decision feedback equalization is that once a data symbol has been detected, the interference induces on the following symbols is estimated and subtracted out before the detection of subsequent symbols. Therefore, knowledge of the desired signal and interference noise is required to make correct decision. Gaussian approximation (GA) is not an appropriate method for finding the interference noise when a few multiple users are present in the system, However, due to its simplicity in the presence of multipath fading, this work assumes GA for the mathematical analysis for different noise terms Giorgetti & Chiani (2005).

3.2 Desired signal energy

The transmitted signal propagates in a multipath channel. To collect these multipath components, each correlator contains several RAKE fingers with different delays and weights. The desired multipath components are correlated with corresponding RAKE fingers, and some of these received energy are the desired or useful signal. In realistic scenarios it is assumed that pulses are orthogonal in a synchronized system. Therefore, the desired signal can be written as Majhi, Madhukumar, Premkumar & Chin (2007a)

$$S_{iq}^{(1)} = \sqrt{E_{tr}^{(1)}} d_m^{(1)} N_s \sum_{p=1}^{L_p} (\alpha_p^{(1)})^2. \tag{7}$$

It is observed that the received energy in the multipath channel increases with increase in the number of RAKE fingers in the correlators, and this improves system performance. However, large number of RAKE fingers increases the system complexity and channel estimation error. Therefore, a minimum number of RAKE finger is used with considerable system performance. Since $d_m^{(1)} \in \{\pm 1\}$, the constellation distance in OPPM-BPSM is far from those in OOK-PSM scheme, which results in better system performance than that in OOK-PSM scheme.

3.3 Inter Symbol Interference (ISI)

In the reality, channel is not perfectly estimated, and each path is not synchronized. So the decision variable is affected by other unexpected signals such as ISI. It occurs when multipath components are not received by their corresponding RAKE fingers, these are received by other RAKE fingers which have different weights and delays. The ISI noise of user 1 can be as Majhi, Madhukumar, Premkumar & Chin (2007a)

$$\sigma_{ISI}^2 = E_{tr}^{(1)} N_s T_f^{-1} \sum_{p=1}^{L_p} \sum_{l=1}^{L_p} \sum_{\substack{p'=1 \\ p' \neq p}}^{L_p} \sum_{\substack{l'=1 \\ l' \neq l}}^{L_p} \alpha_p^{(1)} \alpha_l^{(1)} \alpha_{p'}^{(1)} \alpha_{l'}^{(1)} X(\Delta) \tag{8}$$

where $X(\Delta)$ is the correlation function and Δ is the delay parameter detail given in Majhi, Madhukumar, Premkumar & Chin (2007a). It is observed that ISI is not reduced by orthogonal pulses and their modulation schemes. It depends on channel estimation and its delay spread. The channel delay spread cannot be controlled by modulation schemes or system design. The ISI can be reduced by increasing the duration of pulse repetition interval, which affects system data rates. On the other hand, duration of pulse repetition interval can be shorter when several orthogonal pulses are used in one pulse position but it reduces the system performance. This is one of many practical limitations for higher data rate systems with good system performance.

3.4 Multiple Access Interference (MAI)

In multiple access systems, several users transmit signals over the same channel. Pulses originating in other transmission links may interfere with pulses belonging to a reference transmission giving rise to MAI. The MAI noise of user 1 from $N_u - 1$ users can be written from Majhi, Madhukumar, Premkumar & Chin (2007a) as

$$\sigma_{MAI}^2 = N_s T_f^{-1} \sum_{k=2}^{N_u} E_{tr}^{(k)} \sum_{p=1}^{L_p} \sum_{l=1}^{L_p} \sum_{p'=1}^{L_p} \sum_{l'=1}^{L_p} \alpha_p^{(1)} \alpha_l^{(k)} \alpha_{p'}^{(1)} \alpha_{l'}^{(k)} Y(\Delta') \tag{9}$$

where $Y(\Delta')$ is also a correlation function and Δ' is the delay parameter. The cross correlation value is smaller than the auto correlation values in both synchronous and asynchronous systems. It is observed that MAI depends on correlation properties of orthogonal pulses and number of users in the system. MAI is the sum of interference of one user from all the other users. When all the users use the same set of orthogonal pulses MAI is the sum of expectation of the product of auto correlation and cross correlation. In the conventional system this is expectation of product of auto correlation. Due to the presence of cross correlation term, MAI is reduced. However, when all the users use a different subset of orthogonal pulses, MAI is the sum of expectation of product of cross correlation which is the extreme case for better system performance.

3.5 Performance analysis

It is assumed that the correlator output z_{iq} of ith order pulse and q^{th} pulse position is larger than the other $M/2 - 1$ correlator outputs. As discussed before, each output is corrupted by noise and the total noise at each correlator can be defined as $\sigma_{ISI}^2 + \sigma_{MAI}^2 + \sigma_N^2$ where σ_N^2 is AWGN. The corresponding average probability of a correct decision in the presence of ISI and

MAI can be expressed as Proakis (2001); Zhang & Gulliver (2005a)

$$P_c = \int_0^{\infty} \left(\frac{1}{\sqrt{2\pi}} \int_{-z_{iq}/\sqrt{\sigma_{ISI}^2+\sigma_{MAI}^2+\sigma_N^2}}^{z_{iq}/\sqrt{\sigma_{ISI}^2+\sigma_{MAI}^2+\sigma_N^2}} \exp \frac{-x^2}{2} dx \right)^{\frac{M}{2}-1} p(z_{iq}) dz_{iq} \tag{10}$$

where probability density function of z_{iq} can be written as

$$p(z_{iq}) = \frac{1}{\sqrt{2\pi(\sigma_{ISI}^2 + \sigma_{MAI}^2 + \sigma_N^2)}} \exp \left(-\frac{\left(z_{iq} - N_s \sqrt{E_{tr}^{(1)}} \sum_{p=1}^{L_p} (\alpha_p^1)^2 \right)^2}{2(\sigma_{ISI}^2 + \sigma_{MAI}^2 + \sigma_N^2)} \right) \tag{11}$$

The probability of a symbol error for combined M-ary OPPM-BPSM is given by

$$P_M = 1 - P_c. \tag{12}$$

The BER of OPPM-BPSM scheme can be evaluated as Proakis (2001); Sklar (2001).

$$P_b = \frac{2^{k-1}}{2^k - 1} P_M. \tag{13}$$

3.6 Simulation results

The orthogonal pulse based system has been extensively simulated in different channel conditions. Simulation results of an 8-ary TH-UWB system for different number of pulse positions and orthogonal pulses in the presence of IEEE 802.15.3a UWB multipath channel model Foerster (2003) is discussed in this section. Channel model corresponding to line of sight (0-4m) environment (CM1) is used for this study. The performance is analyzed by using two different sets of orthogonal pulses such as MHPs and PSWFs. Fig. 3 and Fig. 5 show the results of these simulation studies. The theoretical results are also provided for checking the validation of GA with these simulation studies. The solid lines represent performance for MHPs and dashed lines represent the performance for PSWFs.

The performance of an 8-ary scheme is obtained by employing 1 position and 4 pulses, 2 positions and 2 pulses, and 4 positions and 1 pulse. In Fig. 3, the number of significant paths is decided by selecting paths within 10 dB of the strongest path. The performance is also evaluated using multi mode data rates, that is, duration of pulse position is fixed and duration of pulse repetition intervals is selected according to number of positions. The data rates of 1 position, 2 positions and 4 positions are 75 mb/s, 37.5 mb/s, and 18.75 mb/s, respectively for $\delta = 10$ ns and $N_s = 1$. Since duration of pulse repetition interval increases with the increase in the number of positions, the inter frame interference is reduced. Therefore, the system performance with more pulse positions (4 positions 1 pulse) is better than the system performance with 1 position and 4 pulses. However, system with multiple pulse positions reduces the data rate correspondingly. On the other hand, system with 1 position and 4 pulses results in degraded performance due to speculative auto correlation properties of higher order pulses. Therefore, number of pulse positions and pulses can be selected adaptively based on the requirements of data rate and system performance.

Fig. 5 shows the system performance of an 8-ary scheme for the same data rate (50 mb/s). Since pulse repetition interval is fixed for all possibilities of positions and pulses, length of each pulse position is decreased when 4 pulse positions and 1 pulse are considered. The

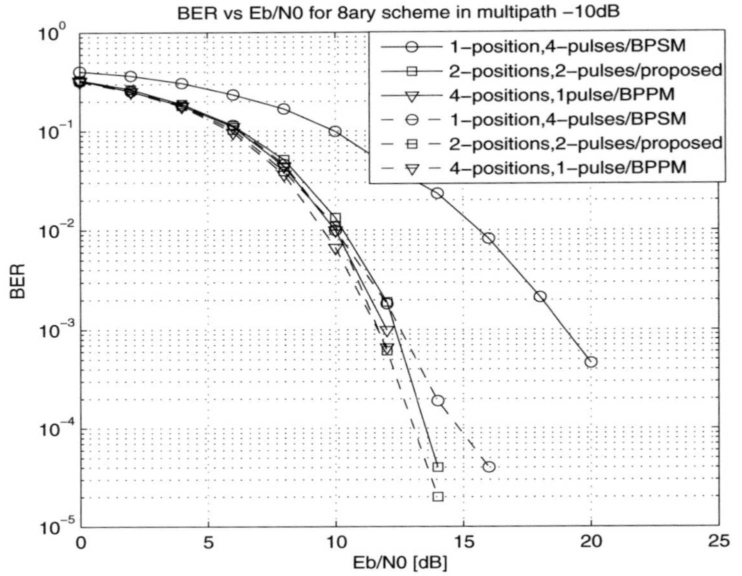

Fig. 3. Performance of 8-ary modulation scheme in different data rate by using modified Hermite and PSWF orthogonal pulses in multipath channel model where upto-10dB path is captured from peak point.

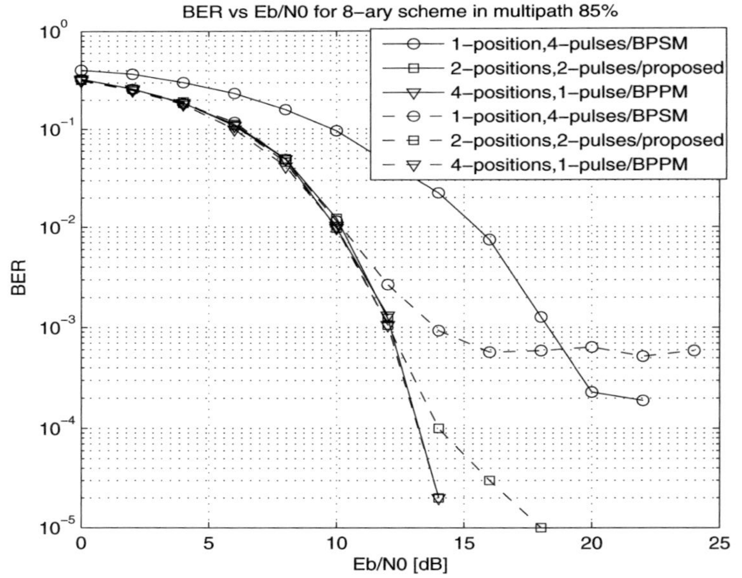

Fig. 4. Performance of 8-ary modulation scheme in different data rate by using modified Hermite and PSWF orthogonal pulses in multipath channel model where 85% energy is captured.

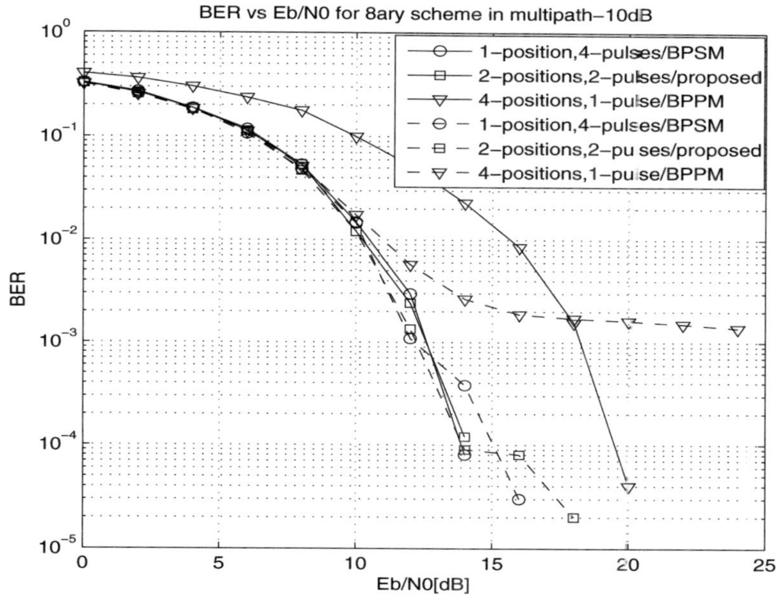

Fig. 5. Performance of 8-ary modulation scheme in the same data rate environment by using modified Hermite and PSWF orthogonal pulses in multipath channel model where upto-10dB path is captured from peak point.

multipath signals of previous pulse positions affect the correlators of the next pulse position resulting in performance degradation. So the noise floor increases with SNR in the presence of ISI and MAI for a large number of pulse positions. The corresponding results is shown in Fig. 5 which shows that the large number of pulse positions (4 positions and 1 pulse) results in performance degradation. It has also been shown that moderate number of pulse positions and pulses (2 positions 2 pulses) is a better choice for an acceptable data rate and system performance.

Since, auto correlation property of 0th order MHPs and 0th order PSWFs is the same, system with 4 positions and 1 pulse gives same performance for both pulses. However, difference in their auto correlation values increases with increase in the order of pulse, which results different performances for the use of higher order orthogonal pulses. Fig. 3 and Fig. 5 show that PSWFs pulses result in better performance than that of MHPs for more number of orthogonal pulses. Therefore, choice of orthogonal pulses also influences the system performance of pulse based modulation schemes.

4. Capacity of TH-UWB systems based on orthogonal pulse waveform

In this section, we provide capacity for different modulation schemes for TH-UWB systems Majhi, Xiang, Madhukumar & Premkumar (2008).

4.1 Capacity of M-ary PSM scheme

For simplicity, a wireless channel is normally assumed to be constant within a coherence time and varies in the next coherence times. A symbol duration for a UWB signal, T_f, is chosen as less than the channel coherence time. In an indoor environment, the delay spread of a UWB channel is much less than the channel coherence time Jiang et al. (2005); Zhang & Gulliver (2005b). The RAKE receiver scheme is adopted at the receiver to combine the total L_p multi-path components (MPC) in following analysis. For a channel with a discrete-value input and a continuous-value output, a piece of k-bit information, $\mathbf{U} = \{U_1, U_2, ..., U_k\}$, is mapped to a set of signals with $M = 2^k$ status of $\mathbf{S} = \{S_1, S_2, ..., S_M\}$, each of which, $\mathbf{s}_{n,PSM}$ for the M-ary PSM scheme, can be simplified from (2) and represented as a M-dimensional signal vector.

$$\mathbf{s}_{n,PSM} = [0, \cdots, \sqrt{E_{tr}}, 0, \cdots, 0] \tag{14}$$

where $\mathbf{s}_{n,PSM}$ is a signal vector with a nonzero and nonnegative value at the nth position. The channel capacity can be derived by maximizing the mutual information between channel input and channel output, which is expressed as,

$$C = \max_{p(\mathbf{S})} I(\mathbf{Y}; \mathbf{S}) \tag{15}$$

where $I(\mathbf{Y}, \mathbf{S})$ is the mutual information between the channel output $\mathbf{Y}$ and the channel input $\mathbf{S}$.

We assume that there is no restriction on input signal other than imposing an average power constraint. As the bandwidth increases, the data rate of a communications system over a multipath fading channel equals to the capacity of a system with an infinite bandwidth over an additive white Gaussian channel with the same SNR. The mutual information approaches close to zero where there are infinite numbers of MPC. This result holds even when the receiver can perfectly track the excess delay of each MPC and the only uncertainty lies on its phase and amplitude. Since $\mathbf{S}$ is the invertible function of $\mathbf{U}$, the capacity becomes

$$C = \max_{p(\mathbf{S})} I(\mathbf{Y}; \mathbf{U}). \tag{16}$$

Let $\mathbf{z}_{PSM}$ be the channel output vector of M correlators which is corrupted by ISI, MAI and AWGN. After maximum ratio combination (MRC), $\mathbf{z}_{PSM}$ can be expressed as a M-dimensional vector shown below.

$$\mathbf{z}_{PSM} = [z_1, z_2, \ldots, z_M] \tag{17}$$

where $z_i = ISI_i + MAI_i + N_i$ for $i \neq n$ and $z_n = S_n + ISI_n + MAI_n + N_n$ for n^{th} correlator. It is assumed that signal is transmitted by n^{th} order pulse waveform where $n \in \{1, 2, \ldots, M\}$.

Since PSM constellations are in orthogonal space, the channel capacity with input signals restricted to an equiprobable M-ary PSM signal and with no restriction on the channel output is given by Li et al. (2005)

$$C_{PSM} = \log_2 M - \frac{1}{M} \sum_{n=1}^{M} \int_{\mathbf{z}_{PSM}} p(\mathbf{z}_{PSM}|\mathbf{s}_{n,PSM}) \log_2 \left(\frac{\sum_{k=1}^{M} p(\mathbf{z}_{PSM}|\mathbf{s}_{k,PSM})}{p(\mathbf{z}_{PSM}|\mathbf{s}_{n,PSM})} \right) d\mathbf{z}_{PSM} \tag{18}$$

where $p(\mathbf{z}_{PSM}|\mathbf{s}_{n,PSM})$ is the probability density function (PDF) of $\mathbf{z}_{PSM}$, for a given $\mathbf{s}_{PSM}$. Since $\mathbf{s}_{PSM}$ is dependant on the modulation scheme, the conditional PDF as well as system capacity vary with modulation scheme.

To find the system capacity of M-ary PSM for TH-UWB systems, it is required to evaluate the PDF of M-ary PSM scheme. $\mathbf{z}_{PSM}$ is a M dimensional joint Gaussian distribution random variable conditioned on $\mathbf{s}_{n,PSM}$ with a PDF given by,

$$p(\mathbf{z}_{PSM}|\mathbf{s}_{n,PSM}) = \left(\frac{1}{2\pi\sigma^2}\right)^{M/2} e^{\left[-\frac{(z_n - E_{b,PSM})^2}{2\sigma^2}\right]} \prod_{\substack{i=1 \\ i \neq n}}^{M} e^{\left[-\frac{z_i^2}{2\sigma^2}\right]}. \tag{19}$$

where the signal energy in multipath channel is obtained from (7) by substituting $d_m = 1$ given as

$$E_{b,PSM} = \sqrt{E_{tr}} N_s \sum_{p=1}^{L_p} (\alpha_p^{(1)})^2 \tag{20}$$

and σ^2 is the total noise and can be written as

$$\sigma^2 = \sigma_{ISI}^2 + \sigma_{MAI}^2 + \sigma_N^2. \tag{21}$$

Using (17) and (19) in (18), the channel capacity for an M-ary PSM scheme on a block fading channel is given as

$$C_{PSM} = \log_2 M - \frac{1}{M} \sum_{n=1}^{M} E_{\mathbf{z}_{PSM}|\mathbf{s}_n} \log_2 \left(\sum_{k=1}^{M} e^{-\frac{z_n^2 - z_k^2 + \left(z_k - E_{b,PSM}\right)^2 - \left(z_n - E_{b,PSM}\right)^2}{2(\sigma_{ISI}^2 + \sigma_{MAI}^2 + \sigma_N^2)}} \right) \tag{22}$$

bits / channel use

It is observed that the bandwidth does not directly influence the mutual information, but it more likely depends on the level of modulation schemes and the total noise σ^2. This noise is directly related to the number of paths is shown in (8) and (9) . And if the underlying number of paths L_p is very large and the delays of these paths spread out, L_p will then increase with increasing bandwidth. The ISI can be reduced by increasing the duration of pulse repetition interval, which affects system data rates for M-ary PSM scheme as well as for M -ary BPSM scheme. It is also observed from (9) that MAI depends on the correlation properties of orthogonal pulses and the number of users. When all the users use pulses from a same set of orthogonal pulses, MAI is the sum of the expectation of the product of auto correlation function and cross correlation function. Hence, MAI is reduced for orthogonal pulse-based modulation due to the presence of cross correlation terms and this is true for all M-ary PSM, M-ary BPSM and M-ary OPPM-BPSM schemes. However, in conventional single pulse approach (pulse position modulation (PPM) and pulse amplitude modulation (PAM)), this is the expectation of the product of auto correlation functions Benedetto & Giancola (2004). It is well-known that a cross correlation function has a smaller value than that of an auto correlation function for orthogonal pulses Dilmaghani et al. (2003).

4.2 Capacity of M-ary BPSM scheme

In M-ary BPSM scheme, two symbols are transmitted by using only one orthogonal pulse. Therefore M-ary BPSM symbol vectors can be generated by using $M/2$-dimensional orthogonal space and expressed as

$$\mathbf{s}_{mn,BPSM} = [0, \cdots, d_m \sqrt{E_{tr}}, 0, \cdots, 0] \tag{23}$$

where $\mathbf{s}_{mn,BPSM}$ is a BPSM signal vector with nonzero value in the nth position, $n \in \{1, 2, \ldots, M/2\}$ and $m \in \pm 1$. Therefore, M-ary BPSM signal vectors are also in orthogonal space and dimension of signal remains the same as PSM scheme. However, the radius of the noise sphere for BPSM scheme becomes larger than the PSM scheme. It increases the system performance of BPSM scheme shown in Majhi, Madhukumar, Premkumar & Chin (2007a). Now the capacity of M-ary BPSM in multipath fading channel can then be expressed as

$$
\begin{aligned}
C_{BPSM} = \log_2 M - \frac{1}{M} \sum_{m=1}^{2} \sum_{n=1}^{M/2} \int_{\mathbf{z}_{BPSM}} & p\left(\mathbf{z}_{BPSM} | \mathbf{s}_{mn,BPSM}\right) \\
& \times \log_2 \left(\frac{\sum_{j=1}^{2} \sum_{k=1}^{M/2} p(\mathbf{z}_{BPSM} | \mathbf{s}_{jk,BPSM})}{p(\mathbf{z}_{BPSM} | \mathbf{s}_{mn,BPSM})} \right) d\mathbf{z}_{BPSM}
\end{aligned}
\tag{24}
$$

where $\mathbf{z}_{BPSM}$ is a $M/2$-dimensional vector defined as

$$
\mathbf{z}_{BPSM} = [z_1, z_2, \ldots, z_{M/2}] \tag{25}
$$

The joint Gaussian distribution of $\mathbf{z}_{BPSM}$ conditioned on $\mathbf{s}_{mn,BPSM}$ is expressed as:

$$
p(\mathbf{z}_{BPSM} | \mathbf{s}_{mn,BPSM}) = \left(\frac{1}{2\pi\sigma^2} \right)^{M/2} e^{\left[-\frac{(z_m - E_{b,BPSM})^2}{2\sigma^2} \right]} \prod_{\substack{i=1 \\ i \neq j}}^{M} e^{\left[-\frac{z_i^2}{2\sigma^2} \right]} \tag{26}
$$

where the signal energy of BPSM in a multipath channel is obtained from (7) as

$$
E_{b,BPSM} = \sqrt{E_{tr}} N_s d_m \sum_{p=1}^{L_p} \left(\alpha_p^{(1)}\right)^2 \tag{27}
$$

Substituting (25) and (26) into (24), the capacity for an M-ary BPSM scheme over a fading channel is given as

$$
C_{BPSM} = \log_2 M - \frac{1}{M} \sum_{m=1}^{2} \sum_{n=1}^{M} E_{\mathbf{z}_{BPSM}|\mathbf{s}_{mn}} \log_2 \left(\sum_{j=1}^{2} \sum_{k=1}^{N} e^{-\frac{z_n^2 - z_k^2 + \left(z_k - E_{b,BPSM} \right)^2 - \left(z_n - E_{b,BPSM} \right)^2}{2(\sigma_{ISI}^2 + \sigma_{MAI}^2 + \sigma_N^2)}} \right) \text{ bits / channel use} \tag{28}
$$

The effects of ISI and MAI of M-ary BPSM are similar as M-ary PSM scheme. However, the complexity of transceiver for BPSM scheme is reduced by half with respect to PSM scheme for $M > 2$. Moreover, BPSM scheme is capable of providing better system performance and capacity by improving the received signal power Majhi, Madhukumar, Premkumar & Chin (2007a).

4.3 Capacity of M-ary OPPM-BPSM scheme

M-ary OPPM-BPSM signal is constructed by using L orthogonal pulse positions and N orthogonal pulses. Therefore, OPPM-BPSM signal can be represented by a $N \times L$-dimensional signal matrix shown below.

$$\mathbf{S}=\begin{pmatrix} 0 & 0 & \cdots & 0 & 0 \\ 0 & \ddots & \cdots & 0 & 0 \\ \vdots & 0 & d_m\sqrt{E_{tr}} & \vdots & 0 \\ 0 & \cdots & 0 & \ddots & 0 \\ 0 & 0 & \cdots & 0 & 0 \end{pmatrix}$$

where $d_m\sqrt{E_{tr}}$ is a signal amplitude at l^{th} row n^{th} column. Since matrix has only one nonzero element, for simplicity, the input signal matrix can be expressed as a NL array signal vector,

$$\mathbf{s}_{mnl} = [0, \cdots, d_m\sqrt{E_{tr}}, 0, \cdots, 0] \tag{29}$$

At the receiver there are NL correlators. Therefore, the received signal can be expressed as a LN array vector

$$\mathbf{z}_{OPPM-BPSM} = [z_{11}, \cdots, z_{1L}, z_{21}, \cdots, z_{NL}] \tag{30}$$

where $z_{ij} = ISI_{ij} + MAI_{ij} + N_{ij}$ for $(i,j) \neq (n,l)$ and $z_{nl} = S_{nl} + ISI_{nl} + MAI_{nl} + N_{nl}$. Similarly, the M-ary OPPM-BPSM modulation scheme for TH-UWB has discrete-valued inputs and continuous-valued outputs, which impose additional constraints on the capacity calculation. Therefore, the capacity for a UWB system using M-ary OPPM-BPSM can be expressed as

$$\begin{aligned} C_{OPPM-BPSM} = \log_2 M - \frac{1}{2NL}\sum_{m=1}^{2}\sum_{n=1}^{N}\sum_{l=1}^{L}\int_{\mathbf{z}_{OPPM-BPSM}} p\left(\mathbf{z}_{OPPM-BPSM}|\mathbf{s}_{mnl}\right) \\ \times \log_2\left(\frac{\sum_{j=1}^{2}\sum_{k=1}^{N}\sum_{p=1}^{L} p(\mathbf{z}_{OPPM-BPSM}|\mathbf{s}_{jkp})}{p(\mathbf{z}_{OPPM-BPSM}|\mathbf{s}_{mnl})}\right) d\mathbf{z}_{OPPM-BPSM} \end{aligned} \tag{31}$$

If it is assumed that signal is transmitted by using n^{th} order pulse at l^{th} pulse position, the joint Gaussian distribution of $\mathbf{z}_{OPPM-BPSM}$ conditioned on $\mathbf{s}_{mnl}$ can be expressed as

$$p(\mathbf{z}_{OPPM-BPSM}|\mathbf{s}_{mnl}) = \left(\frac{1}{2\pi\sigma^2}\right)^{NL}\left(\prod_{\substack{i=1 \\ (i,j)\neq(n,l)}}^{N}\prod_{j=1}^{L} e^{-\frac{z_{jp}^2}{2\sigma^2}}\right) e^{-\frac{(z_{nl}-E_{b,OPPM-BPSM})^2}{2\sigma^2}} \tag{32}$$

where the signal energy over a multipath channel is obtained from (7) as Majhi, Madhukumar, Premkumar & Chin (2007a)

$$E_{b,OPPM-BPSM} = \sqrt{E_{tr}}N_s d_m \sum_{p=1}^{L_p}\left(\alpha_p^{(1)}\right)^2 \tag{33}$$

Substituting (30) and (32) into (31), the capacity is given as

$$C_{OPPM-BPSM} = \log_2 M - \frac{1}{2NL}\sum_{m=1}^{2}\sum_{n=1}^{N}\sum_{l=1}^{L} E_{\mathbf{z}_{OPPM-BPSM}|\mathbf{s}_{mnl}} \times \log_2\left(\sum_{j=1}^{2}\sum_{k=1}^{N}\sum_{i=1}^{L} e^{-\frac{z_{nl}^2 - z_{ki}^2 + \left(z_{ki}-E_{b,OPPM-BPSM}\right)^2 - \left(z_{nl}-E_{b,OPPM-BPSM}\right)^2}{2(\sigma_{ISI}^2+\sigma_{MAI}^2+\sigma_N^2)}}\right) \tag{34}$$

bits / channel use

In M- ary PSM and M-ary BPSM scheme, ISI is reduced only by reducing pulse repetition interval which limits the system data rate. In M-ary OPPM-BPSM scheme, the duration of pulse repetition interval can be reduced by using several orthogonal pulses in one pulse position. However, this process reduces the system performance for higher values of M. This is a practical limitation for high data rate systems and we need to select a suitable value of M in an actual system design.

4.4 Simulation results

In these simulation studies, the capacities of a UWB system adopting M-ary PSM, M-ary BPSM and M-ary OPPM-BPSM modulation schemes are investigated over the modified IEEE 802.15.3a UWB multipath channel model, where M=4, 8, 16 and 32. The capacity is shown in terms of bits per channel use.

For the clear understanding, the system capacity in multipath environment is provided in Fig. 6, Fig. 7 and Fig. 8. The system capacities of M-ary PSM, M-ary BPSM and M-ary OPPM-BPSM for $M = 4$ and 8 are provide in Fig. 6. Although, 8-ary PSM, 8-ary BPSM are the specific schemes of 8-ary OPPM-BPSM, 8-ary OPPM-BPSM outperform 8-ary PSM and 8-ary BPSM. It is because they use different pulse repetition intervals due to their different number of pulse positions and pulses. 4-ary BPSM and 4-ary OPPM-BPSM are identical since both use 1 positions and 2 pulses. It is also observed that at low SNR 4-ary OPPM-BPSM provide better system capacity than the 8-ary PSM scheme. It means that higher level PSM scheme does not provide good system capacity at low SNR.

The capacities of 16-ary PSM, 16-ary BPSM and 16-ary OPPM-BPSM schemes are provided in Fig. 7. 16-ary OPPM-BPSM scheme can be designed in 4 ways such as by using 1 position, 2 positions, 3 positions and 4 positions. 16-ary BPSM is a specific scheme of 16-ary OPPM-BPSM when number of pulse position is 1. The system capacity of the OPPM-BPSM scheme increases with decrease in the number of pulse positions when modulation uses combined of OPPM and BPSM. However, in terms of performance, the system performance decreases with the decrease in the number of pulse positions for a given value of M as shown in our previous work Majhi, Madhukumar, Premkumar & Chin (2007a). Therefore, modulation scheme is to be designed based on capacity and performance requirement of the system.

The capacities of 32-ary PSM, 32-ary BPSM and 32-ary OPPM-BPSM schemes for two different orthogonal pulse sets are provided in Fig. 8. It is observed that PSWFs provide better system capacity than MHPs. It is because of the PSWFs provides less ISI and MAI than the MHPs due to better autocorrelation properties of PSWFs. Therefore, the design of an orthogonal pulse becomes critical to increase the capacity of orthogonal pulse based UWB systems. However, for 32-ary BPPM scheme, PSWFs and MHPs provide the same system capacity. It is because

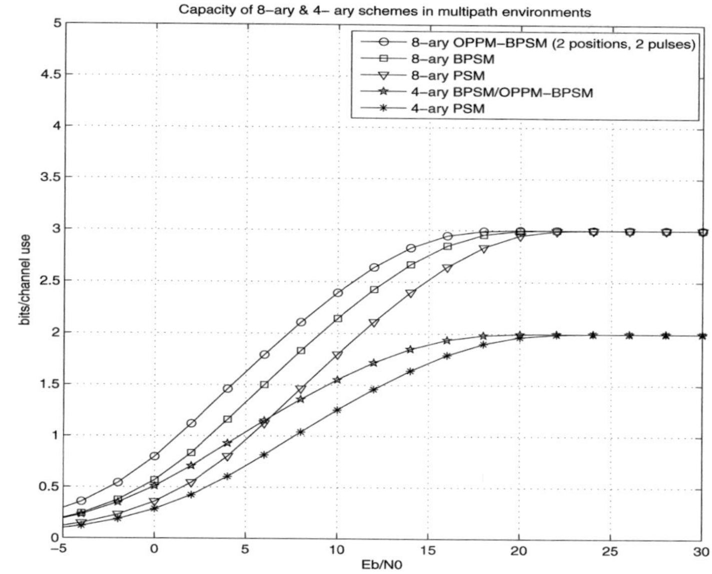

Fig. 6. The capacities of M-ary PSM, M-ary BPSM and M-ary OPPM-BPSM schemes in a multipath environment where M=4 and 8.

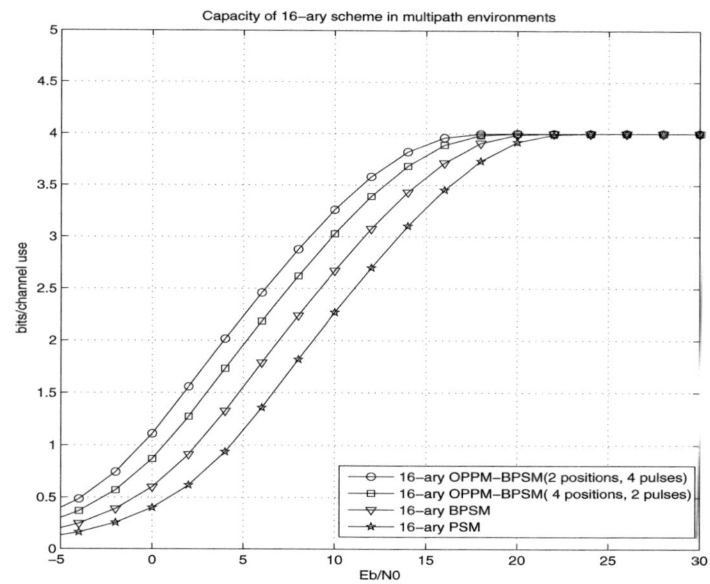

Fig. 7. The capacities of 16-ary PSM, 16-ary BPSM and 16-ary OPPM-BPSM schemes in multipath environment.

we have used 1st order PSWF and 1st order MHP in 32-ary BPPM. It is known that both the pules provide exactly the same correlation properties for the 1st order pulse. Fig. 6, Fig. 7 and Fig. 8 show that the average full capacity for all values of M for M-ary PSM is nearly achieved where the SNR is close to 23 dB, 20 dB for M-ary BPSM and 17 dB for M-ary OPPM-BPSM. It is also observed that M-ary OPPM-BPSM has 3 dB more SNR than M-ary BPSM and 6 dB greater SNR than M-ary PSM at the same capacity. This is because of the use of orthogonal pulses resulting in that ISI and MAI are less for M-ary OPPM-BPSM scheme than M-ary PSM and M-ary BPSM schemes for the same value of M. However, after 25 dB SNR, the capacities are close to the same irrespective of the modulation schemes.
Under the same simulation condition the system capacities of 16-ary BPPM, 16-ary PSM, 16-ary BPSM and 16-ary OPPM-BPSM as a function of number of MPC are provided in Fig. 9. It has been observed that capacities for all schemes decrease with increase in the number of MPC. This is because ISI and MAI increase with the increase in the number of MPC, resulting in the reduction of mutual information. It proves that mutual information is inversely proportional to number of MPC. It is also observed that BPPM and OPPM-BPSM are more sensitive to the number of MPC. When number of MPC is more than 10, the capacities of BPPM and OPPM-BPSM are decreased more gradually than the PSM and BPSM scheme. It is because of involving pulse position modulation in both BPPM and OPPM-BPSM. Indeed, it is known that pulse position modulation is more sensitive in multipath environment. However, OPPM-BPSM still outperforms conventional BPPM scheme for the same values of M.

5. Power spectral analysis of TH-UWB systems

In orthogonal pulse based signal, different symbols are transmitted by different order orthogonal pulses. The continuous spectrum, energy spectral density (ESD), changes with symbol. The discrete spectral component changes with orthogonality of the pulses and TH code. Therefore, a mathematical frame work is essential to understand the orthogonal pulse based PSD in the presence of deterministic TH code Majhi et al. (2010). We assume that the analysis is only for 1 user. For simplicity, the superscript/subscript terms in (35) are omitted/modified. After some modification, sum of M symbol can be written from (2) as

$$s_p(t) = \sum_{l=0}^{M-1} \sum_{h=0}^{N_s-1} a_l w_l(t - lN_pT_f + hT_f - c_{l,h}T_c - \delta_l) \tag{35}$$

where a_l is the amplitude and δ_l is the pulse position. The terms a_l, δ_l and w_l are independent and stationary process. The index p is related to TH code, $c_{l,h}$, and TH period, N_p. To simplify the analysis of the PSD of TH-UWB signal, it is assumed that the number of time frames for a symbol is N_s and it is equal to N_p. Since (35) depends on the time dithering, it can be written in continuous form as

$$y(t) = \sum_l s_p(t - lN_pT_f). \tag{36}$$

The PSD is computed by evaluating the Fourier transform (FT) of the autocorrelation function of $y(t)$ i.e.

$$P_y(f) = \mathcal{F}\Big\{E\left\{y(t)y(t+\tau)\right\}\Big\} \tag{37}$$

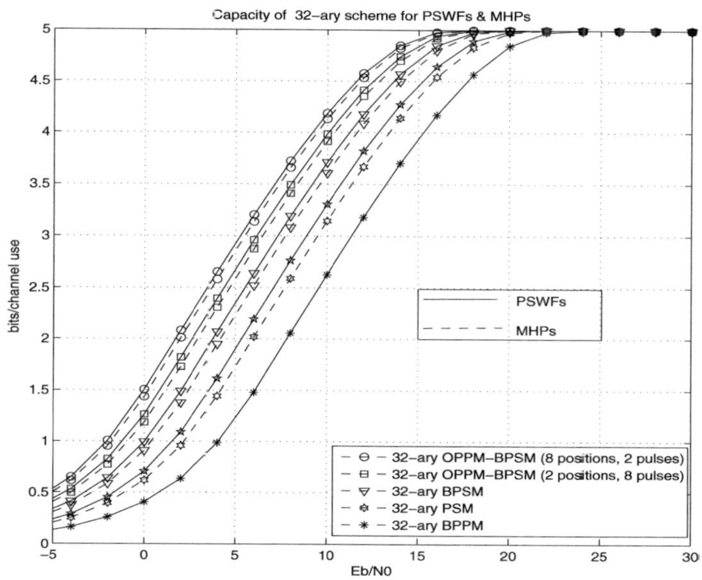

Fig. 8. The capacity of 32-ary PSM, 32-ary BPSM and 32-ary OPPM-BPSM schemes schemes in a multipath environment with different sets of orthogonal pulse waveforms.

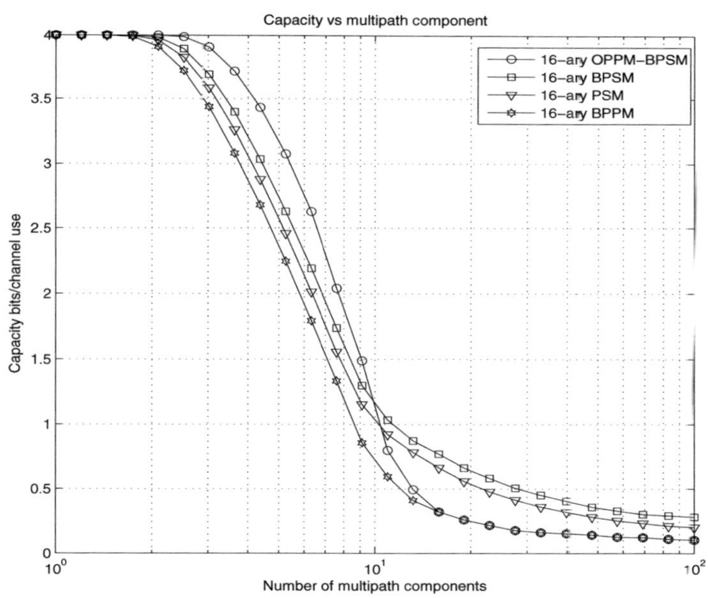

Fig. 9. The capacity versus multipath components is provided for 16-ary BPPM, 16-ary PSM, 16-ary BPSM and 16-ary OPPM-BPSM schemes.

where $\mathcal{F}\{.\}$ denotes the FT and $E\{.\}$ denotes the expectation operator. Therefore, the PSD can be expressed as Padgett et al. (2003)

$$\begin{aligned} P_y(f) = & \frac{1}{N_p T_f}\left[E\left\{|S_p(f)|^2\right\} - E\left\{S_p(f)S_q^*(f)\right\}\right] \\ & + \frac{1}{(N_p T_f)^2}\sum_k E\left\{S_p(f)S_q^*(f)\right\}\delta\left(f - \frac{k}{N_p T_f}\right) \end{aligned} \tag{38}$$

where p and q are two independent random variables with the same probability distribution function. $S_p(f)$ is the FT of $s_p(t)$. It can be expressed as

$$S_p(f) = \sum_{l=0}^{M-1} W_l(f)T_l(f)a_l e^{-j2\pi f\delta_l} \tag{39}$$

where $W_l(f)$ is the FT of the transmitted pulse $w_l(t)$. The time domain representation of $(l+2)^{th}$ order MHPs can be expressed as

$$nw_{l+2}(t) = 2tw_{l+1}(t) - 2(l+1)w_l(t) \tag{40}$$

The FT of $w_{l+1}(f)$ can be expressed as

$$W_{l+1}(f) = j\left[\frac{1}{4\pi}\dot{W}_l(f) - 2\pi f W_l(f)\right] \tag{41}$$

where "˙" stands for derivative with respect to frequency. For MHP, $W_0(f)$ is defined as

$$W_0(f) = 2\sqrt{\pi}e^{-4\pi^2 f^2} \tag{42}$$

The time and frequency domain representation of MHPs are given in Fig. 1.
$T_l(f)$ is the FT of the TH code which transmits the lth symbol

$$T_l(f) = \sum_{h=0}^{N_s-1} e^{-j2\pi f\left(c_{l,h}T_c + (lN_p+h)T_f\right)}. \tag{43}$$

To find the closed form expression of $P_y(f)$ in (38), the expectation of $|S_p(f)|^2$ is to be evaluated. It is given as

$$\begin{aligned} E\left\{|S_p(f)|^2\right\} = E\Bigg\{ & \sum_{l=0}^{M-1}\sum_{n=0}^{M-1} W_l(f)W_n(f)^* T_l(f) \\ & \times T_n(f)^* a_l a_n e^{-j2\pi f(\delta_l - \delta_n)}\Bigg\}. \end{aligned} \tag{44}$$

Since a_l and a_n are independent random variables derived from the same process and δ_l and δ_n are independent random variables derived from different processes. Therefore, (44) can be rewritten as

$$E\left\{|S_p(f)|^2\right\} = \sum_{l=0}^{M-1}\left\{|W_l(f)|^2|T_l(f)|^2E\{a_l^2\}+ \sum_{\substack{n=0\\n\neq l}}^{M-1} W_l(f)W_n^*(f)T_l(f)T_n^*(f) \times E\{a_l\}E\{a_n\}E\{e^{-j2\pi f(\delta_l-\delta_n)}\}\right\}. \tag{45}$$

Similarly, the second expectation in (38) can be expressed as

$$E\{S_p(f)S_q^*(f)\}=\sum_{l=0}^{M-1}\sum_{n=0}^{M-1} W_l(f)W_n^*(f)T_l(f)T_n^*(f) \times E\{a_l\}E\{a_n\}\,E\left\{e^{-j2\pi f(\delta_l-\delta_n)}\right\}. \tag{46}$$

The waveforms $s_p(t)$ and $s_q(t)$ are generated by two i.i.d processes. Therefore, the expectation in (46) is independent of l and n and equal to the case $l \neq n$ of (45) i.e.

$$E\{S_p(f)S_q^*(f)\} = E\{a_l\}E\{a_n\}E\{e^{-j2\pi f(\delta_l-\delta_n)}\} \times \sum_{l=0}^{M-1}\sum_{n=0}^{M-1} W_l(f)W_n^*(f)T_l(f)T_n^*(f) \tag{47}$$

Substituting (45) and (47) in (38), the final PSD can be formulated as in (48)

$$P_y(f) = \frac{E\{a_l^2\} - E\{a_l\}E\{a_n\}E\{e^{-j2\pi f(\delta_l-\delta_n)}\}}{N_pT_f}\sum_{l=0}^{M-1}|W_l(f)|^2|T_l(f)|^2 + \frac{E\{a_l\}E\{a_n\}E\{e^{-j2\pi f(\delta_l-\delta_n)}\}}{(N_pT_f)^2}\sum_{l=0}^{M-1}\sum_{n=0}^{M-1} W_l(f)W_n^*(f)T_l(f)T_n^*(f)\sum_k\delta\left(f-\frac{k}{N_pT_f}\right) \tag{48}$$

Although UWB signals are alike in the frequency domain, they are diverse in the time domain due to their different characteristics of time domain parameters N_p, T_f, a_l and w_l. We see that the PSD of orthogonal pulse-based modulation signals consists of continuous and discrete spectral components which change with the order of pulse waveforms and modulation schemes. The variation of PSD over different orthogonal pulse-based signaling are given in the following section.

5.1 PSD of M-ary PSM scheme

In PSM scheme, symbols are modulated only by the order of orthogonal pulses. The generalized terms in (48) are specified by a_l=1 and $\delta_l = 0$. The expectations of these variables are $E\{a_l^2\} = 1$, $E\{a_l\}E\{a_n\}_{l\neq n} = 0$ and $E\{e^{-j2\pi f(\delta_l-\delta_n)}\} = 1$ respectively. The PSD of the PSM signal can be written from (48) as

$$P_y(f) = p(f) + p_k(f) \tag{49}$$

where

$$p(f) = \frac{1}{N_p T_f} \sum_{l=0}^{M-1} |W_l(f)|^2 |T_l(f)|^2 \tag{50}$$

and

$$p_k(f) = \frac{1}{(N_p T_f)^2} \sum_{l=0}^{M-1} \sum_{n=0}^{M-1} W_l(f) W_n^*(f) T_l(f) T_n^*(f) \times \sum_k \delta\left(f - \frac{k}{N_p T_f}\right) \tag{51}$$

We see that $p(f)$ is continuous spectrum component. It depends on the TH code and the ESD of the lth order orthogonal pulse. Since ESD of different order orthogonal pulses are not identical, the selection of order of the orthogonal pulses plays an important role for continuous spectral component.

$p_k(f)$ is the discrete spectral component which induces UWB interference on the other narrow band systems Majhi, Madhukumar & Ye (2007). The discrete components of the signal appear based on the term $\sum_k \delta\left(f - \frac{k}{N_p T_f}\right)$. It shows that the position of discrete component depends on the TH code and its dynamic range of amplitude depends on the orthogonality of pulses. Since pulses are orthogonal in time and frequency domains, the value of $W_l(f)W_n^*(f)$ is approximately zero, as a result, the dynamic range of amplitude of the discrete spectral components becomes very small. This small dynamic range increases the average transmitted power in pulse and improves the UWB system performance. It helps UWB signal to coexist with other systems without any serious performance degradation. In addition, it facilitates UWB signal to keep its spectrum under the FCC spectral mask without minimizing the average transmitted power in the signal.

5.2 PSD of M-ary BPSM scheme

In BPSM scheme, symbols are modulated by order and amplitude of the pulses, i.e. $a_l \in \{\pm 1\}$ and $\delta_l = 0$. The expectation of these variables are $E\{a_l^2\} = 1$, $E\{a_l\}E\{a_n\}_{l \neq n} = 0$ and $E\{e^{-j2\pi f(\delta_l - \delta_n)}\} = 1$. The corresponding PSD of BPSM scheme can be expressed from (48) as

$$P_y(f) = \frac{1}{N_p T_f} \sum_{l=0}^{M-1} \sum_{h=0}^{N_s-1} \sum_{k=0}^{N_s-1} |W_l(f)|^2 \times \exp\left(-j2\pi f((c_{l,h} - c_{l,k})T_c + (h-k)T_f)\right) \tag{52}$$

The continuous PSD component of BPSM signal is same as PSM scheme. However, the discrete spectral components become zero due to the antipodal pulse. The PSD of the TH-UWB signal for BPSM scheme is smoothed. This allows the signal to coexist with other NB signals. The extensive studies found that any antipodal signal has only continuous spectral component Majhi, Madhukumar & Ye (2007). The continuous component can be easily fitted to FCC by using appropriate MHPs.

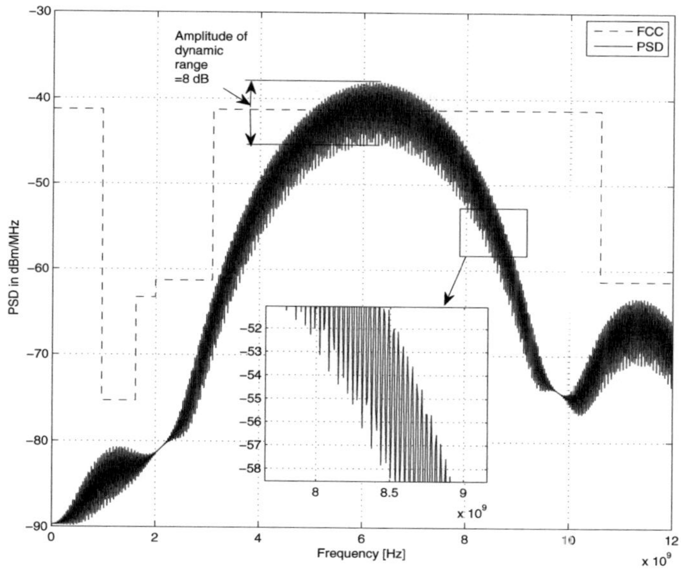

Fig. 10. PSD of 8-ary OPPM scheme with 3rd order MHP and TH code length is 8.

5.3 PSD of M-ary OPPM-BPSM scheme

For OPPM-BPSM scheme, $a_l \in \{\pm 1\}$ and $\delta_l = (l-1)\delta$, where δ is the constant time shift length. This implies, $E\{a_l^2\} = 1$, $E\{a_l a_n\} = 0$ and $E\{e^{-j2\pi f m T_\Delta \delta}\} = (1+\cos(2\pi m f T_\Delta))/2$. The corresponding PSD of OPPM-BPSM signal can be expressed as

$$P_y(f) = \frac{1}{N_p T_f} \sum_{l=0}^{M-1} \sum_{h=0}^{N_s-1} \sum_{k=0}^{N_s-1} |W_l(f)|^2 \times \exp\Big(-j2\pi f\big((c_{l,h} - c_{l,k})T_c + (h-k)T_f\big)\Big) \quad (53)$$

The PSDs of BPSM and OPPM-BPSM schemes are identical. However, OPPM-BPSM can be used for higher level modulation scheme for higher data rate systems. Therefore, OPPM-BPSM modulation is an attractive choice of TH-UWB signal from several aspects.

6. Simulation results and discussions

In this section, PSD is provided for orthogonal pulse-based signaling and compared with conventional OPPM scheme. In simulation, different order of MHPs are used with two different lengths of TH code 8 and 16. The other simulation parameters are set to $T_f = 60$ ns and pulse width is 0.7ns.

Since BPSM and OPPM-BPSM have antipodal signal, they have only continuous spectral component and shape of their spectral is same as continuous component of non antipodal signal. The only difference is that spectral of antipodal signal does not contain any discrete component. The PSD in non antipodal modulation schemes is more complicated. Since OPPM and OPPM-PSM are special cases of OPPM-BPSM, OPPM and OPPM-PSM have been chosen

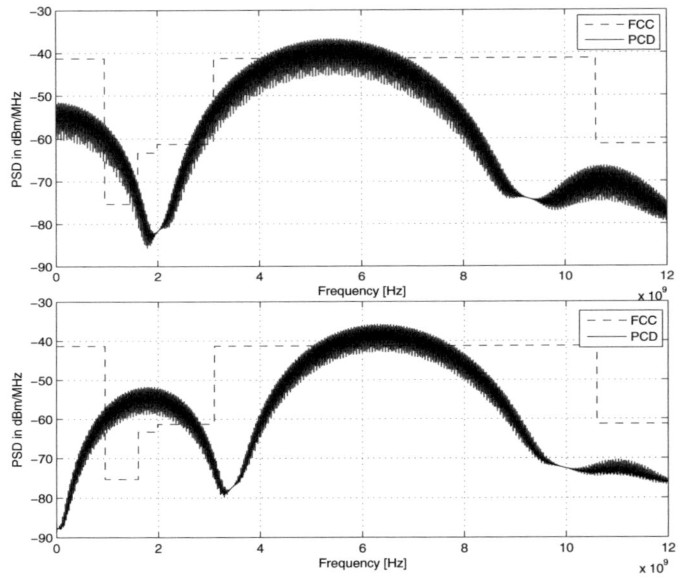

Fig. 11. (a) PSD of 8-ary OPPM scheme with 4th order MHP. (b) PSD of 8-ary OPPM scheme with 5th order MHP and TH code length is 8

to compare the PSD of the signal. The PSD of 8-ary OPPM is given in Fig.10 for 3rd order pulse and in Fig.11 for 4th and 5th order pulses with TH code of length 8 and $T_c = 7.5ns$. Since each time only one pulse is used in OPPM scheme, orthogonality is maintained by position not by pulse. The 3rd order pulse almost satisfy the FCC spectral mask except some discrete components. However, 4th and 5th order pulses do not satisfy the FCC spectral mask shown in Fig.11. The dynamic range of the amplitude of discrete components of OPPM scheme is about 8 dB which is very high. The power of the signal is calculated based on the line where the dynamic range is zero (4 dB below from the pick point). As FCC rules, pick amplitude must be below the -41.25 dBm limit. Therefore, the power of the signal is calculated based on the line which is maximum up to -45.25 dBm. As a result, signal provides low average transmitted power which degrades the system performance. Not that if the dynamic range becomes zero, the maximum limit becomes -41.25 dBm.

Fig. 12 shows the PSD of 8-ary OPPM-PSM for 4 positions and 2 orthogonal pulses with TH code of length 8. We see that that dynamic range of the amplitude of the discrete spectral component of OPPM-PSM scheme is 4 dB which is lower than the OPPM scheme even the same length of TH code is used. It is because of the orthogonality of pulses. So by reducing dynamic range, we can improve the UWB system performance by increasing the average transmitted power in the signal pulse as well as we can reduce the UWB interference over other radio systems. Again by applying TH code over these orthogonal pulse-based modulation, dynamic range of amplitude of discrete component further could be reduced. Fig. 13 shows the PSD of 8-ary OPPM-PSM with TH code of length 16 and $T_c = 3.75ns$. The dynamic range is almost reduced to 1 dB. However, it can not be reduced to zero whatever the length of TH code used. We also see that the average transmitted power in Fig. 13 is more

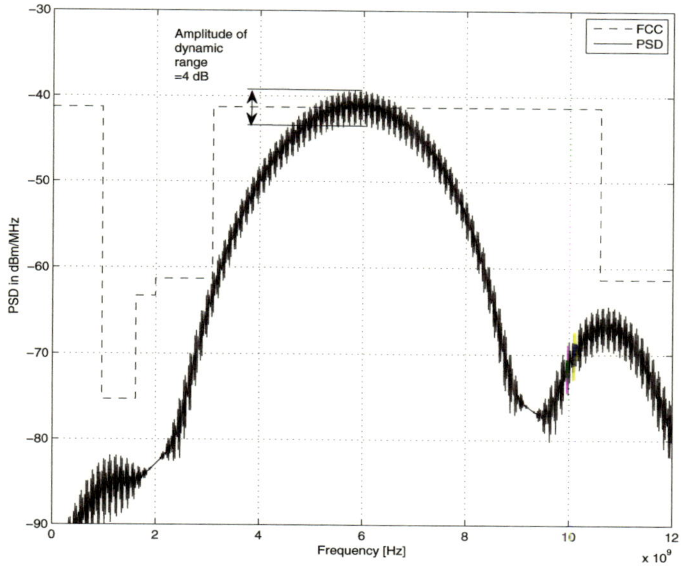

Fig. 12. PSD of 8-ary OPPM-PSM schemes for 4 positions and 2 pulses (0th and 3rd) with TH code of length 8

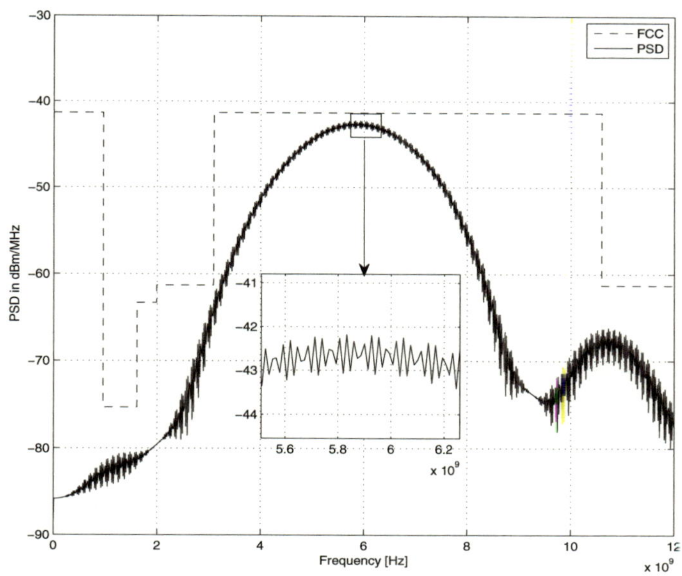

Fig. 13. PSD of 8-ary OPPM-PSM schemes for 4 positions and 2 pulses 0th and 3rd with TH code of length 16

than the previous cases. Therefore, orthogonal pulse-based TH-UWB signaling has several advantages than its complexity burden.

7. Summary

This book chapter provides TH-UWB system model based on orthogonal pulse waveform such as MHPs and PSWFs. The performance of orthogonal pulse based modulation schemes is provided over multipath channel. Several interference issues such as ISI and MAI are provided in the presence of RAKE reception. The system capacity of pulse based modulation schemes over multipath channel is analyzed in details. Finally PSD analysis for PSM, BPSM and OPPM-BPS is drawn by using two different sets of orthogonal pulse waveforms.

8. References

(n.d.).

Benedetto, M. G. D. & Giancola, G. (2004). *Understanding Ultra Wideband radio fundamentals*, Prentice Hall.

Bin, L., Gunawan, E. & Look, L. C. (2003). On the BER performance of TH-PPM UWB using Paa's monocycle in the AWGN channel, *IEEE Conference on Ultra Wideband Systems and Technologies*, pp. 403–407.

Chu, X. & Murch, R. (2005). Multidimensional modulation for ultra-wideband multiple-access impulse radio in wireless multipath channels, *IEEE Transaction on Wireless Communication* 4: 2373–2386.

de Abrue, G. T. F. & Kohno, R. (2003). Design of jitter-robust orthogonal pulse-shape modulation for UWB systems, *IEEE Global Telecommunication Conference*, pp. 739–743.

de Abrue, G. T. F., Mitchell, G. T. & Kohno, R. (2003). On the design of orthogonal pulse-shape modulation for UWB systems using Hermite pulses, *Journal Of Communications And Networks* 5: 328–343.

Dilmaghani, R. S., Ghavami, M., Allen, B. & Aghvami, H. (2003). Novel UWB pulse shaping using Prolate spheroidal wave functions, *The 14th IEEE International Symposium on Personal, Indoor and Mobile Radio Communication Proceedings*, pp. 602 – 606.

Durisi, G. & Benedetto, S. (2003). A general method for SER computation of M-PAM and M-PPM UWB systems for indoor multiuser communications, *IEEE Global Telecommunication Conference*, pp. 734–738.

Foerster, J. (2003). UWB channel modeling sub-committee report final, *IEEEP802.15 Working Group for Wireless Personal Area Networks (WPANs)* .

Gezici, S. & Kobayashi, H. (2005). Performance evaluation of impulse radio UWB systems with pulse-based polarity randomization, *IEEE Transactions on Signal Processing*, pp. 2537–2549.

Gezici, S., Sahinoglu, Z., kobayashi, H. & Poor, H. V. (2006). Ultra-wideband impulse radio systems with multiple pulse types, *IEEE Journal n Selected Areas in Communications* 24: 892–898.

Ghavami, M., Michael, L. B., Haruyama, S. & Kohno, R. (2002). A novel UWB pulse shape modulation system, *Wireless Personal Communications* 23: 105–120.

Giorgetti, A. & Chiani, M. (2005). Influence of fading on the Gaussian approximation for BPSK and QPSK with asynchronous cochanel interference, *IEEE Transaction on Wireless Communications* 4.

Guvenc, I. & Arslan, H. (2003). On the modulation option for UWB systems, *IEEE Military Communications Conference*, pp. 892–897.

Harada, H., Ikemoto, K. & Kohno, R. (2004). Modulation and hopping using modified Hermite pulses for UWB communication, *IEEE Conference on Ultra Wideband Systems and Technologies*, pp. 336–340.

Hu, B. & Beaulieu, N. C. (2004). Pulse shaping in UWB communications systems, *IEEE Vehicular Technology Conference*, pp. 5175– 5179.

Hu, W. & Zheng, G. (2005). Orthogonal Hermite pulses used for UWB M-ary communication, *Proceeding of the International conference on Information Technology*, pp. 97–101.

Hwang, J. H., Kim, S. C., S. Yoon, B. K. & Park, J. S. (2007). Performance analysis of PO-THMA UWB system using mutually orthogonal MHP pulses, *IEEE Transactions on Consumer Electronics* 53.

Jia, T. & Kim, D. I. (2005). Analysis of average signal-to-interference-noise ratio for indoor UWB rake receiving system, *in proceedings of IEEE 61st Vehicular Technology Conference*, pp. 1396–1400.

Jiang, L., , Wang, Y. & Guo, J. (2005). The capacity of M-ary PPM ultra-wideband communication over multipath channels, *IEEE International Symposium on Microwave, Antenna, Propagation and EMC Technology for Wireless Communication Proceedings*, pp. 1606–1609.

Kim, Y., Jang, B., Shin, C. & Womack, F. (2005). Orthonormal pulses for high data rate communication in indoor UWB systems, *IEEE Communication Letters* 9: 405–407.

Kim, Y. & Womack, B. F. (2007). Performance evaluation of UWB systems exploiting orthonormal pulses, *IEEE Transactions on Communication* 55.

Li, W., Gulliver, T. A. & Zhang, H. (2005). Performance and capacity of ultra-wideband transmission with pulse position amplitude modulation over multipath fading channels, *IEEE Global Telecommunications Conference*, pp. 225–229.

Majhi, S., Madhukumar, A. S., Nasser, Y. & Hélard, J.-F. (2010). Power spectral analysis of orthogonal pulse-based th-uwb signals, *VTC Spring*, pp. 1–5.

Majhi, S., Madhukumar, A. S. & Premkumar, A. B. (2006). Reduction of UWB interference at NB systems based on a generalized pulse waveform, *IEICE Electronics Express* 3: 361–367.

Majhi, S., Madhukumar, A. S. & Premkumar, A. B. (2007). Performance of orthogonal based modulation schemes for TH-UWB communication systems, *IEICE Electronics Express* 4: 238–244.

Majhi, S., Madhukumar, A. S., Premkumar, A. B. & Chin, F. (2007a). M-ary signaling for ultra wideband communication systems based on pulse position and orthogonal pulse shape modulation, *IEEE Wireless Communication and Networking Conference (WCNC)*, pp. 2795 – 2799.

Majhi, S., Madhukumar, A. S., Premkumar, A. B. & Chin, F. (2007b). Modulation schemes based on orthogonal pulses for time hopping ultra wideband radio systems, *IEEE International Conference on Communications (ICC)*, pp. 4185–4190.

Majhi, S., Madhukumar, A. S., Premkumar, A. B. & Richardson, P. (2008). Combining OOK with PSM modulation for simple transceiver of orthogonal pulse-based TH-UWB systems, *EURASIP Journal on Wireless Communications and Networking* 2008: 11.

Majhi, S., Madhukumar, A. S., Premkumar, A. B., Xiang, W. & Richardson, P. (2011). Enhancing data rates of TH-UWB systems using M-ary OPPM-BPSM modulation scheme: A system perspective, *Wireless Personal Communications* 56: 583–597.

Majhi, S., Madhukumar, A. S. & Ye, Z. (2007). Coexisting narrowband and ultra wideband systems: Analysis of power spectral density and in-band interference power, *World Scientific and Engineering Academy and Society (WSEAS)* 6: 318–324.

Majhi, S., Xiang, W., Madhukumar, A. S. & Premkumar, A. B. (2008). Theoretical capacity analysis of th-uwb systems for orthogonal pulse based modulation schemes, *VTC Fall*, pp. 1–5.

Michell, C., de Abreu, G. T. F. & Kohno, R. (2003). Combined pulse shape and pulse position modulation for high data rate transmission in ultra-wideband communication, *International Journal of Wireless Information Networks* 10: 167–178.

Mitchell, C. J. & Kohno, R. (2004). Orthogonality and coded modulation for combined pulse position and pulse shape modulation, *International Workshop on UWB Systems, Joint with Conference on UWB Systems and Technologies*, pp. 177–181.

Padgett, J. E., Koshy, J. C. & Triolo, A. A. (2003). Physical-layer modeling of UWB interference, *White Paper of Telcordia Technologies* pp. 1–121.

Parr, B., Cho, B., Wallace, K. & Ding, Z. (2003). A novel ultra-wideband pulse design algorithm, *IEEE Communication Letters* 7.

Proakis, J. G. (2001). *Digital Communications*, New York, NY, McGraw-Hill inc., Fourth Edition.

Ramseier, S. & Schlegel, G. (1993). Bandwidth power efficiencies of trellis coded modulation schemes, *IEEE GLOBAL Telecommunicaiton Conference*, pp. 1634–1638.

Saleh, A. & Valenzuela, R. (1987). A statistical model for indoor multipath propagation, *IEEE Journal of Selected Area in Communication* 5: 128–137.

Sklar, B. (2001). *Digital Communications Fundamentals and Applications*, Singapore, Pearson Education, Second Edition.

Usuda, K., Zhang, H. & Nakagawa, M. (2004). M-ary pulse shape modulation for PSWF-based UWB systems in multipath fading environment, *IEEE Global Telecommunication Conference*, pp. 3498–3504.

Wen, H. & Guoxin, Z. (2005). Orthogonal hermite pulses used for UWB M-ary communication, *Proceedings of the International Conference on Information Technology*, pp. 97–101.

Win, M. Z. & Scholtz, R. A. (1998a). Impulse radio: How it works, *IEEE Communication Letters* 2: 36 – 38.

Win, M. Z. & Scholtz, R. A. (1998b). On the energy capture of ultrawide bandwidth signals in dencemultipath environment, *IEEE Communication Letters* 2: 245 – 247.

Zhang, H. & Gulliver, T. (2005a). Biorthogonal pulse position modulation for time-hopping multiple access UWB communications, *IEEE Transaction on Wireless Communication* 4: 1154–1162.

Zhang, H. & Gulliver, T. A. (2005b). Performance and capacity of PAM and PPM UWB time-hopping multiple access communications with receive diversity, *EURASIP Journal on Applied Signal Processing* 2005: 306–315.

Zhang, L. & Zhou, Z. (2005). Research on orthogonal wavelet synthesized UWB waveform signal, *IEEE International Conference on Communication*, pp. 803–805.

3

A 0.13um CMOS 6-9GHz 9-Bands Double-Carrier OFDM Transceiver for Ultra Wideband Applications

Li Wei, Chen Yunfeng, Gao Ting, Zhou Feng, Chen Danfeng, Fu Haipeng and Cai Deyun
State Key Laboratory of ASIC & System, Fudan University
China

1. Introduction

Since 2002, ultra wideband (UWB) technology has ignited the interests of academia and industry for its potential of achieving high-speed wireless communication in short distance with low power. It is actively investigated today due to the wide available bandwidth for very high data rate up to 480Mb/s and low power service over short distances in 10m range. According to FCC (Federal Communications Commission), the frequency spectrum allocated for UWB is 3.1-10.6 GHz, and the spectrum shape of modulated output power and maximum power level are limited to -41.3dBm/MHz, which ensures that UWB can coexist with existing spectrum users like GSM(Global System of Mobile communication), WLAN(Wireless Local Area Network) and Bluetooth.

Based on MB-OFDM(Multi-Band Orthogonal Frequency Division Multiplexing), WiMedia released the initial version of Physical Layer (PHY) Specification in September 2005. In this proposal, the UWB frequency spectrum from 3.1 GHz to 10.6 GHz is divided into 14 channels with 528MHz for each channel. These sub-bands are grouped into five band groups. It is seen that by increasing the signal bandwidth significantly, ultra-wideband achieves a high channel capacity and becomes an attractive solution to the ever-increasing data rate demands in wireless personal area networks (WPAN). In December 2005, European Computer Manufacturer's Association (ECMA) proposed the standard ECMA 368/369 on high-speed UWB physics layer and media access control layer based on MB-OFDM scheme. This has pushed the industrialization of UWB technology to a new stage again.

In China, UWB technology has also become a hot topic according to the issue of the UWB standard by Chinese Government in 2008. A new UWB scheme named dual carrier-orthogonal frequency division multiplexing (DC-OFDM) has been proposed and applied in China. In China standard, only the band from 6.2GHz to 9.4GHz and the band from 4.2GHz to 4.8GHz are available for UWB applications. These bands are partitioned into 14 sub-bands of 264MHz bandwidth which means the bandwidth is halved in China's DC-OFDM standard compared with the ECMA 368/369 standard. Thus the sampling frequency of the DACs(Digital-to-Analog Converter) and ADCs(Analog-to-Digital Converter) are halved too. The power consumption of the system can be reduced greatly. Moreover, in DC-OFDM

UWB, two bands locating around two different carriers are utilized at the same time to form a bandwidth of 528 MHz for maintaining high-speed communication. In this way, the spectrum usage is more flexible and the spectrum efficiency is enhanced. However, the requirements of less than 9-ns hopping time of the carrier frequency as well as simultaneous dual-carrier outputs challenge the design of dual-carrier frequency synthesizer. Fig.1 shows the frequency spectrum for WiMedia and China UWB standard.

A fully integrated transceiver for DC-OFDM UWB system in the 6-9GHz band is present in this chapter. This chapter will describe the realization of a DC-OFDM UWB transceiver covering 6-9GHz bands in a low cost 0.13um CMOS process. Firstly, the RF receiver design will be described in section 2. Section 3 and 4 introduce respectively the designs of the RF transmitter and the 9-bands frequency synthesizer. The detailed measurement results are demonstrated in section 5, which is followed by the conclusions in section 6.

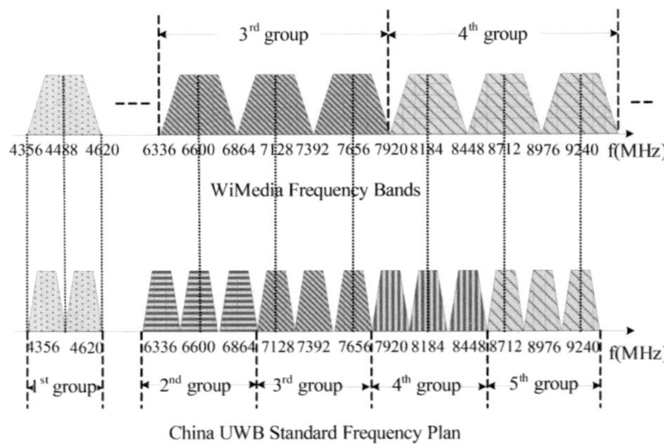

Fig. 1. Frequency spectrum for WiMedia and China UWB standard

2. RF receiver design

Fig.2 shows a block diagram of the proposed UWB receiver. Signals are received and filtered by the off-chip antenna and the RF(Radio Frequency) filter firstly. And then the received signals are amplified and converted to IF(Inter-media Frequency) baseband signal by RF front-end building blocks. After further filtering and amplifying, the analog baseband signals should be large enough to drive the ADC for digital signal processing. The receiver's local oscillator (LO) should be a fast-hopping frequency synthesizer that generates carrier tones according to the band plan in Fig.1. Performances such as in-band phase noise and reference spur are specified as -80 dBc and -40dBc respectively, which are not so stringent. And the I/Q mismatch is designed as 2.5 degree and 0.2 dB.

Normally the noise figure of channel select filter is around 30 dB, thus the conversion gain of RF front-end building blocks should be larger than 30 dB to suppress the noise from LPF(Low Pass Filter). But in that case, the linearity of the receiver will get worse. In order to improve the linearity of the receiver, the conversion gain of the RF front-end building blocks is set to be around 24 dB(average) with variable gain of 12 dB. The NF(Noise Figure) of the LPF is designed to be less than 18 dB to guarantee low noise of the receiver. The LNA(Low Noise Amplifier) utilizes a fully differential structure and presents an input matching to

50ohm for the off-chip antenna. It should provide a maximum gain of 18 dB to suppress noise from mixer and baseband circuits. As LNA sets the baseline for the noise figure of the receiver, the NF of the LNA should be optimized to lower than 5 dB. Following is a quadrature mixer with a fixed gain of 6 dB. The 5th-order Chebyshev type band-selection LPF is implemented after the mixer. Unlike normal channel select filter, the proposed LPF should provide a maximum gain of 30 dB, with a NF less than 18 dB at maximum gain mode.

According to the Friis Equation, the noise of the LPF nearly doesn't contribute to the total input referred noise of the receiver, leading to a very low noise figure. As the back-end block of the receiver, the filter tackles with slightly large signals, leading to stringent linearity requirement for the filter. Since the filter suppresses adjacent channel interferers to some extent, the linearity of the filter is proportionally improved. Sharp rejection of out-of-band signal is also required. Considering the difference between the sub-band's bandwidth of two standards, the cut-off frequency of the filter is switchable between 264 MHz and 132 MHz. Finally, the PGA(Programmable Gain Amplifier) amplifies the signal from the LPF and delivers constant-magnitude signals to the ADC.

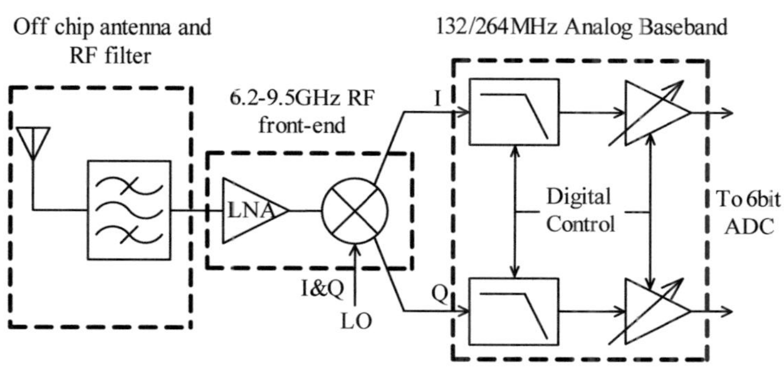

Fig. 2. Architecture of the proposed receiver

2.1 RF front-end design

Attaining an input impedance match for the wide band receiver is particularly difficult because parasitic may dominant the input impedance network. Fig.3 gives a presentation of the LNA for the proposed UWB receiver. A resistive shunt feedback topology is adopted in the LNA design, which achieves a wideband matching with a good balance between area cost and performances. Although there is a slight degradation of the noise figure comparing to other techniques like LC ladder (Bevilacqua A. et al., 2004) and transformer feedback matching (Shin D. H., et al., 2007), quite a large number of inductor coils are avoided. Bonding wire inductance $L_{bonding}$ and the ESD(Electro-Static Discharge) capacitance together with the PAD capacitance C_{pad} are co-designed with other on-chip components. The load stage is an R-L-C tank. The load inductor LL can be replaced by a differential inductor to get a smaller area. However, we split it into two symmetrical inductors for convenience of cascading with mixer in the layout. A fully differential topology is utilized in LNA design to have the input impedance match independent of the bonding wire inductance from the source of M1 to ground. Fig.4 shows the simulated S11 with different bonding wire inductance.

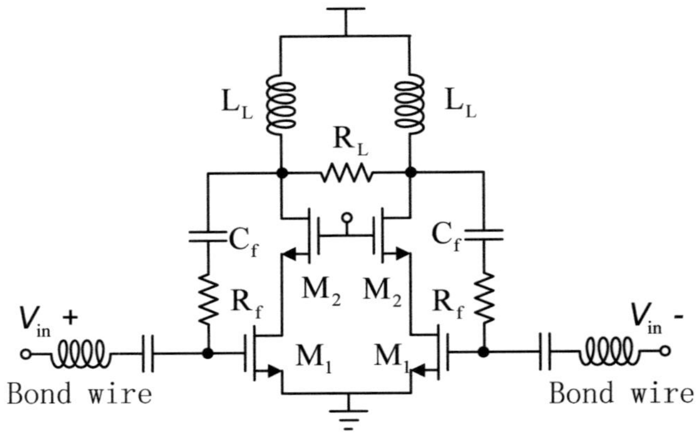

Fig. 3. Schematic Diagram of Low Noise Amplifier

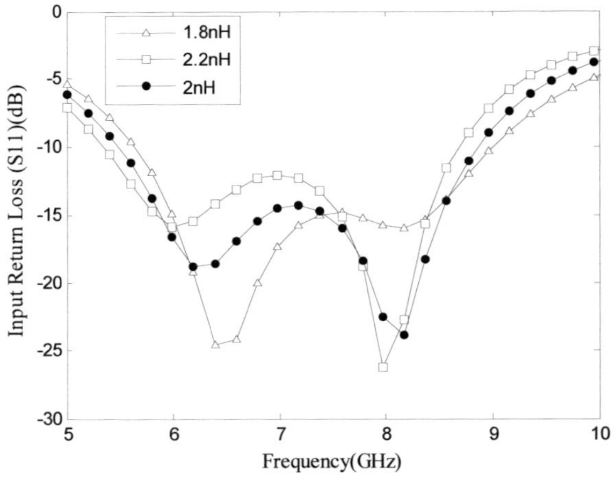

Fig. 4. Simulated S11 with different bonding wire inductance

Fig.5 shows the folded quadarture down- conversion mixer for the UWB receiver. A fully differential Gilbert-cell based structure with I/Q branches sharing the same RF input stage is implemented in the mixer, which eliminates the mismatch present in down conversion topology with separate I/Q mixers. Exploring merged architecture (Sjöland H, et al., 2003) for the quadrature mixer can also minimize the capacitive load to the LNA. Compared with the traditional structure of mixer, the folded structure utilized in this work separates the input stage and switching stage. Thus different bias current can be applied to the input stage and switching stage, better performances are achieved. The bias current of the input stage is bigger to guarantee good performance on conversion gain and noise figure. On the contrary, small current in the switching stage can lower the 1/f noise and dc-offset, which is significantly important in zero-IF receivers.

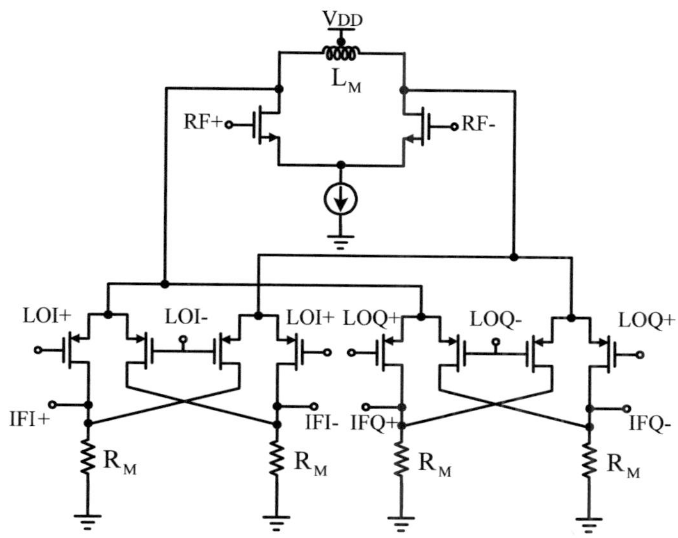

Fig. 5. Quadrature down conversion mixer circuit

2.2 Analog base-band design

The main difference between the two standards is that the intermediate frequency is 4.125MHz-264MHz and 1MHz-132MHz for WiMedia MB-OFDM and China UWB standard respectively. In order to support both standards, the cut-off frequency of the band-select filter should be switchable between 132MHz and 264MHz. Using two different filters to support each standard may be a possible solution, but will sacrifice a lot of die area. Furthermore, as the first stage of IF stage, the NF and linearity of the filter should be optimized. Thus the LPF should provide variable gain to suppress noise substantially at maximum gain mode and meet the linearity requirement when set as minimum gain. In this work, a fifth-order Chebyshev type programmable Gm-C filter is implemented.
The fifth order low pass filter is realized by a cascade of a first order RC filter and two biquads. The proposed architectures of the low pass filter and the biquad are illustrated in Fig.6. Note that the down-conversion mixer's load resistors are utilized to form the first passive RC filter stage. As a result, simulations covering both the mixer and the filters should be taken to make sure that the overall frequency response and gain are optimized.

The modified Nauta Gm cell (as shown in Fig.7) is implemented as the OTA(Operational Transconductance Amplifier) in the filter. The transconductances of the all the OTA are controlled by the digital data.

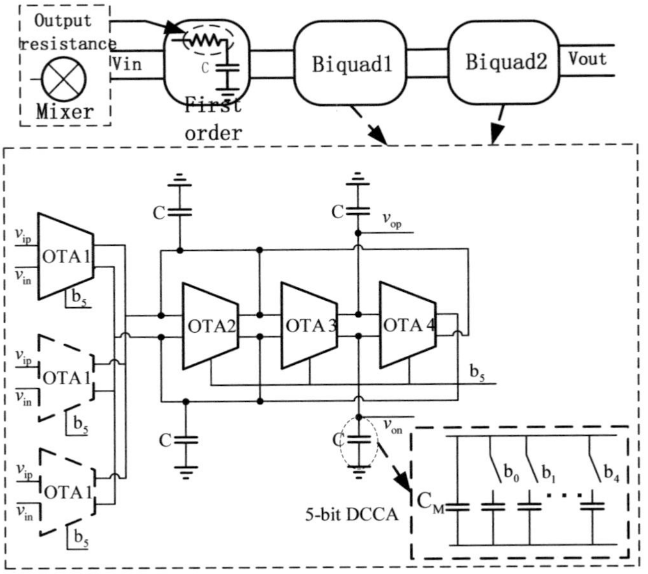

Fig. 6. Structure of the Low Pass Filter

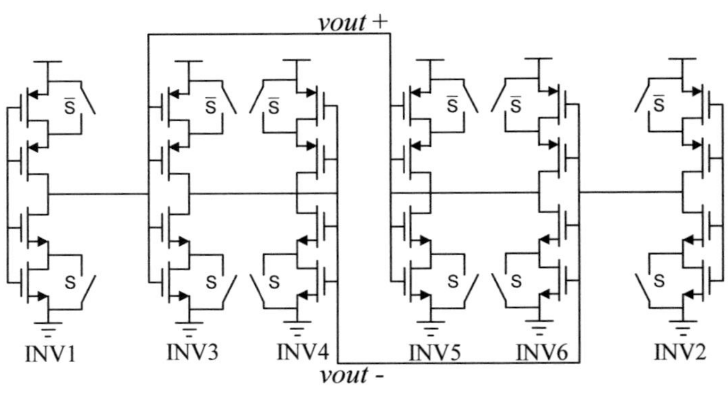

Fig. 7. Modified Nauta OTA

The topology of the PGA (Programmable Gain Amplifier) is based on a source degenerated structure as illustrated in Fig.8. A switched resistor array is implemented to achieve variable gain from 0dB to 18dB with 2dB/step. High-gain amplification easily causes the following stages into saturation due to DC-offset and DC offset also leads to second-order harmonic distortion (HD2) of the received signals, resulting in SNR(Signal-to-Noise Ratio) degradation. Thus the DC-offset cancellation circuits are also included in the PGA design. The amplitude response of the PGA is designed to be flatness within the frequency range of 264MHz.

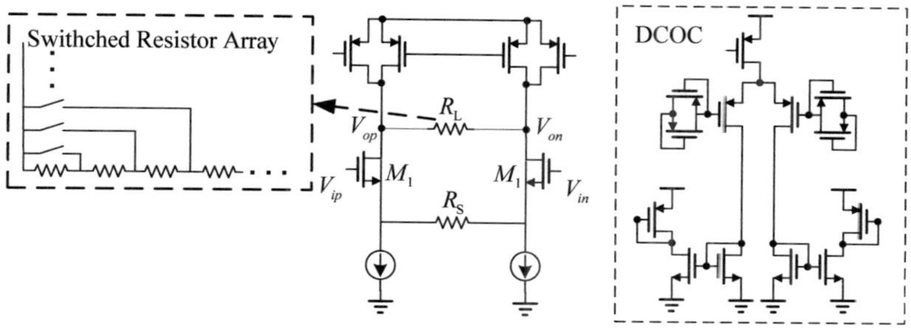

Fig. 8. Topology of the PGA

3. RF transmitter design

The proposed transmitter utilizes the direct conversion architecture for its easiness of integration and low cost. As shown in Fig.9, it consists of a dual-mode I/Q LPF with mode-switch circuits, an I/Q up-conversion mixer with high-linear voltage-to-current (V2I) units, a two-stage power driver amplifier (PA). Besides, the trans-impedance amplifiers (TIAs) are integrated to measure the AC transfer character of the LPF.

The main signal flow of this transmitter is as follows. The ABB(Analog Baseband) voltage signals from the DACs are applied at the inputs of the I/Q LPF. With the correct mode-switch bit as well as the Digital Control Capacitor Array (DCCA) control word, the image signals of the DACs and the unwanted high frequency spurs are all filtered out in both 264-MHz and 132-MHz modes. After the output voltages of the LPF are converted into ABB currents by V2I units, they are up-converted into RF voltages by the switches in the up-conversion mixer at the rate of LO. Lastly, the differential RF voltages are amplified by PA(Power Amplifier) and are converted into single-ended one via the 6-9 GHz off-chip balun, to drive the antenna.

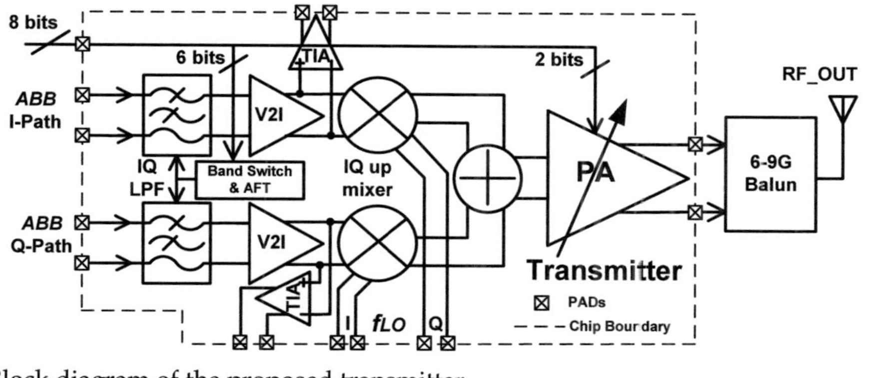

Fig. 9. Block diagram of the proposed transmitter

3.1 Dual-mode I/Q LPF design

The main requirements of this LPF are the attenuation of the out-band signals, the in-band ripple, the dual-mode operation with accurate cut-off frequency controlling and accommodation to the large input ABB voltages. According to the sampling rate of a common UWB DAC, the LPF should have an attenuation of about 45 dB from 264/132 MHz to 600/300 MHz at 264/132-MHz mode. Moreover, an in-band ripple of 0.5 dB is required. To obtain comparably good phase linearity, the 5th-order Chebyshev gm-c LPF is proposed. Besides, to deal with the ABB voltage as large as 300mVpp, the passive sub-filter is placed as the 1st-stage and the high-Q biquad is as the last stage. Also, to improve the linearity of the LPF under low supply voltage with low power, the trans-conductors are built with the Nauta's structure (Nauta B, et al., 1992).

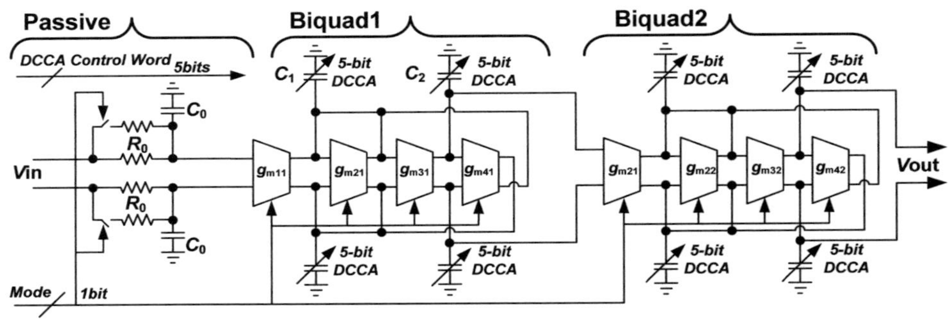

Fig. 10. Architecture of the 5th-order Chebyshev LPF with mode-switch circuits

3.2 Up-conversion mixer design

The simplified I-path schematic of the up-mixer is shown in Fig.11. It utilizes two double balanced Gilbert cells with their outputs summed to realize single-sideband (SSB) up-mixing. Since the I/Q up-mixer acts as I/Q modulator and up-conversion mixer in direct conversion transmitter, the performances of the transmitter are mainly determined by this circuit.

Low spurs, high linearity and wide bandwidth are the main challenges for the design of this up-conversion mixer. The main spurs in the output spectrum of the transmitter are the LO leakage and the sideband signal. The power of the LO leakage is determined by the offset of the I/Q ABB path. In order to reduce the power of LO leakage, an AC coupling is utilized between the V2I unit and the switches of the up-mixer as shown in Fig. 11. Besides, the linearity of the up-mixer is mainly affected by the V2I unit while the impact of the switch stage is of less importance (Zheng Renliang, et al., 2009). Many techniques (Willy Sansen, 2006) have been proposed to improve the linearity of the V2I unit. Although the complete OPAMP-assisted V2I possess better linearity, its application is restricted by the power consumption to achieve sufficient GBW(Gain Bandwidth) of the OPAMP for UWB ABB as well as the limited voltage swing because of the low supply voltage. Instead, the simple OPAMP-assisted V2I unit is preferred. As shown in Fig.11, the V2I unit consists of the input PMOS transistor M1, the source degeneration resistor R1, the current-mirror transistor M2, M3, the AC coupling capacitor CB as well as the bias resistor for eliminating the DC-offset in V2I. The feedback loop is composed of M1, R1, M2, I1 and I2, where M2 acts as the simple single-transistor OPAMP. When applied at the gate of M1, the input ABB voltage is directed transferred to the terminals of R1, because any voltage changes at the gate will be transferred to the source of M1 to maintain a fixed V_{GS} as required by the current source I1 and I2. Thus the input voltage is converted linearly into its current counterpart with a gain of 1/R1. The converted current Δi circulates in M2. Then it is mirrored into the up-mixer by M3. A 400-Ω R1 is used to improve the linearity at the cost of the gain loss in V2I. To compensate it, a 6-dB gain is set at the current mirror. Furthermore, a broadband operation of the mixer is achieved by employing a differential inductor Ld to peak with the parasitic capacitance Cpar and two series resistors Rs to reduce Q of the overall load network.

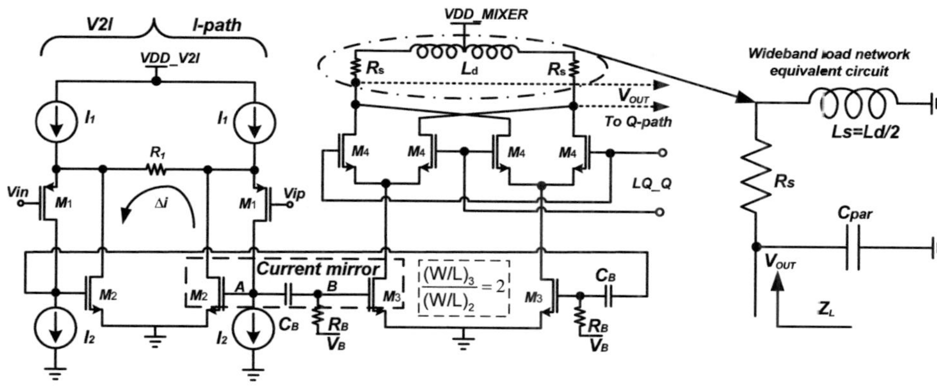

Fig. 11. Simplified I-path schematic of the up-mixer and its wideband load network

3.3 Power driver amplifier

Since the PA is the last stage of the transmitting chain, its linearity determines the output IP3(Input 3rd order Intercept Point) of the transmitter according to the Friis' formula. Moreover, the PA should possess sufficient gain to boost the output power of the up-mixer as well as to reduce the impact of former stages on the linearity of the transmitter. A flat gain of the PA is desired, too. Besides, considerations of the rejection to common-mode interferences should be taken because the tail current sources are eliminated to fit the low supply voltage.

As shown in Fig.12, the 1st stage of the PA is a combination of source follower (M1) and common source (M2) amplifier (Chang-Wan Kim, et al., 2005). The phase shift of the signal passing through the two amplifiers is 0°and 180°respectively. When the input signal Vin is applied at the two amplifiers, the common-mode signals in Vin become out-of-phase and their amplitudes are subtracted at node X/Y while the differential-mode signals in Vin become in-phase and their amplitudes are added at node X/Y. In this configuration, the input differential signals are amplified with the common-mode signals rejected. Therefore, the 1st stage increases the common-mode rejection ratio (CMRR) of the transmitter. In order to obtain a high CMRR, the gain of the two appliers, i.e. the source follower and the common source amplifier, should be equal. The transistors M1 and M2 have the identical size. Under this condition the ideal CMRR is infinite and the differential voltage gain is 6 dB. However, the post simulation of this circuit indicates that the CMRR is improved by 12 dB and the differential voltage gain is about +2 dB because the inherent unbalances between the two amplifiers. Moreover, as the impact of the parasitic capacitors the gain drops at high frequency.

The 2nd stage of the PA amplifies the RF signals to drive the off-chip balun. As the main amplification stage in this PA, its gain and linearity are important. Thus a class-A common source amplifier (Ma) is employed. A differential inductor (LPA) with center tap is used as the load of this stage to resonate with the capacitance including the parasitic capacitance of Mc as well as the PAD. Because the effective 50-Ω input resistors of balun-2 are part of the load network, its Q value is low and the gain is relatively flat. The value of the LPA is optimized according to the PAD capacitance Cpad and the bonding inductance Lb to ensure the peak of the gain is around 9 GHz instead of the middle of 6-9 GHz. Thus it compensates the gain drop of the 1st stage at high frequency. Besides, in this PA the cascode transistors, i.e. M3 and Mc, ease the Miller Effect to reduce the effective loading capacitance to the former stage and avoid the breakdown of the transistors during large signal period. 2-bit digital signals are used to select the required bias voltage for Ma; an 8-dB variable gain is realized.

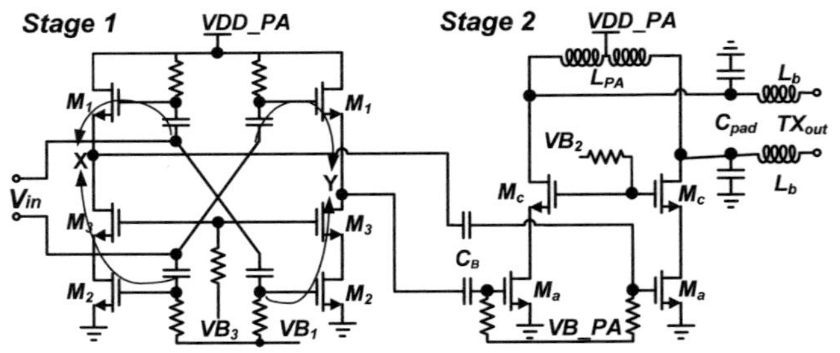

Fig. 12. Simplified schematic of two-stage PA

4. 9-bands frequency synthesizer

According to the band partition for UWB communication system shown in Fig.1, the SSB mixer-based generator for the frequency generations from group2 to group5 is proposed in Fig.13. It is based on the band generation plan (shown in Fig.14), which is designed with the objective of attaining a synthesizer solution that uses a minimum number of components while reducing the generation of spurs.

A PLL(Phase-Locked Loop) with quadrature voltage-controlled oscillator (QVCO) and an external reference of 48 MHz is implemented to generate 8448 MHz I/Q outputs as the fundamental LO frequency. The 8448 MHz in-phase and quadrature phase (I/Q) signals are applied to the quadrature SSB (QSSB) mixer to mix with another input whose frequency is switchable. These switchable input frequencies for QSSB mixer can be derived either from divided-by-2 dividers' output or from a combination of a SSB mixer1 and a divided-by-2 divider. The final output phase accuracy largely depends on the quadrature input signals of the QSSB-mixer. Since divided-by-2 dividers are used to produce I/Q signals for some synthesized frequencies for high phase accuracy. The divider's phase sequence and spectral purity may impact the mixer's phase accuracy. A double balanced quadrature-input divided-by-2 (DBQID) frequency divider is implemented to suppress the third harmonic with high precise quadrature phase sequence.

Two frequency multiplexers are used to choose the right internal frequency for each channel. The band selection is accomplished by switching the capacitor bank of the QSSB mixers to the desired frequency and simultaneously switching its input to the desired frequency and phase. Fast switching can be achieved since they operate simultaneously. To suppress the sidebands caused by nonlinearity and mismatch at the output, the number of SSB mixers has been minimized. The synthesizer's output frequencies are given as f_{fs_out}=8448+/-264*m where m=0,1,2,3. and f_{fs_out}=8448-264*n where n=4,5,6,7,8. The I/Q vectors of the internal frequencies travel through different traces and inevitably suffer from phase and gain mismatches when they reach the QSSB mixers. A Clock buffer is inserted before the QSSB mixer to calibrate the phase and gain mismatches of the input signals coming from different paths.

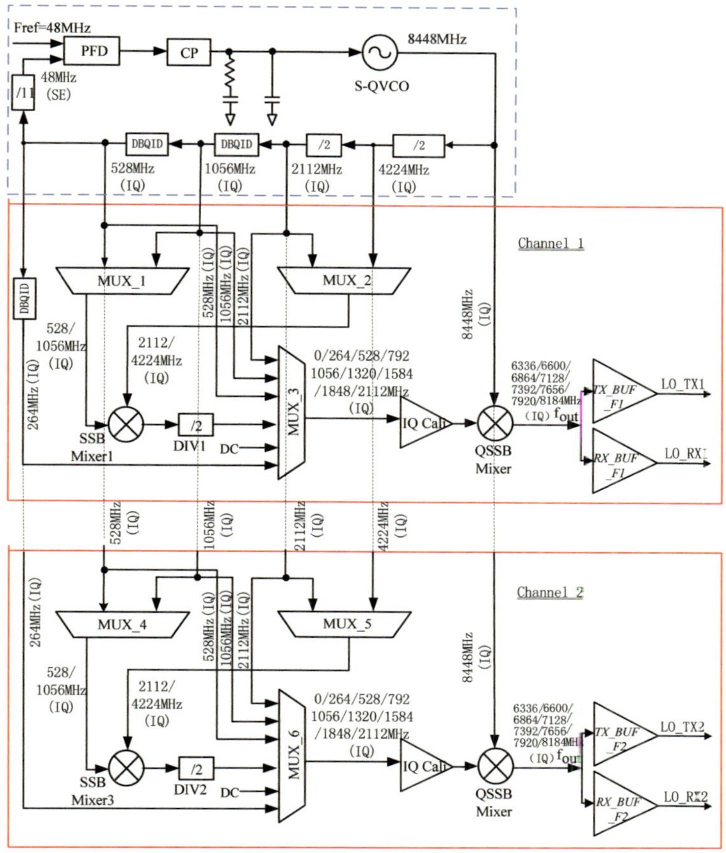

Fig. 13. Architecture of the proposed frequency synthesizer

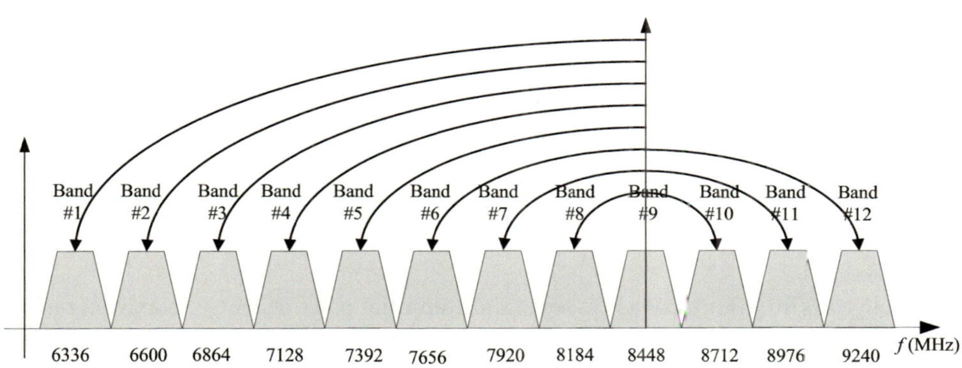

Fig. 14. Frequency plan of the proposed frequency synthesizer

4.1 QVCO design

The QVCO is the most important circuit in a PLL and its phase noise greatly determines the overall PLL output noise performance. Quadrature coupling transistors in parallel quadrature voltage-controlled oscillator (P-QVCO) make a large contribution to the phase noise. A cascode structure can greatly reduce the noise from the cascode device. Better phase noise performance can be achieved by series connection between coupling and switching transistors (Andreani P, et al., 2007).

In the case of P-QVCO, through changing the ratio of the width of coupling transistor to the width of switching transistor, phase noise and phase error can be deal with for each other. The phase error cannot be improved by increasing phase noise for the series quadrature voltage-controlled oscillator (S-QVCO). The phase error of S-QVCO depends on the amount better phase error but worse phase noise performance. To suppress the sideband caused by phase error, top-series QVCO(TS-QVCO) is adopted to generate quadrature LO signal. The width ratio of coupling transistor to switching transistor is 1/2.

As shown in Fig. 15, a linearization technique is used to lower effective K_{VCO} whereas maintain a same tuning range (Kuo C, et al., 2006). By employing this linearization, nearly the whole supply voltage range can be exploited. The varactors biased at different voltages connect with metal-insulator-metal capacitors in series as dc blockers. The dc bias voltages are generated by a resistor ladder. The resonators are both made of a differential inductor, an array of 7 bits two binary weighted switched capacitors and thick oxide MOS varactors. The tuning voltage ranges from 0 to 1.2 V. A small K_{VCO}=60 MHz/V is adopted to achieve low AM-FM noise conversion. To filter the flicker noise from the tail current transistors, a large MOS capacitor is used at the gate of the current mirror.

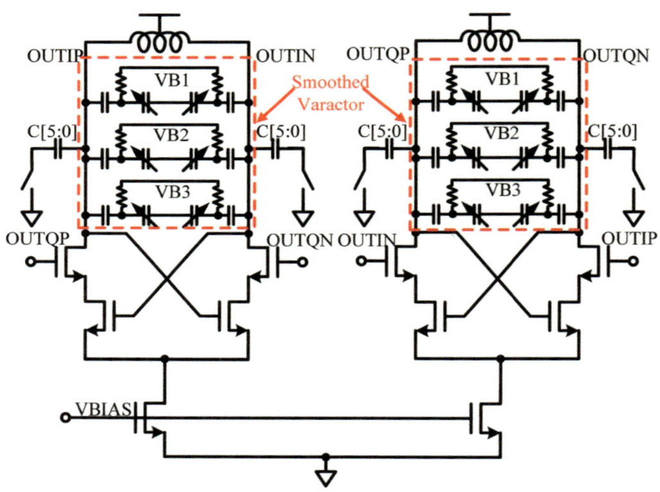

Fig. 15. Schematic of TS-QVCO

4.2 Multiplexer

The multiplexers (MUX) are based on several differential pairs sharing a common resistance load. Their activation or deactivation is through a signal enable or disable the tail current. The port leakage and third harmonic rejection are the key issues. If the unselected input frequency is leaked at the output, it will generate unwanted center frequency, which poisons the output frequency even more than those frequencies not at the center of the bands. For the third MUX, it has as many as six inputs. Thus the port leakage must be solved. There are several methods to suppress the port leakage such as cascode structure. But it is not well suited in a low voltage application.

In this design, a couple of dummy transistors are added to a conventional current-steering MUX to eliminate the unwanted coupling. Fig.16 shows the circuit of the in-phase path of MUX1. Take transistors M1 and M5 for illustration, their input signals are same, but their drains are connected to the opposite output nodes. When Vin1I is not selected and M5 is omitted, Vin1I will couple to the output through the parasitic capacitance of M1. But with the presence of M5, Vin1I will couple to the opposite output node as well. The common response will be suppressed by the differential circuit. Therefore, good isolation is achieved between different inputs. The dummy input pairs consume no extra power. The tail current source of the dummy pairs is zero and the gate of the corresponding transistor is connected to the ground. Fig.17 shows the output spectra of multiplexers with and without the dummy input pairs. The two circuits are simulated with the same operation frequency and power consumption. It shows a port leakage suppression of 46dB better with the dummy input pairs than without them.

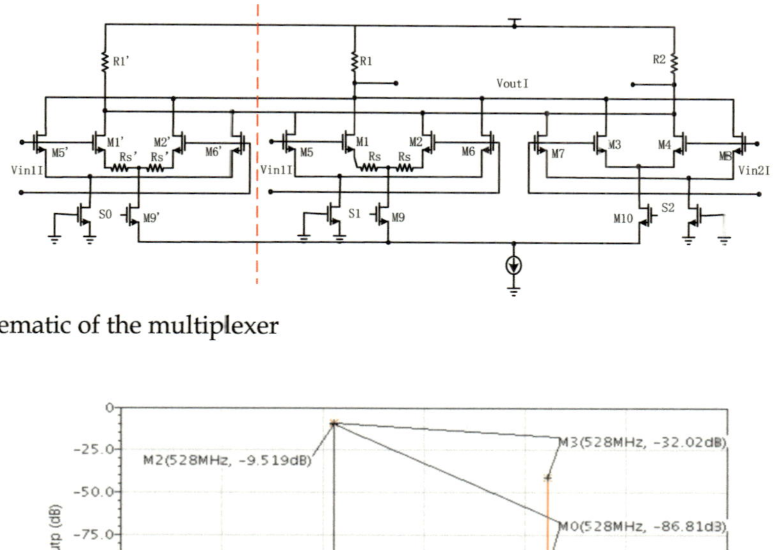

Fig. 16. Schematic of the multiplexer

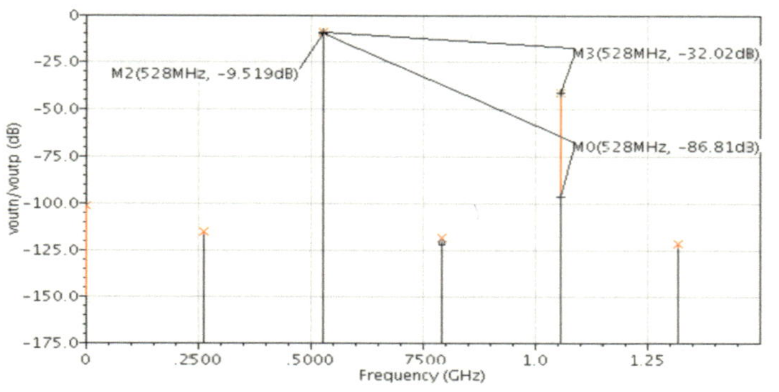

Fig. 17. Simulation result of the MUX with and without the dummy pairs

The output of the MUX is feed to a latter SSB mixer, which functions as up or down conversion according to the input phase sequence. A common way to select up or down conversion is to add a controllable in-phase/opposite-phase buffer before the SSB mixer. In this design, the phase changing is merged in the MUX. As shown in Fig.16, the part in the left of the line is a replication of the right part, excepting the output which is connected to the opposite side. Hence, S0 and S1 change the phase sequences of Vin1, which helps to change between up and down conversion. As a result, such a configuration minimizes the hardware requirement, power consumption, and undesired sidebands.

4.3 IQ calibration

Phase and gain mismatch at the input would cause image signal at the output of the SSB mixers. In order to counterbalance these mismatch, an IQ calibration module is interpolated before the second SSB mixer. Fig.18 shows the circuit of this module. Two identical buffers driven by the in-phase (0°) and quadrature-phase (180°) signals produce two outputs, normally equal to 45° and 135°, respectively. Here, one of the outputs is adjustable, while the other is fixed. These two outputs are quadrature in ideal. If th,

An interesting phenomenon using this method could be found. That is, the phase mismatch at the input will be transformed into the gain mismatch at the output; and the gain mismatch at the input will be transformed into the phase mismatch at the output. This could be illustrated by Fig.19, wherein In,i and In,q represent the input in-phase and quadrature-phase signal, Out,i and Out,q represent the output in-phase and quadrature-phase signal. It can be easily seen that the phase and amplitude error will be transformed into each other at the end. Hence Vbcon which adjusts the output phase will be used when there are some amplitude mismatches at the input, while Icon which adjusts the output amplitude will be used when there are some phase mismatches at the input. This IQ calibration module should be capacitor coupled to the latter stage, since the output common voltage will be changed as the tail current is adjusted. When Vbcon or Icon changes, the tail-current-source transistor should stay in the saturation region. Hence the drain voltage of M7 should be high enough to cover all the changes. Besides that, the Vdsat of M5 should be high enough, because it will determine the range of the phase adjustable. As a result, the Vgst of the input transistor M1 is inevitably small. This will result in a severe third harmonic. A distortion cancellation technique is used here to ease the situation. An additional pair of transistors M3 and M4 is added to the input. They are cross coupled to the output as to the main input pair M1 and M2. The Vgst of the main pair and the additional pair is set to be 1.6:1 while the tail current is set to be 4:1. The third harmonic could be effectively suppressed in this way, with the cost of a degenerated output voltage.

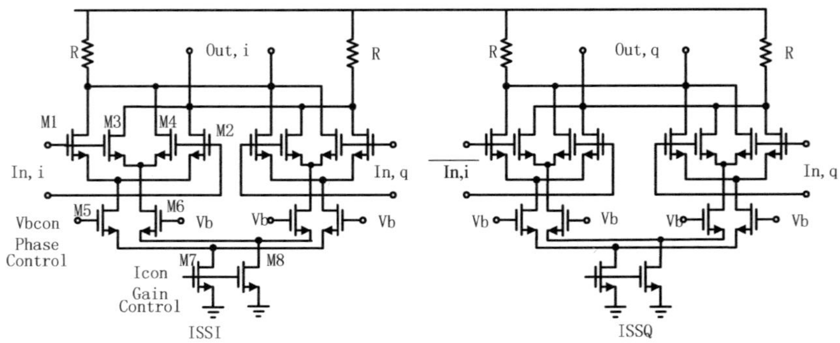

Fig. 18. Schematic of the IQ calibration

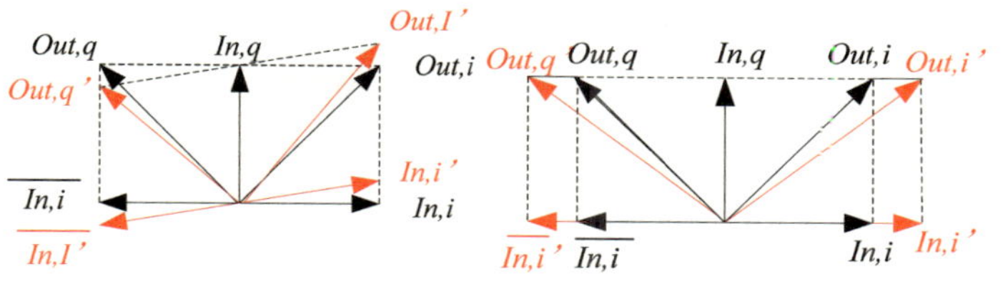

Fig. 19. Illustration of the phase and gain mismatch translation

5. Measurement results

The proposed 6-9GHz DC-OFDM UWB transceiver is fabricated in the 0.13-μm RF CMOS process. Fig.20 shows the die microphotograph. Total employed area including pads is 3.6 mmx4.5 mm.

In order to support both WiMedia MB-OFDM and China UWB standard, the receiver's intermediate frequency is required to be switchable between 4.125-264MHz and 1-132MHz. Fig.21 shows the frequency response of the receiver for two standards with LO signal at 7.656GHz. The receiver achieves variable gain of 18dB to 80dB with 2dB/step for both standards. Gain flatness of less than 3dB is normally achieved. The spectrum mask of the transmitter with 40MHz QPSK modulation signal is shown in Fig.22. The integrated power is calculated at -10dBm.

The measured tuning range of QVCO is from 7.6 to 9.5 GHz. With an external reference of 48MHz, the synthesizer is locked at 8448 MHz with a phase noise is －90 dBc/Hz at 100 kHz offset and －105 dBc/Hz at 1 MHz offset as shown in Fig. 23. A normalized phase noise floor of -226.7 dBc/Hz is achieved. The measured integrated phase noise from 10 kHz to 50 MHz is 1.93°. The reference spur level is -72 dBc seen from the Agilent E4440 spectrum analyzer. Experimental results reveal that the worst-case sideband occurs at 6336 MHz, which is about -30 dBc without I/Q calibration, as shown in Fig. 24, and is suppressed to below -45 dBc when the calibration is on. The band switching behavior is shown in Fig. 25. The bands are switched periodically and the synthesizer output is monitored. The longest switching time approximately equals to 1.4 ns, much less than 9.5 ns which is specified in dual-carrier MB-OFDM UWB systems.

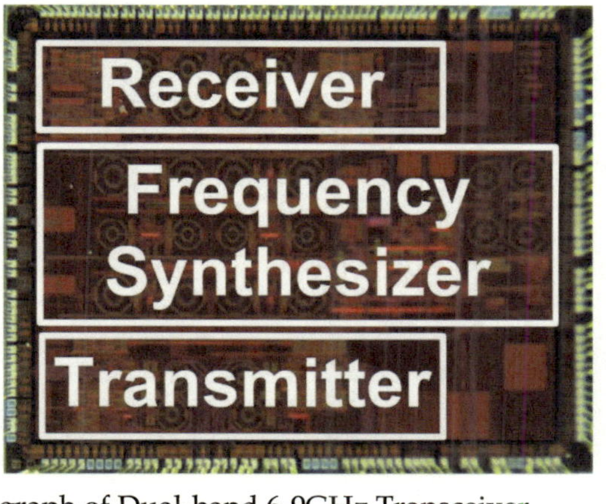

Fig. 20. Die microphotograph of Dual-band 6-9GHz Transceiver

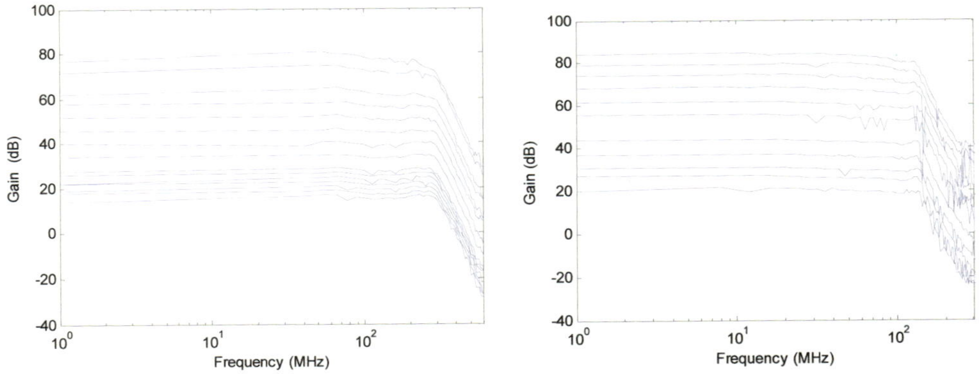

Fig. 21. Receiver frequency response, (left) At cut-off frequency of 264MHz; (right) At cut-off frequency of 132MHz

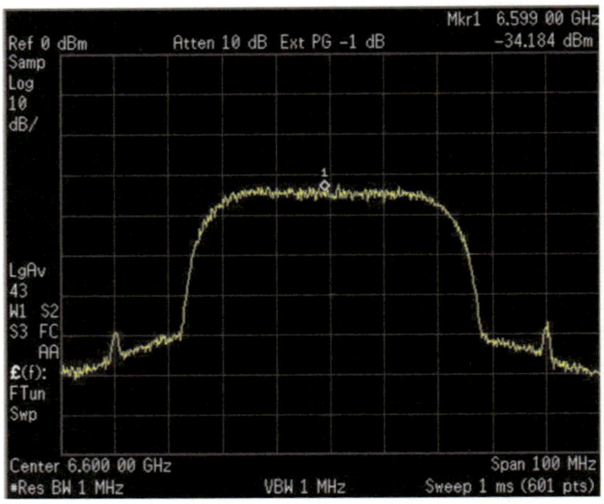

Fig. 22. Spectrum mask of the transmitter with 40MHz QPSK modulation signal

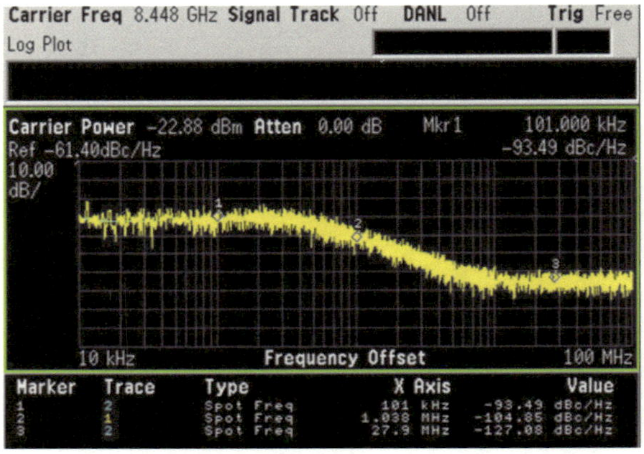

Fig. 23. Measured PLL phase noise and integrated phase noise

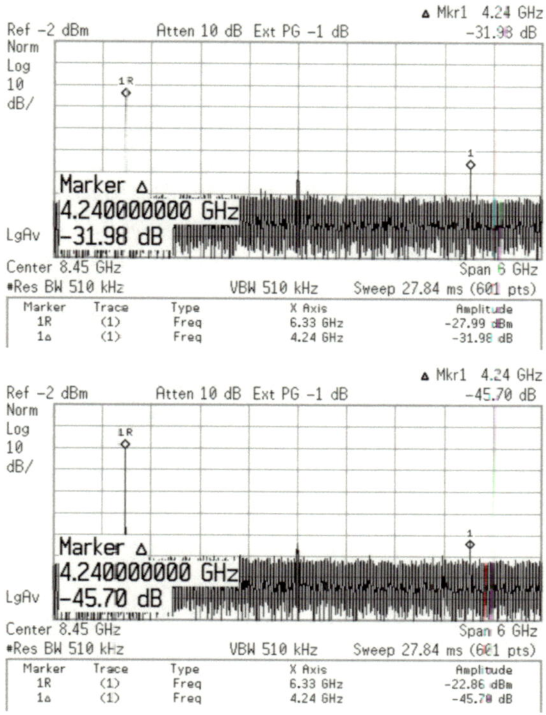

Fig. 24. Output spectrum of 6336 MHz (up) without calibration and (down) with calibration

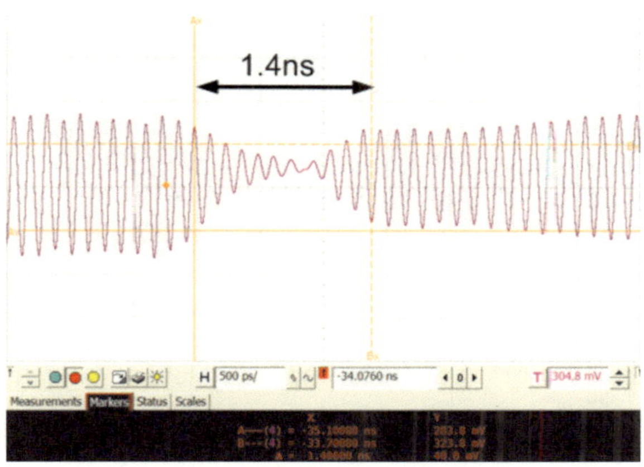

Fig. 25. Measured band switching behaviour

6. Conclusion

A 0.13um CMOS 6-9GHz 9 Band Double-Carrier OFDM Transceiver for Ultra Wideband Application is proposed, which consists of a dual-carrier frequency synthesizer, two 6-9 GHz receiver cores, two 6-9 GHz transmitter cores and the digital control logical circuits, and so on. Only one PLL and two single-sideband mixer have been used in this dual-carrier frequency synthesizer. The VCO in the PLL oscillates at 8448 MHz. All the desired frequencies are generated by mixing the output of the divider chain with that of the VCO. Two carries can be output simultaneously with independent frequency setting and totally 9 carrier frequencies from 6336 MHz to 8712 MHz with a frequency gap of 264 MHz are available. The receiver core is composed of a 6-9 GHz LNA, IQ down-conversion mixer, LPF as well as VGA. The total gain of the receiver chain has been achieved about 80 dB with a minimum NF of 4.5 dB. The transmitter core consists of the IQ LPF, IQ up-conversion mixer and the power driver amplifier. The output IP3 of the transmitter is around +10 dBm and the output 1-dB compression point is about +0 dBm, which reveal high linearity of the transmitter. Besides, the power of LO leakage and sideband signal are less than -35 dB respect to the power of the desired RF signal. The chip size of the DC-OFDM transceiver is 3.6 mm×4.5 mm including the ESD-PADs. The power consumption is around 400mA from 1.2-V supply. The future research will focus on performance optimization, power consumption, and so on.

7. References

Andreani P, A Low-Phase Noise,Low Phase Error 1.8 GHz Quadrature CMOS VCO, ISSCC 2002 : pp.290–291.

Anuj Batra et al., Multi-band OFDM physical layer proposal for IEEE 8.2.15 Task Group 3a, Nov,2003.

A. Mazzanti, E. Sacchi, P. Andreani & F.Svelto, "Analysis and Design of a Double-Quadrature CMOS VCO for Sub-harmonic Mixing at Ka-Band", IEEE Trans. Microwave Theory and Techniques, Vol. 56, No. 2, pp. 355-363, 2008.

A. Mazzanti, L.Larcher & F.Svelto, "Balanced CMOS LC-tank analog frequency dividers for Quadrature LO Generation", IEEE Custom Integrated Circuits Conference, 2005, 9: 575-578.

Bram Nauta, A CMOS Transconductance-C Technique for Very High Frequencies, IEEE Journal of Solid-State Circuits, 1992, 27(2):142.

Bevilacqua A. & Niknejad A.M. An Ultra wideband CMOS Low-Noise Amplier for 3.1-10.6-GHz Wireless Receivers.IEEE J. Solid-State Circuits, 2004, 39(12): 2259-2268

B. Razavi, T. Aytur, F. Yang, R. Yan, H. Kang & C. Hsu, et al. "A 0.13 μm CMOS UWB transceiver", IEEE ISSCC, pp. 216–217, Feberuray 2005.

Chang-Wan Kim, Quoc-Hoang Duong & Seung-Sik Lee et al.,A 0.13μm CMOS UWB RF Transmitter with an On-Chip T/R Switch, ETRI Journal, 2008, 30 (4) :526.

Chang-Wan Kim, Seung-Sik Lee & Bong-Hyuk Park, et al. A CMOS Direct-Conversion I/Q Up-Mixer Block for Ultra-wideband System, Advanced Communication Technology, 2005:23.

Chen Danfeng, Fu Haipeng. Chen Yunfeng, Li Wei, et al., "A Sideband-Suppressed Low-Power Synthesizer for 14-Band Dual-Carrier MB-OFDM UWB Transceivers", IEEE ISCAS, pp.2139-2142, May 2010.

C. Mishra, A. Valdes-Garcia & F. Bahmani, et al., "Frequency planning and synthesizer architectures for multiband OFDM UWB radios", IEEE transaciton on microwave theory and techniques, vol. 53, pp. 3744-3756, Dec. 2005

D. Leenaerts, R. Beek, G. Weide, J. Bergervoet, K. Harish & H. Waite, et al., "A SiGe BiCMOS 1ns Fast Hopping Frequency Synthesizer for UWB Radio", ISSCC 2005, pp. 202-203, February 2005.

H. M. Chien, T. H. Lin & B. Ibrahim, "A 4GHz fractional-N synthesizer for IEEE 802.11a", IEEE VLSI, pp. 46-49, Jun. 2004

Huang Zue-Der, Kuo Fong-Wei, Wang Wen-Chieh & Wu Chung-Yu, "A 1.5-V 3~10-GHz 0.18-μm CMOS frequency synthesizer for MB-OFDM UWB applications", 2008 MTT-S International Microwave Symposium Digest, pp. 229-232, June 2008

Kuo C., Chang J. & Liu S., "A Spur-Reduction Technique for a 5-GHz Frequency Synthesizer", IEEE Trans. Circuits and Systems—I: Regular Papers, Vol. 53, NO. 3, March 2006, pp.526-533.

Liang C.F., Liu S.I., Chen Y.H., Yang T.Y. & Ma G.K., "14-band frequency synthesizer for MB-OFDM UWB application", IEEE Int. Solid-State Circuits Conf. Dig.Tech. Papers (ISSCC), pp. 126-127, Feb. 2006

Nauta B. A CMOS Transconductance-C Filter Technique for Very High Frequencies. IEEE journal of Solid-State Circuits,1992, 27: 142-152

Nikookar, R. Prasad. Introduction to Ultra Wideband for Wireless Communications, Signals and Communication Technology, B.V.: Springer Science & Business Media, 2009.

P.Andreani."A Low-Phase Noise,Low Phase Error 1.8 GHz Quadrature CMOS VCO", ISSCC 2002, pp.290–291,February 2002.

Shin D. H., Park J. & Yue C. P. A Low-Power, 3-5-GHz CMOS UWB LNA Using Transformer Matching Technique.IEEE Asian Solid state circuits conference, 2007, 95-98

Sjöland H, Karimi-Sanjaani A & Abidi A A. A merged CMOS LNA and mixer for a WCDMA receiver [J]. IEEE Journalof Solid-State Circuits, 2003, 38(6): 1045-1050.

Tanaka A, et al. 1.1 V 3.1-to-9.5 GHz MB-OFDM UWB transceiver in 90 nm CMOS. IEEE International Solid-State Circuits Conference, 2006, 120-121

T.Deliyannis, Sun Yichuang & J. Kel Fidleret, Continuous-Time Active Filter Design, Boca Raton: CRC Press LLC, 1999.

Wang C.S., Li W.C., Wang C.K., Shih H.Y. & Yang T.Y., "A 3-10GHz full-band single VCO agile switching frequency generator for MB-OFDM UWB", IEEE Asian Solid-Sate Circuits Conference, pp.75-78, Nov. 2007.

Willy Sansen, Analog Design Essential, Springer Press, 2006.

Werther O, et al. A fully integrated 14-band 3.1-to-10.6 GHz 0.13 _m SiGe BiCMOS UWB RF transceiver. IEEE International Solid-State Circuits Conference, 2008, 122-123

Zheng H., Lou S. H. & Lu D. T., et al. A 3.1 GHz-8.0 GHz Single-Chip Transceiver for MB-OFDM UWB in 0.18_m CMOS Process. IEEE journal of Solid-State Circuits, 2009, 44: 414-426

Zheng Renliang, Ren Junyan & Li Wei, et al., A 3.1-4.8 GHz transmitter with a high frequency divider in 0.18μm CMOS for OFDM-UWB, Journal of Semiconductors, 2009, 30(12):25003-1–8.

Zisan Zhang, Koen Mertens & Marc Tiebout. A 6-9 GHz WiMedia UWB RF Transmitter in 90nm CMOS, IEEE Radio Frequency Integrated Circuits Symposium, 2008:39.

4

Implementation-Aware System-Level Simulations for IR-UWB Receivers: Approach and Design Methodology

Marco Crepaldi[1], Ilze Aulika[1] and Danilo Demarchi[2]
[1]*Center for Space Human Robotics @Polito, Istituto Italiano di Tecnologia, Corso Trento, Torino*
[2]*Dipartimento di Elettronica (DELEN), Politecnico di Torino, Corso Castelfidardo, Torino*
Italy

1. Introduction

Impulse-Radio Ultra-Wide Band technology (IR-UWB) allocates very large bandwidth with short duration pulses. Interest for research started in 2002 when Federal Communication Commission (FCC) normed the power spectral densities allowed for unintentional and unlicensed UWB radiators in the pre-existing full communication band 0-10 GHz FCC (2002). An ultra-wide band pulse has some unique features compared to conventional wireless signals. If on the one hand, narrowband signals envelope is close to a time unlimited continuous function, on the other hand, in a possible conception pulses can be perfect duty cycled tones having limited time support. Pulses with very short duration occupy very large bandwidth and this is in contrast to the narrowband approach, that subdivides the available spectrum into small slices for efficiently allocating radiated power. IR-UWB is then very interesting because it poses these kinds of challenges, i.e. the use of pulses and the coexistence with the existing RF systems.

The use of short duration pulses implies a physical limitation which normally narrowband RF systems are excluded from. These are multipaths, that is reflections from the objects localized in the operating environment. This has conditioned the use of IR-UWB for very high data rates applications because notwithstanding the very large theoretical channel capacity, a very high data rate communication is now almost infeasible with low complexity electronics tackling multipath diversity. IR-UWB has then been proposed for short/medium range Ultra-Low Power (ULP) communication Wireless Sensor Networks (WSN) Bielefeld et al. (2009); IEE (2007); Lecointre et al. (2010); Stoica et al. (2005); Verhelst & Dehaene (2008); Wang et al. (2011). At the transmitters very low average consumed power is possible with aggressive duty cycling, as well as in receivers even if with lower efficiency. Transmitters radiate dBm-order power signals in just 1-3 ns and receivers typically demodulate and synchronize data by detecting the presence of the UWB pulses with time domain computations.

One important key-word for understanding how IR-UWB will possibly impact on new ULP applications is "system-level". The validation of a receiver or a transmitter architecture being aware of the impact of blocks physical implementation prior to full low-level

design can possibly lead to significant performance increase and help lower complexity. Based on these considerations, this book chapter shows a methodology used for IR-UWB receivers simulation, design and conceptualization. A multi-level approach is presented and contextualized with an implementation example, that is an energy detection receiver. This design methodology has been already presented in Crepaldi et al. (2007) and extensively used in Casu et al. (2008). In this book chapter we expand it and provide more comments and considerations based on successive works dealing with IR-UWB system-level design.
Section 2 considers an Energy Detection receiver as a case study and section 3 introduces the design methodology after emphasizing its requirements. Later, section 4 applies the methodology to a specific block of the receiver and section 5 shows the obtained simulation results. Section 6 concludes the chapter.

2. A case study: the Energy Detection receiver

IR-UWB Energy Detection receivers represented mostly the number one choice for WSN and have been widely integrated and researched starting the second half of 2000-2010 decade Crepaldi et al. (2010); Daly et al. (2009); Lee & Chandrakasan (2007). Energy detection receivers are robust and of easy implementation notwithstanding being non-coherent, therefore sub-optimal. In the beginning, research was focused on conceptualized architectures that studied the communication performance of IR-UWB and attempted to solve some system-level issues. An example for non-coherent M-PPM receivers is given in Carbonelli & Mengali (2006). The proposed architectures did not deeply account for circuit-level implementation details. Starting from this first conceptualization mechanism, first energy detection receivers have been proposed Stoica et al. (2005), Lee & Chandrakasan (2007). By then all the required system-level performance figures were validated on silicon for the first time. This, and the successive receivers proposed by then, aimed towards lower energy consumption or to increase performance of some of these reference points. In this book chapter we refer to a somewhat old energy detection receiver scheme, in which an Analog-to-Digital Converter (ADC) is used for data demodulation as well the use of other blocks that differ compared to recent implementations. Here, we explicitly utilize this scheme because it represents a case study, and still, valid ideas can emerge from the analysis of this system from cross-sectional views.
A standard energy detection receiver block scheme is depicted in fig. 1. The complete transceiver is assumed to be fully implemented as a silicon System-on-Chip (SoC) and at this stage the transmitter is assumed to be only behaviorally modeled. The antenna switch commutates the wideband antenna to receiver and transmitter ends, while an external Band-Pass Filter (BPF) ensures that on-chip generated UWB pulses satisfy the FCC mask and, at the same time, filters out-of-band interference from the received ones. The energy detector, depicted in the *front-end* part is composed of a linear amplification block, the Low-Noise Amplifier (LNA), Variable Gain Amplifiers (VGA) a squaring unit and an Integrate&Dump (I&D). The receiver computes the raw pulse energy. By assuming that integration generically starts at t_a and ends at t_b, $Ar(t)$ is signal at the output of the VGA, where A is the gain of the previous blocks, the energy E at the output of the I&D is,

$$E = \int_{t_a}^{t_b} A^2 r(t)^2 dt \tag{1}$$

To run both synchronization and demodulation the receiver circuitry operates on t_a and t_b to detect for example the maximum energy peak and, for 2-PPM receivers, activate

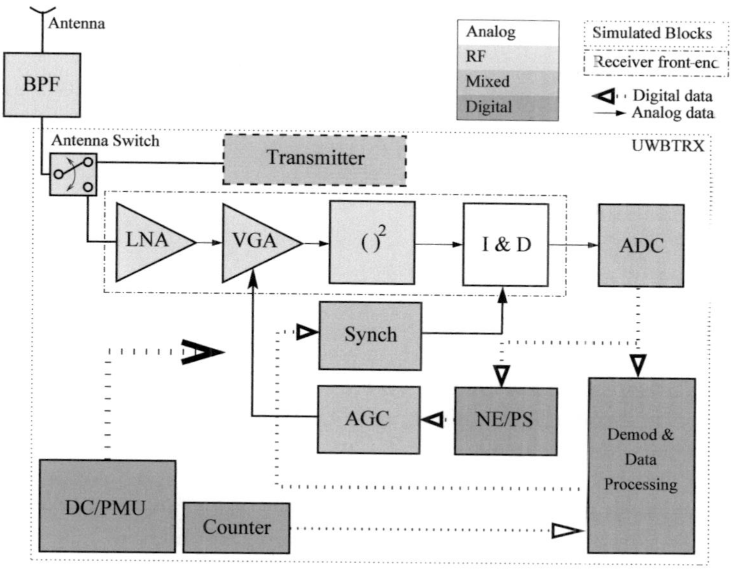

Fig. 1. Energy Detection transceiver block scheme Crepaldi et al. (2007).

integration once pulses timing is acquired at the correct '1' or '0' bins. For gain control the receiver operates on parameter A with an digital-to-analog feedback from the demodulation chain. After energy is calculated it is quantized with an ADC and then processed by the back-end that can implement a threshold based demodulation algorithm for OOK, or a relative comparison as in the case of 2-PPM. Here the receiver operates with 2-PPM modulation. The Data processing block controls also the synchronization unit, that operates similarly to a Delay-Locked-Loop (DLL) for searching the maximum energy peak within a known preamble. The Automatic Gain Control unit (AGC) automatically sets the front-end gain based on the digitized energy. The NE/PS block, namely Noise Estimation&Preamble Sensing block, helps detecting the presence of a preamble once the receiver is activated and collects energy samples from channel when no pulse is transmitted. This helps assessing the clearance of channel as soon as receiver is activated, therefore allowing system shutdown in case no packet is received. Data saved by this digital block is used for adjusting the gain of the receiver front-end for allowing the input range adaptation of the input signal for I&D and consequent A/D conversion. Note that here, receiver sensitivity is defined by the LNA, that shall have the highest gain and the lowest noise figure. The noise figure of the successive VGA units is not as influent as for the first stage because input-referred noise figure is calculated by propagating each amplifier noise figure with Friis formula. Notwithstanding this, the receiver must provide enough amplification to process the UWB pulses, overcome the non-linear law of the squaring unit and the channel path-loss that highly depends on the objects distributed in space. The Counter in the high-level architecture is useful UWB pulses Time-of-Flight calculation, in this case with a Two-Way-Ranging (TWR) packet exchange (defined in section 5). The Duty Cycling/Power Management Unit (DC/PMU) implements receiver duty cycling and deactivates the front-end units to save energy when the receiver is idle. The full implementation of this block requires the definition of the complete packet exchange mechanism as well as detailed information on each single block of the receiver.

Therefore, the complete development of the DC/PMU must be faced at the end of the design but it shall not be considered less important than the others.
It is worth mentioning that our methodological approach is devoted to system-level implications rather than being focused on circuit-level challenges. As recent research shows, we believe that one of the next steps for PHY IR-UWB systems research has to regard both decreasing energy consumption and solving problems from a more general and wide-sense system-level view Gorlatova et al. (2010).

3. The substitute-and-play design methodology

3.1 Simulator and target system

The methodology outlined here is applied on a specific simulation tool called ADVanceMS (ADMS, Mentor Graphics, now Questa ADMS) that allows multi-language descriptions with multi-resolution simulations. It supports VHDL-AMS, Verilog-AMS, VHDL, Verilog, SystemVerilog, SPICE[1] and SystemC in the same simulation environment. The Very High Speed integrated circuit Hardware Description Language (VHDL), similarly to Verilog, is widely used to logically and behaviorally describe digital circuits, modular by construction and based on a very simple math. VHDL is a concurrent language in which every described process works in parallel with the others. Communication among processes is based on events. Before evolving to the next time step, the simulator engine processes a single list in which all process events are queued. While this task is accomplished simulation time is frozen. The VHDL-AMS (AMS is for Analog and Mixed-Signal extensions) language is an extension of the common VHDL IEEE (2007) and adds directives and constructs to support at the same time both digital concurrent and simultaneous statements. These last ones, are used to allow the implementation of the continuous-time nature of analog systems. Continuous-time simulations are not based on events, but on the computation of quantities representing the solution of a continuous mathematical model. In a mixed-signal simulation the inter-communication between these two totally different worlds is ensured by the software tool that handles the different VHDL constructs depending on the cases and interfaces them to a simulation kernel, for example SystemC.
With the same continuous-time granularity the tool can include SPICE-level netlists in the description. Netlists can be directly interfaced to VHDL-AMS, therefore a block can painlessly jump from a behavioral world to the voltage and current domain of silicon devices. Also, other commercial tools such as Cadence IC provide multi-level and multi-resolution descriptions but still they are based on an analog point of view, referring to the system-level use of circuit blocks instead of exploiting the flexibility of a digital description language formalism. Another example is Advanced Design System (ADS, Agilent) that enriches its system-level design flow with low-level electro magnetic simulations. All these tools are frameworks meant to bridge multiple description languages and simulation tools transparently to the user. Here, with this methodology, we believe that that the use of a single and homogenous formalism, with possibly a single simulator, can make the difference.
The evaluation of system-level performance of an IR-UWB system in time-domain is important. As an example, let us consider Duty Cycling (DC). Ideally an IR-UWB receiver has to be kept operating for time durations on the order of few nanoseconds sufficient for receiving pulses from channel and be shut-down for the remaining time to save power

[1] In the following paragraphs we will refer to SPICE descriptions by referring to the name of the Mentor Graphics simulator, ELDO.

consumption. Typically RF front-ends have resonant loads therefore, depending on the implementation, spurious pulses can be erroneously generated whenever a hard digital activation signal operating on active amplification elements is toggled. If the RF amplifiers are simulated only in AC and integrated without a time-domain verification, at the measurements time the system performance can be seriously compromised or even the receiver cannot operate because the successive baseband and backend units are saturated. Therefore, in this methodology we stressed out the time-domain aspect of simulations and to save runtime used the multi-resolution feature to activate only the most important non-ideality required for obtaining figures as much close as possible to the physical verification. Unfortunately running time-domain simulations requires the full large signal expressions of transistors, if simulation includes circuit level blocks, or to solve differential equations whether a high-level behavioral model is conceived. The multi-resolution aspect is then fundamental for obtaining results in a reasonable time because system-level figures of IR-UWB receivers are based on iterative statistical analyses.

Implementation-aware actions on IR-UWB transceivers design require the identification of performance figures that depend on system-level constraints. The most common figures are typically related to Bit-Error-Rate (BER), for communication purposes and, in the case of IR-UWB for ranging applications, to the estimation of the Time-of-Flight (ToF). The UWB channel is statistical, therefore determining these system-level data implies randomizing different multipath realizations according to a specific operating environment, i.e. indoor office, residential, industrial, outdoor, open outdoor, and for Line-Of-Sight (LOS) or Non Line-Of-Sight (NLOS) links IEE (2004). Also, the computation of ToF with TWR schemes requires the modeling of a complete packet transmission mechanism without ideal synchronization. In communications, for bit error-rate tests large random data needs to be tested. Take for example a 10^{-6} BER: theoretically to obtain this single error-rate point at least 100 points are required for high confidence and this implies randomizing an average of 10^8 pulses. Note that from a pure communication point of view all these functionalities can be easily implemented with any high-level modeling language e.g. Matlab but this lacks of flexibility because top-down refinement of heterogeneous blocks is typically not possible. The use of a multi-description modeling tool permits an easy "context switching" between a high-level model to a circuit-level or SPICE post-layout netlists without having to interface the description. This flexibility is not relative only to the simulation tool itself but to the description language and in particular to the use of an homogeneous interface between descriptions. Let us consider an Integrate & Dump unit. Basically, the block shall have an input, an output and an integrate/dump control. Alternatively, if description is at a very high abstraction level control signal can be potentially undefined. These terminals not necessarily convey voltage or current but instead can be, if present, symbolic that only in a successive step are mapped onto a physical counterpart. The use of a priori homogeneous interfacing between different descriptions avoids burdensome conversion times and can be useful for defining electrical interconnections from early design stages.

System-level simulations aiming towards physical implementation predictions, must be enriched with many circuit-level non-ideality concerning silicon integration. Electro-Static Discharge (ESD) protection circuits, bondwire for die soldering on packages and inductive or capacitive parasitic couplings are few of the possible non-ideal effects. These, however, concern circuit-level design and at first design concept phases these can be disregarded, therefore assuming that chip-level integration countermeasures can efficiently tackle them in a next step. For example, if a cascoded tuned amplifier LNA requires a very well controlled to-ground parasitic inductance then this aspect has to be tackled at die-level floorplanning when the number of PAD is decided, therefore at circuit-level design steps. Instead, if the boundary conditions among two or more functional units represents a critical point, this shall be included in system level models. Also, the same parasitic can play different roles if shared among other circuit blocks. For example, if parasitic inductance influences much the operation of a block, for example an UWB coherent correlator, then this shall be included in the system-level model. From this analysis we conclude that the definition of the parameters required in simulation is fundamental.

Non-ideality can depend on many different factors but a flexible high-level simulation requires that they can be effectively modeled as generic parameters. For example, based on circuit-level details, the squarer unit in energy detection receivers, if not differential, can originate additionally to the $()^2$ term a linear by-product that depends on input signal level Han & Sanchez-Sinencio (1998). A high-level parametric behavioral modeling requires the implementation of a mathematical relationship that covers, in the most general conception and with sufficient confidence, the behavior of the circuit-level unit in all the operating conditions. In a high-level methodology this is particularly important because system level simulations are not meant to be a mere verification but instead shall represent a starting point for deriving useful design constraints. The inclusion of circuit-level descriptions at system-level with a uniform and flexbile language serves as inspection and analysis. Successive chip-level integration can be then easily derived by painlessly placing and routing all the blocks at their lowest layout description level.

3.2 Methodological assumptions

Based on the previous analysis, a design methodology for electronics systems shall be referred to at least three important respects: *uniformity*, *partitioning* and *refinement*. Uniformity can be read as the requirement of having an homogeneous formalism to describe the operation of a system. Partitioning can be read as the effort a designer makes for physically mapping the conceptual operation of a system according to very well defined rules. Refinement can be read as the enrichment of physical non-ideality applied to a pure mathematical model to more precisely describe physical behavior. Take for example digital design. Hardware description language as VHDL or Verilog are uniform, because they are completely portable and allow an homogeneous description of a block. The languages permit both gate-level and behavioral-level descriptions at the same time. The logic conception of digital circuits inherently permits a partitioning, that is the identification of input and output signals. Refinement is also possible because, provided that a block has the same inputs and outputs, its description can pass from behavioral to structural, therefore getting closer to single logic gates.

With circuit-level design we have very different aspects. The basic building blocks are not logic gates but devices with a particular electrical interface. In digital domain interface comprises purely logical inputs outputs while here the same input and output terminals are enriched with continuous power by voltage and current. Parasitic are very important in RF design and the well defined input/output paradigm valid for digital circuits is compromised. In the above reading key, couplings between two near blocks on the same silicon chip can generate other inputs and outputs, even if their physical counterpart is a fF order capacitance, a pH order coupling inductance or a GΩ resistor. An RF amplifier having a single input or output, after layout can have more physical interconnections with other blocks that share the same die. In this digital-like input/output key, the effect of parasitic can be also modeled impacting on a given electrical signal, i.e. bandwidth or gain decrease without having to map it as an additional input or output. While the modeling of parasitic effects can be more systematic in digital design (consider for example delay of logic gates), in the analog world this is more complex because it depends on physical design. Filling the modeling gap between analog and digital worlds with a uniform methodology can be possibly obtained by using a description language that forces the same partitioning as in digital domain and at the same time has enough flexibility for being used in the digital simulation domain. Description is not the only aspect that shall be considered. Attention regards also the simulator itself and therefore its inherent capability of accepting hardware described with different languages. Therefore, the design methodology presented here refers to a simulator with which multiple description languages with a uniform formalism are contemplated. Fig. 2 schematizes the interactions between simulation and hardware worlds.

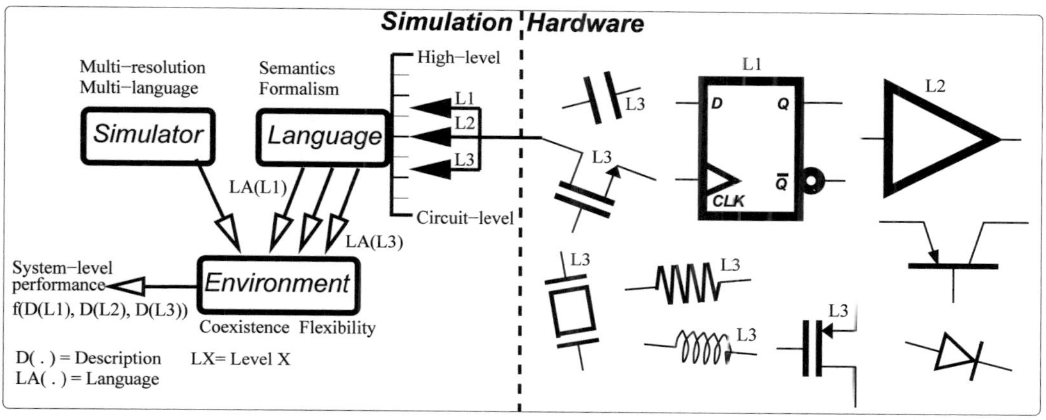

Fig. 2. Simulator and language in a multi-level description.

3.3 Design methodology

The design methodology outlined in this work is organized in four phases. During Phase-I the receiver, or generally the IR-UWB system is behaviorally defined and a first high-level model is generated. This phase is known as *conception*. In the case of our Energy Detection receiver front-end this implies behaviorally modeling e.g. LNA, squaring unit, Integrate and Dump and the Analog-to-Digital Converter (ADC). Note that in the example of figure 3 the front-end is shown but the methodology can be applied to complete systems, even including a dedicated backend for bit and symbol synchronization and demodulation, because VHDL and VHDL-AMS lie on the same domain. At this abstraction level, the description still recalls the formalism of a high-level modeling language e.g. Matlab since an electrical interface is not defined yet and the complete system is packed onto few VHDL-AMS process disregarding the complexity its implementation may imply. Figure 3 (Phase I) shows a single Entity-Architecture (E&A) couple comprising a complete energy detection receiver front-end. At this point, the model is validated by checking consistency with high-level models developed in Matlab or in other high-level languages applied on the system-level figures previously mentioned. Here, from the engineering point of view, the main effort consists of defining the system operation without forcing a design partition that is mandatory towards physical-level implementations.

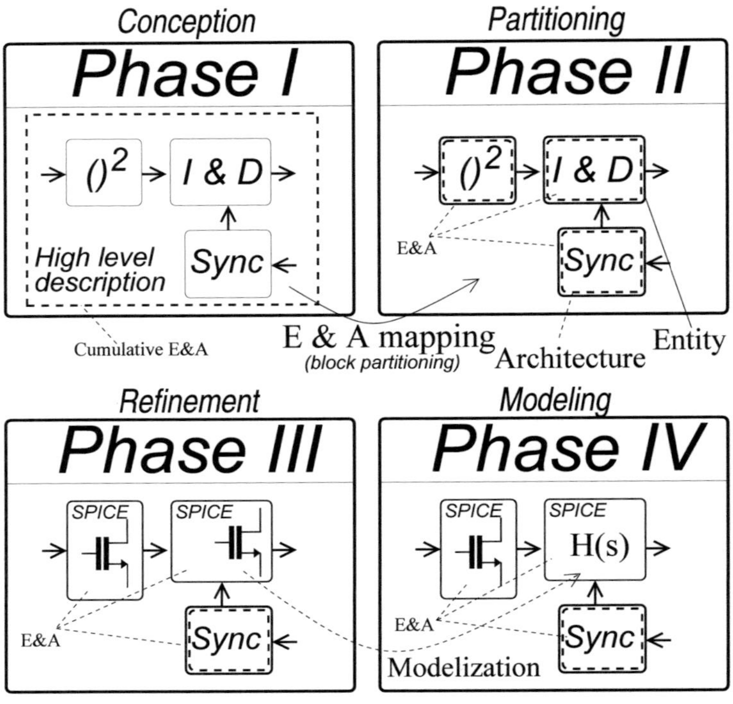

Fig. 3. Design methodology organized in 4 phases.

In Phase-II a first electrical signal definition is forced. We call this very important phase *partitioning*. This implies rearranging the description developed during Phase-I in separate E&A. Here we simply apply the modularity of the VHDL-AMS language on the design to get closer to silicon implementation. Once electrical signals are defined, successive refinement phases applied on a single block are painless provided that electrical interface is the same. Partitioning is the key for efficiently conceiving the system and the later adjustment of system partitioning can be problematic. Here, considering the importance of this phase, no non-ideality are included or modeled in the simulation. The inclusion of non-ideal effects in fact, recalls low-level implementations or, alternatively system-level parameters known to severely impact on system-level performance. The development of a new system, intended not being reported in the state of the art, implies only the partial knowledge of the exact non-ideality that may compromise performance.

The ADC quantization, the AGC look-up table as well as a DAC for AGC gain analog conversion can be all included in this phase not being properly non-ideal effects, rather fundamental circuit features included in normal operation. Bandwidth, saturation and blocks power consumption are not defined at this phase. System partitioning, i.e. electrical interconnection definition, requires the knowledge of lower circuit level constraints. Since the design is simply "rewritten", therefore differently described with the same simulation tool, the result must not change from Phase-I, but consistency with the previous phase needs to be checked. Note that in Phase-II signal electrical partitioning is possible but it is not strictly necessary, while formally only the E&A rearrangement of the conceptual operation is required. Whether this first partitioning does not comprise electrical-level terminals, it can be done in the next phase for each unit by refining each entity declaration.

Once system partitioning is complete, the electrical interface of all the blocks in the IR-UWB system are defined. We are now ready to increase the details in each block. For this reason, Phase-III is called also *refinement*. With refinement, signals partitioned in Phase-II assume a circuit-level meaning. Every important circuit-level non-ideality is modeled according to continuous time or digital statements and included in the architecture. The importance of this phase regards the identification of the non-ideal effects that impact on system-level performance, or, if the system leads the state-of-the-art, even on its basic operation. Efforts in the definition of the number of non-ideality of their description is an important trade-off because very accurate models can severely impact on simulation runtime or excessive efforts on this side can waste time and compromise the overall system-level performance inspection. For energy detection receivers for example, modeling of compression in the front-end is important e.g. for understanding the impact on interference rejection, but still, since the system computes the raw energy of the UWB pulses with a squarer, this is not extremely important. Dedicating weeks of research time on this would avoid taking important decisions next or would block the project at its beginning, while other problems may rise during circuit-level design or chip-level integration.
Phase-III is not only related to the inclusion of non-ideality to the previously idealized blocks. Provided that an homogenous electrical interface derived from Phase-II in the entity declaration of every VHDL-AMS unit is given the complete VHDL architecture can be switched. This enables the replacement of the full VHDL-AMS modeling with transistor-level SPICE models extracted from Cadence Front-end to Back-end or IC Station (Mentor Graphics) other front-end circuit design tools. The description can be also extracted from layout. This Substitute-and-Play (S&P) philosophy allows the identification of the impact of blocks refinement on system-level performance figures. This is very important because it permits architectural analyses by intelligently exploring all the possibilities without focusing on a single abstraction level. Here, a heterogeneous multi-level description can help understanding faster the problems that may arise when solid-state circuits are tested. Provided that refinement is intelligently run, performance e.g. on ranging, demodulation, synchronization, transceiver packet exchange, power consumption, can be forecasted and decision taken whether constraints are not met.
IR-UWB demands time-domain simulations and a complete refined system, even if not for all its blocks, can require very high runtimes especially when statistical tests are executed. Notwithstanding the computational power of workstation and servers keeps increasing as well as code parallelism in software, due to the short duration pulses high simulation accuracy is required and a complete 10 or 100 s packet exchange simulation can require days or even more. This applies also e.g. for PLL, where full SPICE level time-domain simulations are impractical (and in this context also inaccurate) Lai et al. (2005). Moreover, it can result that the effect of some circuit-level blocks severally impacts on system-level performance but cannot be neglected in the description. Therefore, we define a successive Phase-IV, called *modeling* or back-annotation, that aims at the inclusion of the relevant circuit-level non-ideality extracted from the transistor-level description of Phase-III. This can be accomplished in two different ways. The already modeled parameters are refined based on pure circuit level simulation, or, if the non-ideality discovered during Phase-III was not included previously the architecture is redesigned by keeping the same entity definition. The refined models can be used in Phase-III for running again simulations and obtaining further results.
The full design methodology is applied on the I&D unit of our Energy Detection receiver case study as an example. Next paragraph will focus on the design of the block and all

the hypothesis used for its conceptualization will be explained and identified in the outlined methodological key.

4. S&P contextualization: The I&D block design

Fig. 4 shows also the partitioned entity of the I&D and the entity declaration structure. At the highest abstraction level, the I&D electrical boundary is not defined and simply implements the math function $\int x(t)dt$, where $x(t)$ is input signal. $x(t)$ has not a physical counterpart nor it is single-ended or differential and integration output is a quantity that is neither voltage nor a current. A control signal is implicitly defined among the other high-level statements that control the computation of the formula. This integrator has been included in the high-level model and a first consistency check with a Matlab model has been completed. When description enters Phase-II, some circuit level properties must be considered. These are mainly related to 1) power supply, 2) control signals, 3) input and output electrical features (single-ended or differential, AC or DC coupled, current or voltage). By satisfying these constraints, valid for this specific case, the electrical interconnection boundary can be defined. The I&D is a pure analog unit, that has to cope with relatively high frequency signals[2]. Therefore, this block is not critical from the RF point of view and a single power supply and ground connection pin can be considered. Notwithstanding this, the block is critical at system-level. In the case of the LNA for example, having multiple power supplies can help reducing inductance parasitic and, depending on circuit-level design, it can be fundamental for matching. Therefore, modeling multiple power supply pins can be useful even at this abstraction level, and the problems that may arise can be directly tackled here rather than successively, when the floor plan is defined and circuit blocks placed down. A very first constraint we had in the design on the energy detection receiver was that it had to be fully differential, therefore fully differential input and outputs were assumed, in fig. 4 the couples `Inp-Inm` and `Out_intp-Out_intm`, in particular DC coupled voltage signals. For integration control signals the discussion is more complex because the use of a single ended or a differential signal (one, vs. two terminals) depends on the internal implementation of the unit. Homogeneously, we assume also perfectly differential voltage signals `Controlp-Controlm`. In this very first implementation we assume that integrator is the gm-C structure depicted in fig. 4. The transconductor transforms the input voltage into differential current and charges a load capacitor C. When control signal `Controlp-Controlm` is active integration is run, while when it toggles to '0' integration is reset. The biasing circuit is connected to V_{bias1}, V_{bias2} and to V_{bias3}, it consists of two self-biasing stages that generate the required voltages for both transconductor and Common Mode Feedback Network (CMFB), not shown here for sake of brevity. According to the state-of-the-art simpler integrator structures are possible and they can be single ended and much simpler than those depicted here Lee & Chandrakasan (2007). At this point, the target was the replacement of a BiCMOS integrator by then used in a first implementation Stoica et al. (2005) with a lower cost CMOS integrator. Note that at this point the I&D architecture boundary has been fully defined. From an electrical point of view this enables the VHDL architecture switching among different Phase-III domain models. For example, a VHDL-AMS behavioral model, with the given electrical interface can be painlessly substituted with the equivalent circuit-level or layout-description.

[2] Note that after squaring, the useful portion of the spectrum of a UWB signal of bandwidth B is at baseband, $[0, B/2]$, e.g. for a standard UWB pulse having a 500 MHz bandwidth, this corresponds to operating in the band 0-250 MHz.

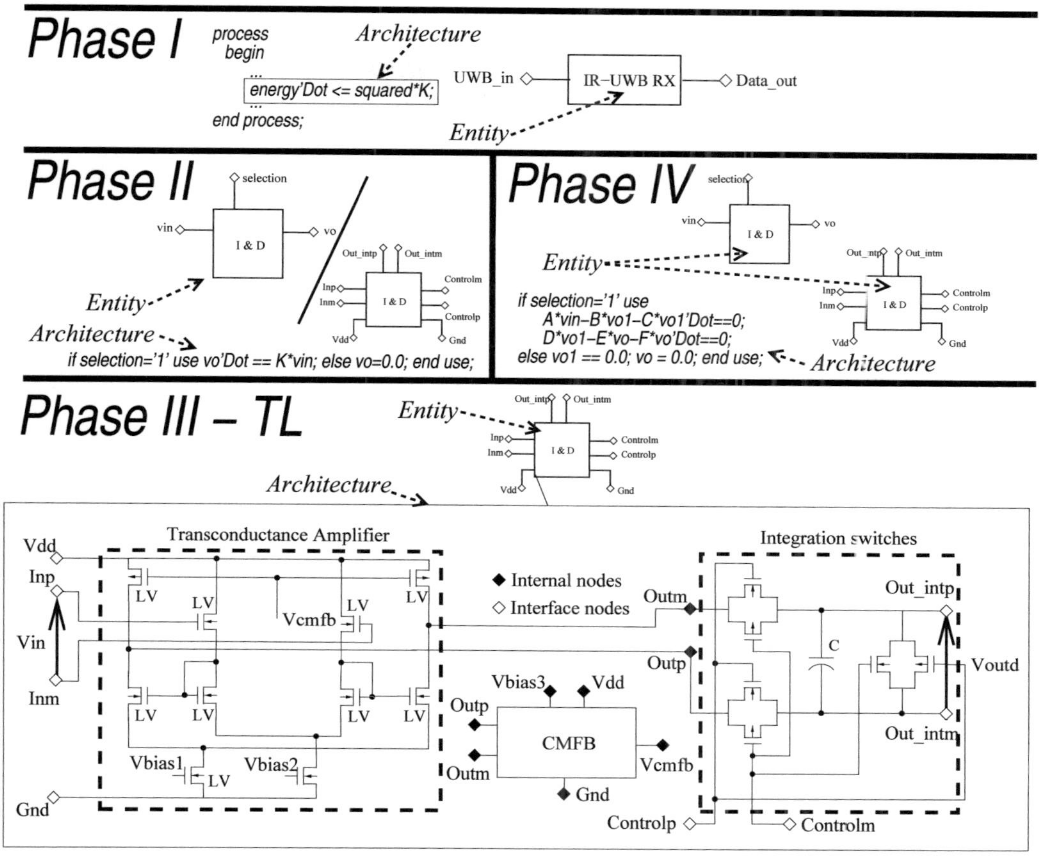

Fig. 4. I&D circuit at the circuit-level design and partitioning level.

With circuit-level simulation the AC behavior of the I&D can be easily extracted. This is reported here on fig. 5 from Crepaldi et al. (2007). The integrator operates from 1 MHz to 1 GHz, has an additional low-pass transfer function, and not ideally infinite DC gain. The second pole at high frequency is due to parasitics of the devices. Note that the useful part of the UWB signal is concentrated from 0 to 250 MHz for a 500 MHz pulse and the behavior of the integrator at very high frequency is not fundamental. The non-infinite DC gain is a loss therefore limiting the maximum length of the integration window. At this point, this AC model can be included in the Phase-III VHDL-AMS models to speed up simulation time. Note that by including the AC model only non-linearities and saturation of the transconductor are not accounted for. This is a clear example of the mandatory requirement of Phase-IV, that is an intelligent inclusion of the relevant non-ideality derived from transistor-level design. In the case the required system-level simulation explicitly requires accounting for this non-ideal effect, then, the backannotation shall be enriched, or alternatively the full circuit shall be included and other blocks non-ideality deactivated to speed-up simulation time. Before applying the substitute-and-play approach, consistency with ideal (Phase-II) and VHDL-AMS models has been checked. As shown in fig. 5, the backannotated model and the AC circuit simulation of Phase-III match.

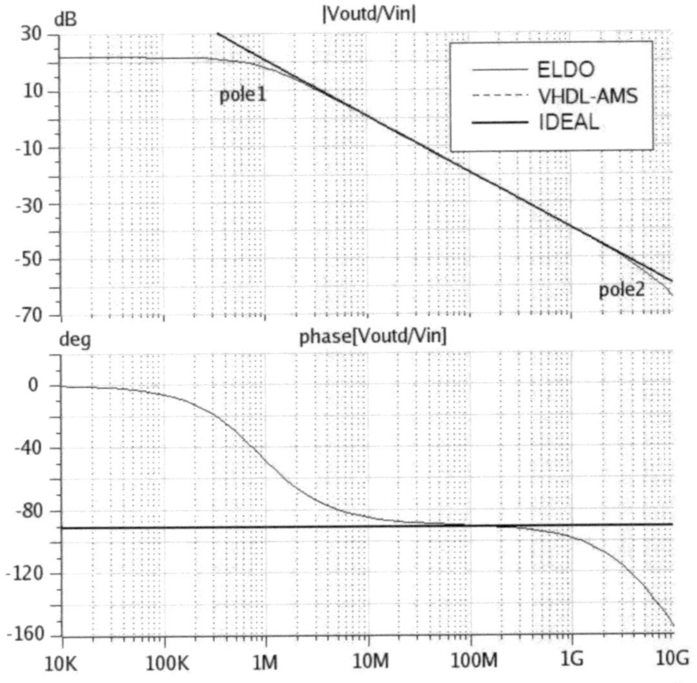

Fig. 5. AC response of the I&D circuit and Phase-II and III models Crepaldi et al. (2007). The IDEAL and VHDL-AMS models overlap.

The connection of transistor level descriptions with ideal blocks can require specific considerations, not only related to the modeling language itself but on the electrical features resulting from blocks interfacing. Take for example a fully ideal Phase-II model of the squarer. A possible VHDL-AMS description can include only the simultaneous statement `vsquare==K*vin**2.0;`, where `square` and `vin` are across quantities defined on two couples of differential terminals. If this is the case, then input and output impedance of the squarer is completely disregarded. If the squarer modeled according to this simple statement is connected to the I&D the resulting integration voltage would be compromised because common mode voltage is disregarded. Therefore, in such cases the inclusion of a boundary element is fundamental for brigding the ideal world to a full custom electrical interface. These boundary elements are inherently included in the surroundings units. In this work, proper boundary elements, operating on the DC level of `vin` have been included.
Fig.6 shows a transient simulation of the integrators during three different modeling phases II, III and IV. Notwithstanding a gain mismatch output is still energy, that is the integral of the squared signal.

5. System-level simulations and results

One very interesting feature of Impulse-Radio UWB regards the possibility of determining the pulses time of flight, that is, the distance between two transceivers. Since UWB pulses are very short, the accuracy with which distance can be estimated can be very high. For example, recent receivers are designed with fine synchronization circuits reaching accuracies of few

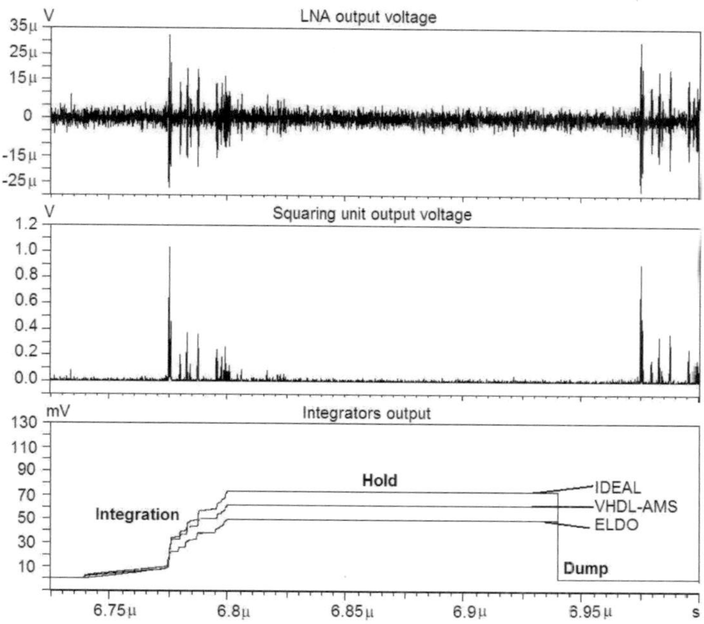

Fig. 6. Transient response of the I&D circuits obtained from different modeling phases Crepaldi et al. (2007).

millimeters Chu et al. (2011). Pulse radio was thought also to serve localization purposes even in space applications Ni et al. (2010). IR-UWB can be easily applied to biomedical devices because pulses are reflected differently depending on dielectric properties mismatches among different mediums. This enables applications in Breast Cancer Detection and wireless biometric parameters sensing. Here, we applied the methodology to Bit-Error-Rate tests for wireless link quality inspection, and to Two-Way-Ranging related to ToF estimation performance.

Figure 7 shows a graphical representation of the effect of the I&D substitution on our energy detection IR-UWB system. The figure shows also a graphical representation of the TWR mechanism implemented between two transceivers[3]. Two-Way-Ranging is a packet exchange mechanism that is based on the transmission of two packets, a request packet and an acknowledge packet between two transceivers A and B. The Time-of-Flight is calculated at the transceiver B, after having received the acknowledge packet from transceiver A. The ToF calculation is based on the determination of the exact leading edge of the UWB pulses with a proper synchronization algorithm. A very common synchronization algorithm, also called window integrator, is based on the determination of the time when the sampling of the maximum UWB energy occurs. It is based on an integration window shift within a fixed pulse repetition period. The shift is realized by a dedicated DLL and phase selector that sequentially shift the control signal of the I&D. After a full exploration within the Pulse Repetition Interval (PRI) the clock phase corresponding to maximum energy is selected. The accuracy of the algorithm depends on the integration window shift, that for coarse synchronization can be

[3] Note that other ranging schemes are possible, for example in Ni et al. (2010) Time-Difference-Of-Arrival (TDOA) is used.

on the order of 5 ns or for fine synchronization even less than 1 ns. Here we applied this windowed integrator for both coarse and fine synchronization. Transceiver B system clock phase is different with respect to transceiver A, therefore the acknowledge packet must include information on both the processing time offset of TRX A and the synchronization phase used for detecting the maximum energy. Transceiver B, processes this information and, according to its synchronization phase, calculates the ToF, therefore distance. Details about the full mechanism can be found in Casu et al. (2008).
Bit-Error-Rate is determined in presence of Additive White Gaussian Noise (AWGN). Its determination implies the inclusion of the Salleh-Valenzuela UWB channel model in the simulation environment with a VHDL-AMS formalism IEE (2004). Natively, the model is implemented in Matlab and here its VHDL-AMS description is based on text files with rendered saved data samples issued with a constant time step. Fig. 8 shows the effect of

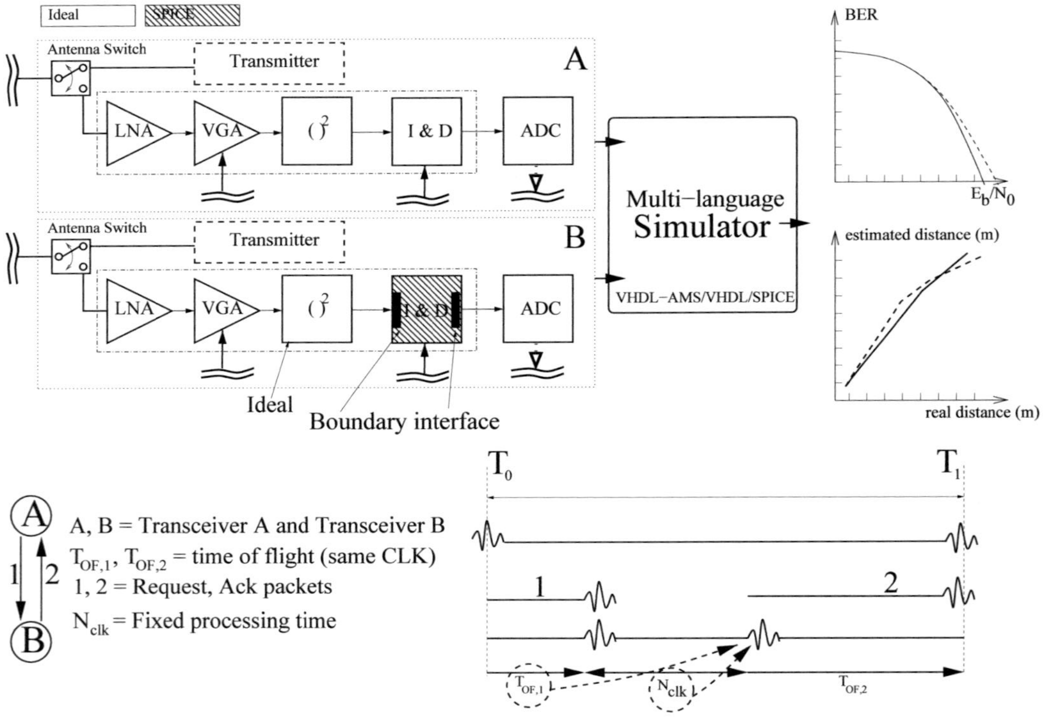

Fig. 7. Deactivation of non-ideal effects during system-level simulation and Two-Way-Ranging.

Phase-III integrator on the BER performance of the system Crepaldi et al. (2007) as a function of E_b/N_0 (proportional to Signal-to-Noise Ratio). The BER curve is slightly shifter because the *pole*2 of the integrator additionally filter input noise out of the squarer. If other blocks are implemented at transistor-level then, noise filtering increases. The results demonstrate that this design methodology permits the determination of transistor-level non-ideality at higher abstraction level.
With the Salleh-Valenzuela channel VHDL-AMS model TWR ranging simulations are also possible. Detailed TWR simulation results are reported in Casu et al. (2008). Two instances of the same IR-UWB transceiver schematized in 1 have been included in the environment

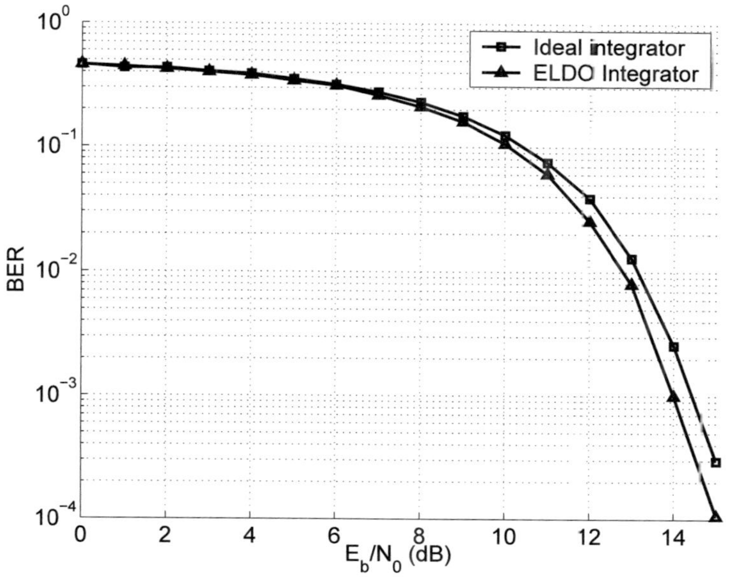

Fig. 8. Bit Error-Rate associated to the circuit-level I&D compared to the ideal system Crepaldi et al. (2007).

as well as specific scripts for enabling batch simulations execution. Channel is residential Line-Of-Sight (LOS) with the recommended path loss. The simulation environment sets its parameters with a parametric constant that models TRX distance. After 10 TWR packet exchanges we can obtain the effect of the I&D refinement on the localization performance of the system Crepaldi et al. (2007). With an ideal integrator at 9.9m distance, the estimated distance is 10.10m and 11.16m with variance 0.49m and 0.10m, for the Phase-II and transistor-level Phase-III models. Thanks to these inherent transient simulations results, the analysis of these two results permits the identification of the circuit blocks influenced in the performance loss. That is, having activated a transistor-level III description, enables the effects of other ideal blocks to influence performance. From the analysis it results that the reason for such a high variation in the estimated distance for Phase-III transistor level implementation depends on the operation of the AGC. The presence of a slightly non-ideal effect on the I&D "excites" the ideal AGC and a incorrect gain adjustment is provided. The incorrect amplification imposed by the AGC loop causes the squared signal to be out of the integrator input range and a lower output voltage is obtained. This causes the ADC quantization to be less effective and the ranging algorithm implemented in the digital back-end fails by few coarse synchronization steps.

Based on successive reasonings, other considerations are possible. The presence of a transistor-level block among other ideal blocks can lead to erroneous simulation conditions. For example, an LNA simply modeled with a perfectly linear amplifier $V_{out} = GV_{in}$, where G is voltage gain, V_{out} and V_{in} are across quantities defined on input and output terminals, does not include saturation. Due to automatic and autonomous system-level operation, an erroneous or partial modeling of some of the other blocks, can force, for example, a gain G on the LNA that leads to output voltage exceeding the allowed signal swing, e.g. 10 times bigger than supply voltage. This problem occurs mainly because the system is conceived

starting from high level models when inputs and outputs miss a physical counterpart. Note that this problem is irrelevant for high-level Matlab simulations in which idealized systems are proven. We conclude that for a consistent and correct system-level modeling, the inclusion of some fundamental circuit-level parameters such as voltage and power ranges limitations, bandwidth and power consumption dependency is extremely important.
The CPU time required to run a 30 μs simulation is an important information that justifies the presence of Phase-IV in our design methodology. As indicated in Crepaldi et al. (2007), on an IBM-Xeon server, 4GB RAM, 3.0 GHz processor with a fixed time step of 0.05 ns, an accuracy EPS=`10e-6` and the Newton/Raphson solving algorithm, the CPU time required with the SPICE netlist is 3 times larger than the time required using the backannotated VHDL-AMS model and 6 times the IDEAL Phase-II description.

6. Conclusion

We have presented a methodology that allows the exploration of the impact of refinement on system-level parameters for an IR-UWB Energy Detection system. The methodology is based on the use of the modular formalism of VHDL, working for the design of digital circuits, properly extended for use in analog continuous-time circuits with AMS extensions. The methodology is based on the use of a multi-language, multi-resolution tool and it is organized in four phases that generally define the main tasks required for a mixed-signal electronic system conception. VHDL-AMS has been conceived for use outside the field of electrical circuit, for example on fluidics, mechanics and all the possible domains governed by linear differential equations. Scientific community endeavors are focused on the efficient integration of any kind of system including MEMS even for IR-UWB Radio Frequency Tiiliharju et al. (2009), smart sensors and energy harvesting powered devices. This design methodology can be utilized also in these contexts, provided that the interface among the different domains is correctly modeled and sufficiently enriched with implementation details. IR-UWB remains, in fact, a valuable ULP wireless technology even for applications in smart sensors.
Based on these results, we believe that to merge both analog and digital design worlds, one interesting topic for successive research can regard a simple, uniform and modular mixed-signal language with a unique simulation tool for both analog and digital circuits disregarding the math they are based on. This language shall allow on-the-fly simulation accuracy directives embedded in each unit description depending on the nature of each block, digital or analog, with a similar semantics. Compared to VHDL-AMS it shall robustly fill the gap between the digital concurrent world and the analog continuous-time paradigm, instead of keeping them separated and making them coexist. In fact, AMS remains still a modeling language, therefore far from being used for automatic low-level synthesis as in digital VHDL design.

7. References

Bielefeld, D., Fabeck, G. & Mathar, R. (2009). Power Allocation and Node Clustering for Distributed Detection in IR-UWB Sensor Networks, *IEEE Vehicular Technology Conference Fall (VTC Fall)*, pp. 1–5.

Carbonelli, C. & Mengali, U. (2006). M-PPM Noncoherent Receivers for UWB Applications, *IEEE Transactions on Wireless Communications* 5(8): 2285–2294.

Casu, M., Crepaldi, M. & Graziano, M. (2008). A VHDL-AMS Simulation Environment for an UWB Impulse Radio Transceiver, *IEEE Transactions on Circuits and Systems I: Regular Papers* 55(5): 1368–1381.

Chu, T.-S., J., R., Chang, S., T., M., C., D. & Hossein, H. (2011). A Short-Range UWB Impulse-Radio CMOS Sensor for Human Feature Detection, *IEEE International Solid-State Circuits Conference (ISSCC), Digest of Technical Papers*, pp. 294–296.

Crepaldi, M., Casu, M., Graziano, M. & Zamboni, M. (2007). An effective AMS Top-Down Methodology Applied to the Design of a Mixed-Signal UWB System-on-Chip, *Design, Automation Test in Europe Conference Exhibition*, pp. 1–6.

Crepaldi, M., Li, C., Dronson, K., Fernandes, J. & Kinget, P. (2010). An Ultra-Low-Power Interference-robust IR-UWB Transceiver Chipset Using Self-synchronizing OOK Modulation, *IEEE International Solid-State Circuits Conference (ISSCC), Digest of Technical Papers*, pp. 226–227.

Daly, D., Mercier, P., Bhardwaj, M., Stone, A., Voldman, J., Levine, R., Hildebrand, J. & Chandrakasan, A. (2009). A Pulsed UWB Receiver SoC for Insect Motion Control, *IEEE International Solid-State Circuits Conference - Digest of Technical Papers*, pp. 200–201,201a.

FCC (2002). Revision of Part 15 of the Commission's Rules Regarding Ultra-Wideband Transmission Systems, Report and order, adopted February 14, 2002, released July 15, 2002.

Gorlatova, M., Kinget, P., Kymissis, I., Rubenstein, D., Wang, X. & Zussman, G. (2010). Energy Harvesting Active Networked Tags (EnHANTs) for Ubiquitous Object Cetworking, *IEEE Wireless Communications* 17(6): 18–25.

Han, G. & Sanchez-Sinencio, E. (1998). CMOS Transconductance Multipliers: A Tutorial, *IEEE Transactions on Circuits and Systems II: Analog and Digital Signal Processing* 45(12): 1550–1563.

IEE (2004). IEEE 802.15.4a Channel Model - Final Report.
URL: *http://www.ieee802.org/15/pub/TG4a.html*

IEE (2007). IEEE 802.15 WPAN Low rate Alternative PHY Task Group 4a (TG4a).
URL: *http://www.ieee802.org/15/pub/TG4a.html*

IEEE (2007). P1076.1/D3.3, Approved Draft Standard VHDL Analog and Mixed-Signal Extensions (Revision of IEEE Std 1076.1-1999).

Lai, X., Wan, Y. & Roychowdhury, J. (2005). Fast PLL Simulation Using Nonlinear VCO Macromodels for Accurate Prediction of Jitter and Cycle-slipping due to Loop Nonidealities and Supply Noise, *IEEE Asia and South Pacific Design Automation Conference*, Vol. 1, pp. 459–464 Vol. 1.

Lecointre, A., Dragomirescu, D. & Plana, R. (2010). Largely Reconfigurable Impulse Radio UWB Transceiver, *Electronics Letters* 46(6): 453–455.

Lee, F. & Chandrakasan, A. (2007). A 2.5nJ/bit 0.65V Pulsed UWB Receiver in 90nm CMOS, *IEEE Journal of Solid-State Circuits* 42(12): 2851–2859.

Ni, J., Arndt, D., Ngo, P., Phan, C., Dekome, K. & Dusl, J. (2010). Ultra-wideband Time-Difference-Of-Arrival High Resolution 3D Proximity Tracking System, *IEEE Position Location and Navigation Symposium (PLANS)*, pp. 37–43.

Stoica, L., Rabbachin, A., Repo, H., Tiuraniemi, T. & Oppermann, I. (2005). An Ultrawideband System Architecture for Tag Based Wireless Sensor Networks, *IEEE Transactions on Vehicular Technology* 54(5): 1632–1645.

Tiiliharju, E., Koivisto, T., Maunu, J., Chekurovy, N. & Tittoneny, I. (2009). Ultra-wideband CMOS-MEMS Radio, *IEEE International Conference on Ultra-Wideband*, pp. 102 –106.

Verhelst, M. & Dehaene, W. (2008). A Flexible, Ultra-Low-Energy 35 pJ/Pulse Digital Back-End for a QAC IR-UWB Receiver, *IEEE Journal of Solid-State Circuits* 43(7): 1677–1687.

Wang, X. Y., Dokania, R. K. & Apsel, A. (2011). PCO-Based Synchronization for Cognitive Duty-Cycled Impulse Radio Sensor Networks, *IEEE Sensors Journal* 11(3): 555–564.

5

Time-Hopping Correlation Property and Its Effects on THSS-UWB System

Zhenyu Zhang[1,2], Fanxin Zeng[2], Lijia Ge[2] and Guixin Xuan[2]
[1]*College of Communication Engineering, Chongqing University*
[2]*Chongqing Communication Institute*
China

1. Introduction

Ultra wideband (UWB) is a promising technology for short-range wireless communications since it potentially combines the reduced complexity with low power consumption, low probability of detection/intercept (LPD/LPI) and immunity to multipath fading (Scholtz, 1993; Win & Scholtz, 1998; Win & Scholtz, 2000). The successful development of UWB technology strongly depends on the advancement of efficient multiple-access techniques. A typical multiple-access format of UWB is time-hopping spread spectrum ultra wideband (THSS-UWB) where data are transmitted by using pulse position modulation (PPM) (Scholtz, 1993) or pulse amplitude modulation (PAM) (Le Martret & Giannakis, 2000) at a rate of many pulses per data symbol. Both of PPM and PAM require good properties of TH sequences.

In terms of TH correlation properties, the known constructions of TH sequences can be mainly classified into two categories, namely aperiodic TH sequences and periodic TH sequences. For aperiodic TH sequences, the chaotic pulse-position modulation (CPPM) which was designed for the transmission of binary information (Sushchick et al, 2000) and the pseudo-chaotic time hopping (PCTH) which exploited the symbolic dynamics of a chaotic map at the transmitter (Maggio et al, 1999; Maggio et al, 2001) were proposed, respectively. In addition, an alternative use of pseudo-chaotic dynamics was proposed as an encoder for the binary data stream while data-independent TH sequences were used to guarantee ease of synchronization and decidability (Erseghe & Bramante, 2002). For periodic TH sequences, the related analyses can be found in the corresponding literatures (Chu & Colbourn, 2004; Corrada-Bravo, 1999; Erseghe, 2002a; Erseghe, 2002b; Iacobucci & Di Benedetto, 2001; Iacobucci & Di Benedetto, 2002; Scholtz et al, 2001). These sequences can be considered as the extensions of existing frequency-hopping (FH) sequences and all of them have good TH correlation properties. In these TH sequences, quadratic congruence code (QCC) (Scholtz et al, 2001) is the best one in terms of TH correlation properties since QCC satisfies $S_{\max} \le 2$ and $C_{\max} \le 4$, where $S_{\max}$ denotes the maximal TH auto-correlation function (ACF) sidelobe and $C_{\max}$ denotes the maximal TH cross-correlation function (CCF) values. For aperiodic and periodic TH sequences, the theoretical bound, namely the relation between four parameters of sequences period L, the number of time slots N, TH sequences family size N_u and maximal TH correlation function values $C_{\max}$ (or $S_{\max}$), plays an important role in construction schemes. So far, several theoretical bounds had been obtained, such as Johnson bounds and new upper bounds (Gao & Chang, 2006; Scholtz et al, 2001).

This chapter mainly focuses on constructions and theoretical bounds of periodic TH sequences. A generalized definition of TH periodic correlation function which can be used to analytically evaluate TH correlation properties is presented. Based on this definition, a method to improve TH correlation properties in practical applications is proposed. By such a method, the maximum correlation function values of TH sequences can be reduced to a half of original values. Consequently, coincident probabilities among TH sequences decrease. In addition, averages of TH periodic correlation function values are investigated, and the relations between the averages and four TH parameters are formulated. From the results, low bounds of maximal TH correlation function values are further obtained.

In terms of the obtained low bounds, the multiple access interference (MAI) of asynchronized THSS-UWB systems is inevitable. Although orthogonal communications will be realized when accurate synchronism is held in the whole system, the accurate synchronization is difficult to be kept and catastrophic collisions will happen when synchronization in the whole system fails to be perfectly kept. In this chapter, a novel of TH sequences with zero correlation zone (ZCZ) is constructed. THSS-UWB systems employing such sequences can be without MAI when the shifts between TH sequences are in range of ZCZ. In other words, when MAI comes as small shifts of cross correlation, the MAI will be eliminated since CCF values are equal to zero in range of ZCZ. As a result, orthogonal communications can be realized while the need of accurate synchronism in whole system is reduced. The idea of approximate synchronization and ZCZ was firstly proposed for direct sequence (DS) in code division multiple access (CDMA) (Suehiro, 1994), and a lot of DSs with ZCZ had been constructed (Fan, 1999; Hayashi, 2009; Hu & Gong, 2010). As for TH sequences with ZCZ, the corresponding studies were also presented (Guvene & Arslan, 2004a; Guvene & Arslan, 2004b; Zeng et al, 2011). Different from the known constructions of ZCZ TH sequences, the method proposed in this chapter can provide a more flexible choice of TH parameters.

Based on the analyses of TH correlation properties, multiple access performance of THSS-UWB systems is presented. So far, most studies on MAI assumed that the interference was a zero-mean Gaussian random variable, namely Gaussian approximation (Canadeo et al, 2003; Durisi & Benedetto, 2003; Hu & Beaulieu, 2003; Mireles, 2001; Scholtz, 1993; Win & Scholtz, 1998; Win & Scholtz, 2000; Zhao & Haimovich, 2002). Multiple access communication performance using UWB waveforms with TH-PPM and DS-BPSK modulations was studied (Canadeo et al, 2003), where the analyses used a fixed sequence, namely Gold codes. Based on Gaussian quadrature rules (GQR) and a characteristic function (CF) technique, two new methods to evaluate the bit error rate (BER) performance of THSS-UWB in the presence of MAI and additional white Gaussian noise (AWGN) channel were proposed (Durisi & Benedetto, 2003; Hu & Beaulieu, 2003), respectively. However, they still considered TH sequences as independent random variables, which will lose the practical characteristics of TH sequences. As a result, the effects of TH correlation properties on the multiple-access performance of THSS-UWB systems are ignored. In order to analyze the practical effect of TH sequences, this chapter derives analytical expressions of the relations between MAI values and TH correlation function values, which can be used to evaluate the BER performance in the presence of MAI.

The organization of this chapter is as follows. After the introduction, the definitions of TH periodic correlation function are provided in Section 2. The definitions are used to obtain theoretical bounds of TH sequences in Section 3 and improve TH correlation properties in Section 4. A novel family of TH sequences with ZCZ is obtained in Section 5. Based on TH correlation properties, the analyses of MAI are presented in Section 6. Finally, Section 7 summarizes the results.

2. TH correlation properties

In this section, we begin with the PPM model of THSS-UWB systems to understand how TH sequences work. We then analyze the correlation property of TH sequences in THSS-UWB systems.

2.1 PPM model of THSS-UWB systems

The PPM model is a kind of hopping format which is studied widely in THSS-UWB systems. In PPM model, the transmitted signal for user i may be expressed as

$$S^{(i)}(t) = \sum_{k=-\infty}^{+\infty} w(t - kT_f - c^{(i)}_{(k)_L} T_c - \delta d^{(i)}_{\lfloor k/N_S \rfloor}) \tag{1}$$

where $w(\cdot)$ represents the transmitted monocycle waveform and $\left\{c^{(i)}_{(k)_L}\right\}$ denotes a TH sequence assigned to user i, where the notation $(\cdot)_L$ denotes a modulo L operation. The TH sequence $\left\{c^{(i)}_{(k)_L}\right\}$ is periodic with period (or sequence length) L and each sequence element is an integer in the range of $0 \le c^{(i)}_{(k)_L} \le N_h$. The notation T_f denotes frame time (or pulse repetition time) and T_c is TH slot time, $T_f = NT_c$, usually $N = N_h + 1$, which represents the number of TH time slots in a frame time. The notation δ is the data shift time. The data sequence $\left\{d^{(i)}_m\right\}$ of user i is a binary stream. One symbol may be conveyed by N_s monocycles. The notation $\lfloor x \rfloor$ denotes the integer part of x.

According to the equation (1), we can see that the properties of TH sequences will play a key role in THSS-UWB systems. They ensure that UWB becomes a multiple access system and these sequences have a significant effect on synchronization and channel estimation. Fig. 1 shows how TH sequences work in THSS-UWB systems. In Fig. 1, $\left\{c^{(1)}_{(k)_5}\right\} = \{2,0,1,4,3\}$ and $\left\{c^{(2)}_{(k)_5}\right\} = \{2,4,3,0,1\}$ represent two TH sequences, respectively, where $N = L = 5$. The two TH sequences control the position of pulse of user 1 and user 2, respectively. In addition, Fig. 1 shows that two collisions between two TH sequences emerge when some shift happens in a period. For more simplicity to be understood, time slots that happened to collision are marked with double-head arrow in Fig. 1.

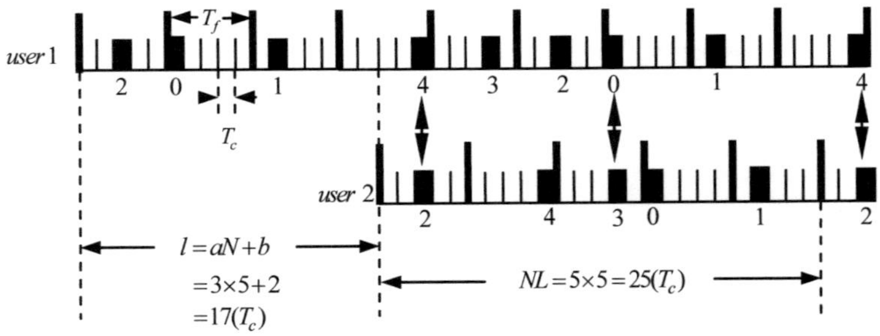

Fig. 1. The collision situation between two TH sequences (two users)

2.2 Definitions of TH periodic correlation function

For TH sequences, TH correlation properties are described by TH correlation function. In terms of chip synchronism (where shift $l \propto T_f$ and $0 \le l \le L-1$), TH periodic correlation function was defined as follows.

Definition 1 (Iacobucci & Di Benedetto, 2002): Let $\{c_{(k)_L}^{(i)}\}$ and $\{c_{(k)_L}^{(j)}\}$ denote two time hopping sequences with period L, and then the Hamming TH periodic correlation function with shift l is given by

$$H_{ij}(l) = \sum_{k=0}^{L-1} h[c_{(k)_L}^{(i)}, c_{(k+l)_L}^{(j)}],\ 0 \le l \le L-1, \tag{2}$$

where

$$h[c_{(k)_L}^{(i)}, c_{(k+l)_L}^{(j)}] = \begin{cases} 1, & c_{(k)_L}^{(i)} = c_{(k+l)_L}^{(j)} \\ 0, & c_{(k)_L}^{(i)} \ne c_{(k+l)_L}^{(j)} \end{cases}.$$

According to the equation (2), we can see that Definition 1 requires $0 \le l \le L-1$ and $l \propto T_f$, that is, the frame time T_f belonging to transmissions of different users must be synchronized. Hence, Definition 1 is the same as the definition of FH correlation function. Specially, we have $l = 0$ when codeword synchronism is held in whole system.

In order to help to understand the collisions situation between the sequences, we give Fig. 2 which is array representation of sequences $\{0,3,8,10,5,1,6,9,2,4,7\}$ with $L = N = 11$. In Fig. 2, columns and rows indicate time frames and time slots, respectively. Also, each column has a unique one (black box) indicating the time slot on which we transmit according to the sequence $\{C_{(k)_L}^{(i)}\}$. Fig. 3 shows how Definition 1 works in the case of chip synchronism. In Fig. 3, we employ QCC sequence as an example. For QCC sequence, $c_{(k)_L}^{(i)} = (k^2 i)_p$, where p is a prime, $L = P$, the number of users $N_u = P-1$, $0 \le k \le L-1$ and $0 \le i \le P-1$. Let $p = 11$, then $L = 11$ and $N_u = 10$. Fig. 3 shows the place of the maximum TH CCF value between $\{c_{(k)_{11}}^{(3)}\}$ and $\{c_{(k)_{11}}^{(5)}\}$ when $l = 1$.

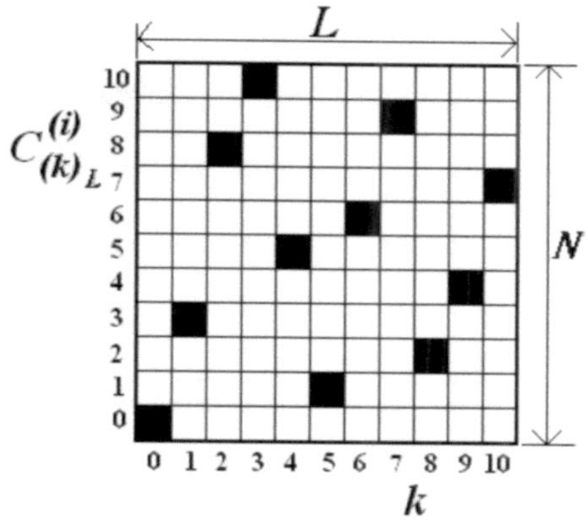

Fig. 2. Array representation of the sequence $\{0,3,8,10,5,1,6,9,2,4,7\}$, where $L = N = 11$

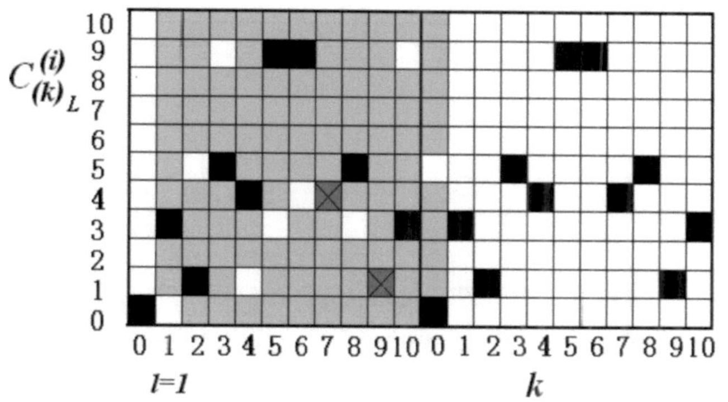

Fig. 3. The illustration of the maximum CCF value between $\left\{c^{(3)}_{(k)_{11}}\right\}$ and $\left\{c^{(5)}_{(k)_{11}}\right\}$ of QCC sequences when $l = 1$, where collisions are denoted by a cross $\times$ and $C_{\max} = 2$

Compared with chip synchronism and codeword synchronism which are difficulty to be kept, the asynchronism format is more interesting. Then, we consider a more general definition of TH periodic correlation function which can be used to analytically evaluate the TH correlation properties in codeword synchronism, chip synchronism and asynchronism in the whole system.

Definition 2: Let $\left\{c^{(i)}_{(k)_L}\right\}$ and $\left\{c^{(j)}_{(k)_L}\right\}$ denote two TH sequences with the period L, and then TH periodic correlation function with shift l can be expressed as

$$C_{ij}(l) = |A \cap B|, \tag{3}$$

where $A = \{x \mid x = (kN + c^{(i)}_{(k+a)_L})_{NL}\}$, $B = \{x \mid x = (kN + c^{(j)}_{(k)_L} + b)_{NL}\}$ and $0 \le a,b,k \le L-1$. The notation $|x|$ represents the number of the elements in set x and $A \cap B$ indicates the intersection between two sets of A and B. The symbol l denotes shift and satisfies $l = aN + b$. Then, we have $0 \le l \le NL-1$.

From the equation (3), we can see that the TH periodic correlation function $C_{ij}(l)$ refers to the number of collisions between the two sequences $\{c^{(i)}_{(k)_L}\}$ and $\{c^{(j)}_{(k)_L}\}$ in a shift period of NL when the shift satisfying $l = aN + b$. The smaller the value of $C_{ij}(l)$ gets, the less the number of collisions of two TH sequences gets. Then, MAI will be reduced. It should be noted that Definition 2 is different from Definition 1. The latter describes the number of collisions in terms of $l \propto T_f$, while the former dose in terms of $l \propto T_c$. In Definition 1, $0 < l \leq L-1$, and in the case, the number of circular shift comparison is equal to L. However, in Definition 2, the number of circular shift comparison is (*N*-1) times more than *L*, namely *NL*.

In order to demonstrate the collision situation between two TH sequences, Fig. 1 is used again. In Fig. 1, *a*=3 and *b*=2, then $l = aN + b = 17$. For $\{c^{(1)}_{(k)_5}\}$ and $\{c^{(2)}_{(k)_5}\}$, the set $A = \{4,8,12,15,21\}$, the set $B = \{4,11,15,17,23\}$, respectively. As a result, their TH periodic correlation function value satisfies $C_{1,2}(17) = |A \cap B| = 2$.

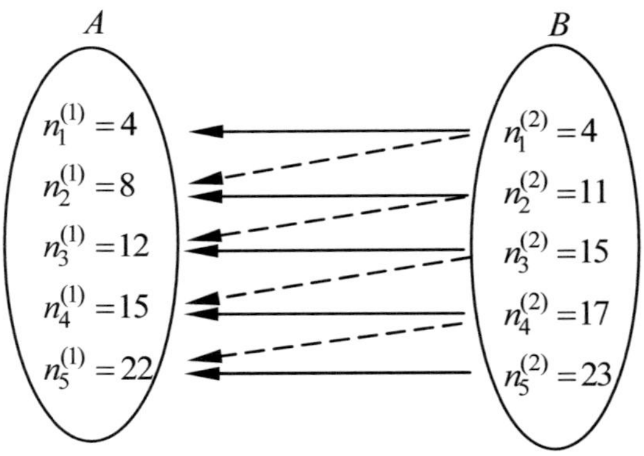

Fig. 4. The corresponding comparison between A and B with no shift (real line arrow) and with one frame shift (broken line arrow)

As for Definition 2, the definition of periodic correlation function of TH sequences is straightforward. However, it is not easy to calculate so that we need to transform the equation (3) into another equation.

We note that in the set B, $0 \leq c^{(j)}_{(k)_L} + b \leq N_h + N - 1$. When $N \leq c^{(j)}_{(k)_L} + b \leq N_h + N - 1$, we have $(kN + c^{(j)}_{(k)_L} + b)_{NL} = ((k+1)N + (c^{(j)}_{(k)_L} + b)_N)_{NL}$. Then, it is possible that the value of $(kN + c^{(j)}_{(k)_L} + b)_{NL}$ in the set B is equal to the value of $((k+1)N + c^{(i)}_{(k+1+a)_L})_{NL}$ in the set A. For example, when $k = 2$ in Fig. 1, we have $c^{(j)}_{(k)_L} + b = c^{(2)}_{(2)_L} + 2 = 5$. Then, $(kN + c^{(j)}_{(k)_L} + b)_{NL} = (2 \times 5 + c^{2}_{(2)_5} + 2)_{5\times5} = 15$, which is equal to the value of $((k+1)N + c^{(i)}_{(k+1+a)_L})_{NL} = ((2+1) \times 5 + c^{(1)}_{(2+1+3)_5})_{5\times5} = 15$. Hence, there exist two cases of collision by means of corresponding comparison between A and B with no shift and one frame shift. The collisions can be depicted in Fig. 4, where the notations $\{n_1^{(1)},\cdots,n_5^{(1)}\}$ and $\{n_1^{(2)},\cdots,n_5^{(2)}\}$ denote time slot values of the first TH sequence and the second TH sequence, respectively.

As a result, an alternative definition of TH periodic correlation function can be obtained as follows.

Definition 3: Let $\{c_{(k)_L}^{(i)}\}$ and $\{c_{(k)_L}^{(j)}\}$ denote two TH sequences with period L, and then TH periodic correlation function with shift l can be given by

$$C_{ij}(l)=\sum_{k=0}^{L-1} h[(kN+c_{(k+a)_L}^{(i)})_{NL},(kN+c_{(k)_L}^{(j)}+b)_{NL}]+\sum_{k=0}^{L-1} h[((k+1)N+c_{(k+1+a)_L}^{(i)})_{NL},(kN+c_{(k)_L}^{(j)}+b)_{NL}], \tag{4}$$

where the notations are the same as that in Definitions 1 and 2.

In Definition 3, it is obvious that the value of $C_{ij}(l)$ contains two parts, namely $\sum_{k=0}^{L-1} h[(kN+c_{(k+a)_L}^{(i)})_{NL},(kN+c_{(k)_L}^{(j)}+b)_{NL}]$ and $\sum_{k=0}^{L-1} h[((k+1)N+c_{(k+1+a)_L}^{(i)})_{NL},(kN+c_{(k)_L}^{(j)}+b)_{NL}]$. The two parts are suitable for $c_{(k)_L}^{(j)}+b<N$ and $c_{(k)_L}^{(j)}+b\ge N$, respectively.

In the case of $c_{(k)_L}^{(j)}+b<N$ and two TH sequences have a corresponding comparison in terms of k, the value of $((k+1)N+c_{(k+1+a)_L}^{(i)})_{NL}$ is impossible to be equal to the value of $(kN+c_{(k)_L}^{(j)}+b)_{NL}$ in the second part of $C_{ij}(l)$. Similarly, in the case of $c_{(k)_L}^{(j)}+b\ge N$ and two TH sequences have a corresponding comparison in terms of k, the value of $(kN+c_{(k+a)_L}^{(i)})_{NL}$ is impossible to be equal to the value of $(kN+c_{(k)_L}^{(j)}+b)_{NL}$ in the first part of $C_{ij}(l)$.

According to the above analyses, we can conclude that the TH periodic correlation function defined in Definition 3 is divided into two parts that do not include each other, and their combination is equal to the quantities of circular shift collisions in the shift period NL. Fig. 5 shows the illustration of the maximum CCF value of QCC sequences on the basis of Definition 3, where p = 11. In comparison with Fig. 3, we can see that TH correlation properties in asynchronism format are obviously inferior to that in synchronism format.

Note that TH periodic correlation functions in previous three definitions are assumed to be discrete functions because the sequences are discrete ones. In real systems, the MAI comes as noninteger shifts of cross correlation, and then we should assume the correlation functions to be continuous.

Continuous sequences, each of which corresponds to a discrete sequence, can be defined in Definition 4. In Definition 4, each term in a discrete sequence is held within a fixed width of a continuous sequence. This allows the continuous correlation function between two continuous sequences to be easily defined, which is similar to the definition of continuous correlation function of DS (Suehiro, 1992).

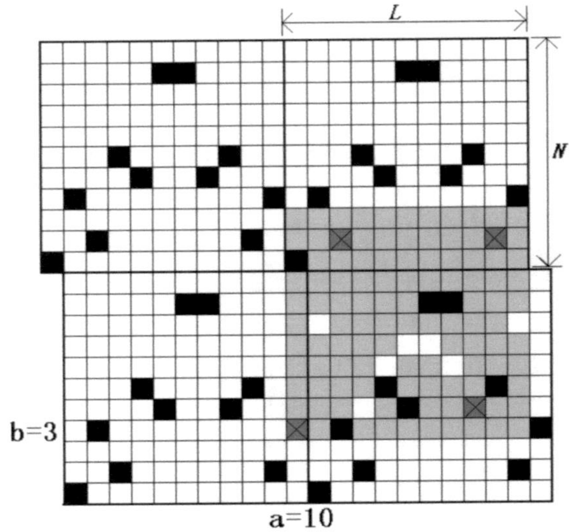

Fig. 5. The illustration of the maximum CCF value between $\left\{c_{(k)_{11}}^{(3)}\right\}$ and $\left\{c_{(k)_{11}}^{(5)}\right\}$ of QCC sequences when $N = N_h + 1 = L = 11$ and $l = aN + b = 113$, where collisions are denoted by a cross $\times$ and $C_{\max} = 4$

Definition 4: Let $\left\{c_{(k)_L}^{(i)}\right\}$ and $\left\{c_{(k)_L}^{(j)}\right\}$ denote two TH sequences with period L. The continuous TH periodic correlation function $Y_{ij}(\tau)$ with shift term τ is defined as

$$Y_{ij}(\tau) = (1-\alpha)\cdot C_{ij}(l) + \alpha \cdot C_{ij}(l+1), \tag{5}$$

where l is an integer, $\tau = l + \alpha$, $0 \le \alpha < 1$, and $C_{ij}(l)$ denotes discrete TH periodic correlation function in Definition 3.

Comparing Definition 4 with Definition 3, we can see that $Y_{ij}(\tau) \le C_{ij}(l)$. We consider the worst scenario, namely $Y_{ij}(\tau) = C_{ij}(l)$, and then we only analyze $C_{ij}(l)$ in the following sections.

3. Theoretical bounds of TH sequences

In this section, TH auto-correlation sidelobes average $\overline{C_{ii}(l)}$ and TH cross-correlation function average $\overline{C_{ij}(l)}$ are evaluated. Based on the obtained averages of $\overline{C_{ii}(l)}$ and $\overline{C_{ij}(l)}$, low bounds of maximal TH auto-correlation sidelobe $S_{\max}$ and maximal TH cross-correlation function value $C_{\max}$ are obtained in the following two theorems, respectively.

Theorem 1: Let $\left\{c_{(k)_L}^{(i)}\right\}$ denote a TH sequence with period L, and then $\overline{C_{ii}(l)}$ can be expressed as

$$\overline{C_{ii}(l)} = \frac{L^2 - L}{NL - 1} \tag{6}$$

and

$$S_{\max} \geq \frac{L^2 - L}{NL-1}. \tag{7}$$

Proof: For the TH sequence $\left\{c_{(k)_L}^{(i)}\right\}$, the sum of TH auto-correlation function values in one shift period NL can be calculated as

$$\sum_{l=0}^{NL-1} C_{ii}(l) = L + \sum_{l=1}^{NL-1} C_{ii}(l) = L + (NL-1)\cdot\overline{C_{ii}(l)}, \tag{8}$$

According to Definition 3, the sum $\sum_{l=0}^{NL-1} C_{ii}(l)$ can be also expressed as

$$\begin{aligned}
&\sum_{l=0}^{NL-1} C_{ii}(l) \\
&= \sum_{l=0}^{NL-1}\{\sum_{k=0}^{L-1} h[(kN + c_{(k+a)_L}^{(i)})_{NL}, (kN + c_{(k)_L}^{(i)} + b)_{NL}] + \sum_{k=0}^{L-1} h[((k+1)N + c_{(k+1+a)_L}^{(i)})_{NL}, (kN + c_{(k)_L}^{(i)} + b)_{NL}]\} \\
&= \sum_{l=0}^{NL-1}\sum_{k=0}^{L-1}\{h[(c_{(k+a)_L}^{(i)})_{NL}, (c_{(k)_L}^{(i)} + b)_{NL}] + h[(N + c_{(k+1+a)_L}^{(i)})_{NL}, (c_{(k)_L}^{(i)} + b)_{NL}]\} \\
&= \sum_{k=0}^{L-1}\sum_{b=0}^{N-1}\sum_{a=0}^{L-1}\{h[(c_{(k+a)_L}^{(i)})_{NL}, (c_{(k)_L}^{(i)} + b)_{NL}] + h[(N + c_{(k+1+a)_L}^{(i)})_{NL}, (c_{(k)_L}^{(i)} + b)_{NL}]\}
\end{aligned} \tag{9}$$

For the equation (9), we have $N \leq N + c_{(k)_L}^{(i)} \leq N + N_h$ and $0 \leq c_{(k)_L}^{(i)} + b \leq N_h + N - 1$ since $0 \leq c_{(k)_L}^{(i)} \leq N_h$ and $0 \leq b \leq N-1$. In order to simplify the equation (9), the value of $c_{(k)_L}^{(i)} + b$ will be divided in two parts.

On the one hand, we have $h[(N + c_{(k+1+a)_L}^{(i)})_{NL}, (c_{(k)_L}^{(i)} + b)_{NL}] = 0$ when $0 \leq c_{(k)_L}^{(i)} + b \leq N-1$. Then, in the equation (9),

$$\begin{aligned}
&\sum_{a=0}^{L-1}\{h[(c_{(k+a)_L}^{(i)})_{NL}, (c_{(k)_L}^{(i)} + b)_{NL}] + h[(N + c_{(k+1+a)_L}^{(i)})_{NL}, (c_{(k)_L}^{(i)} + b)_{NL}]\} \\
&= \sum_{a=0}^{L-1} h[(c_{(k+a)_L}^{(i)})_{NL}, (c_{(k)_L}^{(i)} + b)_{NL}] \\
&= \sum_{a=0}^{L-1} h[c_{(k+a)_L}^{(i)}, (c_{(k)_L}^{(i)} + b)_N]
\end{aligned}$$

When a is from 0 to $L-1$, the sum $\sum_{a=0}^{L-1} h[c_{(k+a)_L}^{(i)}, (c_{(k)_L}^{(i)} + b)_N]$ represents the number of the value of $(c_{(k)_L}^{(i)} + b)_N$ appearing in one period L of TH sequence $\left\{c_{(k)_L}^{(i)}\right\}$.

On the other hand, when $N \le c_{(k)_L}^{(i)} + b \le N_h + N - 1$, it is obvious that $h[(c_{(k+a)_L}^{(i)})_{NL}, (c_{(k)_L}^{(i)} + b)_{NL}] = 0$. Then, in the equation (9),

$$\begin{aligned}
&\sum_{a=0}^{L-1} \{h[(c_{(k+a)_L}^{(i)})_{NL}, (c_{(k)_L}^{(i)} + b)_{NL}] + h[(N + c_{(k+1+a)_L}^{(i)})_{NL}, (c_{(k)_L}^{(i)} + b)_{NL}]\} \\
&= \sum_{a=0}^{L-1} h[(N + c_{(k+1+a)_L}^{(i)})_{NL}, (c_{(k)_L}^{(i)} + b)_{NL}] \\
&= \sum_{a=0}^{L-1} h[N + c_{(k+1+a)_L}^{(i)}, N + (c_{(k)_L}^{(i)} + b)_N] \\
&= \sum_{a=0}^{L-1} h[c_{(k+1+a)_L}^{(i)}, (c_{(k)_L}^{(i)} + b)_N]
\end{aligned}$$

When a is from 0 to $L-1$, the sum $\sum_{a=0}^{L-1} h[c_{(k+1+a)_L}^{(i)}, (c_{(k)_L}^{(i)} + b)_N]$ also represents the number of the value of $(c_{(k)_L}^{(i)} + b)_N$ appearing in one period L of TH sequence $\{c_{(k)_L}^{(i)}\}$.
According to the above analyses, we can obtain that

$$\sum_{l=0}^{NL-1} C_{ii}(l) = \sum_{k=0}^{L-1} \sum_{b=0}^{N-1} num_i((c_{(k)_L}^{(i)} + b)_N), \tag{10}$$

where the notation $num_i((c_{(k)_L}^{(i)} + b)_N)$ denotes the number of the value of $(c_{(k)_L}^{(i)} + b)_N$ appearing in one period L of TH sequence $\{c_{(k)_L}^{(i)}\}$.

For the equation (10), when b is from 0 to $N-1$, the values of $(c_{(k)_L}^{(i)} + b)_N$ just constitute the set $\{0,1,\cdots,N-1\}$. Then, $\sum_{b=0}^{N-1} num_i((c_{(k)_L}^{(i)} + b)_N)$ represents the sum of the number of every element in set $\{0,1,\cdots,N-1\}$ appearing in one period L of TH sequence $\{c_{(k)_L}^{(i)}\}$. Hence, $\sum_{b=0}^{N-1} num_i((c_{(k)_L}^{(i)} + b)_N) = L$, and then $\sum_{l=0}^{NL-1} C_{ii}(l) = \sum_{k=0}^{L-1} L = L^2$. According to the equation (8), we have $\sum_{l=0}^{NL-1} C_{ii}(l) = L + (NL-1) \cdot \overline{C_{ii}(l)} = L^2$, and then $\overline{C_{ii}(l)} = \frac{L^2 - L}{NL - 1}$. As a result, we have $S_{\max} \ge \frac{L^2 - L}{NL - 1}$ since $S_{\max} \ge \overline{C_{ii}(l)}$.

Q.E.D

Similar to TH auto-correlation properties, low bounds of maximal TH cross-correlation function value $C_{\max}$ can be obtained in the following theorem.

Theorem 2: Let $\{c_{(k)_L}^{(i)}\}$ and $\{c_{(k)_L}^{(j)}\}$ denote two TH sequences with period L, respectively. Then $\overline{C_{ij}(l)}$ may be expressed as

$$\overline{C_{ij}(l)} = \frac{L}{N} \tag{11}$$

and

$$C_{\max} \geq \frac{L}{N}, \tag{12}$$

where $i \neq j$.

Proof: According to Definition 3, we have

$$\begin{aligned}
&NL \cdot \overline{C_{ij}(l)} \\
&= \sum_{l=0}^{NL-1} C_{ij}(l) \\
&= \sum_{k=0}^{L-1}\sum_{b=0}^{N-1}\sum_{a=0}^{L-1} \{h[(c^{(i)}_{(k+a)_L})_{NL}, (c^{(j)}_{(k)_L}+b)_{NL}] + h[(N + c^{(i)}_{(k+1+a)_L})_{NL}, (c^{(j)}_{(k)_L}+b)_{NL}]\}
\end{aligned}$$

The analyses of the above equation is similar to Theorem 1, and then we can obtain $NL \cdot \overline{C_{ij}(l)} = \sum_{k=0}^{L-1}\sum_{b=0}^{N-1} num_i((c^{(j)}_{(k)_L}+b)_N) = L^2$ and $\overline{C_{ij}(l)} = \frac{L}{N}$. Also, it is obvious that $C_{\max} \geq \frac{L}{N}$ since $C_{\max} \geq \overline{C_{ij}(l)}$.

Q.E.D

From Theorem 1 and Theorem 2, we can see that TH correlation function averages $\overline{C_{ii}(l)}$ and $\overline{C_{ij}(l)}$ are determined by sequences period L and the number of time slots N. When L and N are fixed, both $\overline{C_{ii}(l)}$ and $\overline{C_{ij}(l)}$ will be fixed for any TH sequence.

In order to explain the conclusions, we give an example. We use linear congruence codes (LCC) (Titlebaum, 1981) and QCC. For LCC sequences, $C^{(i)}_{(k)_L} = (ki)_P$, where $L = P$, $0 \leq k \leq p-1$, $1 \leq i \leq p-1$, and P is a prime. In this example, let $P = 5$ and $N = 5$. Then, we have $\{C^{(1)}_{(k)_5}\} = \{0,1,2,3,4\}$ and $\{C^{(2)}_{(k)_5}\} = \{0,2,4,1,3\}$.

When l is from 1 to 24 (here $NL-1 = 24$), auto-correlation sidelobes of TH sequence $\{C^{(1)}_{(k)_5}\}$ constitute the set {1,0,0,0,0,4,2,0,0,0,0,3,3,0,0,0,0,2,4,0,0,0,0,1}. When l is from 0 to 24, cross-correlation values between $\{C^{(1)}_{(k)_5}\}$ and $\{C^{(2)}_{(k)_5}\}$ constitute the set {1,1,2,0,1,1,1,0,2,1,1,2,1,0,1,1,0,2,2,0,1,1,0,1,2}. Then, the averages of elements in two sets are equal to 5/6 and 1, respectively. The results correspond to $\overline{C_{1,1}(l)} = \frac{L^2 - L}{NL-1} = \frac{5}{6}$ and $\overline{C_{1,2}(l)} = \frac{L}{N} = 1$ in terms of Theorem 1 and Theorem 2.

In addition, for QCC sequences, we have $\{C^{(1)}_{(k)_5}\} = \{0,1,4,4,1\}$ and $\{C^{(2)}_{(k)_5}\} = \{0,2,3,3,2\}$. When l is from 1 to 24, auto-correlation sidelobes of $\{C^{(1)}_{(k)_5}\}$ constitute

{0,1,0,1,1,2,1,1,0,1,1,1,1,1,1,0,1,1,2,1,1,0,1,0}. When l is from 0 to 24, cross-correlation values between $\left\{C^{(1)}_{(k)_5}\right\}$ and $\left\{C^{(2)}_{(k)_5}\right\}$ constitute the set {1,2,0,2,0,0,2,2,1,1,0,0,2,1,2,0,0,1,2,1,0,2,1,0,2}. Then, the averages of elements in two sets are also equal to 5/6 and 1, respectively. As a result, for any sequence, both of $\overline{C_{ii}(l)}$ and $\overline{C_{ij}(l)}$ will be fixed as long as L and N are fixed.

Based on Theorem 1 and Theorem 2, the further result can be also obtained. Two corollaries on TH correlation properties are expressed as follows.

Corollary 1: For a TH sequences family with period L, we have

$$S_{\max}, C_{\max} \geq 1 . \tag{13}$$

Corollary 2: When the period L and the number of time slots N are fixed, in order to obtain good TH correlation properties, correlation function values $C_{ii}(l)$ and $C_{ij}(l)$ should be close to their averages as possible.

In practice, we are also interested in maximal TH correlation function values $max\{C_{ij}(l)\}$ which is the maximum of all correlation function values include cross-correlation function values and auto-correlation sidelobes. Then, the following theorem gives the low bound of $max\{C_{ij}(l)\}$.

Theorem 3: For a TH sequences family with period L and family size N_u, the average of TH correlation function values can be expressed as

$$\overline{C(l)} = \frac{L^2(N_u+1)-2L}{NL(N_u+1)-2} \tag{14}$$

and then

$$max\{C_{ij}(l)\} \geq \frac{L^2(N_u+1)-2L}{NL(N_u+1)-2} . \tag{15}$$

Proof: For a TH sequences family with period L and family size N_u, the number of auto-correlation sidelobes and the number of cross-correlation values are equal to $N_u(NL-1)$ and $\frac{N_u(N_u-1)}{2}NL$, respectively. Then, the number of all correlation function values without auto-correlation peak should be equal to $N_u(NL-1)+\frac{N_u(N_u-1)}{2}NL$.

According to the proof of Theorem 1, the sum of auto-correlation sidelobes for every TH sequence is equal to L^2-L. Then, the sum of auto-correlation sidelobes for TH sequence family is equal to $N_u(L^2-L)$.

Similarly, the sum of cross-correlation values for TH sequence family is equal to $\frac{N_u(N_u-1)}{2}L^2$. Then, the sum of all correlation function values without auto-correlation peak should be equal to $N_u(L^2-L)+\frac{N_u(N_u-1)}{2}L^2$.

In terms of the above analyses, we can obtain that

$$\overline{C(l)} = \frac{N_u(L^2 - L) + \frac{N_u(N_u - 1)}{2}L^2}{N_u(NL - 1) + \frac{N_u(N_u - 1)}{2}NL} = \frac{L^2(N_u + 1) - 2L}{NL(N_u + 1) - 2}.$$

Also, it is obvious that $max\{C_{ij}(l)\} \geq \overline{C(l)}$ and $max\{C_{ij}(l)\} \geq \frac{L^2(N_u + 1) - 2L}{NL(N_u + 1) - 2}$.

Q.E.D

According to Theorem 3, TH correlation function average $\overline{C(l)}$ is determined by three parameters of period L, the number of time slots N and family size M. When L, N and M are fixed, $\overline{C(l)}$ is fixed for any TH sequence family.

4. Improvement of TH correlation properties

In this section, we will provide a method that improves the correlation properties of TH sequences. Before the corresponding analyses, the maximum TH correlation function values are further analyzed according to Definition 3. We give Theorem 4 as follows.
Theorem 4: For TH sequences with period L, the upper bound can be given by

$$S_{\max}, C_{\max} \leq 2max(\sum_{k=0}^{L-1} h[(c_{(k+a)_L}^{(i)})_N, (c_{(k)_L}^{(j)} + b)_N]). \tag{16}$$

Proof: According to the equation (4), we have

$$C_{ij}(l) = \sum_{k=0}^{L-1} h[(c_{(k+a)_L}^{(i)})_{NL}, (c_{(k)_L}^{(j)} + b)_{NL}] + \sum_{k=0}^{L-1} h[(N + c_{(k+1+a)_L}^{(i)})_{NL}, (c_{(k)_L}^{(j)} + b)_{NL}].$$

We first discuss the first part of $C_{ij}(l)$, namely $\sum_{k=0}^{L-1} h[(c_{(k+a)_L}^{(i)})_{NL}, (c_{(k)_L}^{(j)} + b)_{NL}]$. Note that it operates modulo NL. When it operates modulo N, the possibility of collisions between $(c_{(k+a)_L}^{(i)})_N$ and $(c_{(k)_L}^{(j)} + b)_N$ is larger than that of collisions between $(c_{(k+a)_L}^{(i)})_{NL}$ and $(c_{(k)_L}^{(j)} + b)_{NL}$. Then, we have $\sum_{k=0}^{L-1} h[(c_{(k+a)_L}^{(i)})_{NL}, (c_{(k)_L}^{(j)} + b)_{NL}] \leq \sum_{k=0}^{L-1} h[(c_{(k+a)_L}^{(i)})_N, (c_{(k)_L}^{(j)} + b)_N]$.
Similarly, the second part of $C_{ij}(l)$ satisfies

$$\sum_{k=0}^{L-1} h[(N + c_{(k+1+a)_L}^{(i)})_{NL}, (c_{(k)_L}^{(j)} + b)_{NL}] \leq \sum_{k=0}^{L-1} h[(N + c_{(k+1+a)_L}^{(i)})_N, (c_{(k)_L}^{(j)} + b)_N]$$

$$= \sum_{k=0}^{L-1} h[(c_{(k+1+a)_L}^{(i)})_N, (c_{(k)_L}^{(j)} + b)_N].$$

When the shift l is from 0 to $NL - 1$, it is obvious that

$$max(\sum_{k=0}^{L-1} h[(c_{(k+a)_L}^{(i)})_N, (c_{(k)_L}^{(j)} + b)_N]) = max(\sum_{k=0}^{L-1} h[(N + c_{(k+1+a)_L}^{(i)})_N, (c_{(k)_L}^{(j)} + b)_N]).$$

Therefore, we have

$$S_{\max}, C_{\max} = max(\sum_{k=0}^{L-1} h[(c_{(k+a)_L}^{(i)})_{NL}, (c_{(k)_L}^{(j)} + b)_{NL}] + \sum_{k=0}^{L-1} h[(N + c_{(k+1+a)_L}^{(i)})_{NL}, (c_{(k)_L}^{(j)} + b)_{NL}])$$

$$\leq 2max(\sum_{k=0}^{L-1} h[(c_{(k+a)_L}^{(i)})_N, (c_{(k)_L}^{(j)} + b)_N])$$

Q.E.D

Based on Theorem 4, we can obtain another theorem which indicates that the correlation properties of TH sequences will be improved when the number of TH time slot satisfies $N \geq 2N_h + 1$.

Theorem 5: Let $\{c_{(k)_L}^{(i)}\}$ and $\{c_{(k)_L}^{(j)}\}$ denote two TH sequences with period L, respectively. When $N \geq 2N_h + 1$, we have

$$C_{ij}(l) = \begin{cases} \sum_{k=0}^{L-1} h[(c_{(k+a)_L}^{(i)})_{NL}, (c_{(k)_L}^{(j)} + b)_{NL}], & 0 \leq b < N_h + 1 \\ \sum_{k=0}^{L-1} h[(N + c_{(k+1+a)_L}^{(i)})_{NL}, (c_{(k)_L}^{(j)} + b)_{NL}], & N_h + 1 \leq b < N \end{cases}$$

and

$$S_{\max}, C_{\max} \leq max(\sum_{k=0}^{L-1} h[(c_{(k+a)_L}^{(i)})_N, (c_{(k)_L}^{(j)} + b)_N]). \tag{17}$$

Proof: According to the equation (4), we have

$$C_{ij}(l) = \sum_{k=0}^{L-1} h[(c_{(k+a)_L}^{(i)})_{NL}, (c_{(k)_L}^{(j)} + b)_{NL}] + \sum_{k=0}^{L-1} h[(N + c_{(k+1+a)_L}^{(i)})_{NL}, (c_{(k)_L}^{(j)} + b)_{NL}].$$

When $0 \leq b < N_h + 1$, we have $0 \leq (c_{(k)_L}^{(j)} + b)_{NL} < 2N_h + 1$ since $0 \leq c_{(k)_L}^{(j)} \leq N_h$. Similarly, we also have $(N + c_{(k+1+a)_L}^{(i)})_{NL} \geq 2N_h + 1$ when $N \geq 2N_h + 1$. As a result, it is obvious that $\sum_{k=0}^{L-1} h[(N + c_{(k+1+a)_L}^{(i)})_{NL}, (c_{(k)_L}^{(j)} + b)_{NL}] = 0$. Then, $C_{ij}(l) = \sum_{k=0}^{L-1} h[(c_{(k+a)_L}^{(i)})_{NL}, (c_{(k)_L}^{(j)} + b)_{NL}]$ when $0 \leq b < N_h + 1$.

When $N_h + 1 \leq b < N$, we have $(c_{(k)_L}^{(j)} + b)_{NL} \geq N_h + 1$ since $0 \leq c_{(k)_L}^{(j)} \leq N_h$. Combining the result with $0 \leq (c_{(k+a)_L}^{(i)})_{NL} \leq N_h$, we can obtain $\sum_{k=0}^{L-1} h[(c_{(k+a)_L}^{(i)})_{NL}, (c_{(k)_L}^{(j)} + b)_{NL}] = 0$. Hence, $C_{ij}(l) = \sum_{k=0}^{L-1} h[(N + c_{(k+1+a)_L}^{(i)})_{NL}, (c_{(k)_L}^{(j)} + b)_{NL}]$ when $N_h + 1 \leq b < N$.

According to Theorem 4, we have $S_{\max}, C_{\max} \le max(\sum_{k=0}^{L-1} h[(c^{(i)}_{(k+a)_N})_L, (c^{(j)}_{(k)_L} + b)_N])$.

Q.E.D.

To show how Theorem 5 works, we give a simple example using QCC sequences, where $p = 11$ and $L = 11$. Fig. 6 and Fig. 7 show the distributions of correlation function values of QCC sequences when $N = 11$ and $N = 21$, respectively. By comparing two figures, we can see that the maximum TH correlation function values are deceased to a half of original values.

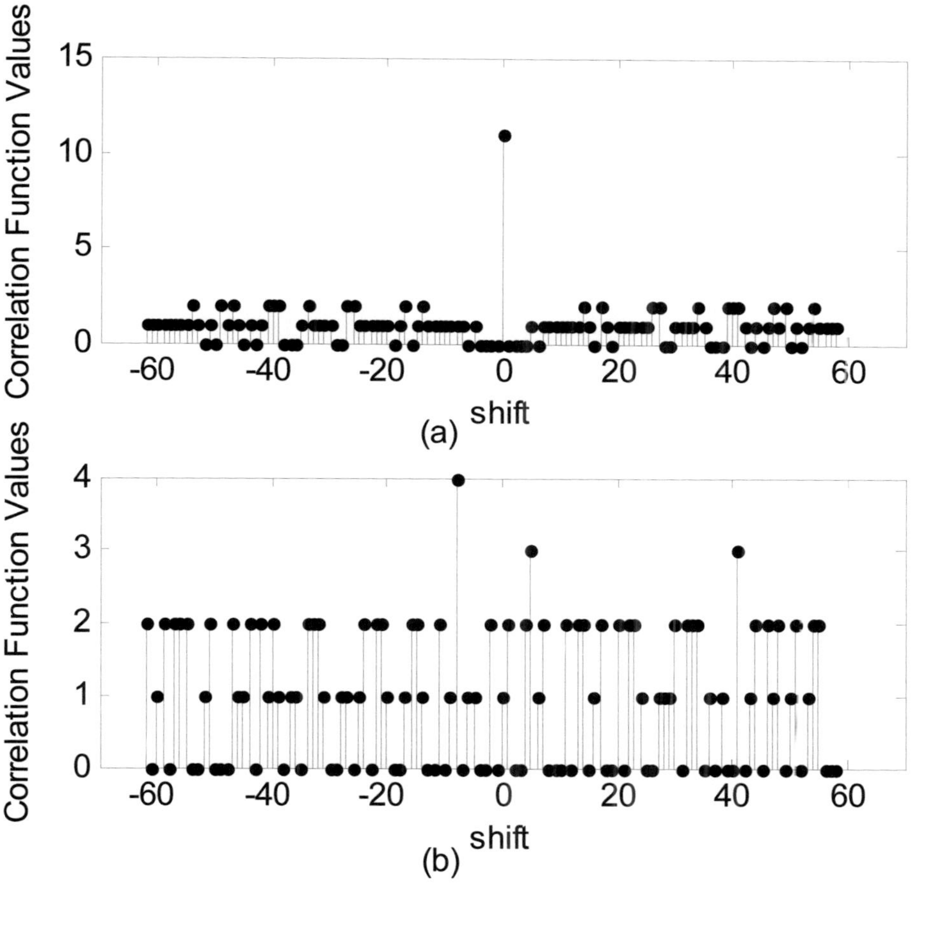

Fig. 6. The distribution of correlation function values of QCC sequences, where $N = 11$. (a). ACF of $\left\{c^{(2)}_{(k)_{11}}\right\}$; (b). CCF between $\left\{c^{(3)}_{(k)_{11}}\right\}$ and $\left\{c^{(5)}_{(k)_{11}}\right\}$

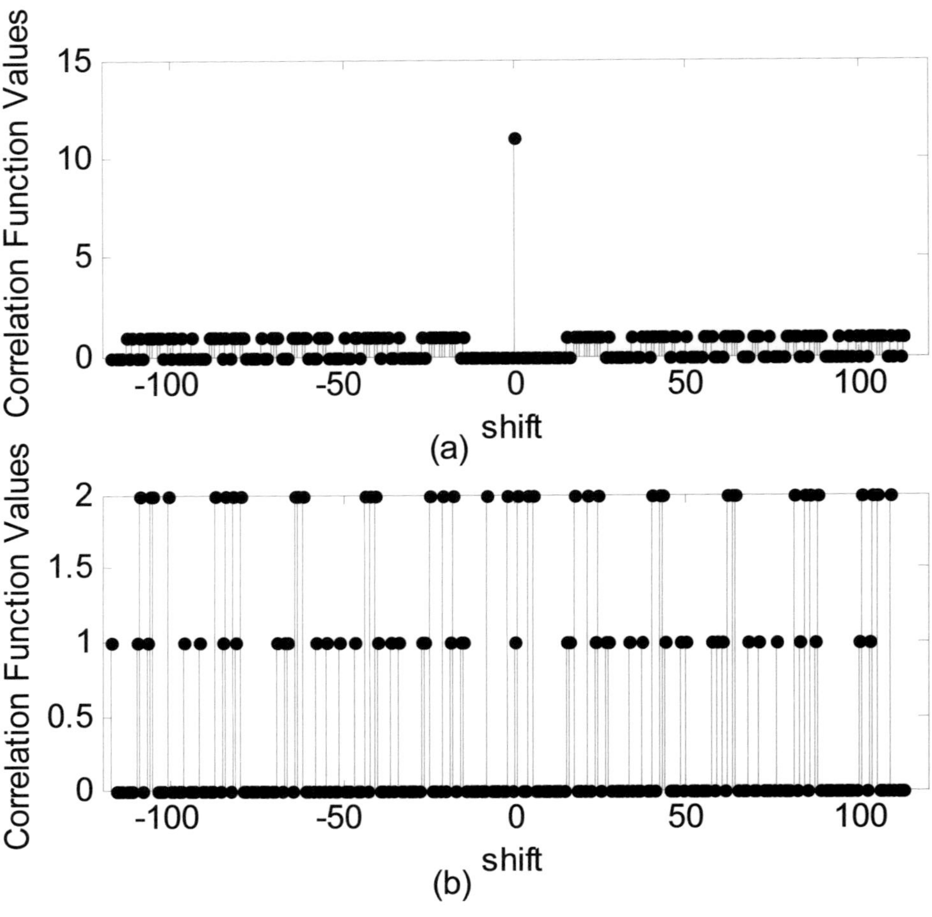

Fig. 7. The distribution of correlation function values of QCC sequences, where $N = 21$. (a). ACF of $\left\{c_{(k)_{11}}^{(2)}\right\}$; (b). CCF between $\left\{c_{(k)_{11}}^{(3)}\right\}$ and $\left\{c_{(k)_{11}}^{(5)}\right\}$

5. TH sequences with ZCZ

In this section, we begin with the definition of ZCZ of TH sequences to understand how ZCZ works. We then construct a class of TH sequences with ZCZ and prove the correlation properties of such TH sequences when the shifts between ZCZ TH sequences are in the range of ZCZ.

5.1 Definition of ZCZ of TH sequences

According to Definition 3 on TH period correlation function, we can define the ZCZ of TH sequences as follows.

Definition 5: Let $C_{ij}(l)$ denotes TH periodic correlation function between two TH sequences $\left\{c_{(k)_L}^{(i)}\right\}$ and $\left\{c_{(k)_L}^{(j)}\right\}$ with period L , and then ZCZ of TH sequences can be expressed as

$$C_{ii}(l)=\begin{cases} L, & l=0 \\ 0, & 0<|l|\le \frac{Z_{ACZ}}{2} \end{cases} \tag{18}$$

and

$$C_{ij}(l)=0,\quad 0\le|l|\le\frac{Z_{CCZ}}{2},\quad i\ne j, \tag{19}$$

where Z_{ACZ} and Z_{CCZ} denote TH zero auto-correlation zone (ZACZ) width and TH zero cross-correlation zone (ZCCZ) width, respectively.
According to definition 5, both of CCF and ACF sidelobes are equal to zero when the shifts between TH sequences are in the range of Z_{CZ}, where $Z_{CZ}=min\{Z_{ACZ},Z_{CCZ}\}$. Then, orthogonal communications can be realized when the approximate chip synchronization is held between users in whole system.

5.2 Construction of ZCZ TH sequences

The principle of construction of ZCZ TH sequences can be depicted in Fig. 8, where $\left\{c^{(1)}_{(k)_L}\right\}=\left\{e^{(1)}_{(k)_L}\right\}$ and $\left\{c^{(2)}_{(k)_L}\right\}=\left\{N_{eh}+1+Z_{CCZ}+e^{(2)}_{(k)_L}\right\}$ respectively denote two TH sequences, and $\left\{e^{(i)}_{(k)_L}\right\}$ is any existing TH sequence satisfying $0\le e^{(i)}_{(k)_L}\le N_{eh}$.

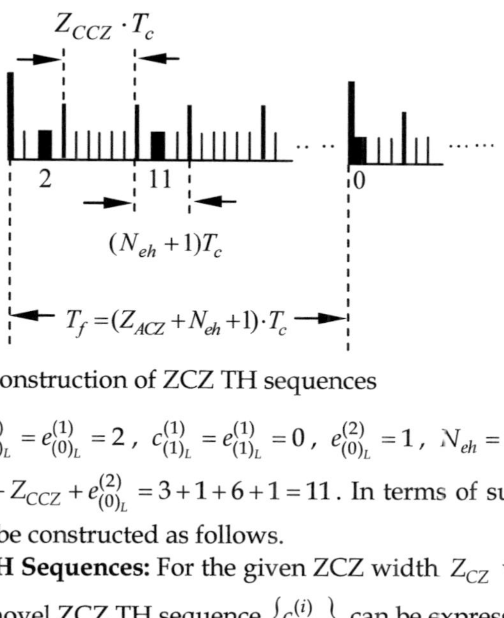

Fig. 8. The principle of construction of ZCZ TH sequences

According to Fig. 8, $c^{(1)}_{(0)_L}=e^{(1)}_{(0)_L}=2$, $c^{(1)}_{(1)_L}=e^{(1)}_{(1)_L}=0$, $e^{(2)}_{(0)_L}=1$, $N_{eh}=3$ and $Z_{CCZ}=6$. Then, we have $c^{(2)}_{(0)_L}=N_{eh}+1+Z_{CCZ}+e^{(2)}_{(0)_L}=3+1+6+1=11$. In terms of such principle, a class of ZCZ TH sequences can be constructed as follows.

Construction of ZCZ TH Sequences: For the given ZCZ width Z_{CZ} which is determined by THSS-UWB systems, a novel ZCZ TH sequence $\left\{c^{(i)}_{(k)_L}\right\}$ can be expressed as

$$c^{(i)}_{(k)_L}=(i-1)(N_{eh}+1+Z_{CZ})+e^{(i)}_{(k)_L}\,. \tag{20}$$

The widths of ZACZ and ZCCZ satisfy

$$Z_{ACZ} = N_u \cdot (N_{eh} + 1 + Z_{CZ}) - N_{eh} - 1 = N - N_{eh} - 1$$

and

$$Z_{CCZ} = Z_{CZ},$$

where $N = N_u \cdot (N_{eh} + 1 + Z_{CZ})$ and $0 \le e_{(k)_L}^{(i)} \le N_{eh}$.

Based on Definition 3, correlation properties of the constructed ZCZ TH sequences can be proved as follows.

Proof: (1). We first consider the case of $i \ne j$, namely CCF.

Let the synchronization error λ of a THSS-UWB system satisfy $|\lambda| \le \frac{Z_{CZ}}{2} \cdot T_c$ when the approximate chip synchronization is held in the whole system. Correspondingly, the shift between two TH sequences is equal to $l = aN + b$, where $0 \le a \le L-1$ and $0 \le b \le Z_{CZ}$. The evaluation of $C_{ij}(l)$ will be carried out in two steps on the basis of its two components.

i. According to the equation (20), the first part of $C_{ij}(l)$ can be expressed as

$$\sum_{k=0}^{L-1} h[(c_{(k+a)_L}^{(i)})_{NL}, (c_{(k)_L}^{(j)} + b)_{NL}] = \sum_{k=0}^{L-1} h[((i-j)(N_{eh} + 1 + Z_{CZ}) + (e_{(k+a)_L}^{(i)} - e_{(k)_L}^{(j)}))_{NL}, (b)_{NL}],$$

where $-N_{eh} \le e_{(k+a)_L}^{(i)} - e_{(k)_L}^{(j)} \le N_{eh}$ since $0 \le e_{(k+a)_L}^{(i)}, e_{(k)_L}^{(j)} \le N_{eh}$. Then, it is obvious that $((i-j)(N_{eh} + 1 + Z_{CZ}) + (e_{(k+a)_L}^{(i)} - e_{(k)_L}^{(j)}))_{NL} \ge 1 + Z_{CZ}$ when $i > j$. If $i < j$, we will have $-(N_u - 1) \le i - j \le -1$. Then,

$$\begin{aligned} -(N_u - 1)(N_{eh} + 1 + Z_{CZ}) - N_{eh} &\le (i-j)(N_{eh} + 1 + Z_{CZ}) + (e_{(k+a)_L}^{(i)} - e_{(k)_L}^{(j)}) \\ &\le -(N_{eh} + 1 + Z_{CZ}) + N_{eh} \\ &= -(1 + Z_{CZ}) \\ &< 0 \end{aligned}$$

We can further obtain that

$$\begin{aligned} ((i-j)(N_{eh} + 1 + Z_{CZ}) + (e_{(k+a)_L}^{(i)} - e_{(k)_L}^{(j)}))_{NL} &= NL - |(i-j)(N_{eh} + 1 + Z_{CZ}) + (e_{(k+a)_L}^{(i)} - e_{(k)_L}^{(j)})| \\ &\ge NL - (N_u - 1)(N_{eh} + 1 + Z_{CZ}) - N_{eh} \\ &= N_u(N_{eh} + 1 + Z_{CZ})L - (N_u - 1)(N_{eh} + 1 + Z_{CZ}) - N_{eh} \\ &= N_u(L-1)N_{eh} + (N_u L - N_u + 1)(1 + Z_{CZ}) \\ &\ge 1 + Z_{CZ} \end{aligned}$$

As a result, when $i \ne j$, we have

$$((i-j)(N_{eh} + 1 + Z_{CZ}) + (e_{(k+a)_L}^{(i)} - e_{(k)_L}^{(j)}))_{NL} \ge 1 + Z_{CZ}.$$

Due to $0 \le b \le Z_{CZ}$, we can obtain that

$$\sum_{k=0}^{L-1} h[(c_{(k+a)_L}^{(i)})_{NL},(c_{(k)_L}^{(j)}+b)_{NL}] = \sum_{k=0}^{L-1} h[((i-j)(N_{eh}+1+Z_{CZ})+(e_{(k+a)_L}^{(i)}-e_{(k)_L}^{(j)}))_{NL},(b)_{NL}]$$
$$=0$$

ii. The second part of $C_{ij}(l)$ can be expressed as

$$\sum_{k=0}^{L-1} h[(N+c_{(k+1+a)_L}^{(i)})_{NL},(c_{(k)_L}^{(j)}+b)_{NL}]$$
$$=\sum_{k=0}^{L-1} h[((N_u+i-j)(N_{eh}+1+Z_{CZ})+(e_{(k+1+a)_L}^{(i)}-e_{(k)_L}^{(j)}))_{NL},(b)_{NL}]$$

Similarly, when $i \neq j$, we can obtain that $((N_u+i-j)(N_{eh}+1+Z_{CZ})+(e_{(k+1+a)_L}^{(i)}-e_{(k)_L}^{(j)}))_{NL} \geq 1+Z_{CZ}$. Due to $0 \leq b \leq Z_{CZ}$, we can obtain that $\sum_{k=0}^{L-1} h[(N+c_{(k+1+a)_L}^{(i)})_{NL},(c_{(k)_L}^{(j)}+b)_{NL}]=0$.

In terms of the above analyses, the CCF values of the constructed ZCZ TH sequences are equal to zero when the shifts are in range of Z_{CCZ}, namely $C_{ij}(l)=0$ when $0 \leq |l| \leq \frac{Z_{CCZ}}{2}$ and $i \neq j$.

(2). Secondly, we consider the case of $i=j$, namely ACF.

For an approximately synchronized THSS-UWB system, when multipath delay is in the range of $Z_{ACZ} \cdot T_c$, the shift of TH sequence $\left\{c_{(k)_L}^{(i)}\right\}$ is correspondingly equal to $l=aN+b$, where $a=0$ and $0 < b \leq Z_{ACZ}$.

Similar to $C_{ij}(l)$, the evaluation of $C_{ii}(l)$ will be carried out in two steps.

i. According to equation (20), the first part of $C_{ii}(l)$ can be expressed as

$$\sum_{k=0}^{L-1} h[((i-i)(N_{eh}+1+Z_{CZ})+(e_{(k)_L}^{(i)}-e_{(k)_L}^{(i)}))_{NL},(b)_{NL}] = \sum_{k=0}^{L-1} h[(0)_{NL},(b)_{NL}].$$

Due to $0 < b \leq Z_{ACZ}$, we have $\sum_{k=0}^{L-1} h[(c_{(k+a)_L}^{(i)})_{NL},(c_{(k)_L}^{(i)}+b)_{NL}] = \sum_{k=0}^{L-1} h[(0)_{NL},(b)_{NL}]=0$.

ii. The second part of $C_{ii}(l)$ can be expressed as

$$\sum_{k=0}^{L-1} h[((N_u+i-i)(N_{eh}+1+Z_{CZ})+(e_{(k+1)_L}^{(i)}-e_{(k)_L}^{(i)}))_{NL},(b)_{NL}]$$
$$=\sum_{k=0}^{L-1} h[(N_u(N_{eh}+1+Z_{CZ})+(e_{(k+1)_L}^{(i)}-e_{(k)_L}^{(i)}))_{NL},(b)_{NL}]$$

Due to $-N_{eh} \leq e_{(k+1)_L}^{(i)}-e_{(k)_L}^{(i)} \leq N_{eh}$ and $N=N_u \cdot (N_{eh}+1+Z_{CZ})$, we can obtain that $N-N_{eh} \leq N_u(N_{eh}+1+Z_{CZ})+(e_{(k+1)_L}^{(i)}-e_{(k)_L}^{(i)}) \leq N+N_{eh}$. Also, since $0 < b \leq Z_{SCZ} = N-N_{eh}-1$, then,

$$\sum_{k=0}^{L-1} h[(N+c_{(k+1+a)_L}^{(i)})_{NL},(c_{(k)_L}^{(j)}+b)_{NL}] = \sum_{k=0}^{L-1} h[(N_u(N_{eh}+1+Z_{CZ})+(e_{(k+1)_L}^{(i)}-e_{(k)_L}^{(i)}))_{NL},(b)_{NL}]$$
$$=0$$

According to the above analyses, the ACF sidelobes of the constructed ZCZ TH sequences are equal to zero when the shifts are in range of Z_{ACZ}.

Q.E.D.

6. Effects of TH correlation properties on MAI in THSS-UWB systems

By transforming the signal model of THSS-UWB communication systems, we obtain expressions for the relation of MAI values and TH correlation function values in this section,.

6.1 Binary model of TH sequences

According to the equation (1), we can see that only one pulse is transmitted to each user within any frame time T_f, i. e. One-Pulse-Per-Frame structure (Erseghe, 2002b; Scholtz et al, 2001). The pulse position is decided by TH sequence $\left\{c_{(k)_L}^{(i)}\right\}$, namely $c_{(k)_L}^{(i)}.T_c$. For more easiness to understand, the structure is depicted in Fig. 9, where elements of a TH sequence are binary ones.

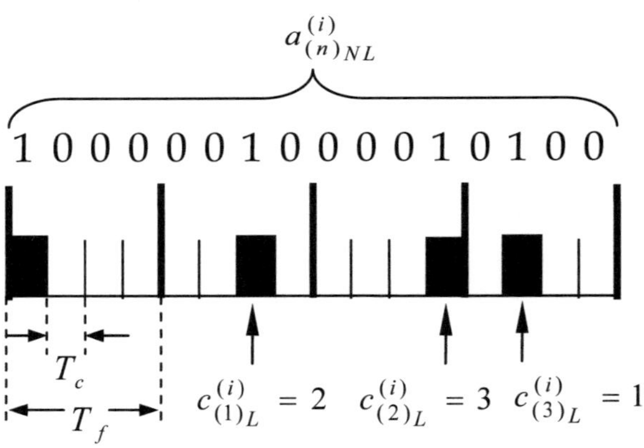

Fig. 9. The hopping format of pulses in PPM

We assume that "1" denotes the time slot where a pulse is modulated, and the other time slots in frame time T_f are "0". As a result, the binary TH sequence $\left\{a_{(n)_{NL}}^{(i)}\right\}$ can be obtained. The sequence $\left\{a_{(n)_{NL}}^{(i)}\right\}$ corresponds to $\left\{c_{(k)_L}^{(i)}\right\}$ and its period is equal to NL. According to the above analyses, the equation (1) may be transformed as

$$S^{(i)}(t) = \sum_{n=-\infty}^{+\infty} a_{(n)_{NL}}^{(i)} \cdot w(t-nT_c-\delta d_{[n/(N \cdot N_s)]}^{(i)}), \tag{21}$$

where

$$a_{(n)_{NL}}^{(i)} = \begin{cases} 1, & n = kN + c_{(k)_L}^{(i)} \text{ for some integer } k \\ 0, & otherwise \end{cases}.$$

Then, the TH periodic correlation function between $\left\{c_{(k)_L}^{(i)}\right\}$ and $\left\{c_{(k)_L}^{(j)}\right\}$ in Definition 3 can be also expressed as

$$C_{ij}(l) = \sum_{n=0}^{L \cdot N-1} a_{(n)_{NL}}^{(i)} a_{(n+l)_{NL}}^{(j)}, \tag{22}$$

where l denotes the shift between $\left\{a_{(n)_{NL}}^{(i)}\right\}$ and $\left\{a_{(n)_{NL}}^{(j)}\right\}$, $0 \le l \le NL-1$.
Note that the equation (22) is different from the periodic correlation function of DSs. The correlation function of binary TH sequences describes the number of agreements to element "1" between sequences, called the number of collisions, where $a_{(n)_{NL}}^{(i)}$ is equal to "0" or "1" instead of " $+1$ " and " -1 " in ordinary correlation function, such as in DS systems. In other words, if both of $a_{(n)_{NL}}^{(i)}$ and $a_{(n)_{NL}}^{(j)}$ are "1" in some time slot where user i and user j collide, then their multiplication $a_{(n)_{NL}}^{(i)} \cdot a_{(n)_{NL}}^{(j)}$ will be also "1".
As a result, the binary TH correlation function in the equation (22) refers to the number of collisions. The smaller $C_{ij}(l)$ gets, the smaller the number of collisions are, and the better TH correlation properties are.

6.2 Multiple-access performance

In THSS-UWB multiple-access communication systems, when N_u links are active, the received signal $r(t)$ may be expressed as follows (Scholtz, 1993),

$$r(t) = \sum_{i=1}^{N_u} A_i S^{(i)}(t - \tau_i) + n(t), \tag{23}$$

where A_i represents the attenuation of transmitter k's signal over the propagation path to the receiver, and τ_i denotes time asynchronisms between the clocks of transmitter k and the receiver. The notation $n(t)$ is white Gaussian receiver noise.
Without loss of generality, we assume that the receiver is interested in determining the data sent by transmitter 1 in the following analyses. We also assume that one data symbol is modulated by L pulses, i. e. $N_S = L$, and the correlation demodulation is employed. When symbol "0" is sent, the shift time is zero, and the shift time is δ when symbol "1" is sent. Then, a template signal can be given by

$$V(t) = \sum_{n=(k-1)\cdot N_s \cdot N}^{k \cdot N_s \cdot N-1} a_{(n)_{NL}}^{(i)} \cdot [w(t - nT_c) - w(t - nT_c - \delta)] \tag{24}$$

For transmitter 1, the demodulation output of the kth bit is

$$T_k^{(1)} = \int_{(k-1)\cdot N_s\cdot N\cdot T_c}^{k\cdot N_s\cdot N\cdot T_c} r(t)\cdot V(t)dt\,. \tag{25}$$

Then, the received bit is decided as "0" when $T_k^{(1)} \geq 0$. Obviously, when $T_k^{(1)} < 0$, the received bit is determined as "1".

The equation (25) is also described as

$$T_k^{(1)} = A_1 N_s (-1)^{d_k^{(1)}} E_P + \sum_{i=2}^{N_u} A_i \cdot [R_{i1}(\tau_i)\cdot(-1)^{d_m^{(i)}} + R_{i1}^*(\tau_i)\cdot(-1)^{d_{m+1}^{(i)}}]\cdot E_P + \int_{(k-1)\cdot N_s\cdot N\cdot T_c}^{k\cdot N_s\cdot N\cdot T_c} n(t)\cdot V(t)dt\,, \tag{26}$$

where $E_P = \int_{-\infty}^{+\infty} w(t)\cdot[w(t) - w(t-\delta)]dt$, $d_m^{(i)}$ and $d_{m+1}^{(i)}$ represent the mth bit and $(m+1)^{th}$ bit of user i, respectively. Transmitter 1 sends the kth bit. $R_{i1}(\tau_i)$ and $R_{i1}^*(\tau_i)$ denote the TH part correlation function between user i and user 1 in continuous time, respectively, namely $R_{ij}(\tau) = \int_0^{\tau} a_{t-\tau}^{(i)} \cdot a_t^{(j)} dt$ and $R_{ij}^*(\tau) = \int_{\tau}^{L\cdot N\cdot T_c} a_{t-\tau}^{(i)} \cdot a_t^{(j)} dt$.

It is obvious that $R_{ij}(\tau) + R_{ij}^*(\tau) = Y_{ij}(\tau)$, where $Y_{ij}(\tau)$ is continuous TH period correlation function in Definition 4.

In the equation (26), the first part is the signal that we desire. The second part represents the MAI that the other users make to user 1, and the last part is the interference made by noise. We are interested in the second part, which will be analyzed in the following. The analysis of THSS-UWB MAI is similar to the performance evaluation for DSSS multiple-access communications (Pursley, 1977).

In order to analyze the second part of equation (26), we define now the TH aperiodic correlation function in discrete time as follows,

$$Z_{ij}(l) = \begin{cases} \sum_{n=0}^{L\cdot N-1-l} a_{(n)_{NL}}^{(i)} \cdot a_{(n+l)_{NL}}^{(j)}, & 0 \leq l \leq L\cdot N - 1 \\ \sum_{n=0}^{L\cdot N-1+l} a_{(n-l)_{NL}}^{(i)} \cdot a_{(n)_{NL}}^{(j)}, & 1 - L\cdot N \leq l < 0\,. \\ 0, & |l| \geq L\cdot N \end{cases} \tag{27}$$

According to the equation (27), in the range of $0 \leq lT_c \leq \tau \leq (l+1)T_c \leq L\cdot N\cdot T_c$, $R_{ij}(\tau)$ and $R_{i1}^*(\tau_i)$ can be expressed as $R_{ij}(\tau) = Z_{ij}(l - L\cdot N)\cdot T_c + [Z_{ij}(l+1-L\cdot N) - Z_{ij}(l - L\cdot N)]\cdot(\tau - lT_c)$ and $R_{ij}^*(\tau) = Z_{ij}(l)\cdot T_c + [Z_{ij}(l+1) - Z_{ij}(l)]\cdot(\tau - lT_c)$, respectively.

Meanwhile, the TH period correlation function $C_{ij}(l)$ in the equation (22) can be expressed as $C_{ij}(l) = Z_{ij}(l) + Z_{ij}(l - L\cdot N)$, and TH period odd correlation function can be also defined as $C_{ij}^*(l) = Z_{ij}(l) - Z_{ij}(l - L\cdot N)$.

From the equation (26), the interference made by user i with user 1 satisfies

$$M_{i1}(\tau_i) = A_i \cdot [R_{i1}(\tau_i)\cdot(-1)^{d_m^{(i)}} + R_{i1}^*(\tau_i)\cdot(-1)^{d_{m+1}^{(i)}}]\cdot E_P\,.$$

Let $l_i T_c \leq \tau_i \leq (l_i + 1)\cdot T_c$. When $d_m^{(i)} = d_{m+1}^{(i)}$, we have

$$\begin{aligned}M_{i1}(\tau_i) &= A_i[R_{i1}(\tau_i)+R_{i1}^*(\tau_i)]\cdot(-1)^{d_m^{(i)}}\cdot E_P \\ &= A_i\{[Z_{i1}(l_i-L\cdot N)+Z_{i1}(l_i)]\cdot T_c+[(Z_{i1}(l_i+1-L\cdot N)+Z_{i1}(l_i+1)) \\ &\quad -(Z_{i1}(l_i-L\cdot N)+Z_{i1}(l_i))]\cdot(\tau_i-l_iT_c)\}\cdot(-1)^{d_m^{(i)}}\cdot E_P \\ &= A_i\{C_{i1}(l_i)\cdot T_c+[C_{i1}(l_i+1)-C_{i1}(l_i)]\cdot(\tau_i-l_iT_c)\}\cdot(-1)^{d_m^{(i)}}\cdot E_P\end{aligned} \tag{28}$$

Similarly, when $d_m^{(i)} \neq d_{m+1}^{(i)}$, we have

$$\begin{aligned}M_{i1}(\tau_i) &= A_i[R_{i1}(\tau_i)-R_{i1}^*(\tau_i)]\cdot(-1)^{d_m^{(i)}}\cdot E_P \\ &= A_i\{C_{i1}^*(l_i)\cdot T_c+[C_{i1}^*(l_i+1)-C_{i1}^*(l_i)]\cdot(\tau_i-l_iT_c)\}\cdot(-1)^{d_m^{(i)}}\cdot E_P\end{aligned} \tag{29}$$

The equations (28)-(29) provide the desired relation between multiple-access interference $M_{i1}(\tau_i)$ and TH correlation function $C_{ij}(l)$ (or $C_{ij}^*(l)$).
Note that $[C_{i1}(l_i+1)-C_{i1}(l_i)](\tau_i-l_iT_c) << C_{i1}(l_i)T_c$ and $[C_{i1}^*(l_i+1)-C_{i1}^*(l_i)]\cdot(\tau_i-l_iT_c) << C_{i1}^*(l_i)T_c$. Hence, the equations (28)-(29) can be respectively expressed as $M_{i1}(\tau_i)=A_i\cdot C_{i1}(l_i)\cdot T_c\cdot(-1)^{d_m^{(i)}}\cdot E_P$ for $d_m^{(i)}=d_{m+1}^{(i)}$ and $M_{i1}(\tau_i)=A_iC_{i1}^*(l_i)\cdot T_c\cdot(-1)^{d_m^{(i)}}\cdot E_P$ for $d_m^{(i)}\neq d_{m+1}^{(i)}$. They describe the interference to user1 from user i. Furthermore, we consider a more general situation, where N_u users are active.
Since $C_{ij}(l)=Z_{ij}(l)+Z_{ij}(l-L\cdot N)$ and $C_{ij}^*(l)=Z_{ij}(l)-Z_{ij}(l-L\cdot N)$, then $C_{i1}(l_i)\le C_{\max}$ and $-C_{\max}\le C_{i1}^*(l_i)\le C_{\max}$. Hence, we consider the worst scenario, which happens when $C_{i1}(l_i)=C_{\max}$ and $C_{i1}^*(l_i)=C_{\max}$. Let I_{SIR} denote signal-to-interference ratio (SIR) which describes the interference to user 1 from the other N_u-1 users and thermal noise, then

$$I_{SIR}=\frac{D_1^2}{N_1^2+\sum_{i=2}^{N_u}M_{i1}^2(\tau_i)}, \tag{30}$$

where $M_{i1}(\tau_i)=A_i\cdot C_{\max}\cdot T_c\cdot(-1)^{d_m^{(i)}}\cdot E_P$, $D_1=A_1\cdot N_s\cdot(-1)^{d_k^{(1)}}\cdot E_p$, and $N_1=\int_{(k-1)\cdot N_s\cdot N\cdot T_c}^{k\cdot N_s\cdot N\cdot T_c}n(t)\cdot V(t)dt$. Then, I_{SIR} can be expressed as

$$I_{SIR}=\frac{1}{\frac{1}{I_{SNR}}+\frac{C_{\max}^2\cdot T_c^2}{N_s^2}\cdot\sum_{i=2}^{N_u}(\frac{A_i}{A_1})^2}, \tag{31}$$

where $I_{SNR}=\frac{D_1^2}{N_1^2}$, which is a convenient parameter and equivalent to the output signal-to-noise (SNR) ratio that one might observe in single link experiments.
Then, the BER can be given by

$$P_e=\frac{1}{\sqrt{2\pi}}\int_{\sqrt{I_{SIR}}}^{\infty}\exp(-x^2/2)dx. \tag{32}$$

According to the equations (30)-(33), we can see that the interference and the BER are determined by $C_{\max}$ when A_1, A_i, N_s, T_c and I_{SNR} are specific. For the construction of TH sequences in THSS-UWB communication systems, TH correlation function values should be as small as possible so that the multiple access interference and the probability of error are small.

Similarly, for another hoping format in THSS-UWB, called PAM, the relation between multiple-access interference $M_{i1}(\tau_i)$ and TH correlation function $C_{ij}(l)$ (or $C^*_{ij}(l)$) may be respectively given as $M_{i1}(\tau_i) = A_i \cdot d_m^{(i)} \cdot \{C_{i1}(l_i) \cdot T_c + [C_{i1}(l_i+1) - C_{i1}(l_i)] \cdot (\tau_i - l_i T_c)\} \cdot E_P$ for $d_m^{(i)} = d_{m+1}^{(i)}$ and $M_{i1}(\tau_i) = A_i \cdot d_m^{(i)} \cdot \{C^*_{i1}(l_i) \cdot T_c + [C^*_{i1}(l_i+1) - C^*_{i1}(l_i)] \cdot (\tau_i - l_i T_c)\} \cdot E_P$ for $d_m^{(i)} = -d_{m+1}^{(i)}$, where the data $d_m^{(i)}$ is a binary stream (" $+1$ "or " -1 ") instead of "0" or "1" in PPM format.

7. Conclusion

The main intention of this chapter is to analyze the correlation properties of TH sequences for THSS-UWB systems. A more general definition of TH periodic correlation function is provided and the definition can be used to analytically evaluate the TH correlation properties in codeword synchronism, chip synchronism and asynchronism in the whole system. Based on the definition, theoretical bounds of TH sequences are presented, which relate four parameters of L, N, N_u and $C_{\max}$ (or $S_{\max}$). The results can be used to evaluate the performance of TH sequences and provide references for the design of TH sequences. Also, based on the definition, a method to improve TH correlation properties in practical applications is proposed. The maximum correlation function values of TH sequences can be reduced to a half of original values by such a method. Specially, in terms of this method, the maximum correlation function values of QCC sequences can be reduced from $S_{\max}=2$ and $C_{\max}=4$ to $S_{\max}=1$ and $C_{\max}=2$, which achieves the best TH correlation properties so far in an asynchronized THSS-UWB system.

A novel TH sequence family with TH ZCZ for approximately synchronized THSS-UWB systems is constructed and its correlation properties are proved in terms of the definition of TH periodic correlation function presented in this chapter. When the approximate chip synchronization is held in the whole system, the MAI of THSS-UWB system employing the proposed ZCZ TH sequences is eliminated, and such THSS-UWB systems are more tolerant to the multipath problem. As a result, orthogonal communications can be realized while the need of accurate synchronism in whole system is reduced.

In addition, the multiple access performance is investigated and the relation between MAI values I_{SIR} and TH correlation function values $C_{\max}$ are formulated by transforming the signal model from decimal sequence $\left\{c_{(k)_L}^{(i)}\right\}$ to binary sequence $\left\{a_{(n)_{NL}}^{(i)}\right\}$. Based on the obtained results, TH correlation function values should be as small as possible so that the multiple access interference and the probability of error are small.

8. Acknowledgement

We thank the National Natural Science Foundation of China (NSFC) under Grant no. 61002034 and 60872164, the Natural Science Foundation Project of CQ CSTC under Grant no. 2009BA2063 and 2010BB2203, Chongqing University Postgraduates' Innovative Team Building Project under Grant no. 200909B1010, and Open Research Fundation of Chongqing Key Laboratory of Signal and Information Processing (CQKLS&IP) under Grant no. CQSIP-2010-01 for supporting this work.

9. References

Canadeo, C.M., Temple, M.A., Boldwin, R.O. & Raines, R.A., (2003). UWB Multiple Access Performance in Synchronous and Asynchronous Networks, *Electron. Lett.*, Vol.39, No. 11, May 2003, pp. 880-882.

Corrada-Bravo, C., Scholtz, R. A. & Kumar, P. V., (1999). Generating TH-SSMA Sequences with Good Correlation and Approximately Flat PSD Level, *Proceedings of UWB'99*, Sept. 28, 1999.

Chu, W. S.& Colbourn, C.J., (2004). Sequence Designs for Ultra-Wideband Impulse Radio with Optimal Correlation Properties, *IEEE Trans. Inf. Theory*, Vol. 50, No. 10, Oct. 2004, pp. 2402-2407.

Durisi, G. & Benedetto, S., (2003). Performance Evaluation of TH-PPM UWB Systems in the Presence of Multiuser Interference, *IEEE Commun. Letters*, Vol. 7, No.5, May 2003, pp. 224-226.

Erseghe, T., (2002). *Ultra Wide Band Pulse Communications*, Ph.D. thesis, Universita' di Padova, Italy, Feb., 2002.

Erseghe, T., (2002). Time-Hopping Patterns Derived from Permutation Sequences for Ultra Wide Band Impulse Radio Applications, *Proceedings of 6th WSEAS International Conf. on Communications*, July 7-14, 2002, pp.109-115.

Erseghe, T. & Bramante, N., (2002). Pseudo-Chaotic Encoding Applied to Ultra-Wide-Band Impulse Radio, *Proceedings of VTC'04F*, Vol. 3, Vancouver, Canada , Dec., 2002, pp. 1711-1715.

Fan, P. Z., Suehiro, N., Kuroyanagi, N. & Deng, X. M., (1999). Class of Binary Sequences with Zero Correlation Zone, *Electron. Lett.*, Vol. 35, No. 10, Oct. 1999, pp. 777-779.

Gao, J. & Chang, Y. (2006). New Upper Bounds for Impulse Radio Sequences, *IEEE Trans. Inf. Theory*, Vol. 52, No. 5, May 2010, pp. 2255-2260.

Guvene, I. & Arslan, H., (2004). TH-Sequences Construction for Centralised UWB-IR Systems in Dispersive Channels, *Electron. Lett.*, Vol. 40, No. 8, April 2004, pp. 491-492.

Guvene, I. & Arslan, H., (2004). Design and Performance Analysis of TH Sequences for UWB-IR Systems, *Proceedings of WCNC'04*, Atlanta, GA, USA, 2004, pp. 914-919.

Hayashi, T., (2009). A Class of Zero-Correlation Zone Sequence Set Using a Perfect Sequence, *IEEE Signal Process. Lett.*, Vol. 16, No.4, April 2009, pp. 331-334.

Hu, B. & Beaulieu, N. C., (2003). Exact Bit Error Rate Analysis of TH-PPM UWB Systems in the Presence of Multiple-Access Interference, *IEEE Commun. Letters*, Vol. 7, No. 12, Dec. 2003, pp. 572-574.

Hu, H. G. & Gong, G., (2010). New Sets of Zero or Low Correlation Zone Sequences via Interleaving Techniques, *IEEE Trans. Inf. Theory*, Vol. 56, No. 4, April 2010, pp. 1702-1713.

Iacobucci, M.S. & Di Benedetto, M.G., (2001). Time Hopping Codes in Impulse Radio Multiple Access Communication Systems, *Proceedings of International Symposium on third generation Infrastructure and services*, Athens, Greece, 2-3 July 2001.

Iacobucci, M.S. & Di Benedetto, M.G., (2002). Multiple Access Design for Impulse Radio Communication Systems, *Proceedings of ICC'02*, Vol. 2, Aug., 2002, pp. 817-820.

Le Martret, C. J. & Giannakis, G. B. (2000). All-Digital PAM Impulse Radio for Multiple-Access through Frequency-Selective Multipath, *Proceedings of GLOBELCOM'00*, Vol. 1, San Francisco, USA, 27 Nov. - 1 Dec. 2000, pp. 77-81.

Maggio, G.M., Rulkov, N., Sushchilk, M., Tsmring, L., Volkovskii, A., Abarbanel, H., Larson, L. & Yao, K., (1999). Chaotic Pulse Position Modulation for UltraWide-Band Communation System, *Proceedings of UWB '99*, Sept., 1999, pp.28-30.

Maggio, G.M., Rulkov, N. & Reggiani, L., (2001). Pseudo-Chaotic Time Hopping for UWB Impulse Radio, *IEEE Trans. on Circuits and Systems-I*, Vol 48, No. 12, Dec., 2001, pp. 1424-1435.

Mireles, F. R., (2001). Performance of Ultrawideband SSMA Using Time Hopping and M-ary PPM, *IEEE J. Sel. Areas Commun.*, Vol. 19, No. 6, June 2001, pp. 1186-1196.

Pursley, M. B., (1977). Performance Evaluation for Phase-Coded Spread-Spectrum Multiple Access Communication—Part I: System Analysis, *IEEE Trans. Commun.*, Vol. 25, No.8, Aug. 1977, pp. 795-799.

Scholtz, R. A. (1993). Multiple Access with Time Hopping Impulse Modulation, *Proceedings of MILCOM '93*, Bedford, MA, USA, Oct. 11-14, 1993, pp. 447-450.

Scholtz, R. A., Kumar, P. V. & Bravo, C. C., "Some Problems and Results in Ultra-WideBand Signal Design, *Proceedings of SETA'01*, May, 2001.

Suehiro, N., (1992). Elimination Filter for Co-Channel Interference in Asynchronous SSMA Systems Using Ployphase Modulatable Orthogonal Sequences, *IEICE Trans. Commun.*, Vol. E75-B, June 1992.

Suehiro, N., (1994). A Signal Design without Co-Channel Interference for Approximately synchronized CDMA System, *IEEE J. Sel. Areas Commun.*, SAC-12, June 1994, pp. 837-841.

Sushchick, M. Jr., Rulkov, N., Tsimring, L., Abarbanel, H., Yao, K. & Volkovskii, A., (2000) Chaotic Pulse Position Modulation: A Robust Method of Communicating with Chaos, *IEEE Commun. Letters*, Vol. 4, No. 4, April, 2000, pp. 128-130.

Titlebaum. E. L., (1981). Time-Frequency Hop Signals, Part I: Coding Based upon the Theory of Linear Congruences, *IEEE Trans. on Aerospace and Elect. Syst.*, Vol.17, No.4, July 1981, pp. 490-493.

Win, M. Z. & Scholtz R. A. (1998). Impulse Radio: How It Works, *IEEE Commun. Letters*, Vol.2, No.2, Feb. 1998, pp. 36-38.

Win, M. Z. & Scholtz R. A. (2000). Ultra-Wide Band-Width Time-Hopping Spread-Spectrum Impulse Radio for Wireless Multiple-Access Communications, *IEEE Trans. on Commun.*, Vol. 48, No. 4, April 2000, pp. 679-690.

Zeng, Q. Peng, D. Y. & Wang, X. N., (2011). Performance of a Novel MFSK/FHMA System Employing No-Hit Zone Sequence Set over Rayleigh Fading Channel, *IEICE Trans. Fundamentals*, Vol. E94-A, No. 2, Feb. 2011, pp. 526–532.

Zhao, L. & Haimovich, A. M., (2002). The Capacity of an UWB Multiple-Access Communications System, *Proceedings of ICC'02*, Vol.3, 2002, pp.1964-1968.

6

Fine Synchronization in UWB Ad-Hoc Environments

Moez Hizem and Ridha Bouallegue
6'Tel/Sup'Com, University of Carthage
Tunisia

1. Introduction

UWB impulse radios (UWB-IR) have attracted increasing interest due to their potential to propose high user capacity with low-complexity and low-power transceivers (Win & Sholtz, 2000). It is approved by the Federal Communications Commission's (FCC) Report in which the UWB spectral mask is released and published in February 2002. Most of these benefits initiate from the distinctive characteristics inherent to UWB wireless transmissions (Yang & Giannakis, 2004). These make UWB connectivity appropriate for indoor and especially short-range high-rate wireless environments, as well as for strategic outdoor communications. However, to harness these benefits, one of the most critical challenges is the synchronization step and more specifically timing offset estimation. Bit error rate (BER) analysis also exposes evident performance degradation of UWB radios due to mistiming (Tian & Giannakis, 2005). The complexity of which is accentuated in UWB owing to the fact that information bearing waveforms are impulse-like and have low amplitude. In addition, compared to narrowband systems, the difficulty of timing UWB signals is increased further by the dense multipath channel that remains unknown at the synchronization step. These reasons give explanation why synchronization has obtained so much importance in UWB research (Fleming et al, 2002; Homier & Sholtz, 2002; Tian & Giannakis, 2003; Yang et al, 2003).

Typically, pulse position modulation (PPM), pulse amplitude modulation or on/off keying (OOK) is employed. PPM modulation transmits pulses with constant amplitude and encodes the information according to the position of the pulse, while PAM and OOK use the amplitude for this purpose. Moreover, PPM is regularly implemented to reduce transceiver complexity in UWB systems. But unlike pulse amplitude modulation (PAM) applied in the context of UWB systems, the difficulty of accurate synchronization is accentuated in PPM UWB systems owing to the fact that information is transmitted by the shifts of the pulse positions.

In the last years, numerous timing algorithms have been studied for UWB impulse radios under various operating environments. Least squares (LS) (Carbonelli et al, 2003) and Maximum-likelihood (ML) approaches (Lottici et al, 2002) are available, but tend to be computationally complex as they need high sampling rates. In (Djapic et al, 2006), a blind synchronization algorithm that takes advantage of the shift invariance structure in the frequency domain is proposed. An accurate signal processing model for a Transmit-reference UWB (TR-UWB) system is given in (Dang et al, 2006). The model considers the channel correlation coefficients that can be estimated blindly. In (Ying et al, 2008), the authors proposed a code-assisted blind synchronization (CABS) algorithm which relies on the discriminative nature of both the time hopping code and a well-designed polarity code. Timing with dirty templates (TDT), which is the starting point of this paper, was introduced in (Yang & Giannakis, 2005) for rapid synchronization of UWB signals and was developed in (Yang, 2006) for PPM-UWB signals with direct sequence (DS) and/or time hopping (TH) spreading. This technique is based on correlating adjacent symbol-long segments of the received waveform. TDT is functional with random and unknown transmitted symbol sequences. When training symbols are approachable, the performance of the TDT synchronizer can be improved by approving a data-aided (DA) mode (Yang & Giannakis, 2005). The DA mode significantly outperforms the non-data-aided (NDA) one. However, the training sequences require an overhead which reduces the bandwidth and energy efficiency. Except (Yang, 2006), all these timing algorithms are developed for PAM-UWB signals. Since their operations greatly rely on zero-mean property of PAM, these presented timing algorithms are not appropriate to PPM-UWB signals.

In this chapter, to address and try to solve the problem of synchronization for Ultra Wideband (UWB) systems in ad-hoc environments, we propose a fine synchronization algorithm for PAM and PPM UWB signals with a spread spectrum involving Time Hopping (TH). We adopt first a blind (or coarse) synchronization technique, which is Timing with dirty templates (TDT). Its principle is to correlate two consecutive symbol-long segments of the received waveform. In particular, synchronization will be asserted when the correlation function reaches its maximum. This allows TDT algorithms to effectively collect the multipath energy even when the spreading codes and the channel are both unknown. However, this technique estimates coarsely (or roughly) the value of timing offset, therefore not precisely, and this may cause a shortfall in performance of our UWB impulse radio systems. To improve synchronization performance of the TDT (Timing with Dirty Templates) algorithm developed in mentioned papers, our contribution will be to implement a new fine synchronization stage and place it after the dirty one, which is TDT approach. The principle of our fine synchronization algorithm is to make a fine research in order to find the exact moment of the beginning pulse (fine estimation of delay time between emitted pulses and those received). This is achieved by correlating two consecutive segments of symbol-length received waveform, but this time in an interval that corresponds to the number of frames included in one data symbol. First, we applied this algorithm in single-user environments and then, have extended the used method in multi-user environments. Simulation results show that this new approach using TDT synchronizer can achieve a lower mean square error (MSE) than the original TDT for both NDA and DA synchronization mode.

The rest of this chapter is organized as follows. We describe our system model (PAM and PPM signals through time hopping spreading) with first stage synchronization (based on TDT) in Section 2, and then we give an outline of the well known TDT approach for UWB TH-PAM and TH-PPM impulse radios to better understand the overall timing synchronization in Section 3. In Section 4, we present the second step or stage of our synchronization approach with UWB time hopping systems in ad-hoc environments. The performance evaluation of our proposed fine synchronization approach with UWB TH-PAM and TH-PPM impulse radio systems in both single-user and multi-user environments is given in Section 5. And finally, we conclude this chapter in Section 6.

2. TH-PAM and TH-PPM UWB system model

Common multiple access techniques implemented for pulse based UWB systems are Time Hopping (TH) and Direct Sequence (DS). Appropriate modulation techniques include OOK (Foerster et al, 2001) and particularly PPM and PAM (Hämäläinen et al, 2002). A given UWB communication system will be a mixture of these techniques, leading to signals based on, for example, TH-PPM, TH-BPAM or DS-BPAM. TH-PAM and TH-PPM are almost certainly the most frequently adopted scheme and will be applied in the following as an example for determining the resources existing in a UWB system in single-user and multi-user environments.

2.1 TH-PAM UWB system model for single-user links

In UWB impulse radios, each information symbol is transmitted over a T_s period that consists of N_f frames (Win & Sholtz, 2000). During each frame of duration T_f, a data-modulated ultra-short pulse p(t) with duration $T_p \ll T_f$ is transmitted from the antenna source. The transmitted signal is

$$v(t) = \sqrt{\varepsilon}\sum_{k=0}^{+\infty} \tilde{s}(k) \sum_{i=0}^{N_f-1} p(t - iT_f - c_{th}(i)T_c - kT_s) \quad (1)$$

where ε is the energy per pulse. $\tilde{s}(k) := s(k)\tilde{s}(k-1)$ are differentially encoded symbols and drawn equiprobably from a finite alphabet. In our case, s(k) are denoting the binary PAM information symbols. User separation is realized with pseudo-random TH-codes $c_{th}(i)$, which time-shift the pulse positions at multiples of the chip duration T_c (Win & Sholtz, 2000). In this paper, we focus on a single user link and treat multi-user interference (MUI) as noise.

The transmitted signal propagates through the multipath channel with impulse response

$$g(t) = \sum_{l=0}^{L-1} \alpha_l \delta(t - \tau_l) \quad (2)$$

where $\{\alpha_l\}_{l=0}^{L-1}$ and $\{\tau_l\}_{l=0}^{L-1}$ are amplitudes and delays of the L multipath elements, respectively. The channel is assumed quasi-static and among $\{\tau_l\}_{l=0}^{L-1}$, τ_0 represents the propagation delay of the channel.

Then, the received waveform is given by

$$r(t) = \sqrt{\varepsilon}\sum_{k=0}^{+\infty} \tilde{s}(k) p_T(t - kT_s - \tau_{l,0} - \tau_0) + \eta(t) \quad (3)$$

where $\tau_{l,0}$ is arbitrary reference at the receiver representing the delay relative to the arrival moment of the first pulse, $\eta(t)$ is the additive noise and $p_T(t)$ denotes the received symbol waveform as

$$p_T(t) = \sum_{i=0}^{N_f-1} p(t - iT_f - c_{th}(i)T_c) * g(t + \tau_0) \quad (4)$$

where * indicates the convolution operation. We define the timing offset as $\Delta\tau := \tau_{l,0} - \tau_0$. Let us suppose that $\Delta\tau$ is in the range of $[0, T_s)$ and we will show in the rest of this paper that this assumption will not affect the timing synchronization.

2.2 TH-PPM UWB system model for single-user links

With PPM modulation (Durisi & Benedetto, 2003; Di Renzo et al, 2005), the transmitted signal in single-user links is described by the following model

$$v(t) = \sqrt{\varepsilon}\sum_{k=0}^{+\infty}\sum_{i=0}^{N_f-1} p(t - iT_f - c_{th}(i)T_c - kT_s - d_i\delta) \quad (5)$$

where ε is the energy per pulse, $d_i \in (0,1)$ represents the i-th information bit transmitted, and δ is the time shift associated with binary PPM. User separation is realized with pseudo-random TH-codes $c_{th}(i)$, which time-shift the pulse positions at multiples of the chip duration T_c (Win & Sholtz, 2000). In this paper, we focus on a single user link and treat multi-user interference (MUI) as noise.

After the transmitted signal propagation through the multipath channel, the received waveform is given by

$$r(t) = \sqrt{\varepsilon}\sum_{l=0}^{L}\alpha_l\sum_{k=0}^{+\infty} p_T\big(t - kT_s - \tau_{l,0} - \tau_0 - d_i\delta\big) + \eta(t) \quad (6)$$

where $\tau_{l,0}$ is arbitrary reference at the receiver representing the delay relative to the arrival moment of the first pulse, $\eta(t)$ is the additive noise and $p_T(t)$ denotes the received symbol waveform as

$$p_T(t) = \sum_{i=0}^{N_f-1} p(t - iT_f - c_{th}(i)T_c) * g(t + \tau_0) \quad (7)$$

where * indicates the convolution operation. We define the timing offset as $\Delta\tau := \tau_{l,0} - \tau_0$. Let us suppose that $\Delta\tau$ is in the range of $[0, T_s)$ and we will show in the rest of this paper that this assumption will not affect the timing synchronization. Let $p_R(t)$ the overall received symbol-long waveform defined as follows

$$p_R(t) = \sum_{l=0}^{L}\alpha_l p_T\big(t - \tau_{l,0}\big) \quad (8)$$

Using (8), the received waveform in (6) becomes

$$r(t) = \sqrt{\varepsilon}\sum_{k=0}^{+\infty} p_R(t - kT_s - \tau_0 - d_i\delta) + \eta(t) \quad (9)$$

2.3 TH-PAM UWB system model for single-user links

The UWB time hopping impulse radio signal considered in this paper is a stream of narrow pulses, which are shifted in amplitude modulated (PAM). The transmitted waveform from the *u*th user is

$$v_u(t) = \sqrt{\varepsilon_u}\sum_{k=0}^{+\infty} s_u(k)p_{u,T}(t - kT_s) \quad (10)$$

where ε_u represents the energy per pulse, $s_u(k)$ are differentially encoded symbols and drawn equiprobably from finite alphabet. In our case, $s_u(k)$ symbolize the binary PAM information symbols and $p_{u,T}(t)$ indicates the transmitted symbol

$$p_{u,T}(t) := \sum_{i=0}^{N_f-1} p(t - iT_f - c_u(i)T_c) \quad (11)$$

where T_c is the chip duration and $c_u(i)$ is the user-specific pseudo-random TH code during the *i*th frame.

After the transmitted signal propagation through the multipath channel, the received waveform from all users is

$$p_{u,T}(t) := \sum_{i=0}^{N_f-1} p(t - iT_f - c_u(i)T_c) \quad (12)$$

where N_u is the user's number, τ_u is the propagation delay of the *u*th user's direct path and $\eta(t)$ is the zero-mean additive Gaussian noise (AGN). The global received symbol-long waveform is therefore given by

$$p_{u,R}(t) := \sum_{l=0}^{L_u-1} \alpha_{u,l} p_{u,T}(t - \tau_{u,l}) \quad (13)$$

Assuming that the nonzero support of waveform $p_{u,R}(t)$ is upper bounded by the symbol time T_s, the received waveform in (3) can be rewritten as

$$r(t) = \sum_{u=0}^{N_u-1} \sqrt{\varepsilon_u} \sum_{k=0}^{+\infty} s_u(k) p_{u,R}(t - kT_s - \tau_u) + \eta(t) \quad (14)$$

3. TDT approach

As mentioned previously, our proposed timing scheme consists of two complementary floors or steps. The first is based on a coarse (or blind) synchronization that is TDT developed in (Yang & Giannakis, 2005). In this section, we will give an outline of the TDT approach to better understand the overall timing synchronization suggested in this paper. The general structure description of our system model with first stage synchronization (TDT) is illustrated in Fig.1.

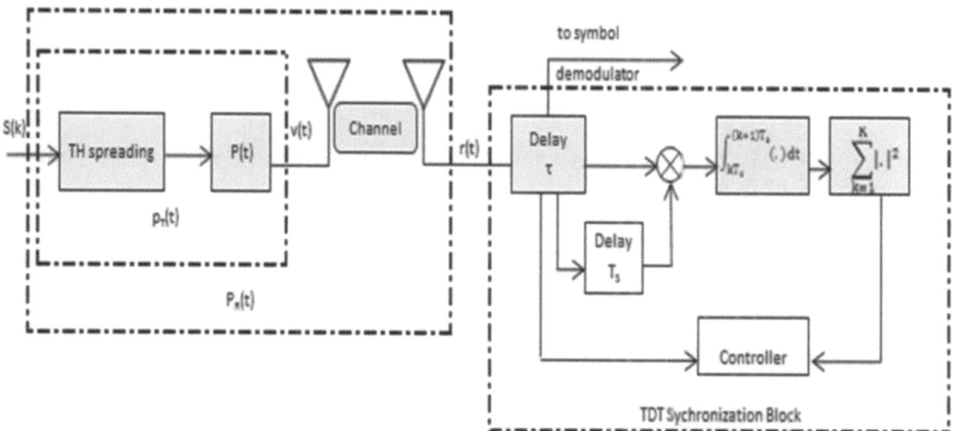

Fig. 1. Description of our model with first stage synchronization

The basic idea behind TDT is to find the maximum of square correlation between pairs of successive symbol-long segments. These symbol-long segments are called "dirty templates" because: i) they are noisy, ii) they are distorted by the unknown channel, and iii) they are subject to the unknown offset τ_0. Then, we will analyze $\tilde{\tau_0}$ representing estimate offset of τ_0 by deriving upper bounds on their mean square error (MSE) in both non-data-aided (NDA) and data-aided (DA) modes.

3.1 TDT approach for TH-PAM UWB system in single-user links

For notational brevity and after setting $p_T(t) := p_R(t-\tau_{1,0})$, the received waveform simplifies to

$$r(t) = \sqrt{\varepsilon} \sum_{k=0}^{+\infty} \tilde{s}(k) p_R(t - kT_s - \tau_0) + \eta(t) \quad (15)$$

Thereafter, a correlation between the two adjacent symbol-long segments $r(t + kT_s)$ and $r(t + (k-1)T_s)$ is achieved. Let $x(k;\tau)$ the value of this correlation $\forall k \in [1, +\infty)$ and $\tau \in [0, T_s)$

$$x(k;\tau) = \int_0^{T_s} r(t + kT_s + \tau)\, r(t + (k-1)T_s + \tau)dt \quad (16)$$

Applying the Cauchy-Schwartz inequality and substituting the expressions of $r(t + kT_s)$ and $r(t + (k-1)T_s)$ to (16), $x(k;\tau)$ becomes

$$x(k;\tau) = \tilde{s}(k-1)[\,\tilde{s}(k-2)\, \varepsilon_A(\tilde{\tau_0}) + \tilde{s}(k)\, \varepsilon_B(\tilde{\tau_0})] + \zeta(k;\tau) \quad (17)$$

where $\varepsilon_A(\tau) := \varepsilon \int_{T_s-\tau}^{T_s} p_R^2\,(t)dt$, $\varepsilon_B(\tau) := \varepsilon \int_0^{T_s-\tau} p_R^2\,(t)dt$, and $\zeta(k;\tau)$ corresponds to the superposition of three noise terms (Yang & Giannakis, 2005) and can be approximated as an additive white Gaussian noise (AWGN) with zero mean and σ_ζ power.

By exploiting the statistical properties of the signal and noise, the mean square of the samples in (17) is given by

$$E\{x^2(k;\tau)\} = \frac{1}{2}\,[\varepsilon_A(\tilde{\tau_0}) + \varepsilon_B(\tilde{\tau_0})]2 + \frac{1}{2}\,[\varepsilon_A(\tilde{\tau_0}) - \varepsilon_B(\tilde{\tau_0})]2 + \sigma_\zeta^2 \quad (18)$$

We notice that $\varepsilon_B(\tilde{\tau_0}) + \varepsilon_A(\tilde{\tau_0}) = \varepsilon\int_0^{T_s} p_R^2(t)dt := \varepsilon_R$ for $\tilde{\tau_0} \in [0, T_s)$, where ε_R represents the constant energy corresponding to the unknown aggregate template at the receiver. Then the mean square of $x^2(k;\tau)$ can be rewritten as follows

$$E\{x^2(k;\tau)\} = \frac{1}{2}\,(\varepsilon_R)^2 + \frac{1}{2}\,[2\varepsilon_A(\tilde{\tau_0}) - \varepsilon_R]2 + \sigma_\zeta^2 \quad (19)$$

Since the term $\varepsilon_A(\tilde{\tau_0})$ reaches its unique maximum at $\tilde{\tau_0} = 0$, then $E\{x^2(k;\tau)\}$ also reached its unique maximum at $\tilde{\tau_0} = 0$. Thus, an estimate of timing offset τ_0 is given by

$$\hat{\tau}_0 = \arg max_{\tau\in[0,T_s]}\, E\{x^2(k;\tau)\} \quad (20)$$

In the practice, the mean square of $x^2(k;\tau)$ is estimated from the average of different values $x^2(k;\tau)$ for k ranging from 0 to M – 1 obtained during an observation interval of duration MT_s. In what follows, we summarize the TDT algorithm in its NDA form and then in its DA form.

3.1.1 Non-data-aided (blind) mode

For the synchronization mode NDA, the synchronization algorithm is defined as follows

$$\begin{cases} \hat{\tau}_{0,nda} = \arg max_{\tau\in[0,T_s]}\, E\,\{x^2(k;\tau)\} \\ x_{nda}(M;\tau) = \frac{1}{M}\sum_{m=0}^{M-1}\left(\int_{mT_s}^{(m+1)T_s} r(t+\tau)r(t+\tau+T_s)dt\right)^2 \end{cases} \quad (21)$$

By using (17), the expression of $x_{nda}(M;\tau)$ can be rewritten as follows

$$x_{nda}(M;\tau) = \frac{1}{M}\sum_{m=0}^{M-1}[s(m-1)s(m-2)\varepsilon_A(\tilde{\tau_0})\; s(m)s(m-1)\varepsilon_B(\tilde{\tau_0}) + \zeta(m;\tau)]^2 \quad (22)$$

From (19) and (20), the estimation of delay τ_0 is made possible due to the presence of the term $\varepsilon_A(\tilde{\tau_0})$ - $\varepsilon_B(\tilde{\tau_0})$. Unfortunately for the estimator $x_{nda}(M;\tau)$, this term exists only if the transmitted sequence presents an alternating sign between the symbols s(m – 2) and s(m). Thus, for the synchronization in NDA mode, the performances of this approach are affected by the sign of the transmitted symbols. To increase the chances that the estimator $x_{nda}(M;\tau)$ is expressed as a function of the energy difference, an increase in the observation interval length is required. However, such an increase leads to increased acquisition delays. Where does the idea of using the data-aided (DA) approach.

3.1.2 Data-aided mode

The number of samples M required for reliable estimation can be reduced noticeably if a data-aided (DA) approach is pursued (Yang & Giannakis, 2003). The delays can be significantly reduced through the use of training sequences with alternating sign between the symbols $s(m-2)$ and $s(m)$, i.e. $s(m - 2) = - s(m)$. This observation suggest that the training sequence $\{s(k)\}$ for DA TDT mode follows the following alternation $[1,1,-1,-1]$ (this by working with a M-ary PAM symbol); i.e.

$$s(k) = (-1)^{\lfloor \frac{k}{2} \rfloor} \tag{23}$$

This pattern is particularly attractive, since it simplifies the algorithm proposed by the TDT approach, for the DA mode, to become

$$\begin{cases} \hat{\tau}_{0,da} = \arg max_{\tau\in[0,T_s]} \{x_{da}(M;\tau)\} \\ x_{da}(M;\tau) = \left(\int_0^{T_s} r(t+\tau)r(t+\tau+T_s)dt\right)^2 \end{cases} \tag{24}$$

with

$$r(t) = \frac{2}{M}\sum_{k=0}^{\frac{M}{2}-1}(-1)^k r(t+2kT_s+\tau).$$

The estimator in (24) is essentially the same as (22), except that training symbols are used in (24). However, theses training symbols are instrumental in improving the estimation performance. This will be approved by the simulation results.

3.2 TDT approach for TH-PPM UWB system in single-user links

For UWB TH-PPM systems, a correlation between the two adjacent symbol-long segments $r_k(t) = r(t + kT_s)$ and $r_{k+1}(t) = r(t + (k+1)T_s)$ is achieved (Yang, 2006). Let $x(k;\tau)$ the value of this correlation $\forall k \in [1,+\infty)$ and $\tau \in [0,T_s)$

$$\begin{cases} x(k;\tau) := \int_0^{T_s} r_{k+1}(t;\tau)\,\tilde{r}_k(t;\tau)dt \\ \tilde{r}_k(t;\tau) := r_k(t+\delta;\tau) - r_k(t-\delta;\tau) \end{cases} \tag{25}$$

By applying the Cauchy-Schwartz inequality and exploiting the statistical properties of the signal and noise (Yang, 2006), the mean square of the samples in (25) is given by

$$E_{s,\zeta}\{x^2(k;\tau)\} \approx \frac{1}{2}\left(\varepsilon_R^2 - 3\varepsilon_A(\tilde{\tau}_0)\varepsilon_B(\tilde{\tau}_0) + 2\sigma_\zeta^2\right) \tag{26}$$

where $\varepsilon_A(\tau) := \varepsilon\int_{T_s-\tau}^{T_s} p_R^2(t)dt$, $\varepsilon_B(\tau) := \varepsilon\int_0^{T_s-\tau} p_R^2(t)dt$, and σ_ζ is the power of $\zeta(k;\tau)$ corresponding to the superposition of three noise terms (Yang, 2006) and can be approximated as an additive white Gaussian noise (AWGN) with zero mean. We notice that $\varepsilon_B(\tilde{\tau}_0) + \varepsilon_A(\tilde{\tau}_0) = \varepsilon\int_0^{T_s} p_R^2(t)dt := \varepsilon_R$ for $\tilde{\tau}_0 \in [0,T_s)$, where ε_R represents the constant energy corresponding to the unknown aggregate template at the receiver.

Similarly to PAM signals, the term $E_{s,\zeta}\{x^2(k;\tau)\}$ reached its unique maximum at $\tilde{\tau}_0 = 0$. In the practice, the mean square of $x^2(k;\tau)$ is estimated from the average of different values $x^2(k;\tau)$ for k ranging from 0 to $M - 1$ obtained during an observation interval of duration MT_s. In what follows, we summarize the TDT algorithm for UWB TH-PPM systems in its NDA form and then in its DA form.

3.2.1 Non-data-aided mode

For the synchronization mode NDA, the synchronization algorithm is defined as follows

$$\begin{cases} \hat{\tau}_{0,nda} = \arg\max_{\tau\in[0,T_s]} x_{nda}(M;\tau) \\ x_{nda}(M;\tau) = \frac{1}{M}\sum_{m=0}^{M-1}\left(\int_0^{T_s} r_k(t;\tau)\tilde{r}_{k-1}(t;\tau)dt\right)^2 \end{cases} \tag{27}$$

The estimator $\hat{\tau}_{0,\,nda}$ in (27) can be verified to be m.s.s. consistent by deriving the mean and variance of the function $x_{nda}(M;\tau)$ (Yang, 2006).

3.2.2 Data-aided mode

For UWB TH-PPM, the training sequence for DA TDT is considered to comprise a repeated pattern (for example (1,0, 1,0)); that is.

$$s(k) = \{k+1\}_2 \tag{28}$$

With this pattern, it can be easily verified that the mean square in (26) becomes

$$E_{s,\zeta}\{x^2(k;\tau)\} = \varepsilon_R^2 - 4\varepsilon_A(\tilde{\tau_0})\varepsilon_B(\tilde{\tau_0}) + \sigma_\zeta^2 \tag{29}$$

With the NDA approach, it is necessary to take expectation with respect to s_k in order to remove the unknown symbol effects; while the DA mode, this is not needed. Hence, the sample mean $M^{-1}\sum_{m=0}^{M-1} x^2(k;\tau)$converges faster to its expected value in (29). This pattern is particularly attractive, since it permits a very rapid acquisition which is a major benefit of the DA mode. Data-aided TDT for UWB TH-PPM signals can be accomplished even when TH codes are present and the multipath channel is unknown, using

$$\begin{cases} \hat{\tau}_{0,da} = \arg\max_{\tau\in[0,T_s]} x_{da} \\ x_{da}(M;\tau) = \left(\frac{1}{M}\sum_{m=0}^{M-1}\int_0^{T_s} r_k(t;\tau)\tilde{r}_{k-1}(t;\tau)dt\right)^2 \end{cases} \tag{30}$$

The estimator in (30) is essentially the same as (27), except training symbols used in (30). However, theses training symbols are essential in improving the estimation performance. This will be approved by the simulation results.

3.3 TDT approach for TH-PAM UWB system in multi-user links

For multi-user UWB TH-PAM systems, a correlation between the two adjacent symbol-long segments $r(t-kT_s)$ and $r(t-(k-1)T_s)$ is achieved. Let $x(k;\tau)$ the value of this correlation $\forall k \in [1,+\infty)$ and $\tau \in [0,T_s)$

$$x(k;\tau) = \sum_{u=0}^{N_u-1}\int_0^{T_s} r(t-kT_s)r(t-(k-1)T_s)\,dt \tag{31}$$

Applying the Cauchy-Schwartz inequality and substituting the expressions of $r(t-kT_s)$ and $r(t-(k-1)T_s)$ to (31), $x(k;\tau)$ becomes

$$x(k;\tau) = \sum_{u=0}^{N_u-1} s_u(k-1)\left[s_u(k-2)\varepsilon_{u,A}(\tilde{\tau}_u) + s_u(k)\varepsilon_{u,B}(\tilde{\tau}_u)\right] + \xi(k;\tau) \tag{32}$$

where $\varepsilon_{u,A}(\tilde{\tau}_u) := \varepsilon_u \int_{T_s-\tilde{\tau}_u}^{T_s} p_{u,R}^2\,(t)dt$, $\varepsilon_{u,B}(\tilde{\tau}_u) := \varepsilon_u \int_0^{T_s-\tilde{\tau}_u} p_{u,R}^2\,(t)dt$, $\tilde{\tau}_u := [\tau_u - \tau]_{T_s}$ and $\zeta(k;\tau)$ corresponds to the superposition of three noise terms (Yang & Giannakis, 2005) and can be approximated as an AWGN with zero mean and σ_ξ power. As mentioned in (Yang &

Giannakis, 2005), the noise-free part of the desired user's samples at the correlator output complies with

$$\chi_0(k;\tau) = \varepsilon_{0,A}(\tilde{\tau}_0) - \varepsilon_{0,B}(\tilde{\tau}_0) \tag{33}$$

Substituting the above equation into (31), we find

$$x(k;\tau) = \chi_0(k;\tau) + \textstyle\sum_{u\neq 0} s_u(k-1)\left[\, s_u(k)\varepsilon_{u,B}(\tilde{\tau}_u) + s_u(k-2)\varepsilon_{u,A}(\tilde{\tau}_u)\right] + \xi(k;\tau) \tag{34}$$

where $s_u(k)$'s are zero-mean information symbols emitted by the $(u \neq 0)$th user. If we calculate the average (without squaring), we obtain $E\{x(k;\tau)\} = \varepsilon_{0,B}(\tilde{\tau}_0) - \varepsilon_{0,A}(\tilde{\tau}_0)$ since $E\{\chi_u(k;\tau)\} = 0$ (Yang & Giannakis, 2005). In what follows, we summarize the TDT approach for multi-user UWB TH-PAM impulse radios in its NDA form and then in its DA form. .

3.3.1 Non-data-aided mode

For the NDA synchronization mode, the timing algorithm is defined as follows

$$\begin{cases} \hat{\tau}_{u,nda} = \arg\max_{\tau\in[0,T_s]} E\left\{x^2(k;\tau)\right\} \\ x_{nda}(M;\tau_u) = \frac{1}{M}\sum_{m=0}^{M-1}\left(x(k;\tau)\right)^2 \end{cases} \tag{35}$$

The estimator can be verified to be m.s.s consistent by deriving the mean and variance of $x_{nda}(M;\tau_u)$. It has been demonstrated that the single-user TDT estimator is operational even in a multi-user environment (Yang & Giannakis, 2005).

3.3.2 Data-aided mode

The pattern described previously in (23) is particularly attractive, since it simplifies the proposed algorithm to become in the DA mode

$$\begin{cases} \hat{\tau}_{u,da} = \arg max_{\tau\in[0,T_s]} \left\{x_{da}(M;\tau)\right\} \\ x_{da}(M;\tau) = \left(\int_0^{T_s} r(t+\tau)r(t+\tau+kT_s)dt\right)^2 \end{cases} \tag{36}$$

With $E\left\{(-1)^{\left\lfloor\frac{k}{2}\right\rfloor} r(t+\tau+kT_s)\right\} = \sqrt{\varepsilon}\left[p_{0,R}(t+T_s-\tilde{\tau}_0) + (-1)^k p_{0,R}(t-\tilde{\tau}_0)\right]$ which signifies that the single-user TDT estimator can also be functional in a multi-user scenario.

4. Proposed fine synchronization approach

In this section, we will develop a low-complexity fine synchronization approach using TDT synchronizer in order to find the desired timing offset. The block diagram of our synchronization scheme is shown in Fig.2. Our approach will be evaluated in both NDA and DA modes, without knowledge of the multipath channel and the transmitted sequence (Hizem & Bouallegue, 2010; Hizem & Bouallegue, 2011, a; Hizem & Bouallegue, 2011, b).

This second floor achieves a fine estimation of the frame beginning, after a coarse research in the first. The concept which is based this floor is extremely simple. The idea is to scan the interval $[\tau_1 - T_{corr}, \tau_1 + T_{corr}]$ with a step noted δ by making integration between the received signal and its replica shifted by T_f on a window of width T_{corr}. τ_1 being the estimate delay deducted after the first synchronization floor and the width integration window value's T_{corr} will be given in Section 5. This principle is illustrated in Fig.3. We can write the integration window output for the nth step $n\delta$ as follows

$$Z_n = \sum_{k=0}^{K-1} \left| \int_{\tau_1+n\delta}^{\tau_1+n\delta+T_s} r(t-kT_s)r(t-(k+1)T_s)dt \right| \quad (37)$$

Where $n = -N+1..0..N-1$, $N = \lfloor T_{corr}/\delta \rfloor$ and K is number of frames considered for improving the decision taken at the first floor. The value of n which maximizes Z_n provides the exact moment of pulse beginning that we note $\tau_2 = \tau_1 + n_{opt}\delta$. Thus, the fine synchronization is performed. Finally, note that this approach will be applied in both NDA and DA modes with UWB TH-PAM and TH-PPM impulse radio systems for both single-user and multi-user environments. We will see later in what mode this approach gives better result compared to those given by the original approach TDT.

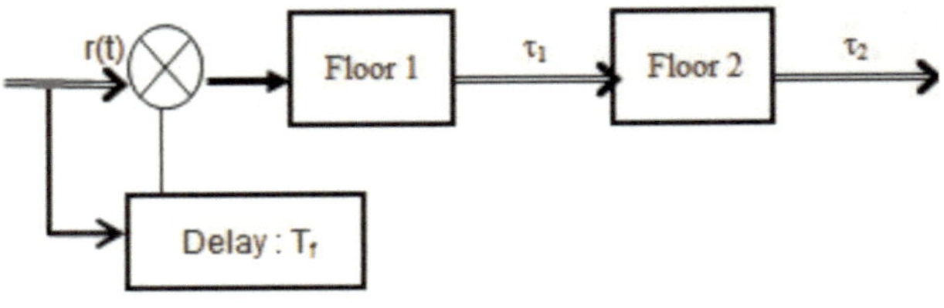

Fig. 2. Block diagram of our synchronization scheme

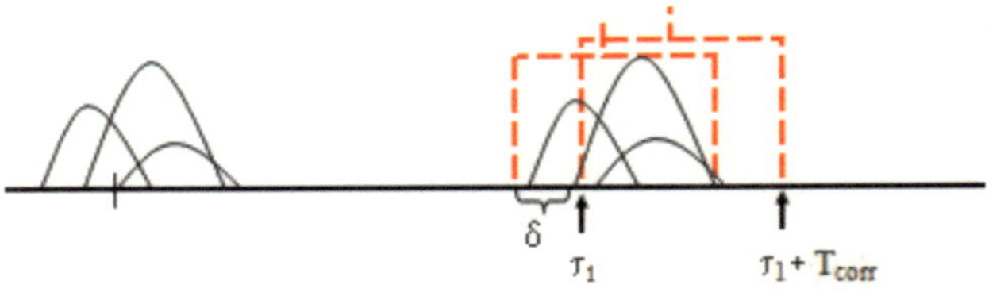

Fig. 3. Principle of second synchronization floor

5. Simulation results

In this section, we will evaluate the performance of our proposed fine synchronization approach with simulations. The UWB pulse is the second derivative of the Gaussian function with duration $T_p \approx 0.8$ns. Simulations are achieved in the IEEE 802.15.3a channel model CM1 (Foerster, 2002). The sampling frequency is f_c = 50 GHz. Each symbol contains N_f = 32 frames each with duration T_f = 35 ns. We used a random TH code over $[0, N_c - 1]$, with N_c = 35 and T_c = 1.0 ns. The width integration window value's T_{corr} is 4 ns. The performance of our synchronization approach is tested for various values of M.

5.1 TH-PAM UWB system in single-user links

In Fig. 4, we first test the MSE of NDA and DA TDT algorithms summarized in Section 3.1. The MSE is normalized by the square of the symbol duration T_s^2. From Fig.4, we note that increasing the duration of the observation interval M leads to improved performance for both NDA and DA modes. We also note that the use of training sequences (DA mode) leads to improved performance compared to the NDA mode.

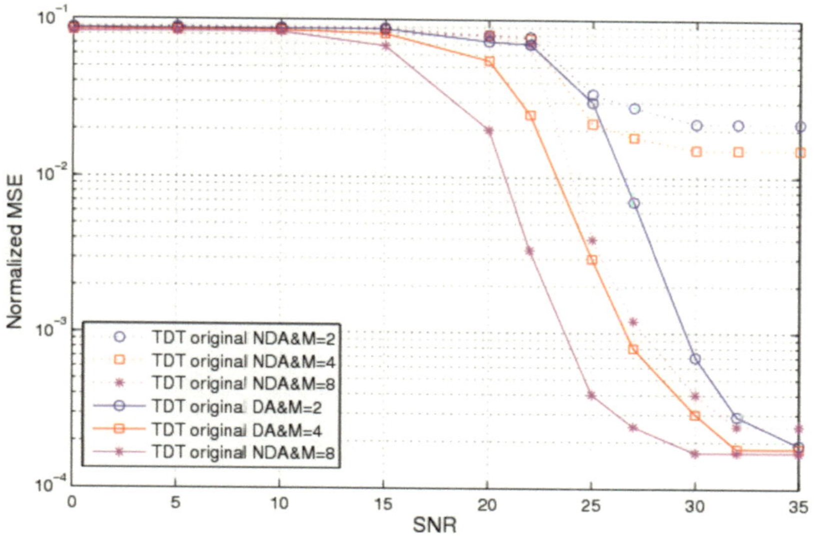

Fig. 4. Normalized MSE of the original TDT synchronizer in both NDA and DA modes

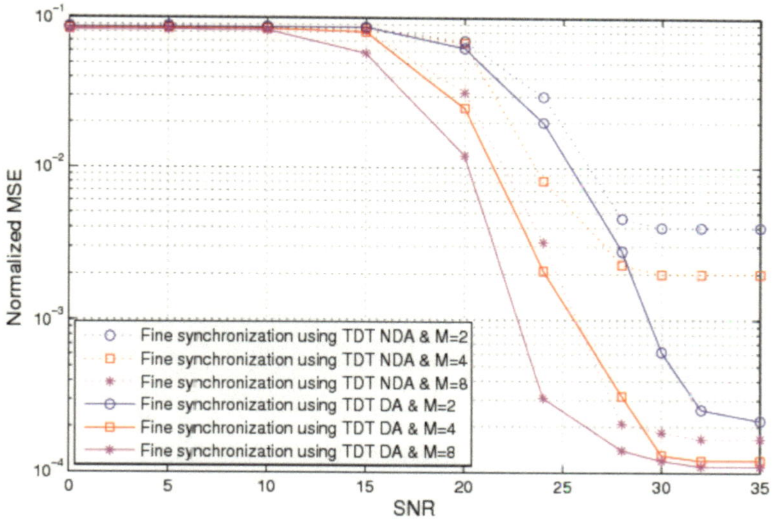

Fig. 5. Normalized MSE of our fine synchronization approach in both NDA and DA modes

In Figs. 5-7, we evaluate and compare by simulation the performance (in term of MSE) of our proposed fine synchronizer with the original NDA and DA TDT algorithms. In Fig. 5, we compare the performances of the new fine synchronization approach in both NDA and DA modes. In Fig. 6-7, we compare the performances of both original TDT and fine synchronization approach proposed for different values of M. In comparison with the original TDT approach, we note that the new approach outperforms the NDA mode and offers a slight improvement in DA mode. Even without any training symbol sequence, our synchronizer can greatly outperform the original NDA TDT especially when M is small.

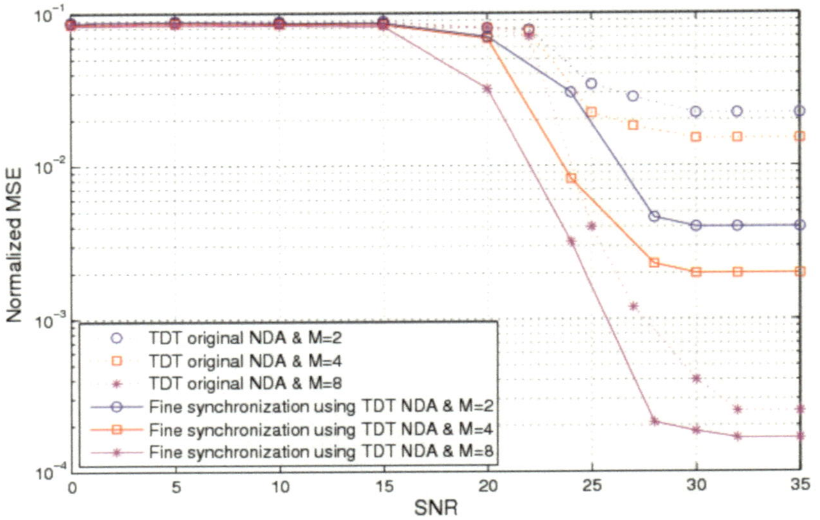

Fig. 6. Performances comparison between original TDT and our fine synchronization approaches in NDA mode

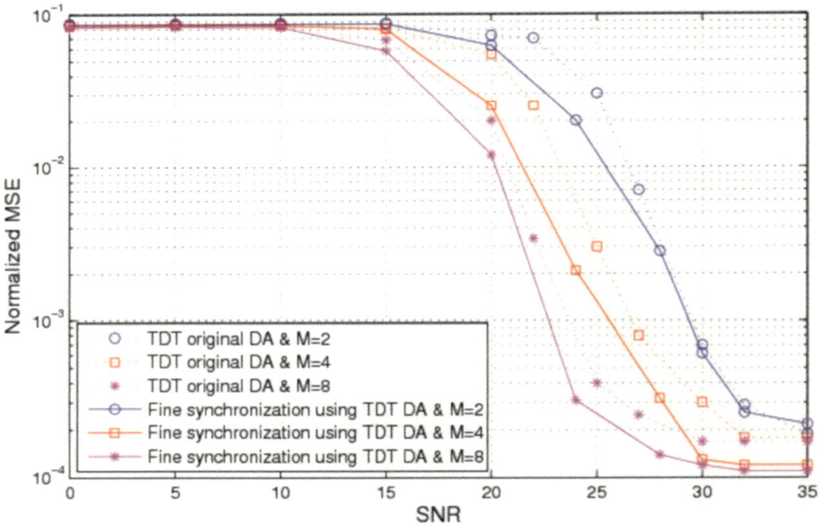

Fig. 7. Performances comparison between original TDT and our fine synchronization approaches in DA mode

5.2 TH-PPM UWB system in single-user links

In addition to other parameters described previously, the time shift associated with binary PPM is δ = 1ns. Then, the performance is tested for various values of M.

In Fig. 8, we first test the MSE of NDA and DA TDT algorithms for UWB TH-PPM systems. We note that increasing the duration of the observation interval M leads to improved performance for both NDA and DA modes. We also note that the use of training sequences (DA mode) leads to improved performance compared to the NDA mode.

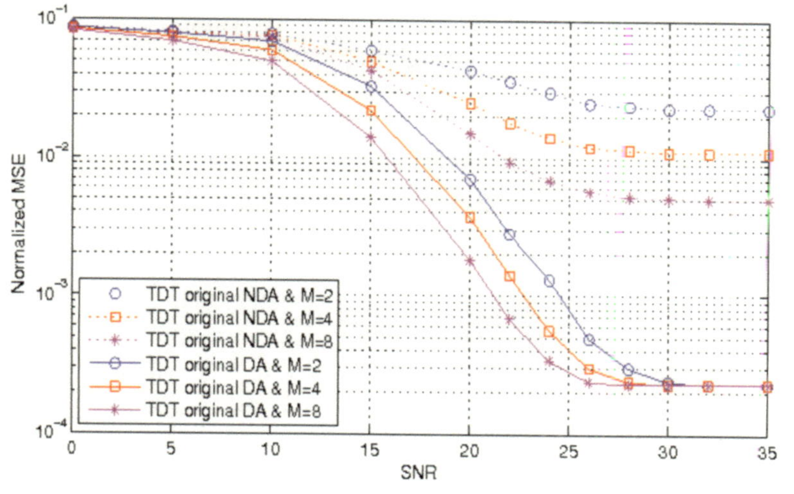

Fig. 8. Normalized MSE of TDT for UWB TH-PPM systems in both NDA and DA modes

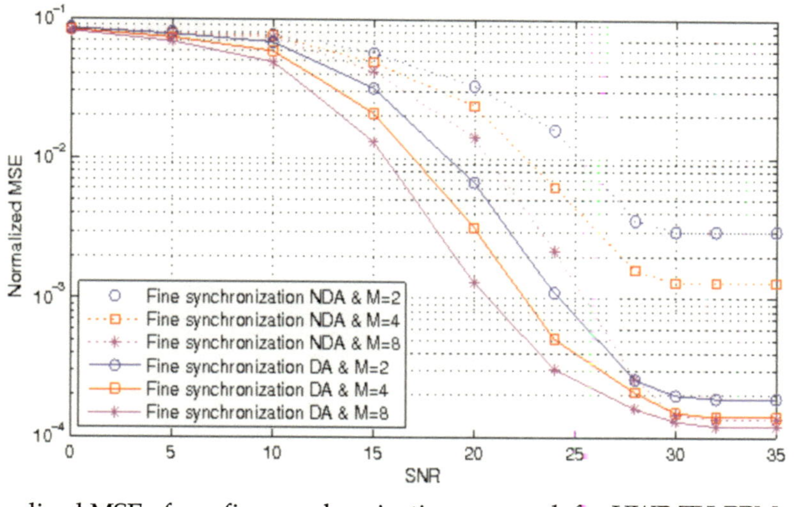

Fig. 9. Normalized MSE of our fine synchronization approach for UWB TH-PPM systems

In Figs. 9-11, we evaluate and compare by simulation the performance of our proposed fine synchronizer with the original TDT algorithms. In Fig. 9, we compare the performances of the new fine synchronization approach in both NDA and DA modes. In Fig. 10-11, we compare the performances of both original TDT and fine synchronization approach proposed in both NDA and DA modes for different values of M. In comparison with the original TDT approach, we note that the new approach outperforms the NDA mode and offers a slight improvement in DA mode. Even without any training symbol sequence, our synchronizer can greatly outperform the original NDA TDT especially when M is small.

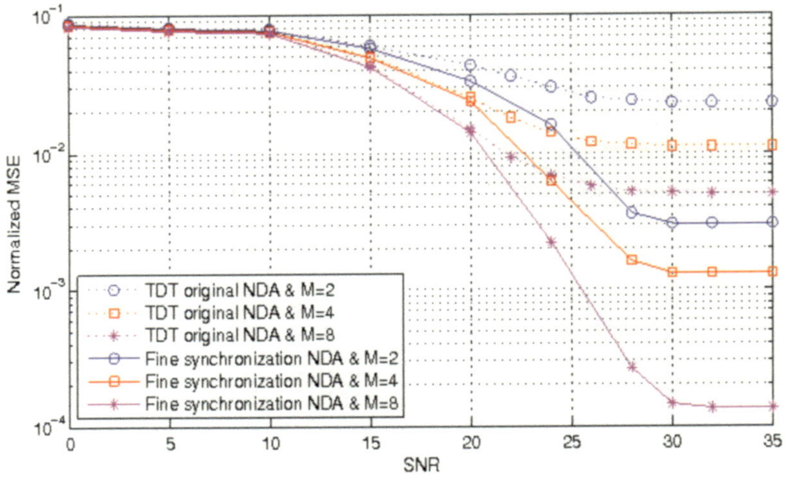

Fig. 10. Performances comparison between original TDT and our fine synchronization approaches for UWB TH-PPM systems in NDA mode

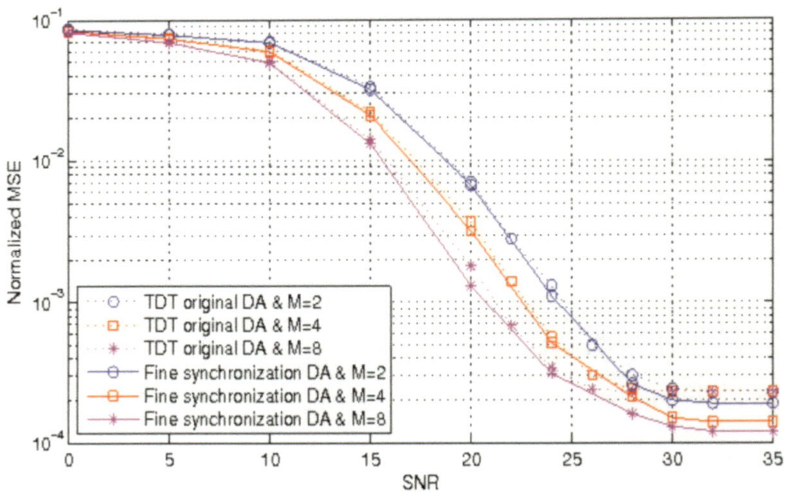

Fig. 11. Performances comparison between original TDT and our fine synchronization approaches for UWB TH-PPM systems in DA mode

5.3 TH-PAM UWB system in multi-user links

In this part, we will evaluate the performance of our proposed fine synchronization approach for UWB TH-PAM signals in ad-hoc multi-user environments. The performance is tested for various values of M.

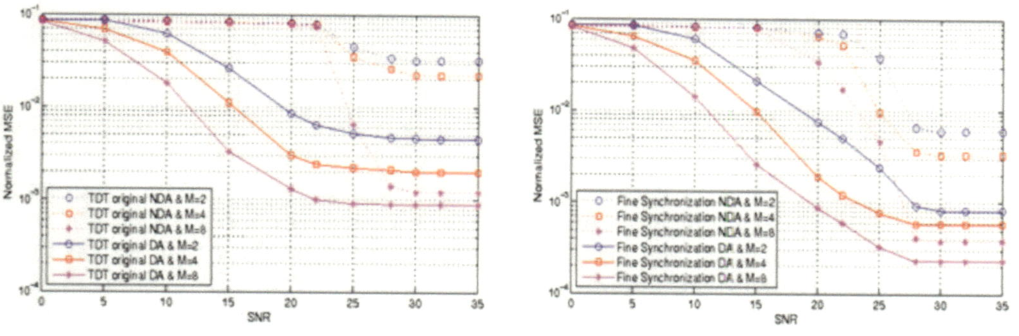

Fig. 12. Normalized MSE of multi-user original TDT synchronizer and our multi-user fine synchronization

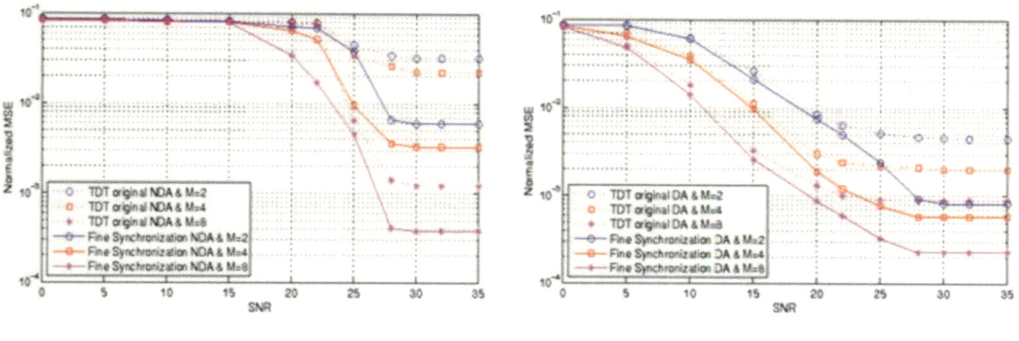

Fig. 13. Performances comparison in NDA and DA modes with multi-user environments

In Fig. 12 on left, we first test the mean square error (MSE) corresponding to (35) and (36). From the simulation results, we note that increasing the duration of the observation interval M leads to improved performance for both NDA and DA modes. We also note that the use of training sequences (DA mode) leads to improved performance compared to the NDA mode. In Fig. 12 on right, we compare the new fine synchronization approach performances in both NDA and DA modes. In Fig. 13, we compare the performances of both original TDT and fine synchronization approach for different values of M. In comparison with the original TDT approach, we note that the new approach greatly outperforms the NDA mode and offers a slight improvement in DA mode. This performance improvement is enabled at the price of fine synchronization approach introduced in second floor which can further improve the timing offset found in first floor.

6. Conclusion

In this chapter, we have discussed the problem of UWB system performance in single-user and multi-user environments. While there is a rich body of literature addressing this problem most of which has emerged recently, this topic is far from being mature. In this context, developing novel approaches with relatively low complexity still represents crucial task in meeting the challenges of UWB communications.

We first describe the TH-PAM and TH-PPM UWB system model in single-user and multi-user environments. Then, we give an outline of the TDT approach. In the rest of this chapter, we propose a novel fine synchronization scheme using TDT algorithm for UWB TH-PAM and TH-PPM radio system in single-user and multi-user links. With the introduced fine synchronization algorithm, we can achieve a fine estimation of the frame beginning. The performance improvement is enabled at the price of fine synchronization approach introduced in second floor which can further improve the timing offset found in first floor (coarse synchronization approach : TDT). The simulation results show that even without training symbols, our new synchronizer can enable a better performance than the original TDT in NDA mode especially when M is small and offers a slight improvement in DA mode.

7. References

Carbonelli, C.; Mengali, U. & Mitra, U. (2003). Synchronization and channel estimation for UWB signals, *Proceedings of GLOBECOM Conference*, San Francisco, CA, vol. 2, pp. 764-768, December 1-5, 2003

Dang, Q. H.; Trindade, A. & Van der Veen, A. J. (2006). Signal model and receiver algorithms for a Transmit-Reference Ultra-Wideband Communication system, *Proceedings of IEEE Journal of Selected Areas in Communications*, vol. 24, No. 4, pp. 773-779, April 2006

Djapic, R.; Leus, G.; Van der Veen, A. J. & Trindade, A. (2006). Blind synchronization in asynchronous UWB networks based on the transmit-reference scheme, *Proceedings of EURASIP Journal on Wireless Communications and Networking*, vol. 2006, No. 2, pp. 65-75, April 2006

Di Renzo, M.; Graziosi, F. & Santucci, F. (2005). A framework for performance analysis for TH-UWB communications, *Proceedings of IEEE International Conference on Ultra-Wideband (ICUWB)*, Zurich, Switzerland, pp. 559-564, September 5-8, 2005

Durisi, G. & Benedetto, S. (2003). Performance evaluation of TH-PPM UWB systems in the presence of multi-user interference, *Proceedings of IEEE Communication Letters*, vol. 5, pp. 224-226, May 2003

Fleming, R.; Kushner, C.; Roberts, G. & Nandiwada U. (2002). Rapid acquisition for ultra-wideband localizers, *Proceedings of Conference on Ultra-Wideband System Technologies*, Baltimore, MD, pp. 245-250, May 20-23, 2002

Foerster, J. R.; Green, E.; Somayazulu, S. & Leeper, D. (2001) Ultra-Wideband Technology for short or medium range wireless communications, *Intel Technology Journal*, Q2, 11p

Foerster, J. R. (2002) Channel Modelling Sub-committee Report Final, *IEEE P802.15-02/368r5-SG3a, IEEE P802.15 Working Group for WPAN*, November 2002

Hämäläinen, M.; Hovinen, V. & Latva-aho, M. (2002) On the UWB System Coexistence with GSM900, UMTS/WCDMA and GPS, *IEEE Journal on Selected Areas in Communications*, Vol. 20, No. 9, (Dec. 2002), pp. 1712-1721, ISSN 0733-8716

Hizem, M. & Bouallegue, R. (2010) Novel Fine Synchronization Using TDT for Ultra Wideband Impulse Radios, *Proceedings of International Information and Telecommunication Technologies Symposium (I2TS)*, Botafogo, Rio de Janeiro, Brazil, December 13-15, 2010

Hizem, M. & Bouallegue, R. (2011a) Fine Synchronization through UWB TH-PPM Impulse Radios, *Proceedings of International Journal of Wireless & Mobile Networks (IJWMN) Vol. 3, No. 1, February 2011*

Hizem, M. & Bouallegue, R. (2011b) Fine Synchronization with UWB TH-PAM Signals in ad-hoc Multi-user Environments, *Proceedings of Progress in Electromagnetics Research Symposium (PIERS)*, Marrakech, Morocco, March 20-23, 2011

Homier, E. A. & Schloltz, R. A. (2002). Rapid acquisition for ultra-wideband signals in the dense multipath channel, *Proceedings of Conference on Ultra-Wideband System Technologies*, Baltimore, MD, pp. 105-110, May 20-23, 2002

Lottici, V.; Andrea, A. D. & Mengali, U. (2002). Channel estimation for ultra wideband communications, *Proceedings of IEEE Journal of Selected Areas in Communications*, vol. 20, pp. 1638-1645, December 2002

Tian, Z. & Giannakis, G. B. (2003). Data-aided ML timing acquisition in ultra-wideband radios, *Proceedings of Conference on Ultra-Wideband System Technologies*, Reston, VA, pp. 245-250, November 16-19, 2003

Tian, Z. & Giannakis, G. B. (2005). BER sensitivity to mistiming in ultra-wideband communications-Part I: Non-random channels, *Proceedings of IEEE on Signal Processing*, vol. 53, No. 4, pp. 1550-1560, April 2005

Yang, L. & Giannakis, G. B. (2003). Low-complexity training for rapid timing synchronization in ultra-wideband communications, *Proceedings of Global Telecommunications Conference*, San Francisco, CA, pp. 769-773, December 2003

Yang, L.; Tian, Z. & Giannakis, G. B. (2003). Non-data aided timing acquisition of ultra-wideband transmissions using cyclostationarity, *Proceedings of International Conference in Acoustics, Speech, Signal Processing*, Hong Kong, China, pp. 121-124, April 6-10, 2003

Yang, L. & Giannakis, G. B. (2004). Ultra-wideband communications: an idea whose time has come, *Proceedings of IEEE on Signal Processing Magazine*, vol. 21, No. 6, pp. 26-54, November 2004

Yang, L. & Giannakis, G. B. (2005). Timing UWB signals using dirty templates, *Proceedings of IEEE Transactions on Communications*, vol. 53, No. 11, pp. 1952-1963, November 2005

Yang, L. (2006). Timing PPM-UWB signals in ad hoc multi-access, *Proceedings of IEEE Journal of Selected Areas in Communications*, vol. 24, No. 4, pp. 794-800, April 2006

Ying, Y.; Ghogho, M. & Swami, A. (2008). Code-Assisted synchronization for UWB-IR systems: algorithms and analysis, *Proceedings of IEEE Transactions on Signal Processing*, vol. 56, No. 10, pp. 5169-5180, October 2008

Win, M. Z. & Scholtz, R. A. (2000). Ultra wide bandwidth time-hopping spread-spectrum impulse radio for wireless multiple access communications, *Proceedings of IEEE Transactions on Communications*, vol. 48, No. 4, PP. 679-691, April 2000

Part 2

Novel UWB Applications in Networks

7

High-Speed Wireless Personal Area Networks: An Application of UWB Technologies

H. K. Lau
The Open University of Hong Kong
Hong Kong

1. Introduction

Recently, a large number of wireless networks are being developed and deployed in the market. According to the communication range, wireless networks can be classified into wireless wide area networks (WWANs), wireless metropolitan area networks (WMANs), wireless local area networks (WLANs), wireless personal area networks (WPANs), and wireless body area networks (WBANs). With the advances in wireless technologies, latest generation of WPANs can provide a data rate of hundreds (or even thousands) of Mbps at a distance of less than 10 meters.

Ultra-wideband (UWB) is an emerging technology that offers distinct advantages, e.g. high bandwidth and small communication ranges, for WPAN applications (Park & Rappaport, 2007; Chong et al., 2006; Fontana, 2004; Intel, 2004; Porcino & Hirt, 2003). One of the 'killer' applications of high-speed WPAN is wireless video area network (WVAN) that offers wireless transmission of high-definition videos (several Gbps) within a small communication distance (Singh et al., 2008; Wirelesshd 2009; Whdi 2009).

This chapter provides a comprehensive summary on the latest development and standardization progress of high-speed WPANs. There are seven sections in this chapter. The first section describes the background of WPANs and introduces the IEEE networking standards for WPAN. The second section discusses characteristics of UWB signals and explains why they are particularly suitable for high-speed WPAN applications. The third section discusses technical challenges and standardization issues. The fourth section reports on the latest development of high-speed WPANs. Standards or systems to be discussed in this section include Certified Wireless USB (WUSB), Bluetooth, TransferJet, WirelessHD, Wireless Home Digital Interface (WHDI), Wireless Gigabit (WiGig), and ECMA-387. The fifth section discusses possible research directions of high-speed WPANs. The sixth and the seventh sections are conclusion and references.

1.1 Background

According to the communication range, wireless networks can be classified into WWANs (e.g. GSM and UMTS), WMANs (e.g. IEEE 802.16), WLANs (e.g. IEEE 802.11a/b/g/n), WPANs (e.g. IEEE 802.15 TG1), and WBANs (e.g. IEEE 802.15 TG6). Among these networks, WLANs have received much attention and achieved great success in recently years. The IEEE 802.11a/b/g/n is now the most popular wireless standard for home networking, small office, and even public Internet access. Table 1 summarizes basic characteristics and Fig. 1 shows the range against peak data rate of various wireless networks.

Classification	Communication range	Examples	Current major applications
WWAN	> 10 km	GSM, UMTS	Mobile Internet access
WMAN	<10 km	IEEE 802.16	Broadband Internet access
WLAN	< 100 m	IEEE 802.11a/b/g/n	Internet access, file sharing
WPAN	< 10 m	IEEE 802.15 TG1	File sharing, headset
WBAN	<1 m	IEEE 802.15 TG6	Body senor network

Table 1. Basic characteristics of wireless networks

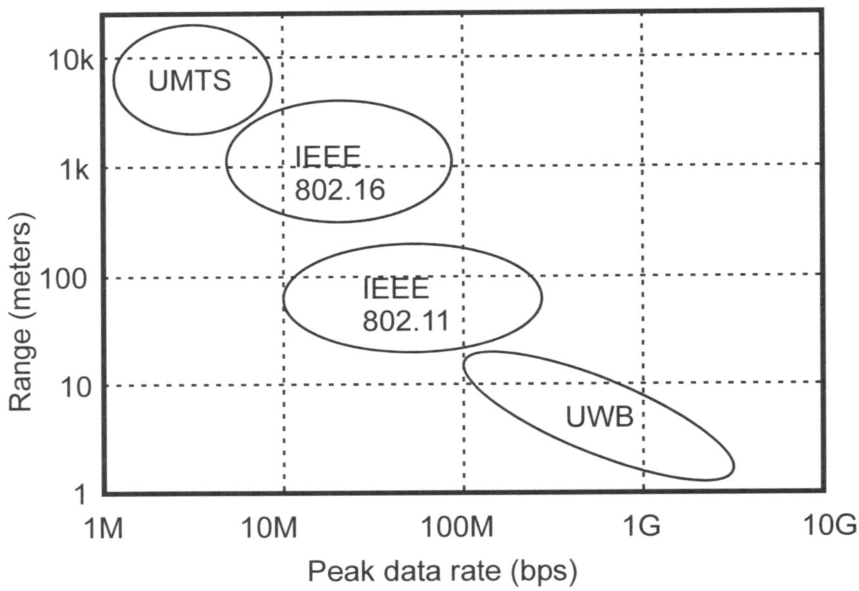

Fig. 1. Communication range against data rate

Recently, high-speed (hundreds of Mbps or several Gbps) WPANs have also received much attention because many innovative ideas and applications (e.g. seamless networking capabilities and HD video streaming) are now becoming a reality and corresponding products are now available in the market. Customer's desires to eliminate cables or complicated connections associated with HDTVs, personal computers or other multimedia systems are not dreams anymore. Obviously, market demands are the major driving force for fast wireless connectivity, especially in WPANs.

1.2 IEEE networking standards for WPAN

Within the IEEE 802 LAN/MAN Standards Committee, the IEEE 802.15 WGs (Working Groups) are responsible for WPAN. The IEEE 802.15.1 (TG1) has derived a WPAN standard based on the Bluetooth v1.1 specifications; while the IEEE 802.15.2 (TG2) has developed a 'Recommended Practices' to facilitate coexistence of WPANs and WLANs. The IEEE

802.15.3 (TG3) and the IEEE 802.15.4 (TG4) are responsible for high and low data rate WPAN, respectively. The IEEE 802.15.5 (TG5) and IEEE 802.15.6 (TG6) focus on mesh networking and WBANs, respectively. The IEEE 802.15.7 (TG7) and IEEE 802.15 IG THZ (IG THZ) are exploring visible light and terahertz communications, respectively. Table 2 summarizes the functions of various TGs in the IEEE 802.15 (IEEE 2011a).

Task group	**Functions/Descriptions**
TG1	*Bluetooth v1.1 specifications*
TG2	*Coexistence of WPANs and WLANs*
TG3	*High rate WPANs*
TG4	*Low rate WPANs*
TG5	*Mesh networking*
TG6	*Wireless body area networks*
TG7	*Visible light communications*
IG THZ	*Terahertz communications*

Table 2. IEEE 802.15 Working groups

Within the IEEE 802.15.3 (TG3), the IEEE 802.15.3a (TG3a) is responsible for WPAN High Rate Alternative PHY. Unfortunately, due to the deadlock between the two available UWB technologies (namely direct sequence UWB (DS UWB) and multiband orthogonal frequency-division multiplexing UWB (MB-OFDM UWB)), the IEEE 802.15.3a (TG3a) was officially disbanded in 2006. The IEEE 802.15.3b (TG3b) aimed to provide amendment and minor optimizations. The IEEE 802.15.3c (TG3c) has developed a high-speed (> 1Gbps) millimeter-wave (57-64 GHz unlicensed band) based alternative PHY for the IEEE 802.15.3 Information about the IEEE 802.15.3 TG3 is summarized in Table 3 (IEEE 2011a).

Task group 3	**Functions/Descriptions**
Task group 3	*High Rate WPAN*
Task group 3a	*WPAN High Rate Alternative PHY (disbanded in 2006)*
Task group 3b	*MAC Amendment*
Task group 3c	*WPAN Millimeter Wave Alternative PHY*

Table 3. IEEE 802.15 Task Group 3 (TG3)

2. Characteristics and benefit of UWB signals

Before the 90's, UWB technologies were restricted to military applications only. In April 2002, the Federal Communications Commission (FCC) issued the first report and order (RAO) and allowed commercial applications of UWB technologies under strictly power emission limits (FCC 2002). According to FCC, UWB is a radio technology that offers a high bandwidth (> 500 MHz) at very low energy levels over a short communication range (< 10 meters).

2.1 UWB signals

UWB technology is very different from other narrowband and spread spectrum technologies. UWB uses an extremely wide band of spectrum to transmit data. According to the RAO from FCC (FCC 2002), UWB technology is not confined to a specific implementation. Instead, any wireless transmission scheme that occupies a bandwidth of more than 20% of a center frequency, or more than 500 MHz can be considered as UWB. Based on their fractional bandwidth, B_f, signals can be classified as narrowband, spread spectrum (or wideband) or UWB as illustrated in Fig. 2 and Table 4.

Two popular approaches to generate UWB signals are single band UWB (often referred as impulse UWB, direct sequence UWB or DS UWB) and multiband UWB (often referred as multiband orthogonal frequency division multiplexing UWB or MB-OFDM UWB). In single band UWB, the concept of impulse radio is adopted and pulses with very short duration (typically between 10 to 1000 picoseconds) that occupy a very wide bandwidth (hundreds of MHz to several GHz) are transmitted. Multiband UWB, on the other hand, divides the whole available UWB frequency spectrum into a number of smaller and non-overlap bands. MB-OFDM UWB signals are transmitted simultaneously over multiple carriers spaced in those non-overlap bands.

Although both approaches can be used to generate UWB signals, they offer different performance degradations. The effect of multipath (Rayleigh) fading on single band UWB is considered to be insignificant; while multiband UWB may suffer from larger performance degradation due to multipath fading. However in multiband UWB, it is possible to avoid the transmission in certain congested bands (e.g. the 5 GHz band currently used extensively in IEEE 802.11a/n or other cordless telephones).

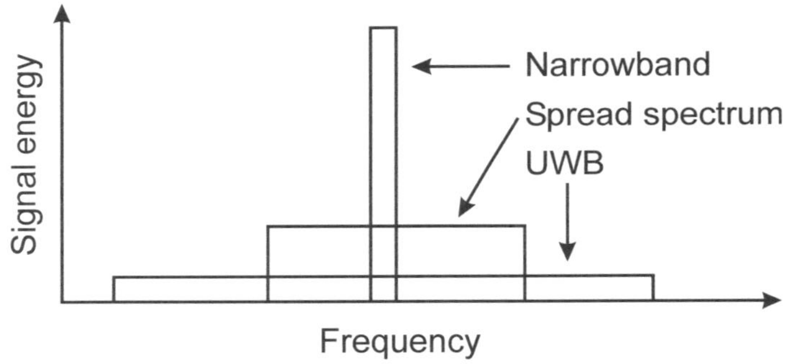

Fig. 2. Spectrum of narrowband, spread spectrum and UWB signals

Signal type	**Fractional bandwidth, B_f**
Narrowband	$B_f < 1\%$
Spread spectrum/wideband	$1\% < B_f < 20\%$
Ultra-wideband	$B_f > 20\%$

Table 4. Fractional bandwidth of narrowband, spread spectrum and UWB signals

2.2 Benefits of UWB technology for WPAN applications

Due to the wide bandwidth and high time resolution characteristics, UWB signals are much more robust to interferences and multipath fading distortion than other narrowband signals. In addition, the large channel capacity and wide bandwidth offer wireless transmission of real-time high quality multimedia files (even uncompressed HD videos in several Gbps). The extremely small transmit power and the very short communication distances result in a large number of other advantages for WPAN applications. Since UWB signals are operating below the noise floor, they provide better security, lower RF health hazards, and lower interference to other systems (which allows the coexistence with current narrowband and wideband systems).

3. Standardization and challenges of UWB WPAN

Although UWB technologies are attractive for WPAN applications, there are standardization and technical issues that need to be addressed.

3.1 Standardization issues

The IEEE 802.15.3a task group is responsible for the WPAN High Rate PHY standardization. The pathway of high-speed WPAN standardization is tough. Due to the deadlock between the two UWB implementations (DS UWB and MB-OFDM UWB), the IEEE 802.15.3a task group was officially disbanded in 2006. Since then, a de-facto standard for high-speed WPAN has emerged in the form of WiMedia Alliance's UWB (Wimedia 2009). However, the WiMedia Alliance announced in March 2009 that all specifications related WiMedia Alliance's UWB will be transferred to the Bluetooth Special Interest Group (SIG), Wireless USB Promoter Group and the USB Implementers Forum. Such a move has big impacts to the specifications and deployment of Wireless USB, Bluetooth and other WPAN systems. Details of Wireless USB and Bluetooth will be discussed later in this chapter.

The use of the FCC approved UWB band (3.1 to 10.6 GHz) avoids the crowded 2.4 GHz band and reduces interferences from Bluetooth, Wi-Fi, DECT phone,…., etc. Currently, the 3.1 to 10.6 GHz band is relatively free for unlicensed used of UWB. As a result, systems that are operating in this UWB band can provide a much larger bandwidth. Fig. 3 shows the worldwide (updated 1-20-1009) spectrum allocation in the 3.1 to 10.6 GHz band (Wimedia 2009). In addition to IEEE 802.15.3a, the IEEE 802.15.3c is a task group which is responsible for the standardization of WPAN millimeter wave alternative PHY. Brief description on millimeter wave PHY will be given later in this chapter.

Although the standardization of UWB technology faced quite a lot of difficulties (including Intel has stopped the development of UWB, missing of UWB technology in Bluetooth 3.0/4.0, keen competition from other WPANs operating in the 60 GHz unlicensed band, …, etc), UWB has been proved to be an effective technology for short range high speed data transmission between devices.

3.2 Challenges

3.2.1 Pulse shaper design

Since the bandwidth of UWB signals is very large and UWB signals are operated as an overlay system, the coexistence of UWB with other narrowband systems must be carefully investigated. Intensive studies are required on three major aspects – (i) interference from UWB systems to other narrowband systems, (ii) interference from other narrowband systems to UWB systems, and (iii) interference from UWB systems to other UWB systems that are operating in the same frequency band. To address this issue, strictly narrowband interference control and accurate out-of-band filter design are required. The FCC emission limits for both outdoor and indoor operations of UWB are summarized in Table 5 (FCC 2002).

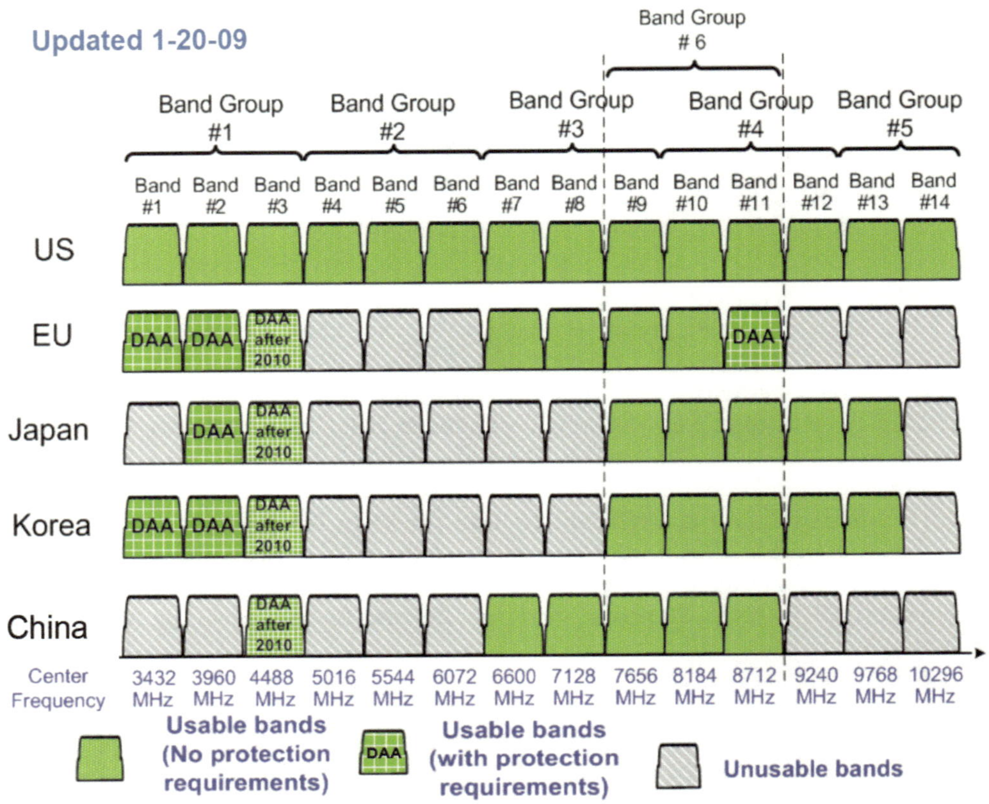

Fig. 3. Spectrum allocation in the 3.1 to 10.6 GHz band (Wimedia 2009)

	Frequency range (MHz)					
	960-1610	1610-1990	1990-3100	3100-10600	Above 10600	1164-1240 1559-1610
Indoor UWB (EIRP)	-75.3 dBm	-53.3 dBm	-51.3 dBm	-41.3 dBm	-51.3 dBm	-85.3 dBm
Outdoor UWB (EIRP)	-75.3 dBm	-63.3 dBm	-61.3 dBm	-41.3 dBm	-61.3 dBm	-85.3 dBm

Table 5. The FCC emission limits for UWB

3.2.2 System design

When MB-OFDM UWB is used (e.g. WiMedia Alliance's UWB), the total transmission power of a UWB signal is distributed over many multipath components. These components are propagating differently and are suffering from different frequency selective fading distortions. To effectively eliminate the effect of multipath fading, accurate channel estimation and synchronization are essential. The choice of modulation techniques for UWB also affects transmission and reception power, data rate and bit error rate performance.

Popular modulation techniques for UWB include pulse position modulation (PPM) and phase shift keying (PSK). Last but not least, effects of multiple access interference (MAI) on system performance must also be investigated.

3.2.3 Wideband RF design

Unobtrusive antennas that can operate effectively under varying propagation conditions are expected in all commercial UWB systems. Due to the nature of UWB signals (very large bandwidth), the design and implementation of wideband RF systems (e.g. antenna and amplifier) are very challenging. Issues related to RF design include impedance matching, radiation patterns, power efficiency, cost and size, …, etc. Recently, the use of multi-input and multi-output (MIMO) in low-cost consumer products (e.g. the IEEE 802.11n Wi-Fi standard) has received much attention. The use of MIMO technology in UWB may further increase the data rate and enhance the interference rejection capability.

3.2.4 Power consumption and battery life

Low power consumption and long battery life are important parameters for all portable and battery-operated devices (especially for consumer products). However, hardware and software complexity play important roles in power consumption. Complex coding and modulation techniques require fast signal processing power, which may increase the power consumption of the devices. In spite of this, UWB-enabled devices can still achieve the lowest power consumption (per Mbps). Table 6 compares the power characteristics of IEEE 802.11g, IEEE 802.11n and WiMedia Alliance's UWB devices (Aiello 2008).

Technology	Range	Throughput	Power
IEEE 802.11g	> 50 m	20 – 30 Mbps	15-20 mW/Mbps
IEEE 802.11n	> 50 m	> 100 Mbps	6-7 mW/Mbps
WiMedia Alliance's UWB	< 10m	> 100 Mbps	1 mW/Mbps

Table 6. Power characteristics of technologies

4. Latest development of high-speed WPANs

This section provides a comprehensive summary on the latest development of high-speed WPANs. Standards or systems reported in this section are (i) Certified Wireless USB (WUSB), (ii) Bluetooth, (iii) TransferJet, (iv) WirelessHD, (v) Wireless Home Digital Interface (WHDI), (vi) Wireless Gigabit Alliance (WiGig), and (vii) ECMA-387.

4.1 Certified Wireless USB (WUSB)

Universal Serial Bus (USB) was originally designed for personal computers, but now has become the most popular de facto standard in connecting peripherals or devices (e.g. digital cameras, scanners, external hard disks, …, etc.). Following the establishment of the Wireless USB Promoter Group in February 2004, the Certified Wireless USB (WUSB) 1.0 specification was released in May 2005. WUSB can be considered as a wireless implementation of USB and is designed to provide high-speed wireless connections between devices that achieving a data rate of 110 Mbps (up to 10 meters) and 480 Mbps (up to 3 meters). WUSB is backward compatible with wired USB. Although the Wireless USB Promoter Group prefers to use the term 'Certified Wireless USB' to distinguish other wireless implementation of USB, Certified

Wireless USB is often referred as Wireless USB or WUSB. Commercial WUSB 1.0 products are available in the market since 2007. Table 7 summarizes the data rate of major USB standards (Wusb 2010).

USB specifications	Date of release	Maximum data rate
USB 1.0	January 1996	1.5 Mbps (Low-speed)
USB 1.1	September 1998	12 Mbps (Full-speed)
USB 2.0 (Hi-Speed USB)	April 2000	480 Mbps
USB 3.0 (Super-Speed USB)	November 2008	5 Gbps
Wireless USB 1.0	May 2005	480 Mbps (up to 2 meters)
Wireless USB 1.1	September 2010	110 Mbps (up to 10 meters)

Table 7. Major USB standards

WUSB is based on the WiMedia Alliance's MB-OFDM UWB radio platform, and is designed to operate in the 3.1 to 10.6 GHz frequency range. The WUSB specification 1.1 released in September 2010 has extended the UWB upper band support for frequencies of 6 GHz and above (Wusb 2010).

4.2 Bluetooth

The Bluetooth v1.0 was announced by the Bluetooth Special Interest Group (SIG) in May 1998. Bluetooth is designed to operate in the 2.4 GHz ISM band, rather than the UWB band (3.1 and 10.6 GHz). Both Bluetooth v1.1 and v1.2 were ratified as IEEE 802.15.1-2002 and IEEE 802.15.1-2005, respectively. The Bluetooth v2.1 adopted in 2007 provides a data rate of 2.1 Mbps. Table 8 summarizes the adopted Bluetooth core specifications (Bluetooth 2010).

Bluetooth specifications	Date of release	Data rate
Bluetooth v1.0a	26 July 1999	721.2 kbps
Bluetooth v1.0B	01 December 1999	
Bluetooth v1.1 (IEEE 802.15.1-2002)	22 February 2001	
Bluetooth v1.2 (IEEE 802.15.1-2005)	05 November 2003	
Bluetooth v2.0 + EDR	04 November 2004	2.1 Mbps
Bluetooth v 2.1 + EDR	26 July 2007	
Bluetooth v3.0 + HS	21 April 2009	24 Mbps
Bluetooth v4.0	30 June 2010	

Table 8. Adopted Bluetooth core specifications

In March 2006, the Bluetooth SIG announced its selection of the WiMedia Alliance's UWB technology for integration with their Bluetooth wireless technology. The most significant improvement in the originally planned Bluetooth v3.0 specification was the adoption of the WiMedia Alliance's MB-OFDM UWB technology that provides a maximum data rate of 480 Mbps. Unfortunately, UWB technology is missing in the final 3.0 specification that was released in April 2009 due to the transfer of WiMedia's technology to other SIGs. The final Bluetooth v3.0 provides a maximum data rate of 24 Mbps through the use of a new High Speed (HS) technology. In June 2010, the Bluetooth SIG also released the Bluetooth v4.0 specification. Two forms of wireless technology systems are adopted in Bluetooth v4.0, namely Basic Rate (BR) and Low Energy (LE). The BR system includes optional Enhanced

Data Rate (EDR) Alternate MAC PHY layer extensions. The BR system provides three different data rates of 721.2 kbps (BR), 2.1 Mbps (EDR) and up to 24 Mbps (High Speed, HS). The HS technology provides better power optimization, better security, enhanced power control and lower latency rate. The LE system is designed for products that require lower power consumption, lower complexity, lower data rates, lower duty cycles and lower cost than BR/EDR. According to the maximum power, Bluetooth devices are divided into three different classes as illustrated in Table 9 (Bluetooth 2010).

Power class	Maximum power	Communication range
1	100 mw (20 dBm)	~ 100 meters
2	2.5 mW (4 dBm)	~ 10 meters
3	1 mW (0 dBm)	~ 1 meter

Table 9. Power classes of Bluetooth devices

4.3 TransferJet

TransferJet is a close-proximity technology developed by Sony and was first presented at the 2008 Consumer Electronics show in Las Vegas (Transferjet 2008). The TransferJet technology is very different from other WPAN technologies that employ electro-magnetic radiation field (e.g. WiMedia Alliance's UWB). TransferJet, on the other hand, is designed to work with longitudinal electric induction fields (Transferjet 2010). It is operating in the UWB band and can achieve a data rate of 560 Mbps (up to 3 cm) with a transmission power of under -70 dBm/MHz. Based on channel conditions, TransferJet is able to determine and adopt the most appropriate data rate for transmission by itself. Sony has also developed a new antenna element for TransferJet called 'TransferJet Coupler' that consists of a coupling electrode, a resonant stub, and ground. Since TransferJet is designed to operate in the near field, which is a non-polarized field, devices are not required to be precisely oriented to initialize communications (Transferjet 2008). Data transfer can be initialized simply by touching the transmitting device to the receiving device.

There are a number of advantages of very short communication distance (within few centimeters) in TransferJet. Firstly, the very short communication distance virtually eliminates the effects of multipath fading and shadowing that commonly exist in other WPANs. It also reduces the interference to other systems and the chance for unauthorized access to TransferJet enabled devices. In addition, the small power requirement can significantly prolong the battery life.

The TransferJet Consortium was established in July 2008 by a group of international companies. The main duties of the consortium include the development of the specification and compliance testing process, management of the certification program and promotion of the TransferJet technology. As of April 2010, there are 18 Consortium members, including Sony, Panasonic, Sharp, and Toshiba. Table 10 summarizes key specifications of TransferJet (Transferjet 2010).

Based on the TransferJet specification, the Technical Committee 50 (TC50) of European Computer Manufacturers Association (Ecma) International has completed the First Edition of its standard titled "Close Proximity Electric Induction Wireless Communications" and is expected to be formally approved by the Ecma General Assembly in June 2011 (Transferjet 2011).

Items	Details
Carrier Center Frequency	4.48 GHz
Transmission Power	At or below -70 dBm/MHz (average)
Transmission Rate	560 Mbps (max) / 375 Mbps (effective throughput)
Communication Distance	A few centimeters (3 cm nominal)
Topology	One-to-one, Point-to-point
Antenna Element	Electric induction field coupler
Modulation	$\pi/2$ shift BPSK + DSSS
FEC	½ Convolutional code + Reed Solomon code

Table 10. TransferJet specifications

4.4 WirelessHD

The WirelessHD specification is one of the industrial standards which are specially designed for wireless transmission of HD videos. A relatively new terminology - wireless video area network (WVAN) - is now commonly used for this special type of WPAN application. The WirelessHD specification v1.0 that provides a data rate of about 4 Gbps was finalized in January 2008 and was the first de facto standard for 60 GHz millimeter wave frequency band applications based on the IEEE 802.15.3c specification (Wirelesshd 2009).

In May 2010, the WirelessHD Consortium released the v1.1 specification that further boosts the data rate to 10 - 28 Gbps and support HD resolution four times beyond that of 1080p. It also defines common 3D formats and resolutions for WirelessHD-enabled devices. Commercial WirelessHD-enabled products, including notebook PCs and TVs, are now available in the market. Table 11 shows the applications supported by WirelessHD v1.1 (Wirelesshd 2010).

Application	Data rate	Target latency
Uncompressed QHD 2560×1440p, 60 Hz, 36 bit color	8.0 Gbps	2 ms
Uncompressed 720p frame sequential 3D A/V 1280×1440p, 60 Hz, 36 bit color	4.0 Gbps	2 ms
Uncompressed 1080p, 120 Hz, 30 bit color	7.5 Gbps	2 ms
Uncompressed 1080p A/V	3.0 Gbps	2 ms
Uncompressed 1080i A/V	1.5 Gbps	2 ms
Uncompressed 720p A/V	1.4Gbps	2 ms
Uncompressed 480p A/V	0.5Gbps	2 ms
Uncompressed 7.1 surround sound audio	40 Mbps	2 ms
Compressed 1080p A/V	20-40 Mbps	2 ms
Uncompressed 5.1 surround sound audio	20 Mbps	2 ms
Compressed 5.1 surround sound audio	1.5 Mbps	2 ms
File transfer	>1.0 Gbps	N/A

Table 11. Applications supported by WirelessHD v1.1 (Wirelesshd 2010)

According to WirelessHD v1.1, the WVAN consists of one *Coordinator* and zero or more *Stations*. The *Coordinator* can be a device that is sink for audio or video data (e.g. a display). A *Station* is a device that has media that it can source and/or sink or has data to exchange. An example of WVAN under WirelessHD is illustrated in Fig. 4 (Wirelesshd 2010).

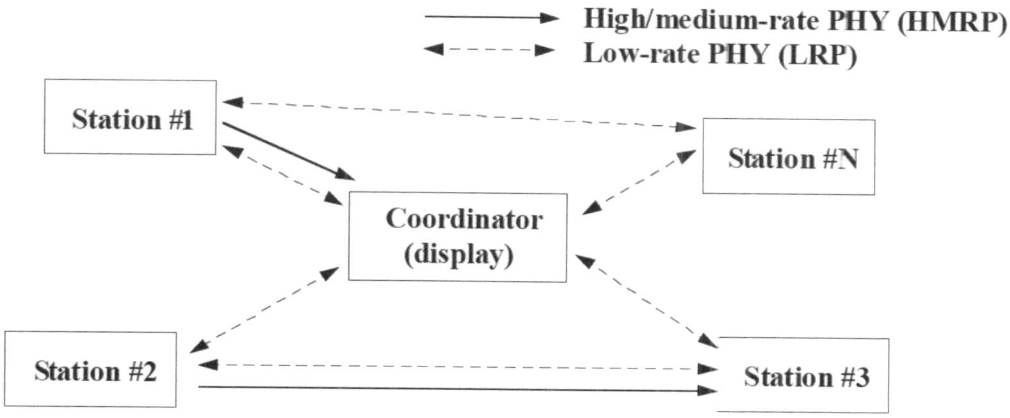

Fig. 4. An example of WirelessHD WVAN (Wirelesshd 2010)

The High and Medium Rate PHY (HRP and MRP) are highly directional and are mainly used for unicast connections (several Gbps). The MRP supports multiple video resolutions with more than one data rates. The Low Rate PHY (LRP) are bidirectional links and can be used for both unicast and broadcast connections (several Mbps). Similar to MRP, the LRP also supports more than one data rates. In a single stream using OFDM modulation with beamform mode, the HRP can achieve a data rate of greater than 7 Gbps. When combined with spatial multiplexing, the HRP may further boost the data rate to greater than 28 Gbps. The transmit masks of HRP and LRP are shown in Figs. 5 and 6, respectively.

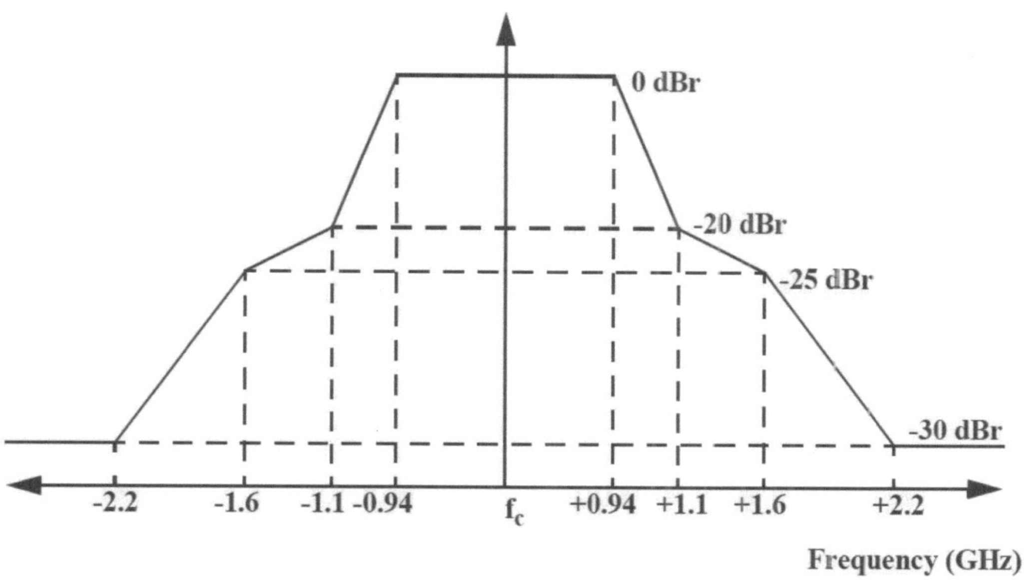

Fig. 5. The HRP transmit mask (Wirelesshd 2010).

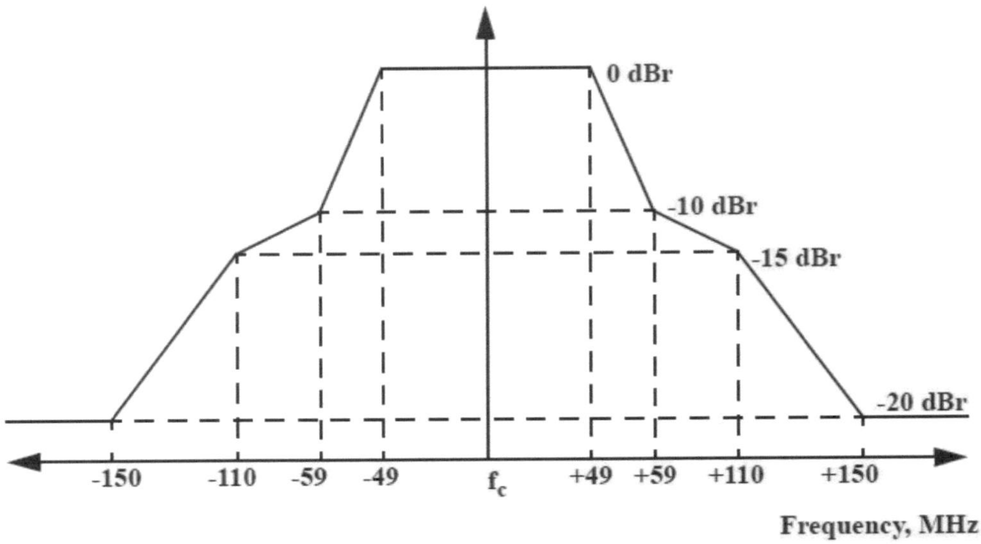

Fig. 6. The LRP transmit mask (Wirelesshd 2010)

Currently, the 57-64 GHz band is allocated in North America and South Korea, the 59-66 GHz band is allocated in Japan and the 57-66 GHz band is allocated in the European Union. The current regulations in the 60 GHz band allow very high effective isotropic radiated power (EIRP) of greater than 10 W for reliable high bandwidth transmission. The use of the 60 GHz band seems to be a good solution to support bandwidth-hungry applications (e.g. uncompressed HD video transmission).

4.5 Wireless Home Digital Interface (WHDI)

Similar to WirelessHD, the Wireless Home Digital Interface (WHDI) is an industrial WVAN standard that offers video transmission (up to 3 Gbps) in the 5 GHz unlicensed band (Whdi 2009). The WHDI v1.0 was released in December 2009 and the communication distance is beyond 30 meters, through walls, and latency is less than one millisecond. Both HDCP revision 2.0 and digital content protection are supported by WHDI.

A multi-input multi-output (MIMO) 20 MHz/40MHz bandwidth channel over the 5 GHz band and the joint source-channel coding (JSCC) approach are used in WHDI (Whdi 2009). There are three major elements in WHDI's JSCC approach. Firstly, video processing and representation are prioritized according to their importance. Secondly, unequal error protection (UEP) is used to protect data with different importance levels. Thirdly, adaptive signal constellation is adopted. According to the report from WHDI, the use of JSCC improves the protection of important components and achieves a better utilization of the available channel capacity (Whdi 2009). Following the WHDI 3D specification update in June 2010, the WHDI v2.0 specification is targeted for release in the second quarter of 2011. Key features in WHDI v2.0 specification include (Whdi 2009):

- Full 3D support (HDMI 1.4a 3D modes, 1080p 60Hz × 2 3D format)
- Support four times the resolution of 1080p (4,096 × 2,160)
- Support WHDI-Wi-Fi integration and same channel co-existence
- Support mobile device integration through reduced power consumption and footprint

Beside WHDI, the IEEE 802.11ac is formed recently for standardization of high throughput WLAN near the 6 GHz band (IEEE 2011b).

4.6 Wireless Gigabit (WiGig) Alliance

The Wireless Gigabit (WiGig) Alliance was formed in May 2009 and aims to establish a unified specification for high-speed (several Gbps) wireless technologies in the 60 GHz band. The WiGig specification is based on the existing IEEE 802.11 standard and was contributed to the IEEE 802.11ad draft standard (Wigig 2011). In May 2010, WiGig Alliance and Wi-Fi Alliance established a cooperation agreement to share technology specifications for the development of certification programs.

Under the WiGig v1.0 specification, WiGig devices with tri-band (2.4 GHz, 5 GHz and 60 GHz) radios are able to seamlessly integrate into existing 2.4 GHz and 5 GHz Wi-Fi networks (e.g. IEEE 802.11a/b/g/n). In addition to uncompressed video transmission, multi Gbps data transfer (e.g. wireless docking station and file transfers between computers/devices) are supported by WiGig. The following key elements are included in the v1.0 specification (Wigig 2011):

- Supports data transmission rates up to 7 Gbps
- Backward compatible with the IEEE 802.11 standard
- Protocol adaptation layers to support specific system interfaces
- Support for beam-forming
- Support for advanced security and power management

4.7 ECMA-387

Ecma International is a standards organization for information and communication systems. Ecma's Technical Committee 48 (TC48) is responsible for the development of standards and technical reports for high rate wireless communications. The ECMA-387 is a standard that specifies the High Rate PHY, MAC, and HDMI (PAL) for the 60 GHz band. The first edition of ECMA-387 has been published by the ISO and IEC as ISO/IEC 13156 in October 2009. Under ECMA-387 2nd edition (revision 2), there are two types of devices, namely Type A and Type B devices (Ecma 2010). Type A devices are equipped with high gain trainable antennas that enable video streaming and other high data rate (0.4 Gbps to 6.4 Gbps) applications within 10 meters communication range under LOS or NLOS environments. Type A devices also support UEP, open-loop and closed-loop antenna training protocols, and transmit switch diversity protocols. Type B devices, on the other hand, are low cost and low power implementation offer a basic data rate of 0.8 Gbps (which can be extended to 3.2 Gbps) within 1-3 meters communication range under LOS environments.

5. Possible research directions

There is a strong customer desire to have a small, portable, cheap, secured and easy to use device that provides wireless transmission of bandwidth-hungry signals. Commercial millimeter-wave transmission technology is a new wireless communications concept that aims to provide multi Gbps transmission in the 60 GHz unlicensed band. In addition to WirelessHD, WiGig, and ECMA-387 that are highlighted in previous section, the IEEE 802.15.3c and the IEEE 802.11ad groups are formed recently to define standardized modifications to the IEEE 802.11 that enable 60 GHz operation (IEEE 2011c).

Apart from commercial millimeter-wave transmission technology, UWB-over-fiber has received much attention recently. The communication range of UWB signal can be significantly increased by connecting the antenna to a fiber, i.e. UWB-over-fiber (Pan & Yan 2010; Guillory et al. 2010).
Other researches (e.g. medical imaging, radar imaging, vehicular radar systems, wireless sensor networks, UWB-based WBANs) are carried out in the areas of high-speed WPANs or UWB technologies. For example the NDSsi ZeroWire technology is a UWB-based medical grade wireless video system that delivers real-time full HD surgical video at a maximum data rate of 480 Mbps within 10 meters (Ayar 2010). Besides imaging applications, UWB technologies are widely used in wireless sensors networks (WSN) and wireless body area networks (Xia et al., 2011). Since UWB technologies can also provide accurate ranging capability and excellent time resolution, other emerging applications are through-wall surveillance radar and vehicular radar systems.

6. Conclusion

Wireless networking products are enjoying great success and high-speed WPANs are undergoing rapid development. Innovative applications like short-range streaming of high-definition video are now possible. This chapter provides a comprehensive summary on the latest development and standardization progress of high-speed WPANs. Although the IEEE 802.15.3a task group was disbanded in 2006, research and development activities on UWB-based WPANs are still carried on. However, approval from regulatory organizations plays an important role in the success of WPANs. The complex mix of standards and technologies introduces barriers in the standardization of high-speed WPANs. When UWB was first introduced, the proposed data rates were attractive (hundreds of Mbps). However, since some regional regulators had posted restrictions on use of UWB in the 3.1 to 10 GHz band, products took a long time to become available in the market. When commercial UWB products are available (e.g. WUSB in mid 2007), their data rates were no longer significantly higher than other completing technologies, like IEEE 802.11n. Obviously, market demands are the major driving force for fast wireless connectivity. High-speed wireless networking would be an important direction of research in telecommunications.

7. References

Aiello, R (2008). Using WiMedia UWB technology to enable future-generation WPANs. EE Times, Visited in July 2011, Available from: http://www.eetimes.com/design/microwave-rf-design/4012949/Using-WiMedia-UWB-technology-to-enable-future-generation-WPANs

Ayar, E. (2010). UWB Wireless Video Transmission Technology in Medical Applications. August 2010, Visited in July 2011, Available from: http://www.ndssi.com/products/surgical/visualization/zerowire/index.php

Bluetooth (2010). Adopted Bluetooth Core Specifications. Visited in July 2011, Available from: https://www.bluetooth.org/Technical/Specifications/adopted.htm

Chong, C.; F. Watanabe F.; & Inamura H. (2006). Potential of UWB Technology for the Next Generation Wireless Communications. *Proceedings of 2006 IEEE Ninth International Symposium on Spread Spectrum Techniques and Applications*, Brazil, August 2006, pp. 422-9.

Ecma (2010). Standard ECMA-387 High Rate 60 GHz PHY, MAC and HDMI PAL: 2nd edition. December 2010, Visited in July 2011, Available from: http://www.ecma-international.org/publications/standards/Ecma-387.htm

FCC (2002). First Report and Order: Revision of Part 15 of the Commissions Rules Regarding Ultra-Wideband Transmission Systems. Visited in July 2011, Available from: http://hraunfoss.fcc.gov/edocs_public/attachmatch/FCC-02-48A1.pdf

Fontana, R. (2004). Recent System Applications of Short-pulse Ultra-wideband (UWB) Technology. *IEEE Transactions on Microwave Theory and Techniques*. Vol. 52, Issue 9, September 2004, pp. 2087-2104.

Guillory, J. et al. (2010). Radio-Over-Fiber Architectures. *IEEE Vehicular Technology Magazine*. Vol 5, Issue 3, September 2010, pp 30-38

IEEE (2011a). Current status of IEEE 802.15 Working Group for WPAN. Visited in July 2011, Available from: http://ieee802.org/15/index.html

IEEE (2011b). Status of Project IEEE 802.11ac. Visited in July 2011, Available from: http://www.ieee802.org/11/Reports/tgac_update.htm

IEEE (2011c). Status of Project IEEE 802.11ad. Visited in July 2011, Available from: http://www.ieee802.org/11/Reports/tgad_update.htm

Intel (2004). Ultra-wideband (UWB) Technology: Enabling high-speed wireless personal area networks. Visited in July 2011, Available from: http://www.intel.com/technology/comms/uwb

Pan, S & Yao, J. (2010). Performance Evaluation of UWB Signal Transmission over Optical Fiber. *IEEE Jounral on Selected Areas in Communications*. Vol 28, Issue 6, August 2010, pp. 889-900

Park, C. & Rappaport, T. (2007). Short-Range Wireless Communications for Next-Generation Networks: UWB, 60 GHz Millimeter-Wave WPAN, and ZigBee. *IEEE Wireless Communications*, Vol. 14, Issue 4, August 2007, pp 70-78.

Porcino, D & Hirt, W (2003). Ultra-Wideband Radio Technology: Potential and Challenges Ahead. *IEEE Communications Magazine*, Vol. 41, Issue 7, July 2003, pp.66-74

Singh, H, et. al., (2008). A 60 GHz Wireless Network for Enabling Uncompressed Video Communication. *IEEE Communications Magazine*. Vol. 46, Issue 12, December 2008, pp. 71-78.

Transferjet (2008). Sony Develops New Close Proximity Wireless Transfer Technology "TransferJet". Visited in July 2011, Available from: http://www.sony.net/SonyInfo/News/Press/200801/08-002E/index.html

Transferjet (2010). TransferJet Overview: Concept and Technology Rev 1.1. February 2010. Visited in July 2011, Available from: http://www.transferjet.org/tj/transferjet_overview.pdf

Transferjet (2011). Ecma completes the First Edition of the TC50 specifications based on TransferJet. February 2011, Visited in July 2011, Available from http://www.transferjet.org/news/index.html

Whdi (2009). WHDI Technology. Visited in July 2011, Available from: http://www.whdi.org/

Wigig (2011). WiGig: Defining the Future of Multi-Gigabit Wireless Communications. July 2010, Visited in July 2011, Available from: http://wirelessgigabitalliance.org/specifications/

Wimedia (2009). Worldwide Regulatory Status. Visited in July 2011, Available from: http://www.wimedia.org/

Wirelesshd (2009). WirelessHD Next Generation Standard Now Available. Visited in July 2011, Available from: http://www.wirelesshd.org/category/news/

Wirelesshd (2010). WirelessHD Specification Version 1.1 Overview. May 2010, Visited in July 2011, Available from: http://www.wirelesshd.org/pdfs/WirelessHD-Specification-Overview-v1.1May2010.pdf

Wusb (2010). Wireless Universal Serial Bus Specification, Revision 1.1. Visited in July 2011, Available from: http://www.usb.org/developers/wusb/wusb1_1_20100910.zip

Xia, L; Redfield, S; & Chiang, P (2011). Experimental Characterization of a UWB Channel for Body Area Networks. EURASIP Journal on Wireless Communications and Networking. Vol 2011, Article ID 703239. Visited in July 2011, Available from: http://www.hindawi.com/journals/wcn/2011/703239/

8

UWB Technology for WSN Applications

Anwarul Azim[1,2], M. A Matin[3], Asaduzzaman[2] and Nowshad Amin[4]
[1]*Dept. of CSE, Faculty of S&E, International Islamic University Chittagong*
[2] *Dept. of Computer Science and Engineering, CUET*
[3] *Dept. of EECS, North South University*
[4]*Dept. of EESE, National University of Malaysia*
[1,2,3]*Bangladesh*
[4]*Malaysia*

1. Introduction

Ultrawide band (UWB) technology has been recognized as a feasible technology for wireless sensor networks (WSNs) applications due to its very good time-domain resolution allowing for precise location, tracking, coexistence with existing narrowband systems (due to the extremely low power spectral density) with low power and low cost on-chip implementation facility. Sensor Nodes (SN) that builds the backbone of such networks is normally micro controller based small devices. As batteries normally supply powers to these nodes that can only provide relatively small and limited processing capabilities. As a result, a number of UWB-based sensor network concepts have been developed both in the industrial and the government domain. For UWB devices, there are three independent bands i.e. the sub-gigahertz band (250–750 MHz), the low band (3.1–5 GHz), and the high band (6–10.6 GHz). Each UWB band has a single mandatory channel and devices that operate independently of the other band. Here, emphasis given on the low band of UWB (3.244-4.742 GHz) that is based on spread spectrum technique for WSN applications. The main feature of the system is the design simplicity having the advantage of using simple binary modulation technique and non-coherent detection scheme. Simulation result shows that, the pulse repetition coder has significant impact on performance as well as controlling data rates and reliable reception. Moreover, data is successfully recovered by an energy detection method (detect and avoid), which facilitates design simplicity at the receiver by avoiding pulse synchronization and coherent detection. We have also analyzed pulse repetition coder in conjunction with spread spectrum technique that facilitates robust and low power transmission system design. The remaining part of this chapter briefly discusses the feasibility of UWB for WSN as a physical layer communication system and then describes the UWB system design, transmission and reception process as well as performance analysis.

2. Applications and overview of WSNs

WSN can be used for many different applications and generally be deployed in an ad hoc manner without stringent advance planning. Therefore, they must be able to organize themselves to form viable single-hop or multi-hop wireless communications networks. After deployment, sensor nodes detect environmental changes and report them to other nodes over their flexible network architecture. Sensor nodes are excellent for deployment in hostile environments, over small, or even large, geographical areas. A WSN is usually deployed on a global scale for information sharing; over a battle field for military surveillance and inspection, in emergent environments for search and rescue, in factories for condition based maintenance, in building for infrastructure health monitoring, in homes to realize smart homes, or even in bodies for patient monitoring. One can retrieve required information from the network by injecting queries and gathering results from the sink. A sink acts like an interface between users and the network. In addition, monitoring environmental conditions extend to irrigation, chemical or biological detection, precision agriculture, forest fire detection, flood detection, bio-complexity mapping of the environment, and pollution study etc. To ensure long-term sustainable economic growth, it is essential to efficiently monitor our environment as well as resources (Land, water etc.). By monitoring the environment we can also protect the environment and people from toxic contaminants that can be released into a variety of sources including air, soil and water from variety of sources.

A WSN is simply defined as a large collection of sensor nodes. Each node equipped with its own sensor, processor and radio transceiver reported by Azim et al (2008). Such networks have substantial data acquisition and processing capabilities that deployed densely throughout the area to monitor specific environmental phenomena. In a multi-hop sensor network, communicating nodes are linked by a wireless medium. To enable global operation of these networks, the chosen transmission medium must be available worldwide. The communication device is used to exchange data between individual nodes. Radio frequency (RF) based communication is commonly used for most WSN applications. The expected feature should be relatively long range, high data rate communications with acceptable error rates at a low energy expenditure that does not require line of sight between sender and receiver. For actual communication, both the transmitter and a receiver are required in a sensor node but can be further optimized to a full or reduced function device as proposed by ZigBee. Generally, each node of a WSN system comprises a transceiver unit, which is in charge of the wireless communication with other nodes. The essential task is to convert a bit stream coming from a micro-controller and convert them to and from radio waves. Recent advancement in wireless communications and electronics has enabled the development of low-cost sensor networks. The IEEE 802.15.4 standard and Zigbee wireless technology are designed to satisfy the market's need for a low-cost, standard-based and flexible wireless network technology, which offers low power consumption, reliability, interoperability and security for control and monitoring applications with low to moderate data rates. The key requirements for transceivers in sensor networks are given in ZigBee discussed by Zhang J, et al (2009).

- Low cost: Since a large number of nodes are to be used, the cost of each node must be kept small. For example, the cost of a node should be less than 1% of the cost of the product it is attached to.
- Small form factor: Transceivers' form factors (including power supply and antenna) must be small, so that they can be easily placed in locations where the sensing actually takes place.

- Low energy consumption: A sensor usually has to operate for several years with no battery maintenance, requiring the energy consumption to be extremely low. Some additional requirements are needed to make the wireless sensor network effective. To evaluate the energy consumption behavior of a radio transceiver, the following parameters need to be considered such as the modes of operation, duty cycle and models for the energy consumption per bit for both sending and receiving. In principle, the sources of energy consumption are RF signal generation, which depends on modulation scheme and target distance as well as on the transmission power (power radiated by the antenna) and the necessities of electronic component for RF front end, amplifier, filter etc.
- Robustness: Reliability of data communication despite interference, small-scale fading, and shadowing is required so that high quality of service (e.g., with respect to delay and outage) can be guaranteed.
- Variable data rate: UWB provides variable data rate although the required data rate for sensor networks is not as high as multimedia transmissions, low data rate is adequate for simple applications while some other applications require moderate data rates.
- Heterogeneous networking: Most sensor networks are heterogeneous, i.e., there are nodes with different capabilities and requirements. In a typical heterogeneous network, clusters are formed around some more capable nodes, usually selected as cluster head (CH), which are responsible for communicating with the sink and the low capability nodes which perform the data collection task, are only responsible for forwarding data to the CH.

3. WSN physical layer and feasibility of UWB

In 2004, the IEEE established the standardization group IEEE 802.15.4a, with the mandate to develop a new physical layer (PHY) for applications such as sensor networks. This UWB PHY provides variable data rates such as: 110 kb/s, 1.70 Mb/s, 6.81 Mb/s, 27.24 Mb/s. In 2005 Reed reported that UWB technology could deliver a large amount of data with low power spectral density (PSD) due to the modulation of extremely narrow pulses. The brief duration of UWB pulses spreads their energy across a wide range of frequencies from near DC level to several GHz. By spreading the energy, UWB signal shares the frequency spectrum with existing radio services. Figure 3.1 illustrates the overlay of UWB devices with some existing radio services, based on the FCC approved emission limits for UWB signals. The UWB signal can be seen as random noise to the IEEE 802.11 WLAN signal whose bandwidth is 22 MHz. The bandwidth of the WLAN interference signal is only a small fraction of the UWB signal bandwidth that means UWB system has robust noise performance. The transmitted average power of the UWB signal is extremely low. Therefore the WLAN and WPAN systems can coexist in the same 2.4 GHz ISM band. Recently, most wireless sensor networks relied upon narrowband transmission schemes such as direct sequence or frequency hopping along with multiple access techniques. Compared to narrowband systems, UWB has several advantages. UWB spreads the transmit signal over a very large bandwidth (typically 500 MHz or more). Due to the combination of wide bandwidth and low power, UWB signals have a low probability of detection facility. Additionally, the wide bandwidth gives UWB excellent immunity to interference from narrowband systems as well as from multi-path effects. FCC regulations limit UWB devices to low average power in order to minimize interference that enables UWB coexists with narrowband systems.

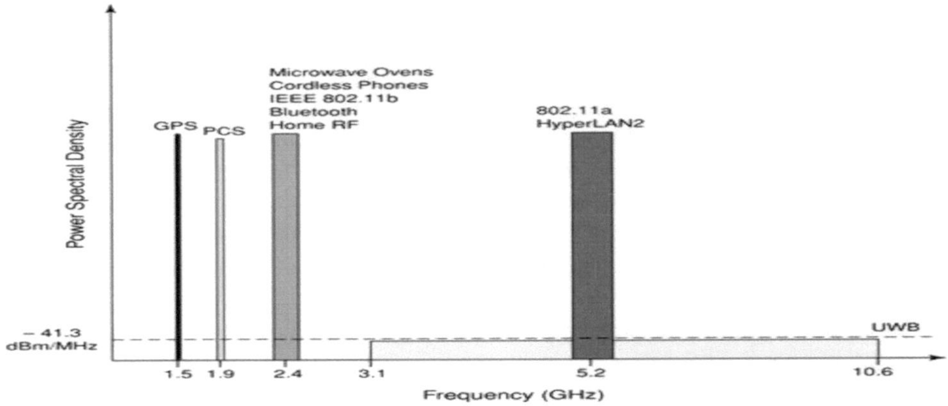

Fig. 3.1. UWB spectrums with some existing radio services (Source: IEEE 802.15).

The PHY is an essential component in any computer network. It is generally used for data transmission and reception, channel sensing, link quality determination, channel selection etc. The UWB PHY was specifically designed to provide enhanced robustness for LR-WPAN applications like WSN. The IEEE 802.15 LR-WPAN specification (2007) is designed to provide robust performance in data and communication system while leveraging the unique capability of UWB waveforms to support precision ranging between devices. The UWB PHY design is intended to make use of the wide bands of spectrum available for UWB operation around the world. The LR-WPANs can operate in multiple independent license-free bands and can be implemented in a single band or multiple bands. In August 2007, IEEE 802.15.4a was released expanding the four PHYs available in the earlier 2006 version, including one PHY using Direct Sequence UWB and another using Chirp Spread Spectrum (CSS). The UWB PHY is allocated frequencies in three bands e.g. below 1 GHz, between 3 to 5 GHz, and between 6 to 10 GHz. The CSS PHY is allocated spectrum in the 2450 MHz ISM band. This standard defines the protocol and compatible interconnection for data communication devices using low data rate, low power and low complexity as well as short-range radio frequency (RF) transmissions within the WPAN. DS-UWB is spectrally efficient that has precision ranging capability. The CSS PHY was added to the standard because it supports communications to devices moving at high speeds and at longer ranges than any of the other PHYs in the IEEE 802.15.4 standard. Basically both new PHYs added scalability to data rates, longer ranges, and lower power consumption into the standard and hence meet the intent of the IEEE 802.15 standard to emphasize very low cost communication system. Table 3.1 represents the standards and technology trend of WPAN technologies.

In 802.15.4a, the UWB PHY layer, which includes modulation, coding, and multiple access schemes, has been designed to achieve optimum performance for WSN applications. At present, Zigbee technology is used as a communication standard for wireless personal area networks like sensor network. UWB technology is more suitable for WSN because it is recommended by the IEEE 802.15.4 standard of Low-Rate Wireless Personal Area Networks (LR-WPANs) that provides low complexity, low cost and low power wireless connectivity among inexpensive devices. The IEEE 802.15.4 specifications according to its upper layers were developed under the ZigBee alliance (a consortium formed in 2002). This standard deals with two PHYs i.e. 868/915 MHz PHY and 2450 MHz PHY where both use the DSSS modulation scheme. This communication standard will be tailored for low power, low data rate, secure wireless communication in the US and European ISM bands. The IEEE 802.15.4/Zigbee technology is specified with a wide range of low-power features at physical and higher levels. The operational power-saving features include low duty cycle operation together with strict power management and low transmission overhead. The implementation of standard-compliant radio-on-a-chip is mainly governed by the PHY specification. The main parameters of the IEEE 802.15.4 PHY are summarized in IEEE 802.15.4a standard (2007). The 2.4 GHz PHY of the IEEE 802.15.4 standard attracts a lot of focus from the wireless industry because the globally available 2.4 GHz ISM band with the largest bandwidth promotes world wide market and flexibility of application design. The IEEE 802.15.4a Task Group has developed an UWB based PHY standard for short-range networks with a precision ranging capability that is optimized for low data rate application. Therefore, comparing to narrow band signals, UWB signal has the advantage of high data throughput, fine range resolution that enables location-aware wireless networking. Moreover, UWB communication system is inherently secure. Since the power density of UWB signals is usually below the environmental noise due to FCC emission limit (i.e. -41 dBm) and with DSSS, signal energy becomes very low which facilitates low probability of detection as well as interference with other radio operating in the same frequency band is negligible. UWB impulse radio is carrier less, so it has only base band processing and no intermediate frequency (IF) processing. This makes impulse radio devices much cheaper than other communication devices.

WPAN (IEEE)	Technology	Data rate	Distance
IEEE 802.15.1	Bluetooth	1 Mbps	10m (Class 3) 100m (Class 1)
IEEE 802.15.2	Coexistence Mechanisms between WLAN and WPAN		
IEEE 802.15.3	High Rate WPAN (UWB)	22, 33, 44, 55 Mbps	30-50m
IEEE 802.15.3a	Alternate 15.3 PHY	>100 Mbps	10m
IEEE 802.15.4	Low Rate WPAN(ZigBee)	250 Kbps	1-100 m
IEEE 802.15. 4a	Low Rate Alternative PHY of 802.15.4 (UWB)	5 Mbps	<1000 m
IEEE 802.15.4b	Revisions and Enhancements IEEE 802.15.4		

Table 3.1. Standards and technology trend of WPAN

4. UWB system design

In recent years, there are a number of implementations of UWB system, such as impulse radio approaches using pulse-position modulation (PPM), binary phase-shift keying (BPSK) modulation, pulse-amplitude modulation (PAM) as well as frequency based approaches using pulsed orthogonal frequency division multiplexing (OFDM) etc. The UWB system can be implemented in a carrier less fashion due to the absence of modulating carrier frequency while conventional narrowband and wideband systems use RF carriers to move the signal from base band to the actual carrier frequency where the system is allowed to operate. So the data transmission as digital pulses substantially simplifies the transceiver circuitry as compared to a traditional RF radio system. The proposed DS-UWB transmitter is shown in Figure 4.2 and the receiver in Figure 4.3, which is simplified greatly as a simple energy detection method reported by Azim et al (2008).

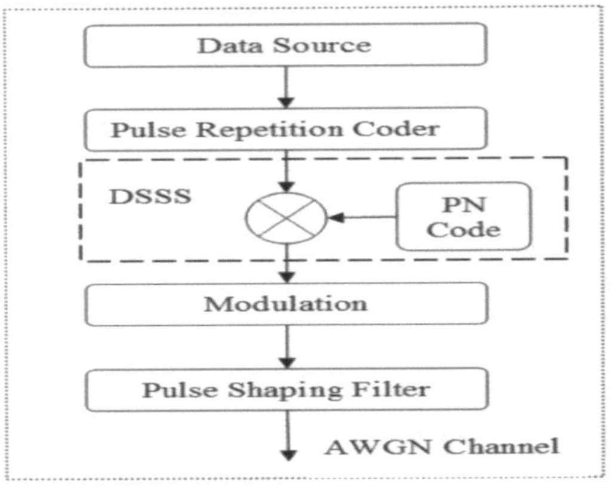

Fig. 4.2. DS-UWB transmitter.

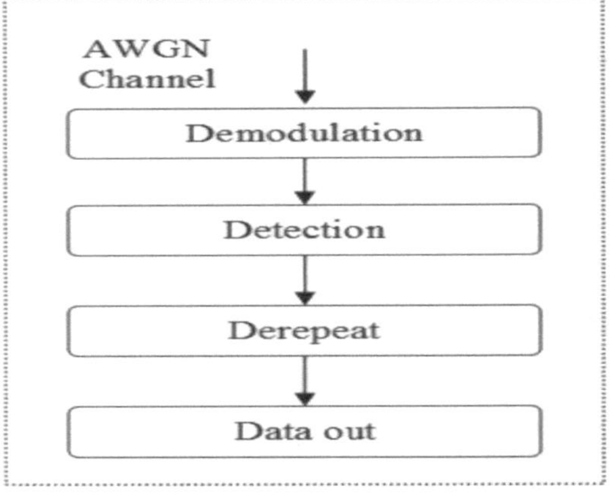

Fig. 4.3. DS-UWB energy detection receiver.

4.1 UWB signal

The FCC rules provide the following definitions for UWB operation:

a. *UWB bandwidth*: The UWB bandwidth (3.244-4.742 GHz or 5.944-10.234 GHz) is the frequency band bounded by the points that are 10 dB below the highest radiated emission.
b. *Center frequency*: The center frequency, f_C, equals $(f_H + f_L)/2$.
c. *Fractional bandwidth*: The fractional bandwidth equals $2(f_H - f_L) / (f_H + f_L)$ and fractional bandwidth equal to or greater than 0.20 and occupies more than 500 MHz of spectrum.
d. *EIRP*: Equivalent isotropic radiated power not greater than -41 dBm or 560 micro-watts.

To generate the UWB signal based on the FCC rules, the following parameter shown in Table 4.1 is considered and simulation performs using matlab.

SN	Parameter	Value	Unit
1.	Pulse width (Pw)	1.3358e-009	Sec
2.	Pulse repetition period (Tp)	2.0035e-009	Sec
	Pulse repetition frequency, Tf = 1/Tp	4.9913e+008	Hz
3.	Lower frequency(fl), Fl = fc-(1/pw)	3.2444e+009	Hz
4.	Higher frequency(fh), Fh = fc+(1/pw)	4.7416e+009	Hz
5.	Centre frequency(fc)=(fh+fl)/2	3.9930e+009	Hz
6.	Bandwidth(Bw) = fh-fl	1.4972e+009	Hz
7.	Fractional bandwidth(fb)=2(fh-fl)/(fh+fl)	0.3750	----
8.	Sampling frequency(fs)=fc*3	1.1979e+010	Hz
9.	Duty cycle (dc) =Pw/Tp*100	66.6733	----
10.	Number of pulse per bit (Ns)	4	----
11.	Length of PN code or chip rate	11	----
12.	Eb/No	5	dB
13.	Coding rate (N = fs/f_chip)	1.0890e+009	Hz/Sec
14.	Average transmitted power in dBm	-30	dBm
15.	Average transmitted power in watt	1.0000e-006	Watt
16.	Energy per pulse (Ex)	2.0035e-015	Jule
17.	Shaping factor	4.0000e-010	Unit less
18.	Pulse Energy after shape (E)	3.3259e-011	Jule
19.	Energy normalization factor (E^0.5)	5.2922e-006	Unit less
20.	Distance (d)	1000	Meter

Table 4.1. Parameters value of UWB system.

Bandwidth, bit rate and symbol rate: Bit rate, $R_b = (N_s/P_w)$, where N_s = Number of pulse per bit and Symbol rate, $T_s = (1/N_sT_p)$ where T_p = pulse repetition period. In the binary case $B = R_b = T_s$ and Bandwidth efficiency can be calculated as R_b/B bits per cycle. Effective bandwidth $B_s = (R_b/N)$, where $N = Log_2M$ (where $M = 2$ in the binary case), in a pulse transmission, signal bandwidth for 90% signal power $B_s = Rb$ Hz and for 95% signal power, $B_s = 2/R_b$ Hz. So depending on the system requirement reliability and accuracy can be achieved while compensating the bit rate and bandwidth utilization.

Capacity: In UWB system, the signal is directly modulated as impulses with a very sharp rise and fall time, thus resulting in a waveform that occupies several GHz of bandwidth. Shannon's capacity formula shows capacity increasing as a function of bandwidth faster than the SNR (signal to noise ratio), $C = Bw \times \log_2(1+SNR)$. Where, C is the channel capacity (bits/sec) and B_w is the channel bandwidth in Hz.

Duty cycle (dc): According to FCC rules, UWB signals consist of very short pulses of energy separated in times but an amount much larger than the length of the pulse. This means that the duty cycle is very low and hence provides low power consumption. The pulse width (or, pulse duration) P_w is much shorter than the pulse repetition time T_p. So duty cycle, $dc = Pw/Tp \times 100$. This means if the duration of pulse increases, the duty cycle decreases and vice versa.

4.2 Transmitter

The first step is to design an information source. Here, we consider the bits "*10110010*". Secondly, A repetition code represents the simplest type of linear block code where a single message bit is encoded into a block of identical n bits, producing $(n, 1)$ block code. This is exercised by means of forward error correction method and acts as a channel coder. Generally the channel coder accepts message bits and adds redundancy according to prescribed rule and exploits the redundancy to decide which message bit was actually transmitted at the receiver end. In our simulation model, repeat bit is the channel encoder and de-repeat bit is the channel de-coder. The goal is to minimize the effect of channel noise. In UWB system we emphasize on reliable transmission than the bandwidth utilization, since we have more bandwidth as required and sensor network needs lower data rates. Therefore, we can adjust the pulse repetition frequency (T_f) to control data rates as required as well as to increase the number of pulse per bit (N_s) provides a lower bit rate, while the redundant pulses improve the processing gain.

Spread spectrum and modulation: The spread spectrum is a means of transmission in which the data of interest occupies a bandwidth in excess of the minimum bandwidth necessary to send the data. In DSSS technique two stage modulations are used reported by Haykin (2006). First the incoming data sequence is used to modulate a wideband code that transform the narrowband data sequence into a noise like wideband signal. The spectrum spreading is accomplished before transmission by using a code sequence that is independent to the data sequence. Usually the same code (PN code) is used in the receiver to de-spread the received signal. But in our system only the length of code is used to estimate the transmitted bits. PN code or chip code spreading the signal bandwidth and its time duration, T_c is called chip interval. So, chip rate, R_c = $1/T_c$ and corresponds to the bandwidth B_w of the transmitter signal is used to make wideband signal. Where B_w>>B_s and B_s is the signal bandwidth. The PN code usually ± 1's sequence is generated at a rate R_c = N_s/T_p bits/s. The ratio of the bit interval P_w to the chip interval (T_c) is usually selected to be an integer in practical spread spectrum system. We consider single user point-to-point communication but in multi-user case DS-CDMA can be used. In our system PN sequence is [-1 -1 1 1 -1 1 1 -1 1 1 1] generated by randsrc (f_chip,1,[1,-1; .6,.4]), where f_chip is the length of PN code. Figure 4.4 shows signal amplitude after spreading.

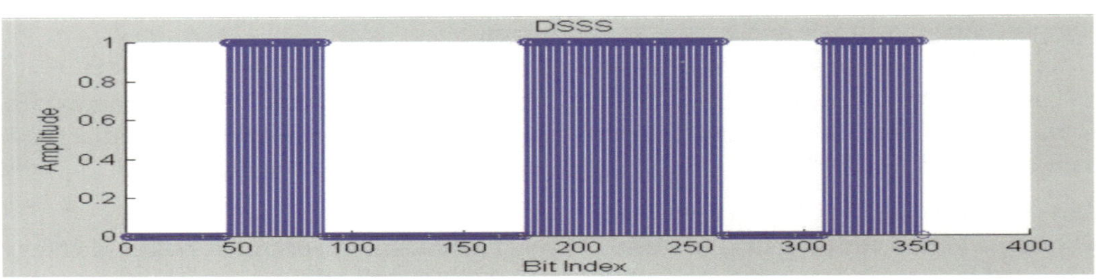

Fig. 4.4. Spreading signal amplitude.

Due to the PN code having a higher rate than the information signal, there will be several chips representing a single information symbol. This adds redundancy to the signal and employs a processing gain due to the increase in the signal bandwidth. It facilitates to resist interference effects and enable secure communication in a hostile environment such that the transmitted signal cannot be easily detected or recognized by unwanted listeners. We consider single user, point-to-point UWB operation. But for multiple users, spread spectrum can be used as a multiple-access communication system where a number of independent users are required to share a common channel without an external synchronizing mechanism. Here DSSS technique is used prior with modulation, which greatly reduced the noise sensitivity (i.e. noise immunity). Spreading creates a lower power spectral density than the original signal; however the total transmitted power remains the same. This allows the SNR of the signal to be below the noise floor level. It has several advantages for the system, as the signal will be less likely to interfere with other users on the same spectrum. Also other unauthorized users are unable to detect the signal, as the signal amplitude will appear as a slight increase in noise, so adds security to the system.

Modulation format: In this UWB system lower order modulation format is used for the transmission of sensor information. Table 4.2 shows the BPSK and PAM modulation format discussed by Haykin (2006).

Polarity of data sequence b(t) at time t		+		-	
		PSK	PAM	PSK	PAM
Polarity of PN sequence c(t) at time t	+	0	1	π	-1
	-	π	-1	0	1

Table 4.2. BPSK and BPAM modulation format.

Pulse shaping: The choice of the pulse is critical as its impulse response affects the PSD of the transmitted signal. Zeng (2005) has proposed several UWB pulse shapes where Gaussian pulse is more suitable for UWB transmission. To increase the derivative of the pulse, the relative bandwidth decreases while the center frequency increases for a fixed value of pulse width. The Nth order Gaussian pulse can be generated by $p(t) = A\frac{(2\pi f\sigma)^{10} e^{-(2\pi f\sigma)^2}}{n^n e^{-n}}$ and Figure 4.5 shown different pulse shapes. We used Gaussian doublet (2nd order Gaussian pulse) because it is the most currently adopted pulse that meet the appropriate UWB operation with regulation explained by Benedetto and Giancola (2004), which is usually generated by the equation. $p(t) = (1 - 4\pi\frac{t^2}{pw^2})e^{-\frac{2\pi t^2}{pw^2}}$

Here *p(t)* is a Gaussian pulse (Gaussian doublet) where pulse duration or width is much smaller than pulse repetition period, i.e. $T_p >> P_w$, so it can produce low duty cycle operation.

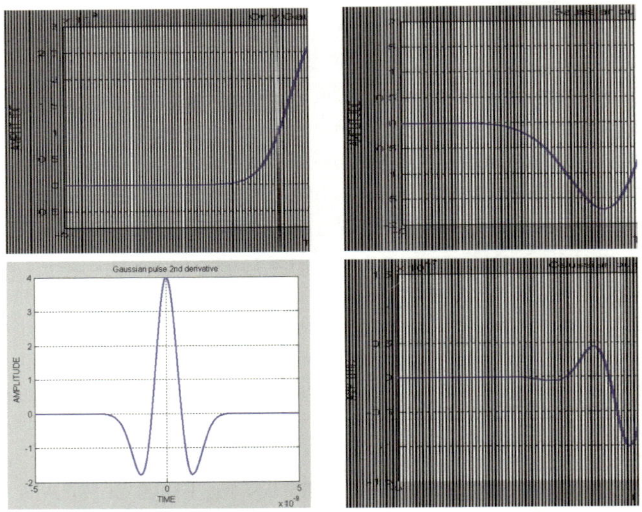

Fig. 4.5. Gaussian pulse shape.

The output of the modulator enters the pulse shaper filter, which acts as a low pass filter and after convolution operation between the modulated data and Gaussian pulse. Signal amplitude is shown for BPSK and BPAM in Figure 4.6 and transmitted pulse after shaping is shown in Figure 4.7.

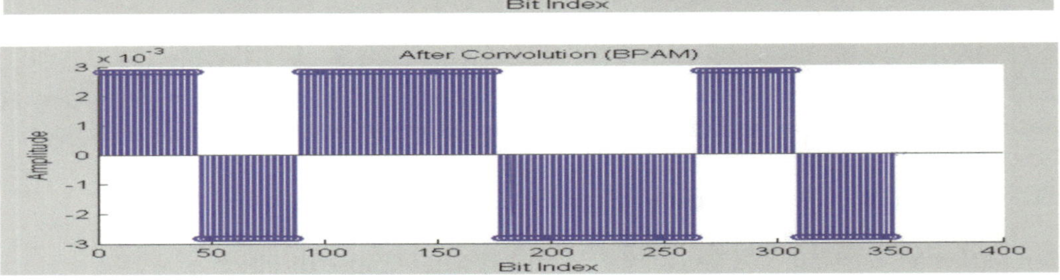

Fig. 4.6. Transmitted signal amplitude (BPSK & BPAM).

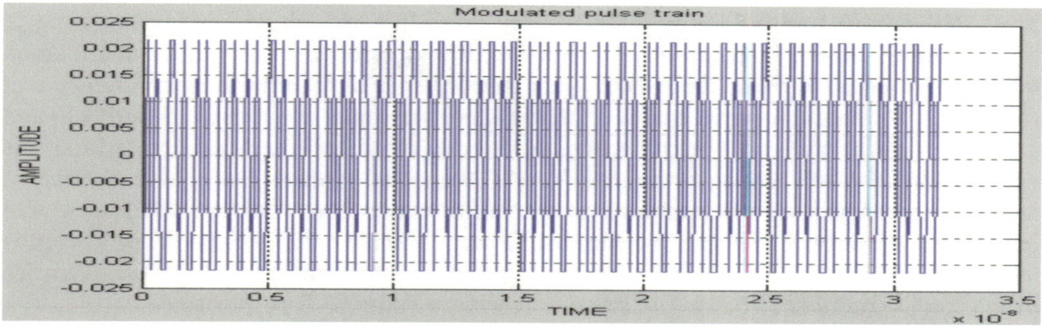

Fig. 4.7. Transmitted pulse train after shaping.

4.3 Channel

The UWB radio signal is ideally composed of a sequence of pulses that do not overlap in time. Each pulse is confined within a specific time interval and the pulse itself has finite duration. The received signal can be expressed as $r(t) = s(t)+n(t),\ 0 \le t \le T$. where $n(t)$ denotes a sample function of the additive white Gaussian noise (AWGN) process with power spectral density of $N_0/2$ W/Hz. Here single user point-to-point communication system is considered with the absence of inter symbol interference (ISI) and multi-user interference (MUI) phenomenon. Figures 4.8 and 4.9 show the channel output of BPSK and BPAM respectively.

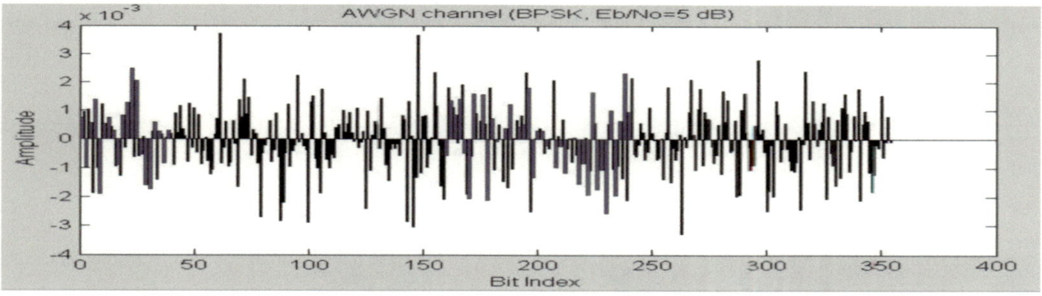

Fig. 4.8. AWGN channel output (BPSK), where Eb/No=5 dB.

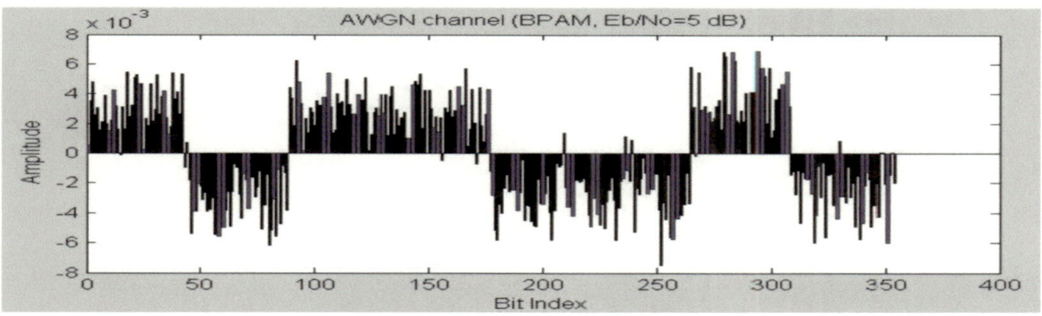

Fig. 4.9. AWGN channel output (BPAM), where Eb/No=5 dB.

The BPSK output shown in Figure 4.8 is more noise like and undetectable comparing to BPAM output shown in Figure 4.9. The probability of error depends on the modulation scheme and Signal to Noise Ratio (SNR). The performance of the impulse radio signal over the AWGN channel can be realized with the BER performances as shown in Figure 4.10 and 4.11, where number of pulse per bit is one and four, while different modulation technique is used. In the DS-UWB propagation through AWGN channel, transmitted pulses are delayed and attenuated due to thermal noise, but multi path effect, ISI and MUI were not considered. Here by increasing the number of pulses per bit (N_s), the received energy is increased by a factor N_s, without increasing the average transmitted power (P_{av}). To increasing the number of pulses per bit we can achieve better SNR performance.

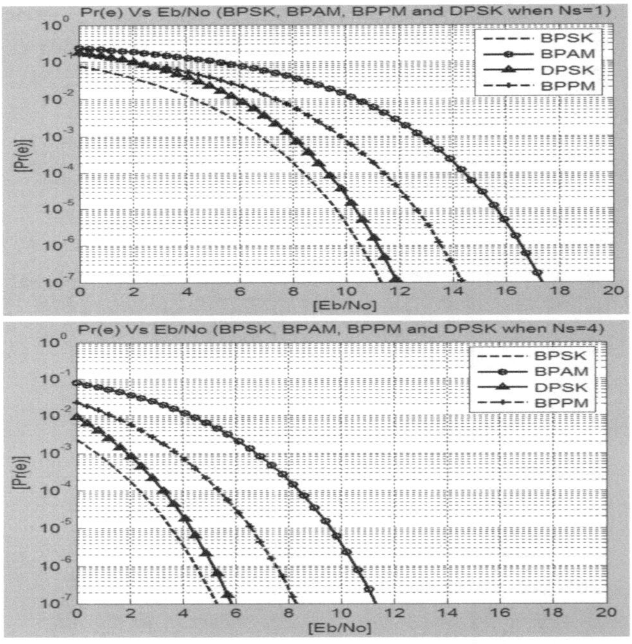

Fig. 4.10. BER performance BPSK, BPAM, DPSK, BPPM (*Ns=1, 4*).

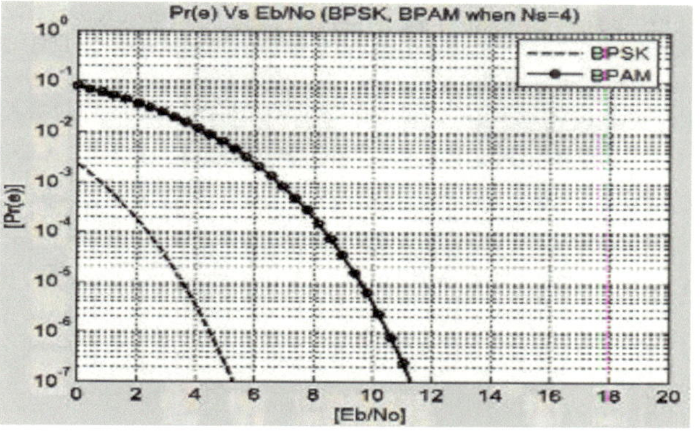

Fig. 4.11. BER performance BPSK, BPAM (*Ns*=4).

4.4 Receiver

At the receiver shown in Figure 4.3, de-modulation operation is performed with the noisy signal. The constellation diagram is shown in Figure 4.12 and the signal after demodulation is shown in Figure 4.13. The received signal is successfully recovered by using an energy detection method. A sample of matlab code for detection is shown in Figure 4.14.

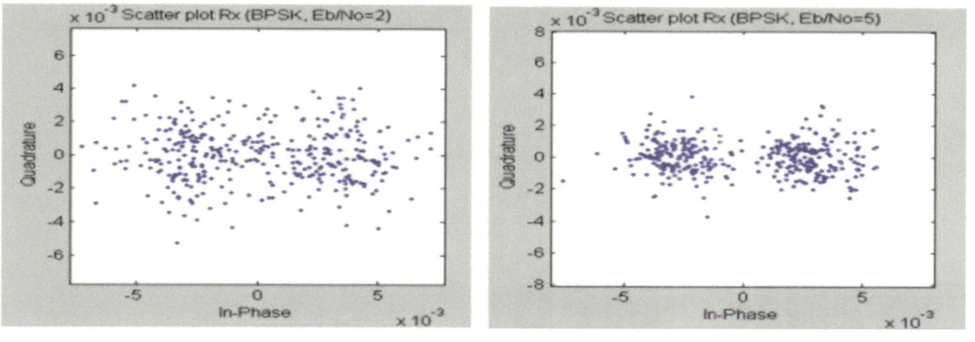

Fig. 4.12. Received signal constellation (BPSK, Eb/No=2, 5)

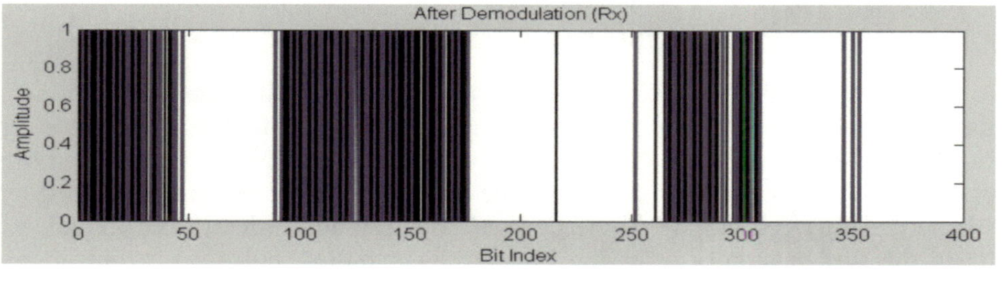

Fig. 4.13. Received signal amplitude after demodulation.

The decision is obtained by applying a simple majority criterion. Given the number of pulses falling over a threshold and comparing this number with the number of pulses falling below the same threshold, the estimated bit corresponds to the higher of these two numbers. An error occurs if more than half of the pulses are misinterpreted. So this decision factor achieves accurate reception and by increasing the number of pulses per bit provides more efficiency. The length of PN code (f_chip) is used to correlate with the received bits after demodulation while f_chip/2 decision metrics provides the estimated repeat bits at the receiver shown in Figure 4.15. Finally $N_s/2$ decision threshold facilitates to recover bits in the de-repetition process, which are compared to the transmitted bits for error estimation. For large number of transmitted data, no error is found as shown successfully by the simulation results.

```
%Detection
 j=1;
 for i = 1:length(tx_repbits)
   sum=0;
   for k = j:j+f_chip
     sum=sum+rx_demod(k);
   end;
   if (sum >=f_chip/2)
     msg_rep(i)=1;
   else
     msg_rep(i)=0;
   end;
 j = f_chip*i+1;
 end;
```

Fig. 4.14. Detection code.

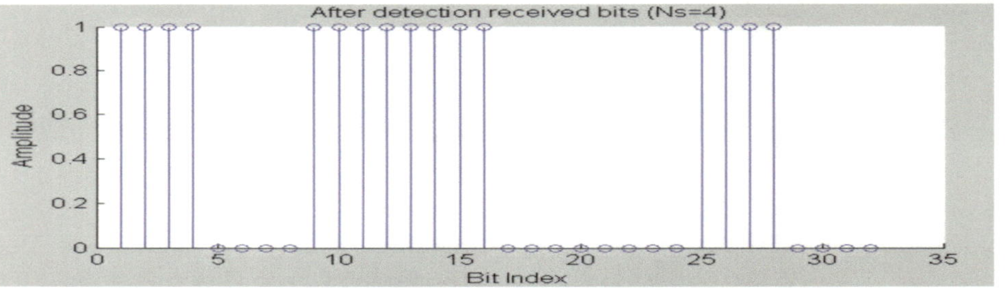

Fig. 4.15. Output after detection (*10110010*), *Ns* = *4*.

The proposed transceiver model is efficient and ensures reliable transmission, so it is suitable for sensor network communication system. Here, by increasing the number of pulses per bit (*Ns*), the received energy is increased by a factor Ns, without increasing the average transmitted power but at the same time compensating the bit rate of dividing by *Ns*. Data is successfully recovered by energy detection technique (detect and avoid), which facilitates the design simplicity at the receiver by avoiding pulse synchronization and coherent detection. Moreover having 50% of data corruption during the propagation, the system still recovers the bit stream accurately (Ns/2, bit=8, Tx bit=8×4, Sum> Ns/2) . Also power emission and consumption are very low .(Power = 794 μW and Energy per pulse = 280 nW).So it's a noise like signal, which is difficult to detect by unwanted user and immune to interference with other existing radio operating in the same band.

5. Summary

UWB technology is feasible for the implementation of sensor networks as it offers high robustness to interference and provides low complexity receivers and transmitters with low energy consumption. The IEEE 802.15.4a standard enables UWB-based sensor networks, which offer a high degree of flexibility and includes modulation, coding, and multiple access schemes that permit non-coherent receiver design. The specification for UWB LR-WPAN devices incorporates a number of optional enhancements to potentially improve performance, reduce power consumption and enhance coexistence characteristics. In particular, DS-UWB is a suitable communication platform for wireless sensor networks where accuracy and reliability is more important factor than bandwidth utilization. Due to the ability of noise immunity and low probability of detection and interference rejection, DS-UWB is a good choice for wireless sensor networks. Pictorial signal behavior shown in the simulation process helps to realize the above-mentioned facts. The UWB information rates as a function of transmission distance over AWGN and other channels can be considered for further development. Moreover, in future, multiple access interference on transceiver design can be investigated in a multi user environment. It might be interesting to explore the coding-spreading tradeoffs, channel estimation and design of optimum transceiver architecture.

6. References

Allen, B. (2004). Ultra wideband wireless sensor networks. IEE Seminar on Ultra Wideband Communications Technologies and System Design, King's College, London. Pp: 35-36

Azim M A, et al., (2008). *Direct Sequence Ultra Wideband System Design for Wireless Sensor Network.* Proceedings of the International Conference on Computer and Communication Engineering (ICCCE'08). Kuala Lumpur, Malaysia. Pp: 1136 to 1140

Azim M A, et al., (2008). *Development of Low-cost Sensor Interface for Wireless Sensor Network Monitoring Application.* 5th International Conference on Information Technology and Applications (ICITA 2008), 23 - 26 June 2008, Cairns, Queensland, AUSTRALIA.

Benedetto, M. D. and Giancola, G. (2004). *Understanding ultra wide band radio fundamentals.* Prentice Hall. Communications Engineering and Emerging Technologies Series. Pp: 121-234

Haykin, S. (2006). *Digital communications.* John Wiley & Sons, Inc. New York, NY, USA. Page 445 to 471

IEEE802.15.4 specifications. (2003). Online article, Retrieved June 22, 2006, from http://www.ieee802.org/15/pub/TG4.html

IEEE 802.15.4a. (2007). *IEEE Standard for PART 15.4: Wireless MAC and PHY Specifications for Low-Rate Wireless Personal Area Networks (LR-WPANs):* Amendment 1: Add Alternate PHY. Retrieved July 2, 2007, from

http://standards.ieee.org/getieee802/download/802.15.4a-2007.pdf
Oppermann, I., Hamalainen, M., and Iinatti, J. (2004). *UWB theory and applications*. Wiley Press.
Reed, J. H. (2005). An introduction to Ultra wideband communication systems. Prentice Hall.
Zeng, D. (2005). *Pulse Shaping Filter Design and Interference Analysis in UWB Communication Systems.* Dissertation Submitted to the Department of Electrical and Computer Engineering, Virginia Polytechnic Institute and State University.
Zhang J, et al (2009). *UWB Systems for Wireless Sensor Networks.* Research article by Mitsubishi Electric Research Laboratories. Available online at http://www.merl.com

Green Femtocell Based on UWB Technologies

Moshe Ran[1] and Yossef Ben Ezra[2]
[1]*MostlyTek Ltd. 58 Keshet St., Reut*
[2]*H.I.T - Holon Institute of Technology, Holon,*
Israel

1. Introduction

The rapid evolution of mobile communications through four generations of mobile communication, envisages the operation at 100Mb/s for mobile users and at 1Gb/s for stationary applications in the near future. The tremendous increase of data rates must be considered in the context of four decades of the mobile cellular technologies progress since its first introduction by the Nippon Telephone and Telegraph Company (NTT) in the late 70's Rappaport (2002). On the other hand, fixed wireless communications are already available to provide over 300 Mbps raw data rates through wireless local area networks (LAN) protocols as 802.11n, and over 1Gbps through Ultra Wideband (UWB) in wireless personal area networks (PAN), see (ECMA-368), (ECMA-387).

With the introduction of the femtocell concept Zhang (2010), new opportunities have been opened for approaching the 4G mobile vision through fixed mobile convergence (FMC). Femtocell Access Point (FAP), are low power access points that connect mobile terminal to the mobile core network using wired broadband or fixed broadband wireless technologies. The FAP provides viable opportunities for mobile operators, to meet the indoor coverage challenges for most demanding applications at low cost.

We propose a novel concept of 4G femtocell, denoted a "Green Femtocell", and high level network architecture to support the new paradigm of FMC, in which convergence of 4G cellular with short-range wireless and wired are realized. The proposed approach paves the way of green framework in which increase by x100 in energy efficiency and x100 reduction of human exposure to wireless radiation become feasible.

Our approach relies on radio-over-fiber and all-optical solutions that can already be considered "green" in offering reduced energy consumption to alternative wireless access solutions, see CELTIC Purple Book (2011). The new concept is based on the following novel technical and business entities (Fig. 1):

- We introduce a green remote Home Access Node (HAN) that relays range of radio protocols, including UWB, WLAN, LTE-A, and IEEE 802.16m as radio signals over hybrid wireless-fiber media from 1.8 GHz to 10.6 GHz; with strict limitation of radiated power. Wireless radiation for indoor environments is reduced by 2-3 order of magnitudes, while potentially support target **1Gbps end-user data rates, by using dual-mode cellular-UWB** for most common indoor applications. Indoor HAN should support mobile users at distances ranging from 0.3m to 30m over-the-air. For outdoor and longer range indoor topologies, we enable protocol-transparent architectures capable of relaying range of radio protocols from 30 to 300 m.

- Processing for multiple HANs is centralized with optical Multi-cell Base Station (O-McBS) capable of performing parallel multiple input multiple output (MIMO) processing of radio-over-fiber (ROF) links over 100's of GRANs. Unlike the McBS approaches suggested in Foshini (2006) and Gambini (2010), our O-McBS approach involved with **optical MIMO over multi mode fiber (MMF)**. The multicell processing performed at McBS enables clear benefits of centralized approach to interference management over the hybrid wireless-fiber medium and efficient radio resource managements (RRM). We note that MIMO over MMF is a very recent enabling technology that has been shown to attain 400Gb/s signalling rate over several hundreds of meters of MMF at 10^{-10} BER Greenberg (2007). A promising solution for radio signalling and multiple access over hybrid wireless-MMF is based on orthogonal frequency-division multiple access (OFDMA). However, most of the works have addressed only the indoor wireless channels Perez (2009).
- O-McBS are connected through optical femto gate way (O-FemtoGW) to core network through Tb/s optical links. O-FemtoGW multiplex data from 10's of O-McBS, and forms through all-optical real-time processing an optical OFDM (O-OFDM) signal carrying 100Gb/s. Recently, works on all optical FFT schemes to implement efficiently O-OFDM to enable 1 Tb/s have been published by Hillerkuss (2010). Recent survey on O-OFDM with MIMO can be found in Shieh (2010), and general aspects of O-OFDM in Armstrong (2009) and Gidding (2009).

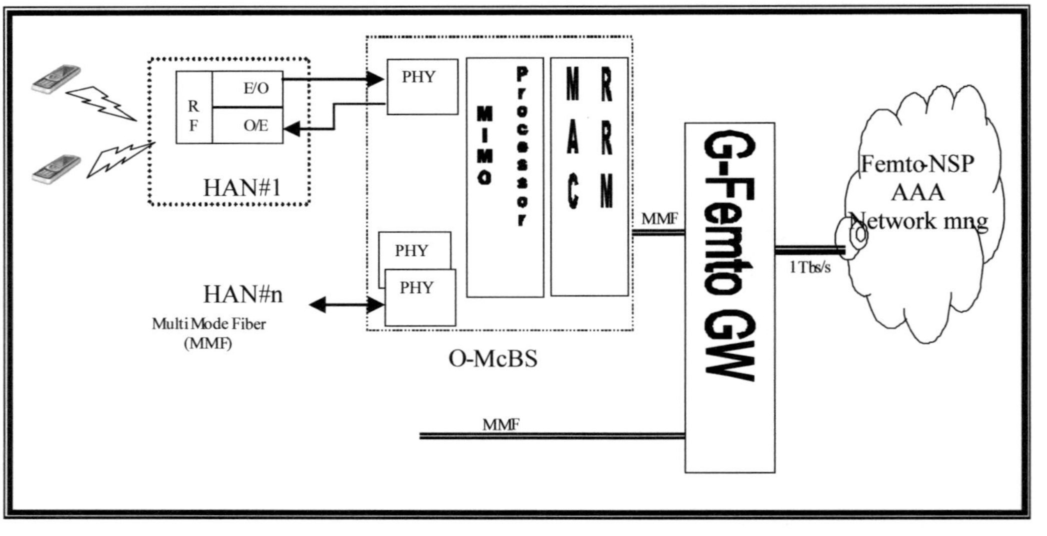

NSP: Network service provider
O-McBS: optical multicell BS
MAC: Medium access control
O/E: Optical/Electrical converter
MMF: Multimode Fiber
RRM: radio resource management
E/O: Electrical/Optical converter

Fig. 1. Green Femtocell Access Network (high-level architecture).

The proposed architecture leverages and extends the concepts and technologies of UWB radio over optical fiber (UROOF) Ran (2010a, b), Ben-Ezra (2010), and further investigated in the context of future mobile technologies in Ran (2009) and Altman (2010). Our technical approach is directed to solve the crucial problem of interference management in local area environments in femtocells deployment. The femto-to-macro co-channel interference is solved by allocating unlicensed frequencies for indoor (e.g., UWB, WLAN). The femto-to-femto interference is addressed through centralized approach within the O-McBS.

In this chapter we focus on the local area coverage through HAN and elaborate possible indoor and outdoor architectures that can support 4G femto mobile vision in the near future. The chapter is organized into the following sections. In Section 2, we review state of the art of femtocell technology. In section 3, we review the very-low radiation distributed antenna system (VLR-DAS) concept. We elaborate indoor architectures and performances of green femtocells, and further extend the discussion to outdoor architectures in section 4. In section 5, we provide interference analysis for single cell and multiple cell scenarios and discuss mitigation techniques to enable co-existence with other systems. Theoretical and experimental investigation is provided in Section 6. Conclusions are presented in Section 7.

2. Femtocells technologies

The use of densely deployed many low-power, low cost and high performance base stations, e.g., FAPs, seems a promising approach to cope with the ever-growing indoor coverage demand. Since 70% of voice and more than 90% of data services occur indoors Roche (2010), it becomes promising to deploy in-house FAP, which connects standard mobile devices to a mobile operator's through existing broadband Internet connection. According to a recent market prediction, Jarich (2010), leading femtocells vendors expect 2013 to be the year in which LTE and possibly WiMax femtocells will be commercially deployed. **Femtocells are potentially industry-changing disruptive shift in technology for radio access in cellular networks.** Femtocells should provide small scale functions of BS, and also certain functions of the customer premises equipment (CPE). Therefore, to gain benefits for network operators and consumer, the interaction between femto and macro radio layers should be carefully managed.

2.1 State of the art of femtocells

FAP's can be classified into two categories: home FAP, also called home base station (HBS), or enterprise FAP. HBS typically supports 2- 8 simultaneous users, where enterprise FAP can support 4 – 16 users, and can easily configured in cascade architecture to support 32- or 64-simultaneous users. FAP can further be classified according their underlined cellular technologies: GSM -, UMTS -, LTE-, WiMax -, FAP etc.

Femtocells can be configured in three ways to restrict and control their usage by certain users. In **open access** (or public access) all users, including outdoor users, or neighbouring femtocells are able to make use of nearby femtocells. In **closed access** (or private access) only a list of registered users are allowed to access a femtocell. In **Hybrid access** nonsubscribers use only limited amount of the femtocell resources, as e.g., emergency call services.

Open access benefits outdoor users, who are making use of nearby indoor femtocells and thus clearly improve the overall capacity of the network. From interference point of view, open access is superior to closed access, since it allows customers to connect to nearest access point. Thus it enables reducing the overall use of system resources (power-frequency-time). Possible drawback to this approach is the increase number of handoffs. Additionally, the femtocell customer pays by himself for the FAP and the broadband Internet connection, and is likely to reject the sharing of his own resources with users passing by his neighbourhood.

Some key technical challenges to large scale femtocell deployment are the following Yongho (2009):

- **Interference to/from other femtocells and macrocell BS.** Massive deployment will pose serious issues on the radio interference management with the surrounding cells (both femtocelss and macrocells). Since femtocells are planned to be installed in an ad-hoc manner by end-users and in large numbers, it will be challenging to do centralized and coordinated radio planning as in legacy macrocell system.
- **Seamless Handover between a femtocell and macrocell or other femtocells.** The conventional broadcast mechanism to advertise neighbour BS information may not be viable and scalable to include information about femtocells due to the excessive overhead needed. In the absence of this information, the macrocell-to-femtocell handover becomes challenging. In particular, handouts (femto to macro) and handin (macro to femto) efficient algorithms are required.
- **Lack of standard solutions for scalability, redundancy and traffic partioning.** For the femtocells to be widely used, it is essential that femto BSs interface with the rest of the network, both control and management planes, be fully standardised. No current guarantee that the fixed broadband connection will prioritize the traffic originating from the FAPs for a service without call blocking or dropping. It is highly desired that implementation of standard solutions be verified for multivendor interoperability.
- **Synchronization and location.** Inter-cell synchronization and femtocell location are critical for proper operation of femtocells, but GPS cannot be used in many indoor cases. Therefore, solutions for timing synchronization and location are needed.

Many aspects of current state of the art of femtocells technologies were published within the special issues of IEEE communications magazine (September 2009 and January 2010). Several research projects within the seventh EU framework program for research and technological development (FP7), and industry-driven research initiative CELTIC-Plus framework (www.celtic-initiative.org) are addressing some key aspects of future femtocells technologies.

HOMESNET project, supported in part by CELTIC (CP6-009), is focused on three key challenges in HBS: dense deployment of self-organizing networks (SON), self-optimization and low radio emissions, Altman (2010). To achieve very low radio emissions in the house or constrained environment like hospitals, an architectural network option for HBS based on radio-over-fiber approach is investigated. **The HOMESNET vision can certainly provide a good starting point for the new paradigm of Green Femtocell concept described in this chapter.**

FREEDOM project (Femtocell-based Network Enhancement by Interference Management and Coordination, see www.ict-freedom.eu) aims at improving the efficiency of networks with massive femtocell deployment. The focus is on addressing the key question: How much the whole system efficiency can be improved by exploiting the available quality of the IP-based backhaul link?

The solutions addressed in FREEDOM include the two main flavours of the 4G femtocell paradigm, namely IEEE 802.16m and LTE-Advanced. In both cases the core concepts investigated are: Interference management and cooperation; dense femtocell-specific RRM; scalability and effectiveness and femtocell-based network planning.

BeFEMTO (Broadband Evolved FEMTO Networks, see www.ict-befemto.eu) is a recent integrated FP-7 project aiming to develop femtocell technologies based on LTE-A. The project is targeting ambitious objectives such as:

- Minimum system spectral efficiency of 8 b/s/Hz/cell.
- A maximum averaged transmit power of less than 10mW for indoor femto nodes.
- Seamless convergence between fixed broadband and mobile cellular systems.

Rocket (Reconfigurable OFDMA-based Cooperative Networks Enabled by Agile Spectrum Use, see www.ict-rocket.eu) is another FP7 project aimed at providing solutions for LTE-A and 802.16m to reach data rates 100Mbps and peak throughput higher than 1Gbps. The technical approach is based on advanced opportunistic spectrum usage, multi-user cooperative transmissions and ultra-efficient MAC design.

2.2 Standardization of Femtocells

Since a femtocell is a small scale cellular BS, it transmits over RF bands using licensed spectrum granted by the appropriate government authority. This requires that mobile operator be responsible for the control of radio transmission in a strict way, and follows regulations and standards. Standardization is certainly important for femtocells market to reach massive deployment. There are several standard development organizations (SDOs) and non-SDO forums that play an important role in the standardization of femto technology.

SDOs:

3GPP (www.3GPP.org) was created in 1998. The 3rd Generation Partnership Project (3GPP) unites 6 telecommunications standards bodies from Asia, Europe and North America. Over 350 companies participate in 3GPP through their membership of one of the 6 partners. The scope of 3GPP is to produce Specifications for a Mobile System based on evolved GSM core networks and the radio access technologies that they support. The femtocell concept applies as modifications for 2G/3G/4G mobile cellular generations described below in TABLE 1. Recent success with the creation of LTE and Systems Architecture Evolution (SAE) Specifications has made 3GPP the focal point for Mobile Broadband systems and a genuine contender as point of convergence for future Specifications for mobile networks. The standardization is defined in series of Technical Specification (TS) and Technical Reports (TR). The work is done through several working groups RAN2-RAN4 and SA1-SA5.

3GPP Radio Interfaces	2G radio: GSM, GPRS, EDGE 3G radio: WCDMA, SSPA, HSPA, LTE 4G radio: LTE Advanced	Rel.99 Rel.4 - 7 Rel. 8 /9 Rel. 10
3GPP Core Network	2G/3G : GSM core network 3G/4G: Evolved packet Core (EPC)	Rel. 8
3GPP Service layer	GSM, IMS, Multimedia Telephony (MMTEL),	Rel. 9

Table 1. 3GPP key releases and areas in which Femtocell concepts apply

Recent reference model for stage 2 UTRAN architecture for 3G Home NodeB (HNB) was finalized in 3GPP RAN3 within **TS 25.467**. The basic elements of Iu-h interface are given in Fig. 2. The HNB connects to the mobile core network through HNB-GW, which acts as a concentrator to aggregate large number of HNB's. The Iu-h interface goes through a security gateway (SeGW) to HNB-GW. The HNB Management System (HMS) is based on TR-069 family of standards and provides authentication of HNBs and access from HNB to HNB-GW.

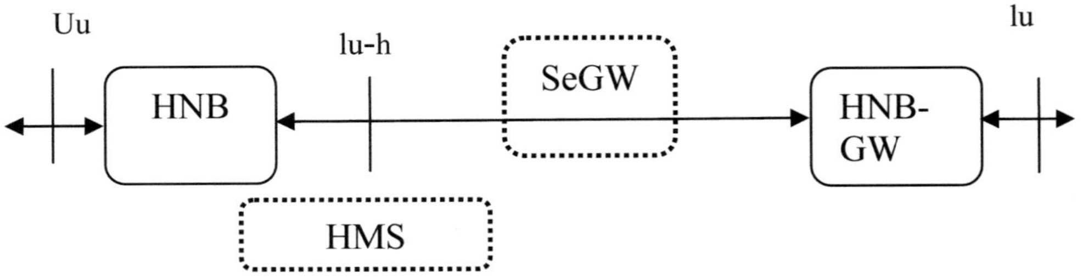

Fig. 2. Iu-h reference point in 3GPP for 3G Home NodeB.

Evolution to IMS/HSPA+/LTE is addressed in **3GPP TR R3.020**. In January 2009, the 3GPP published overall architecture of LTE HeNB architecture within **TS 36.300-870**.
The 4G aspects of 3GPP are studied within the 3GPP approach to address the ITU's IMT-advanced system. Key 4G based on LTE-Advanced features, as defined within 3GPP Rel.10 and are summarized below.

- **Channel BW**. Support for wider bandwidth (up to 100MHz).
- **Downlink transmission** scheme will support data rates of 100Mb/s with high mobility and 1Gb/s with low mobility. It will be improvement to LTE along the evolution path by using 8x8 MIMO.
- **Uplink transmission scheme** supports data rates up to 500Mb/s.
- **Relay functionality.** Improvement to cell edge coverage and more efficient coverage in rural areas.
- Coordinated multiple point transmission and reception (CoMP) in both downlink and uplink
- **LIPA** (local IP Access) & **eHNB** (enhanced HNB) to allow traffic off-load

3GPP2 (www.3GPP2.org) was created in 1999 as a partnership among SDOs from US, Korea and Japan and more recently from China to facilitate the CDMA based radio technologies for mobile cellular evolving from the IS-95 CDMA family of standards. The 3GPP2 architecture for femtocells is heading toward all-IP architecture for voice services based on Session Initiation Management Protocol (SIP), (see RFC3261), and IP multimedia sub-system (IMS) (see TS23.238)
BBF (Broadband Forum, previously DSL Forum, see www.brodbandforum.org) The TR-069 was originally created to manage DSL gateway device, and has been grown over the years for supporting new devices including femtocells. These modifications were published as amendments in (BBF TR-069), (BBF TR-098) and (BBF TR-106). In 2009 the forum published its data model for femtocells (BBF TR-196), supporting interoperability between FAP and network equipment.

Non- SDO bodies and forums relevant to the femtocells paradigm:

The Femto Forum (www.femtoforum.org) has around 100 members from all parts of femtocell industry: major operators, major infrastructure vendors, vendors of components and subsystems. Femto Forum works with SDO and regulators worldwide to provide an aggregated view of femtocells market. The forum is focused on building and maintaining an eco-system that delivers the most commercial and technically efficient solutions based on femtocells. It has now four working groups (WGs): marketing & promotion, radio & physical layer, network & interoperability and regulatory. Recently four special interest groups (SIGs) were founded: LTE, WiMAX, Interoperability and Services.

Femto Forum started discussions on femto architectures with 15 variations early 2008, and soon converged into only one based on Iu-h that has led to the proposal to 3GPP.

WiMAX forum (www.wimaxforum.org) is an industry-led organization that certifies and promotes the compatibility and interoperability of broadband wireless products based upon IEEE Standard 802.16. The forum has hundreds of members, comprising the majority of operators, component vendors and equipment vendors in the communications ecosystem. The WiMAX Forum's primary goal is to accelerate the adoption, deployment and expansion of WiMAX technologies across the globe while facilitating roaming agreements, sharing best practices within our membership and certifying products. WiMAX products are interoperable and support broadband fixed, nomadic, portable and mobile services. WiMAX has two phases for femtocell evolution:

Phase 1 is "femto aware" version based on IEEE802.16-Rev2 network release 1.6 and system profile release 1.0/1.5 with basically no change in the air interface standard to enable basic femtocell deployment. This version was completed late 2010.

Phase 2 "femto enhanced" version is based on network release 2.0; system profile release 2.0 and the air interface defined in 802.16m. This version is expected to be completed by 2012 with target deployments in 2012-2013.

NGMN (next Generation Mobile Networks http://www.ngmn.org) Alliance was founded by leading international mobile network operators in 2006, and joined recently (May 2011) the 3GPP as a market representation member. Its goal is to ensure that the standards for next generation network infrastructure, service platforms and devices will meet the requirements of operators and, ultimately, will satisfy end user demand and expectations.

GreenTouch (www.greentouch.org) is a recently established consortium dedicated to fundamentally transforming communications and data networks, including the Internet, and significantly reducing the carbon footprint of ICT devices, platforms and networks. By 2015, its goal is to deliver the architecture, specifications and roadmap — and demonstrate key components — needed **to increase network energy efficiency by a factor of 1000 from current levels.**

3. Green femtocell and Very-Low Radiation Distributed Antenna System (VLR-DAS)

The basic idea of distributed antenna system (DAS) is replacing an antenna radiating at high power with N small antennas using low-power. Passive DAS use only passive elements as splitters, taps, terminators, circulators, filters and coaxial cables to split the RF signals into N antennas. Active DAS use different active elements (amplifiers, convertors E/O and O/E). Radio-over-fiber (ROF) DAS are most common active DAS techniques currently used.

3.1 Passive DAS

The basic features of Passive DAS are well known Saleh (1987). The benefits of passive DAS using Omni antennas are discussed in Chow (1994). In particularly, the following useful features are evident:

- **Maximizing coverage area.** For a given radiated power, wireless channel with path loss exponent γ, N antennas system will have an increased coverage area over a single antenna system by factor $N^{1-\frac{2}{\gamma}}$.
- **Minimizing the radiated power.** For a given coverage area, an N antenna system will have a reduction in minimum required radiated power in downlink channel by the factor of $N^{1-\frac{\gamma}{2}}$ compared to single antenna system. Furthermore, the maximum radiated power in the uplink channel will be reduced by the factor $N^{-\frac{\gamma}{2}}$.
- **Minimizing maximum path loss.** For a given coverage area, an N antenna system will have reduced maximum path loss by factor of $N^{-\frac{\gamma}{2}}$.
- **Minimizing far-field interference.** For a given coverage area, an N antenna system will have reduced far field interference by factor of $N^{-\frac{\gamma}{2}}$

Thus for example, for N=8 and γ =5 and given coverage area, radiated power reduction of 22.6 (13.5dB) will be achieved for the downlink and 181 (22.6dB) in the uplink compared to a single antenna system. With such deployment, radiation from MS is reduced by 22.6dB and thus the C/I is improved by the same factor.

3.2 Active DAS

A typical active DAS uses master unit, which is the intelligent part of the active DAS, to distribute the RF signals from the BS antenna to multiple expansion units over an optical fiber of lengths up to 6km. Each expansion unit is connected to multiple remote radio units (RRUs) with thin coax, CAT5 or fiber of lengths up to 400m. Unlike passive DAS, active DAS has the ability to automatically compensate for the losses in the system by using internal calibrating signals and amplifiers. Active DAS is the preferred solution for large building and can provide monitoring and alarms in the event of malfunction.

DAS can be combined with MIMO communications concepts by treating the RRU's as a distributed antenna array, see Heath (2010). The multi-user MIMO DAS system model is given in Fig. 3

The model considers R RRUs in each cell, each BS and RRU are equipped with N_t antennas and mobile user with N_r =1 receive antennas. Some aspects of MIMO DAS using beam-forming have been addressed in Li (2009). However, the overall benefits of multiuser MIMO DAS have not been established yet in DAS deployments Heath (2010).

Recently, there has been much research interest on the cooperative (or collocated) antenna system (CAS). It is shown You (2011) that theoretically CAS can alleviate major problems of the cellular systems as intercell interference (ICI) and cell edge effect problem. The performance of CAS depends very much on the degree of cooperation. In one extreme, the CAS consists of several distributed antennas which are connected to a central processing unit. This type of CAS is the DAS case, which has the best performance. In the other extreme, the BTS only exchange limited information, normally in order to boost the performance of the UEs at the edge of the cell. This case is also known as the Coordinated Multiple Point Transmission and Reception (CoMP) system. In this case, the transmission of the data is coordinated in time, frequency or space so as to minimize the inter-cell interferences and propagation loss effects. The CoMP concept for both uplink and downlink is central concept in LTE-A systems defined through 3GPP rel. 10 and future versions.

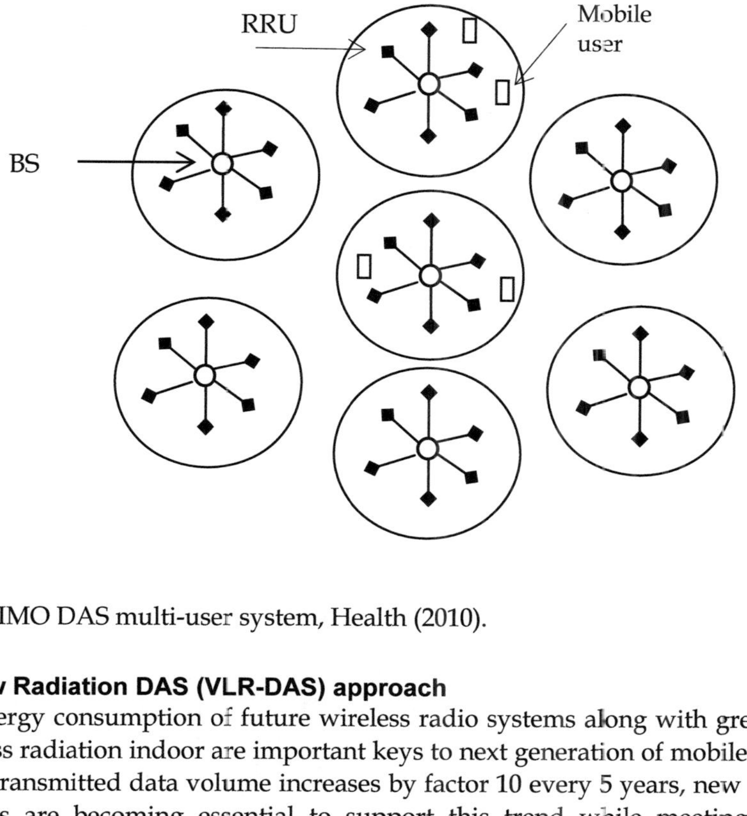

Fig. 3. The MIMO DAS multi-user system, Health (2010).

3.3 Very Low Radiation DAS (VLR-DAS) approach

Lowering energy consumption of future wireless radio systems along with great reduction of the wireless radiation indoor are important keys to next generation of mobile radio access systems. As transmitted data volume increases by factor 10 every 5 years, new technologies and solutions are becoming essential to support this trend while meeting the energy consumption along with the wireless radiation. The VLR-DAS concept, Ran (2010c), targets the distribution of various radio-protocols including 3G, LTE, IMT-Advanced and WiMax at a very-low radiation based on radio-over-X (optical fiber, Coax, Cat-5, power-line etc.). The focus is on scalable hybrid wired-wireless topologies to achieve a remarkable reduction of the emitted power to minimum level (order of 1mW and less) needed for coverage of small cells.

VLR-DAS approach in the context of 3G femtocell paradigm is described in Fig. 4, Ran (2010c). The new functional device, called Home Access Node (HAN) is a special version of RRU with strict limitation of radiated power to below 1mW over the air. HAN also used to adapt the RF signal to the wired media with minimum distortion, and perform initial identification and filtering of target radio signal.

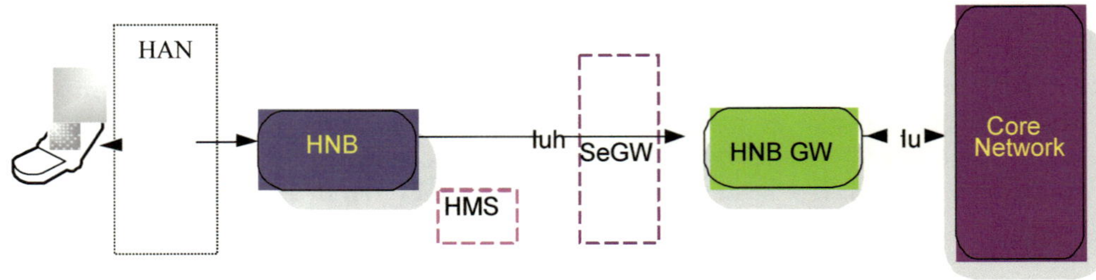

Fig. 4. 3G femtocell logical architecture Deployment configuration/Options with HAN function

Some general requirements for the successful operation of HAN with 3G femtocell are the following.

- The introduction of the HAN as an intermediate unit between HNB and the UE should not affect the 3G protocol stack between the HBS and the core network.
- Coexistence of the relayed radio signals with other radio signals (adjacent HBSs, macro BS, WLAN etc.) is required both for wired and for the wireless indoor infrastructure.
- The transmission of the various radio signals from the HBS through the HAN and the wired-wireless infrastructure to the UE should introduce minimum delay and negligible distortion to the performance of the communication link between HBS and UE.

VLR-DAS basic scenarios

Case 1: Residential in-house distribution network. For home users the VLR-DAS concept implies using the distribution of target wireless signals inside small number of rooms (say less than 8) based on existing and future home infrastructure. The radio signals are relayed transparently from one HBS to several rooms as "radio-over-wired" signals (with possibly frequency translate over wired home infrastructure). The wired part is serving as a "range extension" unit over 10's of meters to a set of HANs. Each HAN transmits at minimum power (less than 1mW) to cover a single room of less than 10m.
This case, shown in Fig. 5, is based on the simplified serial concatenation of several HAN's connected through a multimode fiber (MMF). The use of a separate fiber for the uplink and downlink conversion aimed to simplify and reduce the cost of the HAN implementation. In this example, HAN#1 serves as the Master Unit (MU) that communicates with the HBS (or external macro BS either directly, or through a repeater) over the air to simplify the installation of the home network. The other HAN's transmit the radio signals over the air to the target UE.
Case 2: Enterprise "Green Hospital". The Green Hospital represents a corporate use case of citizen-to-authority and authority-to-citizen (C2A-A2C) where the exchange of large amount of medical data over 100s of femtocells is considered. Here the range extension over the wired media is 100-3000m, and hundreds of users are supported through highly dense multi-femtocells architecture.
In hospitals the need for wireless is growing exponentially because all kinds of "vital signs" functions are monitored continuously. In hospital networks, there is an inherent inconsistency between the wish for mobility of the monitored patients and the interference of the radio signals with the sensitive diagnostic equipment. At present, some 8-10 functions per patient are wirelessly recorded. The number of monitored functions will grow rapidly in

the coming years. In many cases the monitoring of patient functions is done wireless because of the need for improved "quality of life" for the patients. In essence, the use of "wireless" in a hospital is conflicting with the safety and security demands for the sensitive diagnostic and patient treatment equipment. Green Femtocell would be a great benefit by enabling unique capabilities to medical centres and health management offices (HMO).

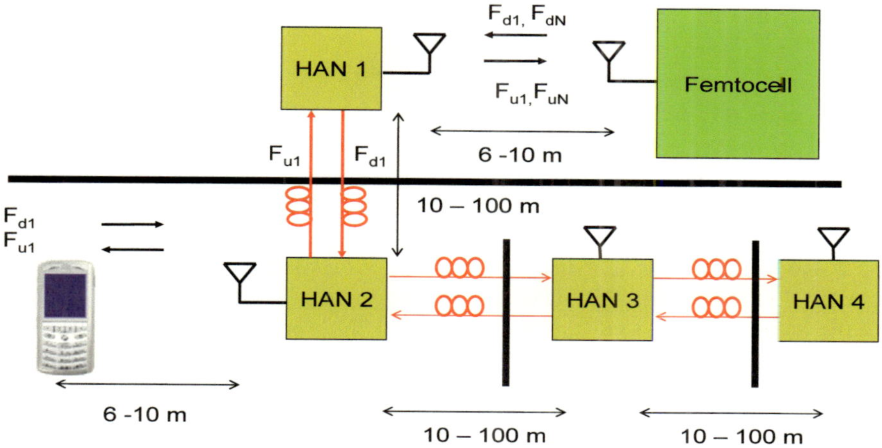

Fig. 5. Typical residential configuration of VLR-DAS over 4 rooms

Green hospital scenario for radiological department: The relevant patient data including previous examination images is transmitted within 1-2 seconds to the HAN sensor located near the desk. The data includes also medical information such as allergy to contrast media or renal failure, which will pop-up automatically on the screens of the staff in the department. The images are transferred from the HAN on the front desk immediately to the Picture Archiving and Communication System (PACS) via the wired optical fiber backbone. The PACS consists of Host Communication Controllers (HCCs) and the central Hospital Information System (HIS). HIS data would appear automatically on the computer screen on the desk. Then, the secretary has just to show the data to a visitor in order to get his confirmation that the data are correct.

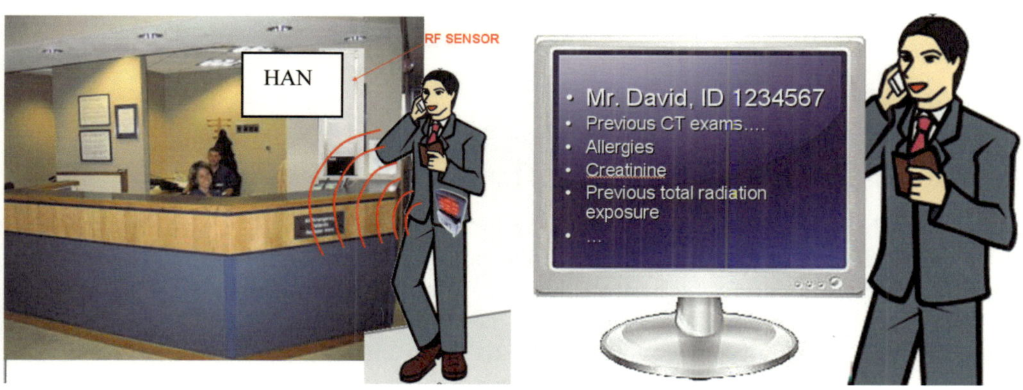

Fig. 6. Green Hospital scenario for radiological department.

Derivatives of this scenario for fast exchange of medical records such as part/all the medical history of a patient may include: summary and reports of medical visits in different clinics, all laboratory results as blood tests, urine, previous ECG records, etc. obtained during the patient's life. It may include large files of diagnostic imaging procedures, pathology and histology results (including digitalized images), summary or lists of previous and currently active diseases, side effects and hypersensitivity reactions to drugs and contrast media, allergies, family history, a list of drugs taken by the patient, etc.

3.4 Experimental results of VLR-DAS for 3G UMTS

Some results of UMTS Green Femtocells were published in Ran (2010c). Fig. 7 shows the experimental setup for the testing of the VLR-DAS concept in Case 1. The system comprised a femto BS (Agilent E4432B) transmitting "green" W-CDMA FDD signals of -20dBm at band I (UL: 1920 -1980 MHz and DL: 2110 -2170MHz). We used the Test Model 1 containing 64 DPCH signal. The optical sub-system contained 10Gbps 850nm Vertical-Cavity Surface-Emitting Laser (VCSEL) for the direct E/O conversion of the W-CDMA signal into radio-over-fiber (ROF) signal. The ROF UMTS signal propagated through a standard MMF (type OM3) of the length of 30m, and was detected by a PIN diode. Then, the detected signal was onward transmitted through the tested channel to the W-CDMA receiver. The purpose of this experiment was the study of the wired channel performance. The measured gain of the wired channel is about 3dB. The performance of the wired channel strongly depends on the RF signal power level at the VCSEL input due the VCSEL strong nonlinearity.

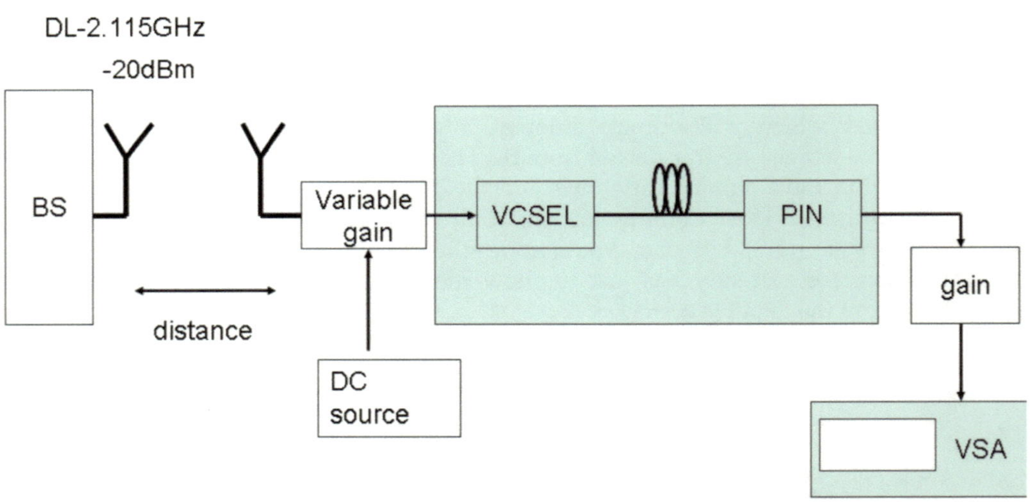

Fig. 7. Experimental set-up for UMTS in VLR-DAS scenario.

As shown in Fig. 8, the optimal value for the input RF power is about -30dBm. The input power values above -25dBm lead to distortions expressed by higher order intermodulations and link gain degradation which affect the error vector magnitude (EVM). EVM and link gain vs. input signal power were recorded for the various scenarios. The EVM degradation, denoted ΔEVM, is obtained by comparing the EVM for the "best case" where BS is directly connected to the HAN, i.e., no optical or wireless segment, and the wired case where HAN is directly connected to the BS, and through the optical segment to the UE.

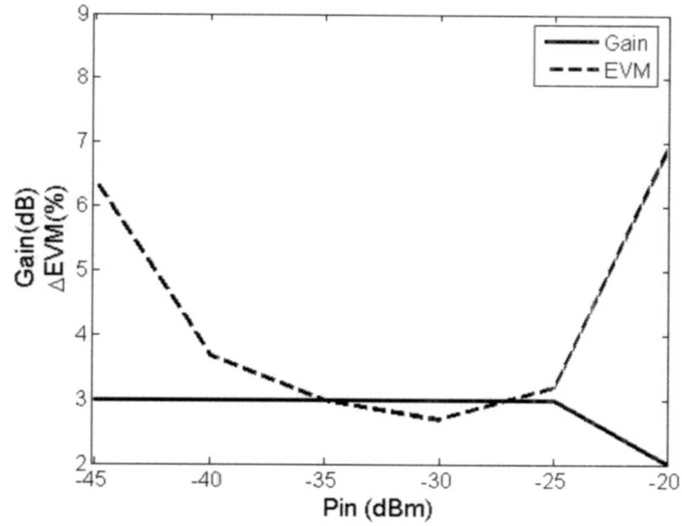

Fig. 8. ΔEVM and link gain for the wired case. The HAN is connected to the BS through a wired connection and to the user through VLR-DAS optical segment. The link shows negligible performance degradation over wide range of input power.

Fig. 9 shows the gain and ΔEVM for a combined wireless-wired channel and different lengths of wireless segments. The measured gain of the combined channel is about 15dB. Although the combined wireless-wired channel causes the degradation of EVM in the range of 8% -15%, still the satisfactory performance of EVM values below 11% is observed.

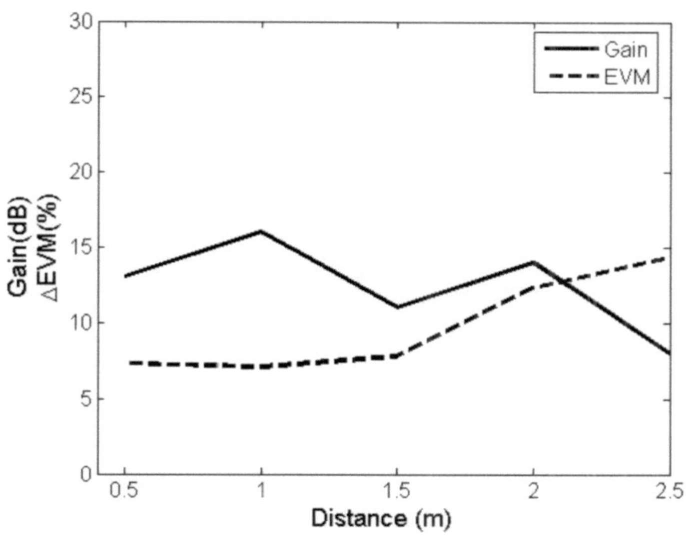

Fig. 9. ΔEVM and link gain for the wireless-wired case. The HAN is wirelessly connected to BS and connected to the user through VLR-DAS optical segment.

4. 4G green femtocell: Indoor architectures based on MB-OFDM UROOF technologies

The UWB radio over optical fiber (UROOF) was investigated recently in series of works, see (UROOF), Ran (2010a, b, c), Ben-Ezra (2010), Kraemer (2009) and Ran (2009). In this section we propose a framework, based on UROOF technologies, to elaborate **indoor architectures** for implementing the 4G green femtocell paradigm (see Fig. 1). We start with the proposed wireless air interface based on MB-OFDM UWB. We then provide some experimental data on transmitting multi-gigabits data over hybrid fiber-wireless channels and review MB-OFDM with MIMO as a promising approach for the compound indoor wireless-MMF channel.

4.1 MB-OFDM UWB over wireless indoor channel

MB-OFDM has been proposed for the draft standard 802.15.3a (withdrawn on January 2006) and by the WiMedia Alliance in ECMA-368. MB-OFDM UWB is a combination of extremely broadband OFDM signal and frequency hopping, in which 528 MHz channels are selected over the entire 7.5GHz bandwidth between 3.1 to 10.6 GHz. Time frequency codes (TFCs) of length 6 are used to select a sequence of "logical channels" from a band group. Unique logical channels are defined by using up to seven different TFC codes for each band group. TFCs for band group 1, according to (ECMA-368) are given for example in Table 2.

TFC Number	BAND_ID for Band Group 1					
1	1	2	3	1	2	3
2	1	3	2	1	3	2
3	1	1	2	2	3	3
4	1	1	3	3	2	2
5	1	1	1	1	1	1
6	2	2	2	2	2	2
7	3	3	3	3	3	3

Table 2. Time Frequency codes patterns for band group 1

There are 5 Band Groups (see Fig. 10):

- Band group #1 is mandatory, remaining (#2 - #5) are optional.
- Only two Time-Frequency coded Logical Channels for Band group #5.

Band group #2 can be avoided when interference from U-NII bands is present.

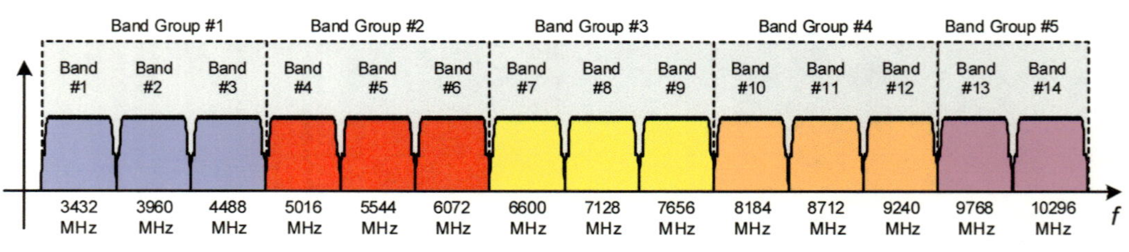

Fig. 10. Band plan for 3.1-10.6 GHz MB-OFDM with centre band frequencies shown.

The MB-OFDM signal can be expressed by

$$x_{MB-OFDM}(t) = \mathrm{Re}\left\{\sum_{k=0}^{N-1} x_{OFDM,k}(t - kT_{OFDM})\exp\left[j2\pi f_{(k \bmod 6)}t\right]\right\}$$

Where, $x_{OFDM,k}(t)$ is kth the OFDM symbol of duration T_{OFDM}, N is the number of transmitted OFDM symbols and f_k is the carrier frequency over which the symbol is transmitted. The designed value for T_{OFDM} is 312.5ns, where information length is 242.4ns, 9.5ns kept for guard time, and 60.6ns are length of the cyclic prefix, providing guard against multipath of length up to 60.6ns.

UWB wireless channel models were developed within IEEE 802.15.3a group during 2002 for the range 0-10m. The model defines four radio environments, based on simplified Saleh-Valenzuela (S-V) channel model, Saleh (1987a):

CM1: near line-of-sight (LOS) with distance of 0-4 m between Tx and Rx.

CM2, 3: non-LOS for a distance 0-4m, 4- 10 m, respectively.

CM4: heavy multipath environment.

Realistic simulations over CM1-CM4 were carried out in which: losses due to front-end filtering, clipping at the DAC, DAC precision, ADC degradation, multi-path degradation, channel estimation, carrier tracking, packet acquisition, overlap and add of 32 samples (equivalent to 60.6 ns of multi-path protection), etc. were considered. The distance at which the MB-OFDM system can achieve a PER of 8% for a 90% link success probability is tabulated in Table 3 below:

Range	AWGN	CM1	CM2	CM3	CM4
110 Mbps	20.5 m	11.4 m	10.7 m	11.5 m	10.9 m
200 Mbps	14.1 m	6.9 m	6.3 m	6.8 m	4.7 m
480 Mbps	7.8 m	2.9 m	2.6 m	N/A	N/A

Table 3. Simulation results of MB-OFDM over practical indoor channel models

An improved version to this model was published within IEEE 802.15.4a, based on more detailed field measurements. The model allows a larger number of environments; treats the number of clusters of multipath components in the S-V as a random variable and allows frequency dependence of the path loss, according to the following formula.

$$PL(f) = E\left\{\int_{f-\Delta f/2}^{f+\Delta f/2} |H(f')|^2 df'\right\}$$

$E\{\ \}$ is taken over large enough area to allow averaging out the small scale fading as well as shadowing. Δf is chosen small enough so that the dielectric constant, diffraction coefficients are constant within that bandwidth. An updated survey on UWB propagation channels is given in Chapter 3 of Kraemer (2009).

4.2 MB-OFDM UWB radio over mixed wireless-MMF

Channel impairments of MMF fibers are overviewed in Shieh (2010). When addressing the mixed wireless-MMF with MB-OFDM UWB technology, both frequency selective fading due to time delay spread over indoor channel, and multimode dispersion in MMF links should be considered. In particular, the cyclic prefix design for OFDM symbol should be longer than maximum delay due multipath wireless propagation and multimode dispersion spread in MMF.

A simplified model for UROOF "optical relay system" is given in Fig. 11, Ran (2010a), consisting of MMF, directly modulated VCSEL and a photodetector (PD) PIN diode.

It is shown based on series of experiments with UROOF platform (UROOF) that range extension by two orders of magnitude can be achieved for all MB-OFDM UWB RF signals.

A highly efficient method of RF and optical signal mixing to achieve optical OFDM transmission of MB-OFDM beyond 40Gbp/s was presented recently by Ben-Ezra (2010). The first concept is based on parallel–RF/serial optics architecture shown in Fig. 12. Basically, this architecture takes 128 conventional WiMedia/ECMA baseband channels of 528MHz and uses all-optical mixing to multiplex them over SMF. One of the key advantages of such approach is the ability to provide hybrid fiber-wireless solution, where the wireless segment at the available ultra-wideband (UWB) transmission is fully compliant with UWB regulations.

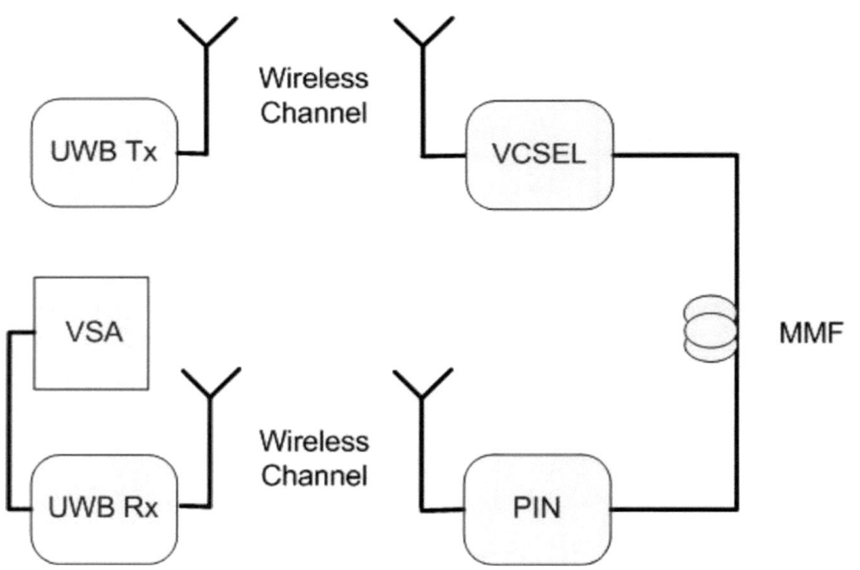

Fig. 11. UROOF channel consisting of an optical link (VCSEL-MMF-PD) and a wireless channel

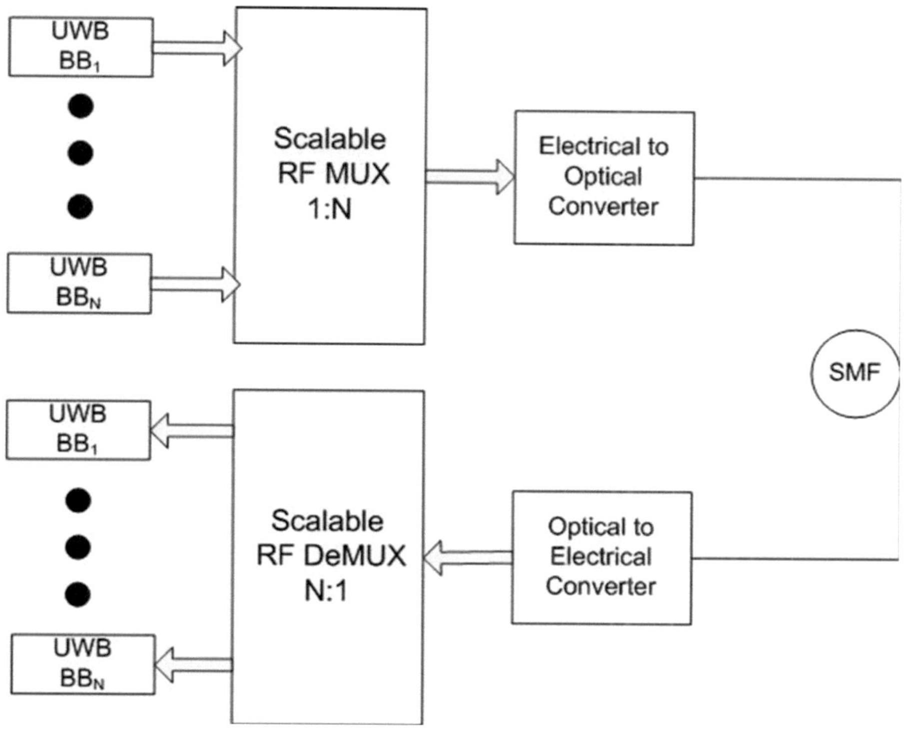

Fig. 12. 61.44 Gb/s based on parallel RF/Serial optics with N=128 channels

Another architecture based on parallel-RF/parallel-optics based on 12 low cost VCSELs array that transmits over bus of 12 MMF fibers to array of 12 photo detectors s is analysed in Ben-Ezra (2010).

4.3 MB OFDM UWB with MIMO over the wireless channel

Dense deployment of low-power femto BSs offers significantly higher capacity per area than legacy macrocells due to more efficient spatial re-used and smaller cell size. MIMO techniques permit to exploit the rich scattering of short range indoor wireless channel to increase the spectral efficiency linearly. The UWB-MIMO studies can be classified into four categories Kaiser (2009):

- **MIMO channel modelling and measurement**. Spatial characterization of UWB MIMO channel is given in Malik (2008).
- **Channel capacity.** It was shown in Zheng (2006) and Martini (2007) that the system capacity is limited by N_T, the number of Tx antennas, N_R number of receiving antennas, and η, the number of spatial degrees of freedom of scattered field. A fundamental result by Telatar (2000) showed that for achieving channel capacity at low SNR input signals must be spiky in time or frequency.
- **Space-Time Coding (STC) for MB OFDM MIMO** is presented in Siriwongpairat (2006). By applying space-time-frequency coding across K OFDM symbols maximum achievable diversity order of KLN_TN_R can be achieved, where L is the number of resolvable paths.

- **Beam forming for MB OFDM UWB** is presented in Malik (2006). It is shown that the signal bandwidth has little impact on beam width or direction

The high level scenario of MB-OFDM UWB with MIMO processing over the wireless channel is shown in Fig. 13. It is assumed that HAN nodes are separated 10m apart from each other, and a latency control to support the MIMO algorithm will be needed. Each cluster should support 10-100 femtocells. The concentrator achieves macro diversity

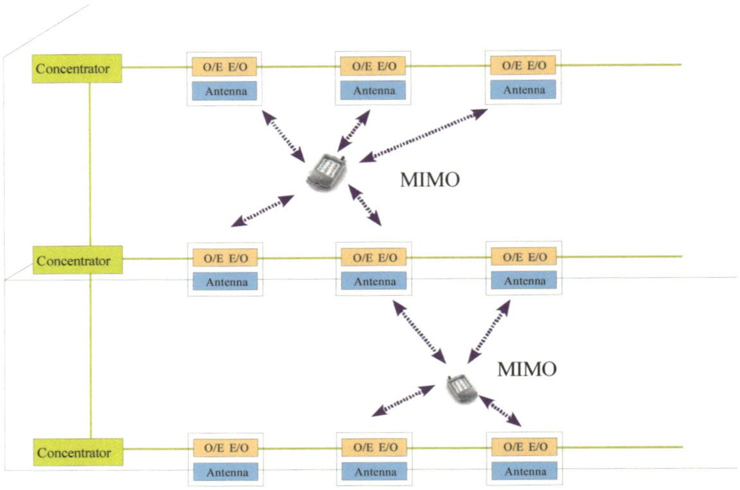

Fig. 13. Multimcell BS processing with MIMO over the air.

4.4 MB OFDM UWB with MIMO over the wireless-MMF channel

Recently, the MIMO approach to exploit the modal dispersion in MMF has been address by several researchers Targihat (2007), Greenberg (2007), Greenberg (2008), Armstrong (2009), and Shieh (2010).

The application of MIMO processing to the hybrid wireless-MMF may provide solution for implementing the VLR-DAS paradigm and approaching the challenging goal of green femtocell. The proposed approach paves the way for a green framework in which increase of x100 in energy efficiency and x100 reduction of human exposure to wireless radiation become feasible.

The fundamentals and challenges of this hybrid wireless-optical are yet to be explored.

5. Conclusion and further prospective

The research on UWB-MIMO is still in its early stage. Further studies, especially on its MB OFDM implementation, are necessary to bring this technology into the market. The research on MB OFDM UWB over fiber-wireless channels with MIMO presents a promising solution for achieving data rates needed for 4G indoor systems at extremely low power.

6. Acknowledgment

The authors would like to thank the partners of HOMESNET project (CELTIC CP6-009) for the helpful discussions on future femtocells technologies and trends.

7. References

Altman, Z.; Balageas, C.; Formet, E.; Hämäläinen, J.; Marce, O. ; Mutafungwa, E.; Zheng, Z.; Ran, M.; Ben-Ezra, Y.; Beltran, P. & Perales S. (2010). Femtocells: The HOMESNET Vision, *21st Annual IEEE Int'l. Symposium on Personal, Indoor and Mobile Radio Communications (PIMRC 2010)*, Istanbul, Turkey, 26-29 September 2010

Armstrong, J. (2009), OFDM for optical communications, *J. Lightwave Technology* Vol. 27 N.o.3 pp.189-204

Ben-Ezra, Y.; Ran, M.; Lembrikov, B.I.; Mahlab, U; Haridim, M. & Leibovich, A. (2010). Optimal transmission of OFDM ultra-wideband signals beyond 40 Gb/s, *Special issue on transparent optical networking, J. of Networks*, Vol. 5, No.2, (February 2010), pp. 140-151 [Invited]

CELTIC Purple Book by CELTIC-Plus Version 2011, http://ebookbrowse.com/celtic-plus-purple-book-2011-web-pdf-d110054593

Chow, P; Karim, A; Fung, V & Dietrich C. (1994). Performance advantages of distributed antennas in indoor wireless communications systems. *IEEE VT conference*, vol.3 pp. 1522-1526, June 1994.

ECMA-368 http://www.ecma-international.org/publications/standards/Ecma-368.htm

ECMA-387 http://www.ecma-international.org/publications/standards/Ecma-387.htm

FCC 2002 www.fcc.gov/Speeches/misc/statements/knapp060502.pdf

Foschini, G. J.; Karakayali, K. R. & Valenzuela, A. (2006). Coordinating multiple antenna cellular networks to achieve enormous spectral efficiency, *Communications, IEE Proceedings*, vol. 153, no. 4, (August 2006), pp. 548 – 555.

IEEE802.16-2009 IEEE Std. 802.16-2009, IEEE Standard for Local and Metropolitan area network part 16: Air Interface for Broadband wireless access systems, May 29, 2009.

Gambini, J. & Spagnolini, U. (2010). Radio over telephone lines in femtocell systems, *in Proc.IEEE 21st International Symposium on Personal Indoor and Mobile Radio Communications* (PIMRC, Sept. 2010), pp. 1544-1549.

Gidding, R.P.; Jin, X.O. & Tang, J.M. (2009). First experimental demonstration of 6Gb/s real-time OFDM transceivers incorporating chanel estimation and variable power loading. *Opt. Express* Vol. 17 No. 2 pp. 19727-19738.

Greenberg, M.; Nazarathy M. & Orensein M. (2007). Data parallelization by optical MIMO transmission over multimode fiber with intermodal coupling. *J. of lightwave technology* vol.25, no.6 (June 2007), pp.1503-1514.

Greenberg, M.; Nazarathy M. & Orensein M. (2008). Performance of high-bitrate multiple-output links over multimode fiber with intermodal dispersion J. *of lightwave technology* Vol.26, No.14 (July 2008), pp.2192-2200.

Heath, R.W; Wu, T; Kwon, Y.H & Soong, A.C.K (2010). Multiuser MIMO in Distributed Antenna System. *Signals, Systems and Computers, Asilomar 2010.*

Hillerkuss D. et al., (2010), OFDM for optical communications, *OPTICS EXPRESS*, vol.18, No.9 (April 2010), pp. 9324-9340.

Jarich, P. (2010) *Femtocells Circa 2011: Surveying the Ecosystem Players*, in www.currentanalysis.com

Kaiser, T.; Zheng, F. & Dimitinov, E (2009). An overview of ultra wideband systems with MIMO. *Proceedings of the IEEE*, Vol. 97, No.2 February 2009, pp.285-312.

Kraemer, R and Katz, M.D (editors) (2009), Short-Range wireless communications – Emerging Technologies and Applications, Chapter 24, pp274-327: Ultra-Widedband Radio-Over-Optical Fiber Technologies. *Wiley-WWRF series* ISBN 978-0-47-69995-9

Li, X; Luo, M et al. Downlink performance and capacity of distributed antenna system in multi-user scenario, *in Wicom*, 2009 pp1-4.

Malik, W.Q (2008) Spatial correlation in UWB channels. *IEEE Transaction on Wireless communications* vol. 7 no. 2, 604-610.

Malik, W.Q.; Allen, B. & Eswards, D.J. (2006), A simple adaptive beam former for UWB wireless systems, in IEEE conferece on ultra-wideband, pp.453-457.

Martini, A. et al. (2007). Capacity of wideband MIMO channels via Space Time diversity of scattered fields, *in IEEE Asilomar Conference on Signals, Systems and Computers* pp.138-142.

Perez, D.L.; Valcarce, A.; De La Roche, G. & Zhang, J. (2009). OFDMA Femtocells: a roadmap on interference avoidance. *IEEE Comm. Magazine* (Sep 2009), pp41-48

Rappaport, T. S. (2002). *Wireless Communications Principles and Practice,* Prentice Hall PTR, ISBN 0-13-042232-0, New Jersey, USA

Ran, M.; Lembrikov, B.I. & Ben-Ezra, Y. (2010a). Ultra-Wideband Radio-Over-Optical Fiber Concepts, Technologies and Applications, *IEEE Photonics Journal,* Vol. 2, No.1 (February 2010), pp. 36-48, [Invited]

Ran, M.; Ben-Ezra, Y. & Lembrikov, B.I. (2010b). High Performance Analog Optical Links Based on quantum dot devices for UWB signal transmission, in: *in: Ultra WidebandBoris Lembrikov (ed.) Sciyo, Croatia,* 2010, pp. 75-96, ISBN 978-953-307-139-8

Ran, M.; Lebovitch, A.; Yurchenko, Y. & Ben-Ezra Y. (2010c). Green Femtocell: The VLR-DAS Approach, *2nd Int'l. Conf. on Evolving Internet (INTERNET 2010)*, pp. 226-228 Valencia, Spain, September 20-25, 2010

Ran, M; Ben Ezra, Y. & Lembrikov B.I. (2009). Ultra-wideband Radio-over-optical-fiber Technologies, *In: Short-Range Wireless Communications*, Kraemer, R. & Katz, M. D. (Eds.), pp.271-327,Wiley, ISBN 978-0-470-69995-9 (H/B), Chichester, England

RFC3261, SIP: Session Initiation Protocol, IETF, June 2002

Roche, G.; Valcarce, A.; Lopez-Perez, D. & Zhang, J. (2010). Access Control Mechanisms for Femtocells, *IEEE Comm. Magazine,* (January 2010), pp. 33-39

Saleh, A.A.M; Rustako, A.J. Jr & Roman, R.S. (1987). Distributed antennas for indoor radio communications. *IEEE Trans. on Communications*, vol. 35 no. 12, pp.1245-1251.

Saleh, A and Valenzuela, R.A (1987a). A statistical model for indoor multipath propogation, IEEE Journal on selected Areas Communications, Vol. 5, No.2 pp.128-137

Siriwongpairat, W.P. et al. Multiband-OFDM MIMO coding framework for UWB communication system. IEEE transaction on Signals Processing Vol. 54. No. 4 pp. 214-224.

Tarighat, A. et. al (2007). Fundamental and challenges of optical MIMO multimode fiber links. *IEEE communications Magazine,* May 2007, pp.57-63

Telatar, I.E. and Tse , D.N.C (2000). Capacity and mutual information of wideband multipath fading channels. IEEE trans. On Information Theory vol. 46, No. 4 pp.1384-1400.

TS23.238, 3GPP (2006) Technical Specification Group Services and Systems Aspects, IP Multimedia Subsystem (IMS), Stage 2, VS.15.0

[UROOF] http://www.ist-uroof.org/

Yongho, R.; Kwak. J.S. & Etemad, K., (2009). WiMax Femtocell: Requirements, Challenges, and Solutions, *IEEE communication Magazine,* (September 2009), pp.84-91

Zheng, F. and Kaiser T. (2006) Channel capacity of MIMO UWB indoor wireless systems, *in UWB Communication Systems – A comprehensive overview.* Hindawi Publishing Corp. pp.376-409.

Zhang, J. & De la Roche, G. (2010). *Femtocells: technologies and deployment,* John Wiley & Sons Ltd., ISBN 978-0-470-74298-3

10

A Telematics System using In-Vehicle UWB Communications

I.J. Garcia Zuazola[1], J.M.H. Elmirghani[2] and J.C. Batchelor[3]
[1]*Deusto Institute of Technology - DeustoTech, University of Deusto*
[2]*School of Electronic & Electrical Engineering, University of Leeds*
[3]*School of Engineering and Digital Arts, University of Kent*
[1]*Spain*
[2,3]*UK*

1. Introduction

Owing to the rapid evolution in wireless and cellular communications usage, and the resulting huge demand, new communications technologies are needed to support the cost effective growth of mobile telephony and the tremendous increase in Internet data traffic in new applications such as Telematics. New wireless communication technologies are therefore needed for the delivery of enhanced applications, for example telematics services to motorway users. This is targeted towards transforming wireless techniques to operate on: fast moving vehicles with high impairments encountered at high data rates; small cellular cells which are key to high data rate operation; and, more frequent cell handovers dictated by speed and small cell size. This scenario can make use of data from inductive loops for traffic (vehicular and data) monitoring, prediction, grooming and control.

Central base stations driving distributed antenna units will facilitate this new bandwidth provision and Radio-over-Fibre (RoF) is a technology proposed to achieve transparent data distribution. The replacement of coaxial cables with optical fibres is desirable. Fibre transmission offers sufficient bandwidth with reduced loss over long distances and decreased electromagnetic interference. Low loss Single Mode Fibre (SMF) is commonly deployed outdoors while Multimode Fibre (MMF), with higher losses, is often pre-installed in buildings. Unfortunately MMF is not of a quality that lends itself to broadband Radio-over-Fibre.

Two data transmission approaches meet the data rate requirements criteria, (i) high performance Worldwide Interoperability for Microwave Access (WiMAX) approaches that use an unlicensed spectrum band (5.470-5.725 GHz) for the motorway-to-vehicle communication and (ii) the Ultra Wide Band (UWB) communications standard for communication exchange from the vehicle to the in-vehicle applications. UWB in the unlicensed lower band (3.168-4.752 GHz) offers multipath immunity.

WiMAX technology, based on the IEEE 802.16m standard, is of great interest to Intelligent Transport Systems (ITS) due to the expected need for the delivery of peak rates of up to 1 Gbit/s at fixed speeds and 100Mbit/s to mobile users over micro-cells. Theoretically, antennas are reciprocal, which means they are able to transmit and receive electromagnetic waves with similar properties. Antennas play a key role in wireless communications; the antenna performance is crucial to efficient electromagnetic propagation. Cost-effective while

efficient antenna units have been assessed (Garcia et al., 2008) for Intelligent Transport Systems (ITS) in a highway scenario. The antennas designed and tested served the WiMAX standard over full-duplex bi-directional optical links.
UWB technology is of global interest in modern communication systems due to its potential to deliver high speed data rate transmissions with excellent immunity to multipath interference at low power consumption, low cost (Elmirghani et al., 2006) and simple hardware configuration. The deployment of UWB inside vehicles can provide mobility and connectivity to a host of passenger devices while significantly reducing the costs associated with wiring. Vehicle chasses may contain large steel plates and antennas over ground planes are favoured for ceiling mounts (Garcia et al., 2009).
Key potential applications include advanced ultra-high speed cellular communications to vehicles, emergency services, ramp entry and variable speed control, and roadside assistance notification. The systems can also provide live traffic information and triple play services on in-car traffic information terminals or bespoke terminals.
In this work the following topics are discussed. First, a Telematics system, based on a published patent GB2436909 is presented. In essence, WiMAX is used here in highways to provide communication services to moving vehicles incorporating an in-vehicle UWB network of 480 Mbps. Second, a detailed motorway-to-vehicle wireless communication system using recent developments in RoF systems is introduced. This utilizes centralized base stations driving a host of antenna units along a highway through Single Mode Fibre (SMF). Third, a UWB wireless propagation analysis in an in-vehicle scenario is analysed. The analysis is performed in the lower unlicensed band (3.168-4.752 GHz) of the allocated Federal Communications Commission (FCC) spectrum that has an assigned band from 3.1 to 10.6 GHz (FCC, 2002). Finally, the design of a UWB antenna for an in-vehicle application is described.

2. The telematics system

An attractive nature of a Telematics system is its capability to revolutionise transport using the latest wireless communication technologies including computer networking and communications infrastructure for delivering enhanced services to motorway users, Fig. 1. This provides the users with live traffic information and multi-play services on in-car traffic information terminals or bespoke systems.
The introduced Telematics system utilises RoF as a technology to achieve the modulated carrier distribution. RoF brings the opportunity to simplify typical Base Station (BS) deployments by making them part of the system backbone while being able to transmit over long distances. In addition, among the advantages of using optical fibres are: bandwidth, security, lack of interference and an extreme reduction in maintenance costs that is achieved by centralisation.
The RoF term is used when light is modulated using a radio frequency (RF) information carrying signal and is transmitted over optical fibre. The electromagnetic spectrum that can in theory be used to modulate the light is typically of several GHz bandwidth for the semiconductor lasers and several MHz for light emitting diodes (LEDs). Microwaves are commonly used due to their reasonable coverage, high data rates and capacity using moderate cells with low frequency reuse.

(a)

(b)

Fig. 1. A highway communication infrastructure (a) and an in-vehicle femtocell (b)

High capacity, Quality of Service (QoS) and data transfer communications are in demand. Methods to meet these demands include techniques that range from the use of smaller radius cells and the use of sectorisation, to the most recent use of multiple antennas in spatial and/or frequency diversity configurations. For these systems it is desirable to have easy integration and small package antenna units. Multiple Input Multiple Output (MIMO) Antennas composed of three 120° sectors exploiting frequency diversity gain have been reported in (Garcia et al., 2008). In the same way, wireless systems devices for the vehicle to road intercommunication are usually of small profile. A tri-sector configuration antenna to meet this demand has been reported in (Garcia et al., 2010).

Low cost RoF distribution deployed in-vehicles is of great interest, Fig. 1b. Unlike coaxial, fibre lines are capable of providing electromagnetic interference free transmissions over the distance. Centralisation of transceivers to a system backbone might complement with lower engineering costs, both in deployment and maintenance.

Antennas play a key role in wireless transceivers as they are the last step to radiating energy into the space. Miniature antennas are required for the in-vehicle application; the prototypes will have to ensure a good system performance and be platform tolerant.

In the next section, the motorway-to-vehicle wireless communication system, the in-vehicle wireless communication system including a UWB in-car wireless channel analysis, and finally the design of an UWB antenna for in-vehicle applications are introduced.

2.1 Motorway-to-vehicle communication system

In Telematics applications, an autonomous data collection and processing systems can exploit recent advances in UWB, Worldwide Interoperability for Microwave Access (WiMAX) and radio fibre technologies to provide data and multimedia services for vehicles along sophisticated highways (Kerner et al., 2005). Traffic prediction systems are receiving increased attention and are being evaluated and adopted in a variety of contexts (Gunter & Grosmann, 2005). Communication between vehicles and vehicle to infrastructure has also been explored using a range of technologies (Lee & Williams, 2000). Refined highways hold sophisticated telematics infrastructure that includes fibre, inductive loops buried, routers, switches and digital video cameras with number plate recognition systems among others.

An example of a highway sensing and communication infrastructure is depicted in Fig. 1.

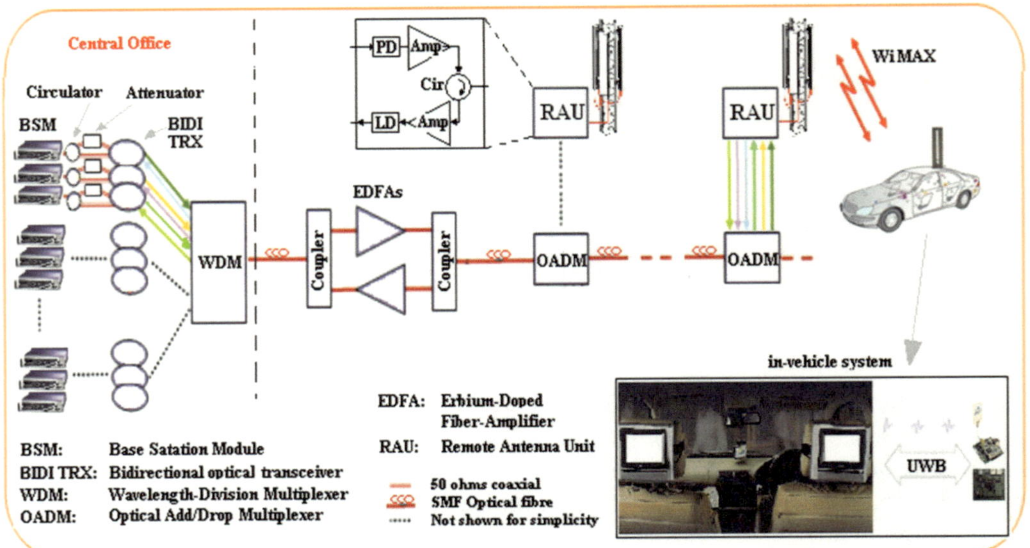

Fig. 2. The Telematics system architecture

The motorway-to-vehicle wireless communication system is realised using various components of the system architecture in Fig. 2, the RoF architecture.

In essence, the system architecture comprises a (virtual) central office (CO) which consists of base station modules (BSM), circulators, attenuators, bidirectional optical transceivers (BIDI TRX) and wavelength division multiplexers (WDM). Erbium-doped fibre amplifiers (EDFAs) provide bi-directional amplification to compensate for losses due to long runs of the single mode fibre (SMF). Optical add drop multiplexers (OADM) provide interfaces between the fibre and the remote antenna units (RAUs).

The functional view of the RoF system is now detailed. The system employs a high performance unlicensed 5.470-5.725 GHz WiMAX band directly modulated over a bi-directional SMF link using remote units composed of three sector antennas (Garcia et al., 2008) which aims to achieve a better capacity and coverage performance over micro-cells, and is intended to support around 100 Mbps. The radio frequency (RF) signals are directly modulated using a distributed feedback laser (DFL) contained in the BIDI TRX. The signal generated is carried over the downlink SMF to a remote RAU site which converts this optical signal to an RF signal by positive-intrinsic-negative photodiode with a transimpedance amplifier (PIN-TIA) receiver with high sensitivity. For uplink transmission, the wireless signal received from an RAU is amplified, converted into an optical form by a DFB laser (chosen as an ideal device for SMF and long haul communications), and transported uplink by the SMF to the CO where another PIN-TIA receiver in BIDI TRX demodulates the optical signal into the RF.

2.2 In-vehicle communication system

UWB technology has gained huge interest globally due to its potential to deliver high data rate and spatial capacity, with multipath immunity (Das et al., 2006; Way, 1989) and low power, low cost design. The deployment of this wireless technology in vehicles will provide mobility and connectivity to a host of passenger devices while reducing significantly the costs associated with wiring.

In addition, large vehicles can benefit from the use of low cost optical fibre communications. A RoF system can be used as part of a distributed antenna system (DAS), (with centeralised control) which supports the deployment of femtocellular access networks at 480Mbps within airplanes, buses, coaches, cars, lorries, trains, trams and other transport vehicles. Such a system can assist in the minimisation of radio frequency (RF) inference when compared to coaxial cable links, simplifies the infrastructure and reduces engineering cost. A high-level block diagram for the in-car system is depicted in Fig. 3.

Next, a study of a UWB system over RoF in an in-vehicle scenario is described in Section 2.2.1. Experimental results of the radio propagation within the car in a realistic environment validate the system and are described in Section 2.2.2.

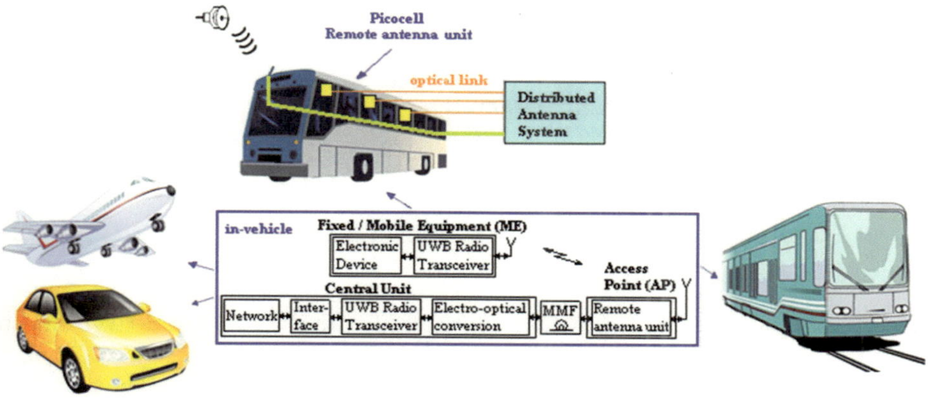

Fig. 3. The in-vehicle distributed antenna system

2.2.1 UWB RoF transmission in-vehicle

A feasibility study of an IEEE 802.15.3a UWB system based on Multi-Band Orthogonal Frequency Division Multiplexing (MB-OFDM) transmitted over RoF inside a vehicle is described. The in-vehicle system set-up includes a DV9110 Development Kit (DVK) from Wisair Ltd and a Renault Extra Van that was used as a base for the experiments. Two UWB transceivers were placed in the vehicle; one was used as an Access Point (AP) and the other as fixed/Mobile Equipment (ME). Each transceiver had an integrated monopole antenna of 2 dBi gain. The transceiver emits a short pulse of output power 80μW (Power Spectral Density PSD of -42 dBm/MHz max) containing the WiMedia/MBOA Group 1 sub-band (3.168–4.752 GHz) and using a modulated signal MB-OFDM quadrature phase shift keying (QPSK) at a varying physical data rate between 53.3 Mbps and 480 Mbps.

The in-vehicle system design is depicted in Fig. 4 which shows the UWB RoF architecture.

A UWB radio at the central unit was directly modulated over an optical signal using a Vertical Cavity Surface Emitting Laser (VCSEL) and then distributed over a MultiMode Fibre (MMF) to a remote antenna unit where the transmitted modulated optical signal was demodulated back to radio using a PhotoDetector (PD) and then propagated by the antenna into the space. A multimode fibre network was used for the UWB radio distribution due to the large bandwidth, low loss and the ability of centralisation at a relatively low cost (Garcia et al., 2005).

Results using Agilent ADS software demonstrate the feasibility of the RoF system. A waveform generator was used to transmit/receive through the system. It used a signal spectrum at 1 MHz narrow resolution bandwidth (RBW) filter within a Federal Communications Commission (FCC) mask. The transmitted and received signals are depicted in Fig. 5. They overlapped and showed good agreement.

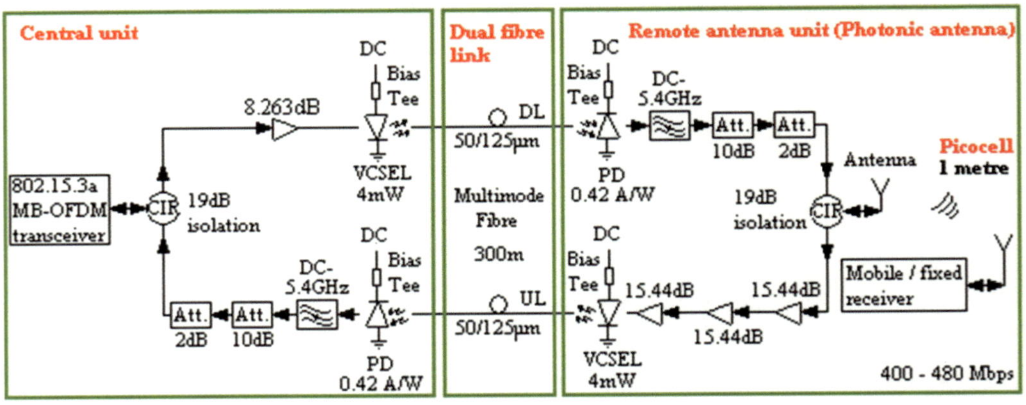

Fig. 4. The complete UWB RoF femtocell

The measured output power was -41.3dBm; this agreed with the maximum UWB Effective Isotropic Radiated Power (EIRP) allowance and the receiver sensitivity was -79.55 dBm. The system Transmit (Tx) power budget was measured near 0dBm after compensating for a power penalty (attributed to optical and RF devices loses) of 8.26dB.

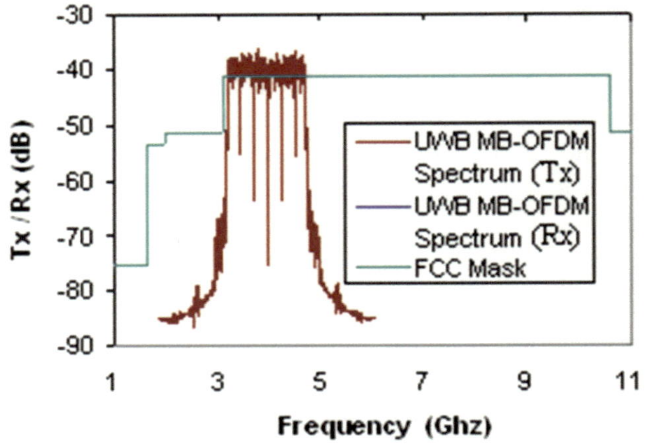

Fig. 5. Transmitted/received UWB MB-OFDM signal

A Bit Error Rate (BER) test evaluated the performance of the system. The transmitter produced OFDM UWB symbols that a receiver was capable of analysing; with/without multipath over an in-door channel. The resulting test for the down link is depicted in Fig. 6, where an irrelevant BER difference between a referenced transmitter and the UWB RoF full-duplex system is observed.

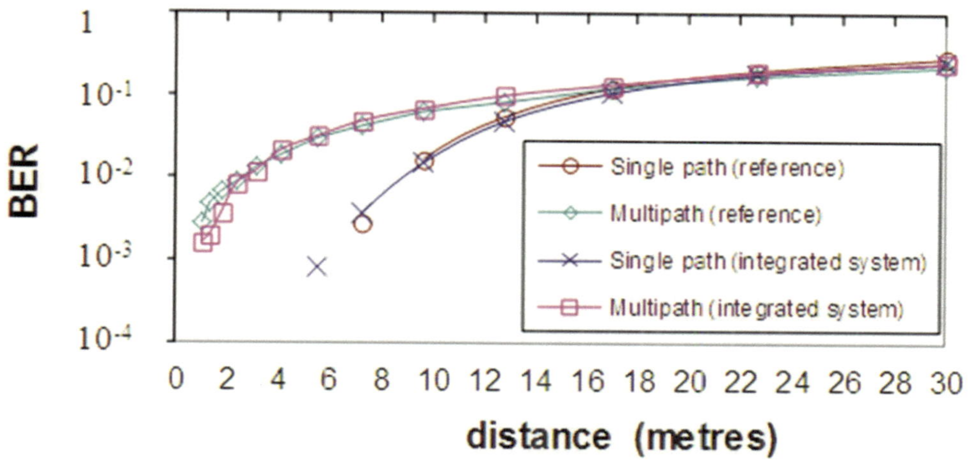

Fig. 6. The system Bit Error Rate (BER)

The results show the feasibility of the RoF system; this would allow extending UWB radio signals over hundreds of meters distances, well enough for in-vehicle applications.

2.2.2 UWB in-car wireless propagation

Based on the MultiBand technique, the multiband UWB (MB-UWB) splits the spectrum into sub-bands and uses conventional narrow band techniques, such as Orthogonal Frequency-Division Multiplexing (OFDM), to transmit the information in each sub-band (Elmirghani et al., 2006).

In this section, the propagation of a MB-UWB wireless system is studied within the in-car environment.

Using the same wireless set-up as in Section 2.2.1 without the optical fibre distribution, each transceiver was connected to a laptop for control and datalogging.

The cell size which is limited by the in-vehicle dimension was 0.9 metres in radius. The access point was set in the middle of the ceiling of the car as the preferred location in vehicles (Garcia et al., 2009); this ensures good power distribution and minimises field exposure to occupants.

The maximum channel path loss (where the antennas are considered as part of the channel) was measured at 33.53dB at 3.8GHz and the antenna mismatch or de-tuning is best at about 4cm away from the metallic in-car ceiling. A -91.30dBm noise floor was observed and an Access Point (AP) UWB Development Kit (DVK) power of -43.42dBm (1MHz resolution bandwidth, RBW) was employed.

Several channel analysis have been reported in (Garcia et al., 2009), this includes a rich multipath in this application, high reflections, path delay changes in open and closed environments with a possible Doppler dispersion, and the Inter-Symbol Interference (ISI) for moving vehicles. A Bit Error Rate (BER) measurement was made to validate the system.

The BER was set up having the Access Point (AP) and Mobile Equipment (ME) intercommunicating reciprocally with each other. To predict the maximum achievable data rate at the allowed BER, packets of certain known length were sent over the in-car channel from the AP to the ME when the vehicle was stationary and then at different vehicle velocities using the driver closed environment scenario. The received data is analysed and recorded as BER in Figs. 7a and 7b respectively. Throughputs (capacity) for the same set up are shown in Figs. 8a and 8b. Although there are fluctuations in the BER performance, an average of 2.5×10^{-4} is obtained while the vehicle is stationary and BERs up to 3.2×10^{-2} are measured at a speed of 120km/h.
The vehicle in motion (system closed) affects the BER results. The ISI mainly arising from high reflections within the small car metallic chamber is conjectured to be aggravated by the antenna instability due to the mobile vehicle vibration and interaction through the restricted window area. This is translated into a collection of received variations in the amplitudes and phases of differently delayed waves caused by further fading and multipath. The interference of direct path and the reflected waves results in higher BERs.

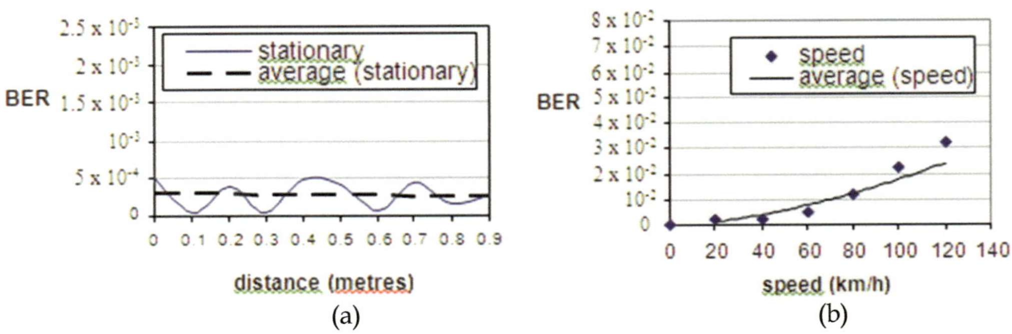

Fig. 7. BER as a function of distance (a) and BER as a function of mobility (b).

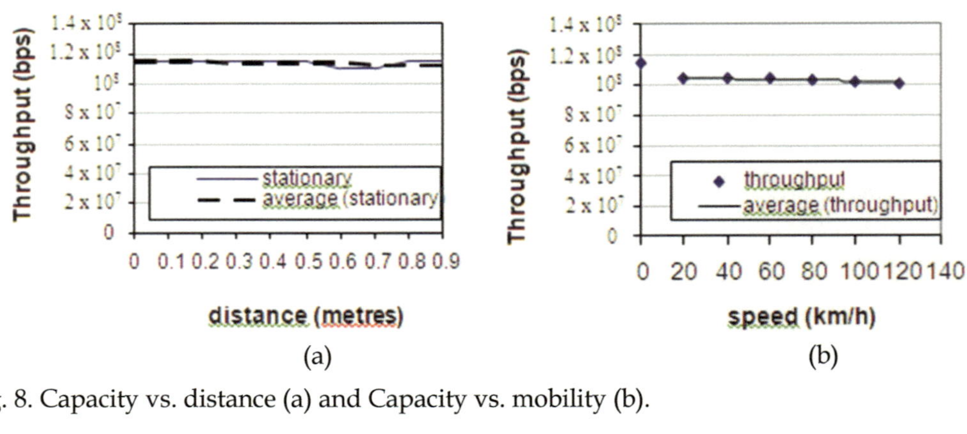

Fig. 8. Capacity vs. distance (a) and Capacity vs. mobility (b).

An average of 115Mbps throughput is measured when stationary and up to 102Mbps at a speed of 120km/h. The vehicle in motion affects the BER results due to the antenna instability that is created while in motion. There was high multipath in-vehicle, and this, in the moving vehicle, resulted in ISI and caused higher BER measured at higher speeds. In the same way, a lower data rate was achieved in motion.

3. UWB Antenna for in-vehicle applications

Planar Inverted-F Antennas (PIFAs) are well suited for integration inside vehicles. Their chasses may contain large steel plates and antennas over ground planes are favoured for ceiling mounts (Garcia et al., 2009). A UWB PIFA incorporating two shorting posts with coupling gaps is presented. The antenna operates at the lower UWB band (3.168-4.752 GHz) with a 3.57:1 VSWR and has a tailored impedance bandwidth and roll-off comparable to a standard frontend Band Pass Filter (BPF). To bring down unit cost, there has been a drive to simplify the hardware of UWB systems (Mohammad & Ismail 2008) and hardware could be further reduced by the adoption of the UWB PIFA proposed here, because the commonly deployed front-end BPFs would not be required. For the antenna to be implemented in car, the antenna performance was verified when in close proximity to a large conducting plate. Additionally, the elimination of the BPF with its associated insertion loss can offer power savings from the vehicle's battery especially when stationary.

Fig. 9 depicts the geometry of the UWB PIFA antenna. It consists of two planes, an etched upper layer (A) and a bottom ground (B) separated by an air substrate ($\varepsilon_r \approx 1$).

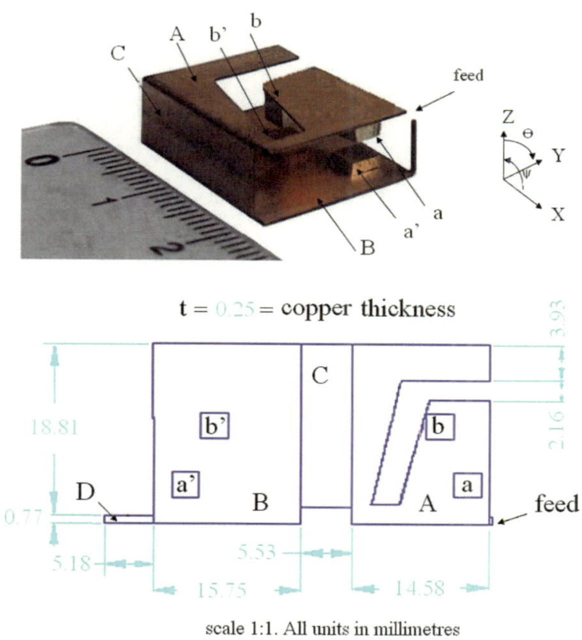

Fig. 9. Geometry of the UWB PIFA

The A and B planes are capacitively coupled via the two pairs of pins (a & a′ and b & b′). The dimensions of both posts a & b are 2.9 x 2.9 x 2.9 mm^3 while posts a′ & b′ are 2.9 x 2.9 x 1.45 mm^3. Coupling between planes A and B is achieved across the gaps in the posts.

The antenna is fed at the upper plane A using the inner core (0.51 mm) of a 50Ω rigid coax cable with a total diameter of 2.16mm and 62mm length. The outer shield of the cable is

attached to a grounding strip, D, electrically connected to B. The total volume of the antenna is 19.58 x 15.75 x 5.53 mm^3. The maximum dimension is smaller than 0.21λ at the lowest frequency of operation.
A simulated parametric study of the capacitive gap is reported in (Garcia et al., 2010b) where decreasing the gap between a and a' (or b and b') tended to improve the band-notch depth and impedance roll-off. Therefore, adjusting the gap capacitance of the electrically unconnected shorting posts allows a BPF like characteristic to be defined. An optimum length value of a = b = 2.9 mm and, a' = b' = 1.45 mm was found to give a band-notch at 5.5GHz, a return loss (RL) of -5dB, roll-off of 0.18 and 0.03 dB/MHz and a -5dB S11 fractional bandwidth of 40%. The optimal value of the gap corresponds to 1.18mm.

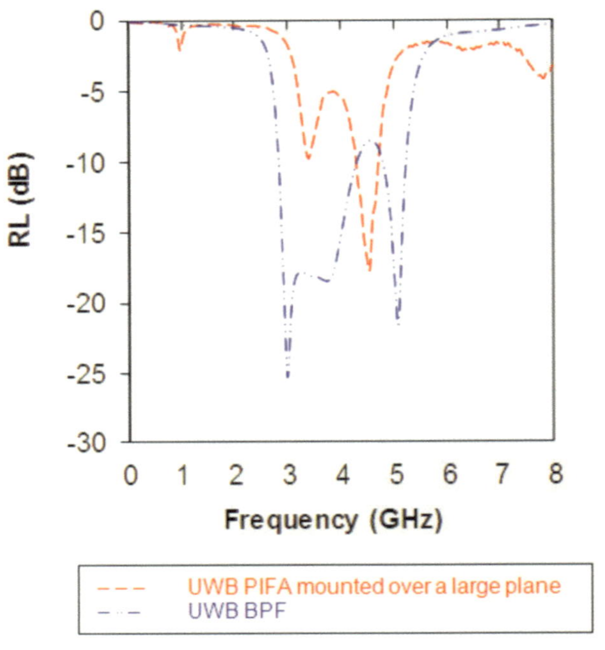

Fig. 10. The UWB PIFA antenna

The reflection coefficient of the UWB antenna is shown in Fig. 10 compared to a standard frontend BPF (RFlambda, n.d.). Compared to the commercially available BPF (2441MHz pass-band rejection and 2dB insertion loss; roll-offs of 0.050dB/MHz and 0.031dB/MHz for the lower and upper bands respectively), the UWB antenna has a lower 1108MHz pass-band rejection and improves the roll-offs to 0.024dB/MHz and 0.030dB/MHz.
An antenna having a VSWR of 3.57 (5dB RL) can be calculated to present an equivalent mismatch loss of 1.65dB (Kraus & Marhefka, 2001). Therefore, if the BPF and its associated mismatch loss of 2dB was removed, then there will still be an overall reduction in loss of 0.8dB.
The measured 5dB return loss bandwidth of the proposed PIFA is 42.15% for the 3.168-4.860 GHz FCC UWB. To investigate the effect of attaching the antenna to a large conducting plate in a car chassis, a larger ground plane of dimensions 510 x 800 x 0.75 mm^3 was placed ¼ wavelength below the PIFA. Fig. 10 shows the S11 response.

The far-field radiation patterns of the antenna including the large plane in polar form (measured at 4.752 GHz) are depicted in Fig. 11. The patterns are essentially directional, presenting a 120° half power beam-width (HPBW) and 1.53/1 front-to-back ratio in the H-azimuth plane and similar value for the E-elevation. A gain of 7.11dBi was measured with the E plane present.

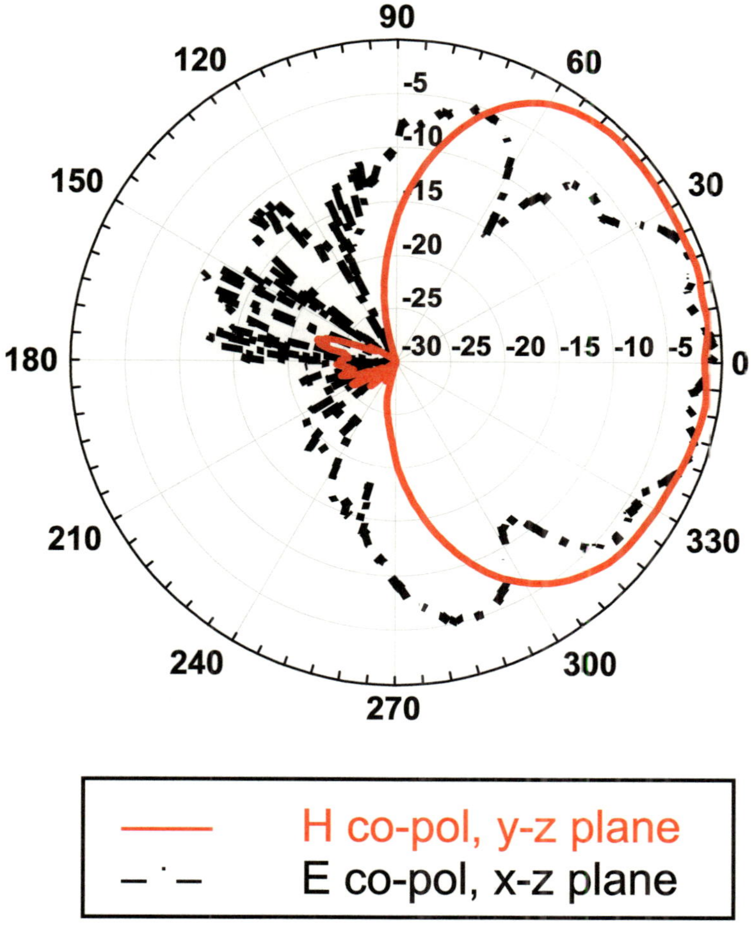

Fig. 11. Radiation patterns of the antenna including the large ground plane

4. Conclusion

This chapter has introduced new techniques and methodologies to support the cost effective growth of mobile telephony and the tremendous increase in Internet data traffic in Telematics for the delivery of enhanced services to highways users.
A Telematics system has been described based on a high performance WiMAX spectrum using an unlicensed band (5.470-5.725 GHz) and recent developments in RoF systems as a base for the delivery of wireless communications to motorway-to-vehicle applications. This approach results in a relative low cost deployment and maintenance, extends the radio over long distances and delivers peak rates of at least 1 Gbit/s to fixed users and 100Mbit/s to mobile users over micro-cells.
Cost-effective while efficient narrow band tri-sector antenna units have been assessed for Intelligent Transport Systems (ITS) in the presented highway scenario. The antennas served the WiMAX standard over full-duplex bi-directional optical links (Garcia et al., 2008). These antennas seem to be reasonably proficient for use in ITS due to their potential higher gains and reduced spatial limitations. The low VSWR performance achieved by the use of these narrow band antennas can improve the system link budget, which is translated into a relatively higher coverage/throughput.
The Robustness to multipath interference offered by the unlicensed lower band (3.168-4.752 GHz) UWB communication is to be exploited for in-vehicle communications. Within this work a promising low cost RoF link to extend the UWB radio over relatively long distances (i.e.: trains, trams and airplanes) has been introduced. The transmission was assessed using a relatively inexpensive multimode RoF link. The transmission network was capable of providing high data rates of 400-480Mbps at picocells of about a metre radius with inconsiderable SNR degradation performance over fibre links of several hundred of meters.
In addition, a wireless propagation of UWB radio inside a vehicle is analysed. The analysis of the UWB radio channel in-vehicles demonstrates that UWB is a very suitable and promising technology for transmission networks able to provide high data rates of 400Mbps within cars. Path loss was not of a significant level due to the short ranges that are encountered within cars. However, the main attenuation might perhaps be due to shadowing effects. High data rates were achieved in closed environment scenarios (Garcia et al., 2009). As many new cars include air conditioning, it is not unreasonable to expect the environment to be closed for the majority of the time.
A UWB antenna design example for an in-vehicle application has been introduced. The results indicate that the presented antenna works satisfactorily in the unlicensed UWB band and that the antenna element can be mounted on a large ground plane without degrading its performance. Owing to the low volume of the design it can be easily integrated inside vehicles in close proximity to the body.

5. Acknowledgment

This work was partially funded by the European Union.

6. References

Biagi, M. & Baccarelli, E. (2003). A simple multiple-antenna ultra wide band transceiver scheme for 4th generation WLAN, *IEEE 58th Vehicular Technology Conference*, pp. 1903 – 1907, Volume 3, Orlando, Florida, USA, 2003.

Das, A., Nkansah, A., Gomes, N. J., García Zuazola, I. J., Batchelor, J. C. & Wake, D. (2006). Design of Low Cost Multimode Fiber-Fed Indoor Wireless Networks, *IEEE Transactions on Microwave Theory and Techniques,* pp. 3426-3432, Vol.54, No.8, August 2006.

Elmirghani, J.M.H., Badic, B., Li, Y., Liu, R., Mehmood, R., Wang, C., Xing, W., García Zuazola, I.J. & Jones, S. (2006). IRIS: An Intelligent Radio-fibre Telematics System, *proc. of 13th ITS World Congress and Exhibition in London,* UK, 8-12 October 2006.

FCC - Federal Communications Commision, (2002). *First Report and Order on Ultra-Wideband Technology,* fCC 02-48, Washington, DC, 22nd April, 2002.

García Zuazola, I.J., Batchelor, J.C., Langley, R.J., Das, A., Nkansah, A., Wake, D. & Gomes, N.J. (2005). Photonic Antenna Units containing Bi-directional Amplification for TDD and FDD in Picocell Systems, *Proc. LAPC Conference,* pp. 217-220, Loughborough, UK, April 2005.

García Zuazola, I.J., Elmirghani, J.M.H. & Batchelor, J.C. (2008). WiMAX Antennas for Intelligent Transport Systems communications, *Proc. LAPC Conference,* pp. 133-136, Loughborough, UK, 17-18 March, 2008.

García Zuazola, I.J., Elmirghani, J.M.H. & and Batchelor, J.C. (2009). High-speed ultra-wide band in-car wireless channel measurements, *IET Communications.,* pp. 1115–1123, Volume 3, Issue 7, 2009.

García Zuazola, I.J., Batchelor, J.C. & Elmirghani, J.M.H. (2010). Sectorized WIMAX Antenna for future Vehicular Communications Systems, *Microwaves, Antennas & Propagation, IET,* pp. 210 – 218, Volume 4, Issue 2, Feb. 2010.

García Zuazola, I.J., Batchelor, J.C., Elmirghani, J.M.H. & Gomes, N.J. (2010b). UWB PIFA Antenna for simplified transceivers, *Electronics Letters,* pp. 116–118, Volume 46, Issue 2, January 2010.

Gunter, Y. & Grosmann, H.P. (2005). Usage of Wireless LAN for Inter-Vehicle Communication, *Proceedings of the 8th International IEEE Conference on Intelligent Transportation Systems,* pp. 296-301, Vienna, Austria, September 13-16, 2005.

Kerner, B. S., Rehborn, H., Aleksi, M. & Haug, A. (2005). Traffic Prediction Systems in Vehicles, *Proceedings of the 8th International IEEE Conference on Intelligent Transportation Systems,* pp. 251-256, Vienna, Austria, September 13-16, 2005.

Kraus, J. D. & Marhefka R. J. (2001). *Antennas for all applications, 3rd edition,* McGraw-Hill, ISBN 0072321032, Boston, 2001.

Lee, K.F. & Williams, D. B. (2000). A Space-Frequency Transmitter Diversity Technique for OFDM Systems, *IEEE Globecom,* pp. 1473-1477, Volume 3, San Francisco, Nov. 2000.

Mohammad, N.H. & Ismail, W. (2008) System-level integration and simulation of ultra wideband receiver front-end, *Communications, Propagation and Electronics, MIC-CPE Mosharaka International Conference,* pp. 1-6, Jordan, 6-8 March 2008

RFlambda (n.d.) Available from: www.rflambda.com

Way, W. I. (1989). Subcarrier multiplexed lightwave system design considerations for subscriber loop applications, *Journal of Lightwave Technology*, pp. 1806–1818, vol. 7, no. 11, November 1989.

Part 3

Novel UWB Applications in Cognitive Radio Systems

11

UWB Cognitive Radios

Sithamparanathan Kandeepan[1,2], Gianmarco Baldini[3]
and Radoslaw Piesiewicz[4]
[1]*RMIT University, Melbourne*
[2]*CREATE-NET, Trento*
[3]*Joint Research Center, European Commission, Ispra*
[4]*Wroclaw Research Center, EIT+, Wroclaw*
[1]*Australia*
[2,3]*Italy*
[3]*Poland*

1. Introduction

In this chapter we present UWB communication as a potential candidate for cognitive radio technology. Cognitive radios are intelligent radios that could adopt itself by sensing and learning the radio environment and optimize its transmission strategies to maximize the utilization of the scarce radio resources such as the radio spectrum. This has been motivated by the radio regulatory bodies around the world (EC, 2007; FCC, 2003) to utilize unused radio spectrum known as white space in the spatio-temporal domain. In the recent years UWB communication has emerged as a potential candidate for the CR technology due to its ability to share the spectrum with others for short range wireless communications. In this context we present the concept of cognitive radios and the necessary techniques to adopt UWB as cognitive radios in this chapter. Especially, we enhance on the fundamentals of cognitive radios and spectrum sensing which enable the UWB radio to learn the radio environment. We also touch upon other cognitive radio related topics that are related to UWB communications such as dynamic spectrum access, interference mitigation and localization techniques. Furthermore, we present some potential applications for the use of UWB based cognitive radios which are derived from the European Union funded projects EUWB (EUWB, 2008) which is one of the biggest UWB projects that the world has seen so far, and the C2POWER project (C2POWER, 2010) which is related to energy efficiency in short range wireless communications with the use of cognitive radios. In this chapter we do not consider the technological aspects related to the use of cognitive radios for energy efficiency but only consider the use of cognitive radios for dynamic spectrum access. However, at the end of the chapter we present a scenario for the use of cognitive radios for energy efficiency derived from (C2POWER, 2010).

In the material presented in this chapter we mainly consider the high data rate UWB radios based on the Multi-Band Orthogonal Frequency Division Multiplexing (MB-OFDM) technique following the Wimedia specifications (Wimedia-PHY, 2009). The OFDM based transceiver design makes it feasible for the UWB radio to sense the radio environment and dynamically change the transmission parameters accordingly. This makes the UWB radios much more attractive and to suit cognitive radio technology that require having intelligence and adoptability in the radio itself. Moreover, the low transmit power in UWB communications also makes it feasible to have secondary user access to the spectrum without interfering with the primary users of the spectrum. The concepts of secondary users and primary users are treated subsequently in this chapter.

2. Cognitive radio fundamentals

The term cognitive radio was coined by Joseph Mitola (Mitola, J. & Maguire Jr. G.) considering ideal context aware radios with embedded intelligence. Mitola's vision of cognitive radios spans across all the layers of the communication protocol stack emphasizing on the need for optimum utilization of the radio resources by adopting its transmission policies and strategies. The adaptation of the local policies is based on sensing and learning the environment or by being informed about the radio environment by an information broker in the network. Haykin (Haykin, S. 2005) then adopted Mitola's ideal cognitive radio concept to wireless communications by defining the corresponding communications and signal processing problems associated with cognitive radios in the lower layers of the protocol stack. Here we present the fundamentals of cognitive radios explaining the cognitive engine and the cognitive cycle as described by Mitola and Haykin. We present the concept of white-space in the spatio-temporal domain in regards to spectrum utilization and the underlay and overlay technologies for dynamic spectrum access.

2.1 Spectrum classification in a broader sense

First let us classify the spectral usage in the spatio-temporal domain. By computing the power spectra of the received radio stimuli at a particular point and time one could broadly classify the spectra into three types (Haykin, S. 2005), as given below.

Black Spaces: spectra occupied by high-power 'local' interferers.

Gray Spaces: spectra occupied partially by low power interferers.

White spaces : spectra free of radio frequency interferers except for ambient natural and man-made noise.

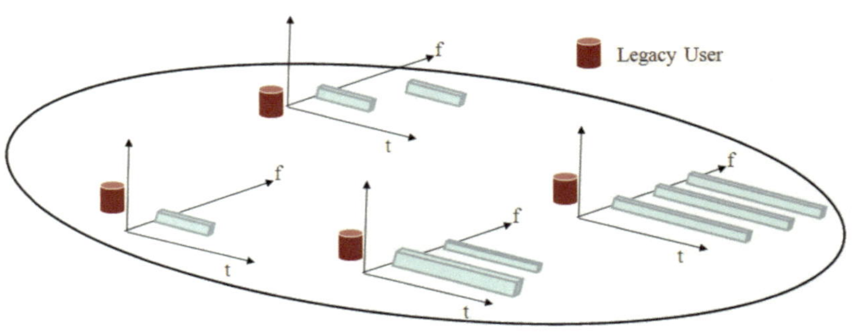

Fig. 1. The evolution of 'spectrum holes' in the spatio-temporal domain

One could clearly see that the above classification is a function in the spatio-temporal domain. For example, 'black', 'gray' and 'white' spaces could appear and disappear back and forth at a particular location over time. Therefore it is necessary to sense and learn the radio environment in order to maximize the spectral usage opportunistically. In other words detecting 'spectrum holes' as it is termed is quite crucial for dynamic spectrum access. Figure-1 depicts the concept of 'spectrum hole' evolution in the spatio-temporal domain.

2.2 Spectrum sharing in cognitive radio networks: 'Underlay' and 'Overlay' techniques

With cognitive radio technology the concept of 'primary users' and 'secondary users' of the spectrum are developed. The primary users are the incumbent users with the exclusive rights to use the spectrum at anytime and the secondary users, also known as the cognitive radio users, are the users that use the spectrum without interfering with the primary users. There are basically two spectrum sharing techniques considered for cognitive radio networks for maximizing the spectral efficiency between the primary and the secondary users. First is the 'spectrum underlay' technique and second is the 'spectrum overlay' technique.

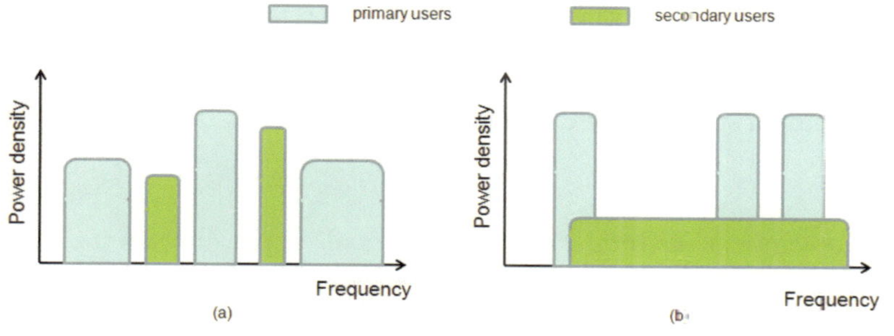

Fig. 2. Spectrum sharing in cognitive radio networks, with (a) overlay and (b) underlay sharing techniques.

In the 'spectrum underlay' method the secondary users can utilize the spectrum simultaneously with the primary users without exceeding a predefined interference level to the primary users. Secondary users in this case can share the spectrum such that the total interference power from the secondary users to the primary users are controlled below the interference limit set by the relevant regulatory authorities. The characterization of such interference limit is given in the next subsection. Figure-2 depicts the concept of spectrum underlay technology. UWB radio technology due to the low powered transmissions in the ultra wide band frequency range is therefore a potential candidate for deploying spectrum underlay technology for spectral sharing. Using the low powered transmissions and making sure that the interference limit is not exceeded UWB radios can potentially share the spectrum with the primary users and coexist.

In the 'spectrum overlay' method the cognitive radios can identify the spectrum holes in the spatio-temporal domain and opportunistically utilize them by giving higher priority to the primary users. Whenever a primary user is not using the spectrum secondary users (cognitive radios) are allowed to transmit however when a primary user is detected in that particular band then secondary users need to immediately vacate the band by stopping transmitting in that particular band. In this sense spectrum sensing and primary user detection become a crucial functionality for reliably detecting the primary users in the environment in the spatio-temporal domain. Figure-2 depicts the concept of spectrum overlay technology at a particular time in some space. The MB-OFDM based UWB technology is considered as a potential candidate for spectrum overlay technology for spectrum sharing by inherently making use of the OFDM transmission technique. By using OFDM, UWB devices can dynamically turn on and off the corresponding subcarriers depending on whether any primary users exist or not in a particular band in the environment. In other words the transmission spectrum of UWB radios can be sculpt according to the presence of the primary users in the respective frequency bands in the environment.

2.3 The interference temperature limit

The interference temperature T_I is a measure of the interference power level due to wireless transmissions at a particular location as defined by the FCC in (FCC, 2002). The interference temperature follows a similar definition as to the thermal noise temperature in receivers. It is well known that the thermal noise power P_n in receivers is given by (Sklar, B.),

$$P_n = kT_N B \tag{1}$$

where $k = 1.38 \times 10^{-23}$ is the Boltzmann's constant, B (Hz) is the receiver operating frequency bandwidth and T_N (in degrees Kelvin unit) is the noise temperature. Likewise the total interference power P_I due to the transmissions of wireless devices and natural interferences at a particular point in space can be characterized by,

$$P_I = kT_I B \tag{2}$$

The interference temperature limit $T_I^{\max}$ therefore is an upper limit on the value of T_I that can be used to control and limit the interference in the radio environment. Such limits for the interference temperature can be used to enable the underlay spectrum sharing technique by coordinating or policing the interference level in the environment generated by the secondary users to the primary users.

2.4 The cognitive cycle

The cognitive cycle is the term describing the activities involving the intelligence of the radio device such as sensing, learning and adopting. In (Mitola, J. & Maguire Jr. G.), Mitola had presented a generic cognitive cycle that corresponds to his view of ideal cognitive radios. By adopting this model, Haykin then presented a similar cognitive cycle model in (Haykin, S. 2005) by mainly describing the PHY and MAC layer aspects of the radio device considering the communications and signal processing functionalities. Here we explain both the cognitive cycles described by Mitola and Haykin.

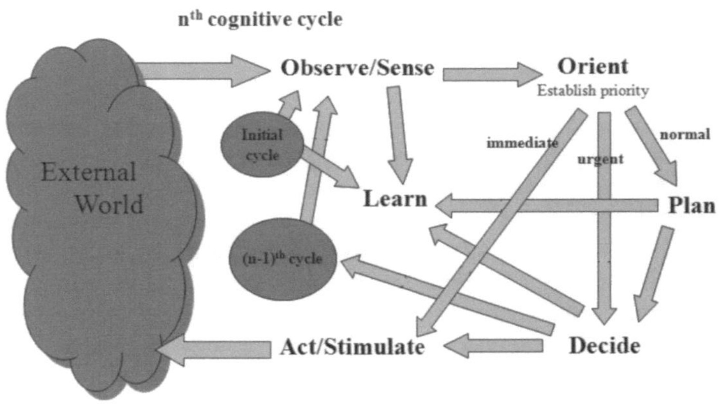

Fig. 3. Cognitive Cycle described by (Mitola, J. & Maguire Jr. G.)

Figure-3 depicts the cognitive cycle described by Mitola. In the figure the radio observes and senses the external world and orients itself according to the internal policies and plans before making a decision on how to act upon that situation. Once the decision is made the radio then acts accordingly. Then in the next and the consecutive cycles it goes through a similar process until it decides not to operate. The cognitive radio learns from the observations as shown in the figure. The core of the cognitive cycle that lies inside the radio is known as the cognitive engine.

The equivalent cognitive cycle presented by Haykin is depicted in Figure-4. In this figure, the corresponding signal processing and communications functionalities associated with the radio is presented within the cognitive engine. As shown in the figure, the cognitive radio observes the radio environment using the sensed radio stimuli and creates a radio environment map of the potential radios in the environment considering the spatio-temporal usage of the frequency bands. This information is then used together with the channel state estimation by the transmitter to adopt its transmissions accordingly.

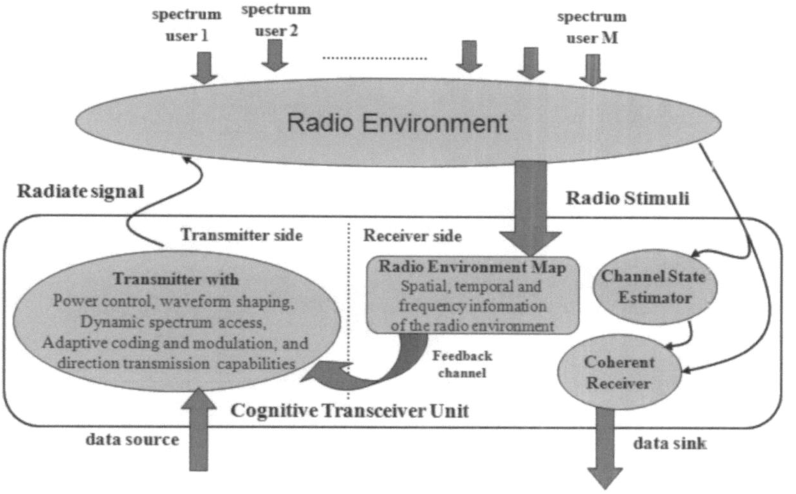

Fig. 4. Cognitive Cycle corresponding to the communications and signal processing aspects in the radio, as described by (Haykin, S. 2005)

The cognitive engines presented in Figure-3 and Figure-4 are only the conceptual ones which include the basic functionalities required for the radio to have intelligence. The spectrum sensing functionality helps the radio to observe the radio environment, and is one of the hot topics in the field of cognitive radios. Spectrum sensing is covered later in Section-5. The radio environment map is then created with the use of spectrum sensing information and the radio-localization functionality (if available) of the cognitive radio. Localizing a radio in the environment is not always feasible given the fact that the localization task needs to be performed blindly. Once the cognitive radio nodes have a good understanding of the radio environment it would then perform power control with appropriate interference mitigation techniques in the spatio-temporal domain to transmit its data. Furthermore, other functionalities also can be added into the cognitive engine depending on the applications and any specific requirements appropriately.

3. Dynamic spectrum access

The radio spectrum can be utilized by considering various access strategies, methodologies or policies. In this section we provide a quick background on the spectrum access models to explain how cognitive radios are used for dynamic spectrum access. Spectrum access models can be classified as command and control model, exclusive-use model, commons model and the shared model (Hossain, E. et. al.). In the shared use model the secondary user of the spectrum will opportunistically access the spectrum without interfering with the primary user of the spectrum, in the exclusive use model the unlicensed secondary user can be granted access to the spectrum by the licensed primary user, and in the commons model the secondary user can access the spectrum without any restrictions. In Figure-5 we present a taxonomy of the different spectrum access models.

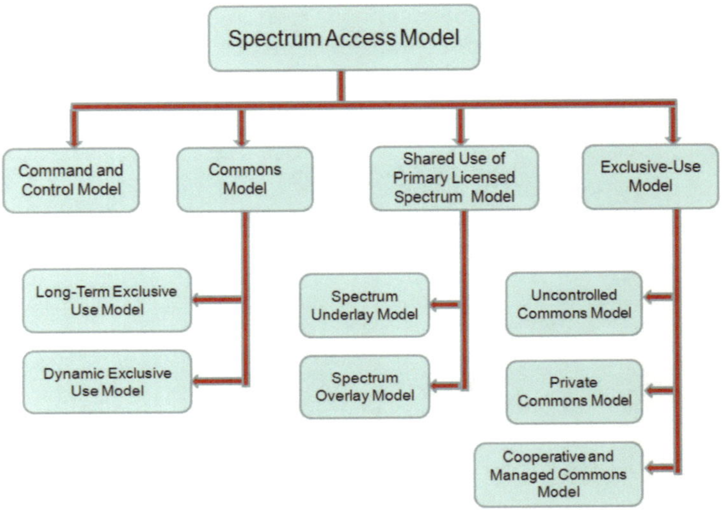

Fig. 5. Classification of spectrum access models

For a detailed description of the different access models the reader is referred to (Hossain, E. et. al.). In the previous section we briefly described the access model that is of interest to us which is the shared spectrum access model that includes the spectrum underlay and overlay techniques. In the 'shared' model the concept of primary and secondary users of the spectrum are derived and the spectrum can be shared simultaneously between the primary and the secondary users of spectrum. The primary users are the incumbent users of the spectrum however the secondary radios also can use the spectrum. In this case the secondary radios need to make sure that they do not interfere with the primary radio transmissions, and as long as the interference constraint is met the secondary users can use the spectrum transparently to a primary user.

4. UWB as cognitive radio, and coexistence

As described previously cognitive radio nodes require intelligence and self adoptability in order to dynamically adopt its strategies based on the time varying radio environment. In this section we see how UWB devices can suit such requirements and be considered as a potential candidate for cognitive radio technology. Based on MB-OFDM transmission, Figure-6 depicts how UWB radios could be used as secondary radios based on cognitive radio technology for sharing the spectrum with the other users. In this section we further describe how the UWB radios can be considered to adopt both underlay and overlay spectrum sharing models.

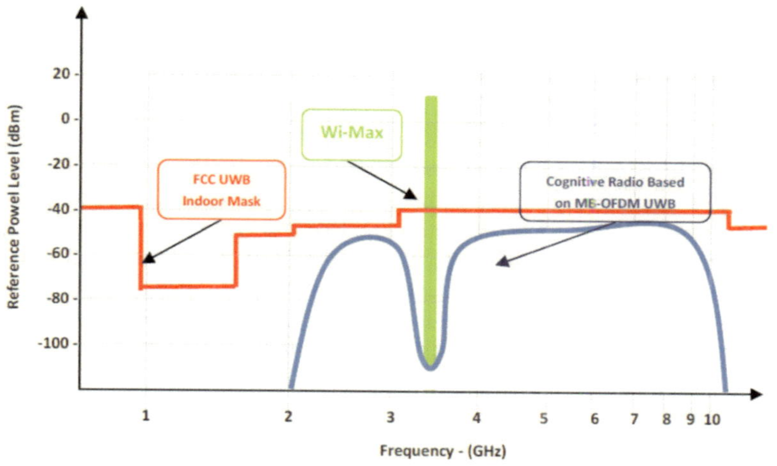

Fig. 6. Intelligent spectrum sharing mechanism based on underlay and overlay policies by UWB radios using cognitive radio technology

As we know there exist many radio technologies in the UWB frequency range operating in the licensed as well as the unlicensed frequency bands. The UWB radios therefore need to coexist with all the radios in the frequency range which makes UWB as a potential candidate for cognitive radio technology which maximizes the usage of the scarce spectrum. In this section we further consider the coexistence of various spectrum users in the UWB frequency range with the MB-OFDM based UWB radios. In particularly, we discuss the policies and requirements for the UWB radios to coexist with the other radios and utilize the spectrum as a secondary user considering both underlay and overlay spectrum access methods. Power controlling with the concept of interference temperature limit, spectrum sculpting together with detect-and-avoid (DAA) techniques are some of the strategies used by UWB radios in order share the spectrum with the primary users. Below we provide some background on spectrum sculpting and power control in UWB radios in the context of spectrum sharing, and later we present detect-and-avoid technique in detail.

4.1 Spectrum sculpting

For the UWB radios to share the spectrum with the primary users using the overlay method it needs to shape its transmission spectrum in such away that the primary users are not interfered. Spectrum sculpting techniques are used for shaping the spectrum in UWB radios (Wang, Z.; Yamaguchi, H.). The two most common spectrum sculpting methods are the spectrum shaping in time domain using shaped pulses and spectrum shaping in the frequency domain using tone nulling (in OFDM systems). The time domain method in general may not be possible to shape the spectrum in all the cases, the frequency domain tone nulling method on the hand can provide better performances in terms of shaping the spectrum. The tone nulling technique can cause spectral overshoots in the transmission band and hence various derivatives of this method are also considered such as enhanced active interference cancelation as proposed in (Wang, Z.). The example shown in Figure-6 clearly depicts how the spectrum sculpting technique is used in UWB radios in order to coexist and share the radio spectrum with the primary user radios in the environment.

4.2 Power control

Power control in wireless and mobile communications is a well studied topic for more than twenty years. It has attained more attention in the recent years for potential spectrum sharing in cognitive radio networks. Traditionally power control was considered for maximizing the transmission rate with fare-scheduling without degrading the QoS of the other users in the environment. In a similar context power control is also considered for cognitive radio networks as presented in (Gu, H.; Radunovic, B.; Xing, Y.; Zhang, L.). Here we briefly explain the concept on power control for dynamic spectrum sharing with underlay technology in cognitive radio networks by having the total interference power as a constraint.

Suppose P_I is the interference power limit corresponding to the interference temperature T_I as explained in (2). If there exist K number of cognitive radios in the environment sharing the spectrum with the incumbent users, then the total interference caused to the lth primary user is given by,

$$I_l = \sum_{k=1}^{K} h_{kl} P_k \tag{3}$$

where, P_k is the transmitted power from the kth cognitive radio node, and h_{kl} is the channel gain from the kth cognitive radio node to the lth primary user in the environment. In order to comply with the interference regulatory level, for the interference caused from the secondary users to the primary users, the following constraint should be met,

$$I_l = \sum_{k=1}^{K} h_{kl} P_k \leq P_I \forall l \tag{4}$$

The cognitive radio nodes on the other hand would like to achieve the highest possible transmission rate which is related to the received signal to interference ratio γ_{km} at the mth secondary receiver where $m = 1 \dots K$ and $m \neq k$, given by,

$$\gamma_{km} = \frac{h_{km} P_k}{\sum_{u=1, u \neq k,m}^{K} h_{um} P_u + \sigma_m^2} \tag{5}$$

where, h_{km} is the channel gain from the transmitter k to the intended receiver m, h_{um} is the channel gain from the transmitter u to the unintended receiver m, P_u is transmitted power from the transmitter u, and σ_m^2 is the receiver noise power at the receiver node m. Then in order for the secondary communication pair $\{k, m\}$ to have the best possible transmission rate, considering the constraint in (4), the simplest optimization strategy for power control is given by,

$$\hat{P}_k = \max_{P_k} \gamma_{km}, \text{ such that } I_l \leq P_I \tag{6}$$

It might be difficult to measure the interference power at the primary user node unless the primary user cooperates. In such situations there can be a power controller or a monitor serving the purpose of controlling the power by measuring the total interference power at some central location. In literature one could find various cooperative and distributed power controlling methods using game theoretic approaches which we do not cover in this chapter.

5. Spectrum sensing

Spectrum sensing is one of the crucial functionalities of a cognitive radio in order to learn the radio environment. Various spectrum sensing techniques exist (Kandeepan, S. et. al; Yucek, T. and Arslan,H.) and in general could be classified as 1) energy based sensing, 2) cyclostationary feature based sensing and 3) matched filter based sensing. The energy based sensing is the simplest method to sense the environment in a blind manner, the cyclostationary based sensing may require some information about the spectral-user signal characteristics, and the matched filter based sensing requires the complete information of the spectral-user signal. In this section we elaborate in detail on the various spectrum sensing techniques and their related detection performance for MB-OFDM based sensing. Moreover, we present collaborative sensing techniques in order to address the 'hidden node problem'.

Let us provide some background on spectrum sensing prior to presenting the related techniques. Spectrum sensing and detecting the presence of a radio in the environment is treated as a classical statistical detection problem (Kay, S.). We define the two binary hypotheses H_0 and H_1 to indicate the absence and the presence of the primary users in the environment respectively. In the discrete signal domain this could be represented as,

$$r(n) = \begin{cases} \nu(n), & H_0 \\ s(n) + \nu(n), & H_1 \end{cases} \tag{7}$$

where $\nu(n)$ is the additive Gaussian channel noise and $s(n)$ is the received signal. If the test statistic that is used for the detection is given by $\xi(r(n))$, which is a function of the sensed signal $r(n)$ with $n = 1, 2 \ldots N$, then the detection criteria is given by,

$$d = \begin{cases} 0; \ \xi < \lambda \\ 1; \ \xi \geq \lambda \end{cases} \tag{8}$$

where, λ is known as the detection threshold. The probability of detection and the probability of false alarm are then defined as,

$$P_D = Pr[d = 1 | H_1] \tag{9}$$

$$P_{FA} = Pr[d = 1 | H_0] \tag{10}$$

The probability of miss detection on the other hand is defined by $Pr[d = 0|H_1]$, and thus is given by $P_M = 1 - P_D$. In general the detection threshold λ is chosen in order to trade off between the detection and false alarm probabilities. Different criteria can be used in order to find the optimal threshold which is a well treated topic in the literature of statistical detection, which we do not present in this chapter. In the subsequent sections we provide various ways to derive the test statistic ξ used for the detection of primary users.

5.1 The hidden terminal problem

Prior to presenting the spectrum sensing techniques we present why spectrum sensing is treated as an important topic in cognitive radio literature. We mentioned that the detection performance is characterized by the probability of successfully detecting the radio and the probability of false alarm. In cognitive network applications the regulatory bodies are quite strict on secondary nodes causing any interference to the primary users, in this sense the primary users need to be reliably detected by the secondary users with a high detection probability (close to 100% or $P_D \simeq 1$). The detection probability usually depends on the signal to noise ratio of the received signal, the received signal power depends on how far the transmitting node is from the sensing node characterized by the path loss. Moreover, channel fading is also a factor that affects the received signal power. In this sense, radio nodes (primary users) closer to the cognitive radios are easily detected with a higher probability of detection compared to the radio nodes that are further away from the cognitive radios. When the primary user radios are not detected by the cognitive radio nodes they do not appear in the radio environment map created by the cognitive radio nodes, and hence the primary user nodes become hidden to the cognitive radio nodes. This is known as the 'hidden terminal problem'. Figure-7 depicts a typical hidden terminal problem scenario. In the figure, CR-1 is unable to detect the PU and hence the PU node is hidden from the CR-1 node. The hidden node problem can create interference from the secondary nodes to the primary nodes and hence therefore harming the communication rights of the primary radios in the allocated band and violating the regulatory requirements. The hidden terminal problem also can harm the performance of secondary user communications interfered by the primary user in this case. Therefore, various spectrum sensing and detection techniques are considered to solve the hidden terminal problem to increase the detection probability for detecting the primary users in the environment.

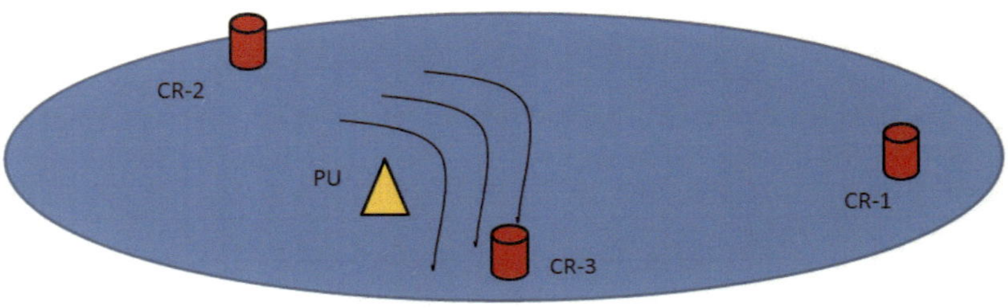

Fig. 7. An example of the hidden terminal problem, where the PU node is hidden from CR-1.

5.2 Spectrum sensing with energy detection

The energy detector is the simplest spectrum sensing method for detecting primary users in the environment in a blind manner (Urkowitz, H.). It is computationally efficient and also be used conveniently with analog and digital signals (or in other words at the RF/IF stages or at the base band). It also has a well known drawback in the detection performance when the noise variance is unknown to the sensing node. When the signal to noise ratio is very low the knowledge of the noise power can be used to improve the detection performance of the energy detectors. In energy detectors, the energy of the received signal is computed over a time period T or equivalently over N samples in the discrete domain and used as the test statistic, where $T = NT_s$ and T_s is the signal sampling period. The test statistic at the base band considering the complex envelope of the received signal is therefore given by,

$$\xi = \int_{t_1}^{t_2} r(t)\tilde{r}(t)dt \tag{11}$$

where, $\tilde{r}(t)$ is the complex conjugate of $r(t)$. The signal to noise ratio (SNR) is then defined based on the received signal $s(t)$ for $t_1 < t \leq t_2$ for some $t_1, t_2 \in \mathbb{R}^+$, given by,

$$\rho = \frac{1}{\sigma_i^2[t_2 - t_1]} \int_{t_1}^{t_2} s(t)\tilde{s}(t)dt \tag{12}$$

Note that based on the transmission pattern of the primary user the instantaneous signal to noise ratio would vary, here however we assume ρ to be a constant. For the discrete signal on the other hand, the energy based test statistic is given by,

$$\xi \approx T_s \sum_{n=0}^{N-1} r[n]\tilde{r}[n] \tag{13}$$

where, N is the total number of complex samples and is also known as the time-bandwidth product (Urkowitz, H.). Note that in (12) there are essentially N number of real component samples and N number of imaginary component samples. Considering the discrete domain test statistic the detection criteria is then given by,

$$d = \begin{cases} 0; \ \xi < \lambda \\ 1; \ \xi \geq \lambda \end{cases} \tag{14}$$

In order to compute the detection probability and the false alarm probability we consider the distribution of the test statistic ξ. The energy based test statistic ξ follows a non-central and a central chi-sqaure distribution under H_0 and H_1 respectively with $2N$ degrees of freedom. Using the distributions of the test statistic under H_0 and H_1 we can derive the detection probability and the false alarm probability using equation (9) and (10) and in closed form expressions as (Dingham, F.,F.),

$$P_D = Q_N(\sqrt{2N\rho}, \sqrt{\lambda}) \tag{15}$$

$$P_{FA} = \Gamma(N, \lambda/2) \tag{16}$$

where, $\Gamma(a,b) = \frac{1}{\Gamma(N)} \int_b^{\infty} u^{a-1} \exp(-u) du$ is the regularized upper incomplete Gamma function, $\Gamma(.)$ is the Gamma function, $Q_N(a,b) = \int_b^{\infty} u^N \exp(-(u^2+a^2)/2) I_{N-1}(au)/a^{N-1} du$ is the generalized Marcum Q-function, and $I_{N-1}(.)$ is the modified Bessel function of first kind with order $N-1$.

Let us look at some results for the detection performance of the energy detector in the additive Gaussian noise channel by plotting the complementary receiver operating characteristics (C-ROC) curve. The C-ROC depicts the probability of false alarm in the x-axis and probability of miss detection in the y-axis. Figure-8 shows the C-ROC curves for the energy detector for various values of signal to noise ratio levels ρ. As we observe from the figure, the detection performance improves with increasing values of ρ by achieving lower miss detection probabilities for lower false alarm probabilities when ρ increases. Figure-9 on the other hand shows the C-ROC curves for various values of N, and again we observe that the detection performance improves with increasing values of N.

Note that the analytical results presented here do not consider the wireless channel effects such as fading or shadowing, authors in (Dingham, F.,F.) and (Atapattu. S., et. al.) have presented closed form expressions for the detection probability for the energy detector considering various wireless channels which we do not cover in this chapter.

5.3 Spectrum sensing with cyclostationary feature detection

The cyclostationary feature analysis is a well developed topic in the literature of signal processing (Gardner, W.). In wireless communications, depending on the modulation type, data rate and carrier frequency etc. the transmitted signals show very strong cyclostationary features, especially when excess bandwidth is utilized. Therefore identifying the unique set

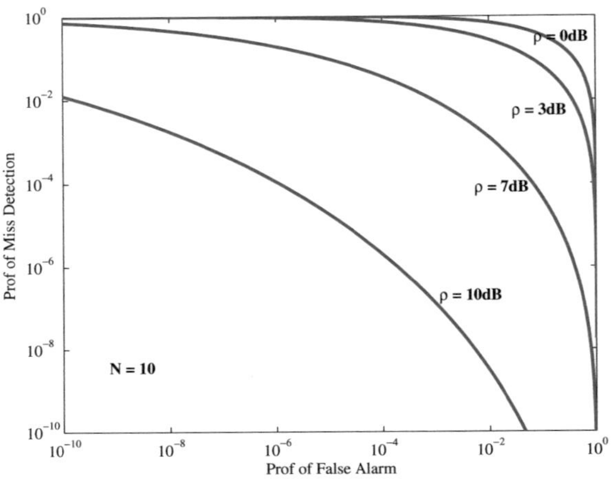

Fig. 8. Complementary ROC curves for the energy detector for various signal to noise ratio levels.

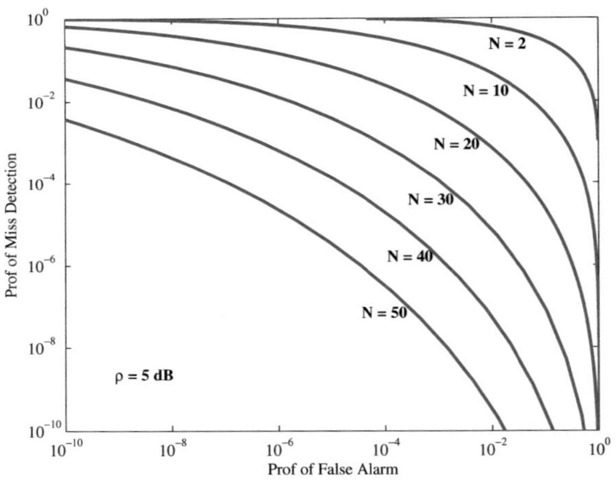

Fig. 9. Complementary ROC curves for the energy detector for various values of time-bandwidth product N.

of features of a particular radio signal for a given wireless access system can be used to detect the system based on its cyclostationary features. In the context of spectrum sensing some studies have been performed in using the cyclostationary features to detect the primary users in the environment (Kandeepan, S.; Kyouwoong, K.). For sufficient number of samples, when the cyclostationary features are properly identified, this method can perform better than the energy based detection method. However the main drawbacks with this method are the complexity associated with the technique and the requirement of some a-priori knowledge of the primary user signal.

Here we provide some of the fundamentals of cyclostationary feature analysis and show how it can be used to detect primary users in the environment for cognitive radio networks.

A random process $x(t)$ can be classified as wide sense cyclostationary if its mean and autocorrelation are periodic in time with a period of T. Mathematically the mean and the autocorrelation are respectively given by,

$$E_x(t) = E_x(t + mT) \tag{17}$$

and

$$R_x(t,\tau) = R_x(t + mT, \tau) \tag{18}$$

where, t is the time, τ is the lag associated with the autocorrelation function, and m is an integer. The periodic autocorrelation function can be expressed in terms of the Fourier series given by,

$$R_x(t,\tau) = \sum_{\alpha=-\infty}^{\infty} R_x^{\alpha}(\tau) \exp(2\pi j\alpha t) \tag{19}$$

where,

$$R_x^{\alpha}(\tau) = \lim_{T\to\infty} \frac{1}{T} \int_T x(t + \frac{\tau}{2}) x^*(t - \frac{\tau}{2}) \exp(-2\pi j\alpha t) dt \tag{20}$$

The expression in (20) is known as the cycle autocorrelation, and for a cyclostationary process with a period T_0, the function $R_x^{\alpha}(\tau)$ will have component at $\alpha = 1/T$. Using the Wiener relationship, the Cyclic Power Spectrum (CPS) or the spectral correlation function can be defined as,

$$S_x^{\alpha}(f) = \int_{-\infty}^{\infty} R_x^{\alpha}(\tau) \exp(-j2\pi f\tau) d\tau \tag{21}$$

The CPS in (21) is a function of the frequency f and the cycle frequency α, and any cyclostationary features can be detected in the cyclic frequency domain. An alternative expression for (21), for the ease of computing the CPS, is given by.

$$S_x^{\alpha}(f) = \lim_{T_0\to\infty} \lim_{T\to\infty} \frac{1}{T_0 T} \int_{-T_0/2}^{T_0/2} X_T(t, f + \frac{1}{\alpha}) X_T^*(t, f - \frac{1}{\alpha}) dt \tag{22}$$

where, $X_T^*(t,u)$ is the complex conjugate of $X_T(t,u)$, and $X_T(t,u)$ is given by,

$$X_T(t,u) = \int_{t-T/2}^{t+T/2} x(v) \exp(-2j\pi uv) dv \tag{23}$$

Expression in (22) is also known as the time-averaged CPS which achieves the theoretical CPS. Figure-10 depicts an example of a CPS plot for BPSK modulated signal. In the figure N is the number of samples per block corresponding to the time interval T and M is the total number of blocks used for time averaging. The figure clearly depicts the cyclic frequency components centered at $\alpha = \pm 40$ MHz, and the additive noise component appearing at $\alpha = 0$. Therefore, using the CSD one could detect the presence of the primary user.

Now, we present how the CSD can be used to generate a test statistic to perform primary user detection. Considering the CSD to detect the primary user, we can re-write equation (7) with respect to the CSD as,

$$\begin{aligned} H_0 &: S_r^{\alpha}(f) = S_v^{\alpha}(f) \\ H_1 &: S_r^{\alpha}(f) = S_s^{\alpha}(f) + S_v^{\alpha}(f) \end{aligned} \tag{24}$$

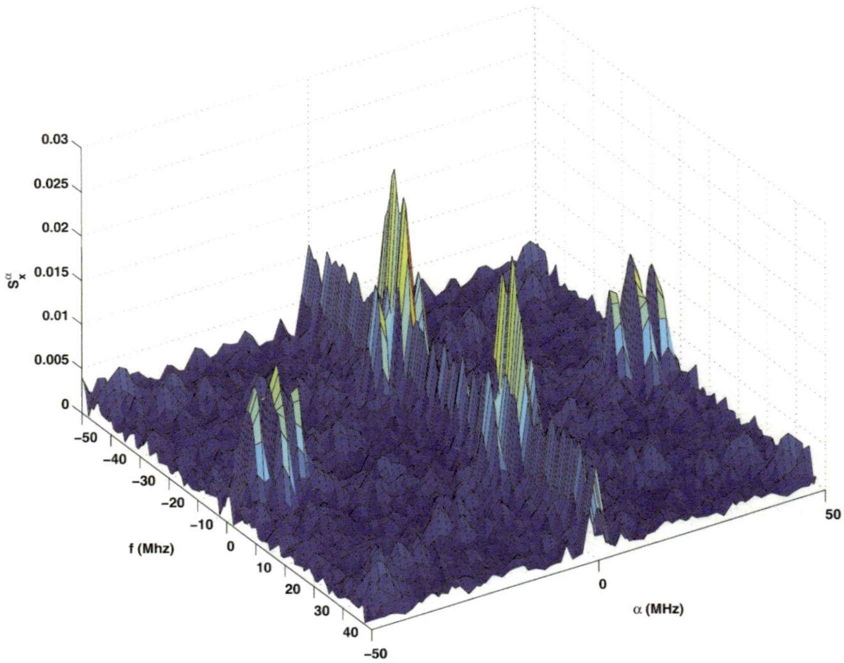

Fig. 10. Cyclic Spectral Density for BPSK with a signal to noise ratio of -13.3dB with $N = 50$ samples averaged over $M = 40$, with $f_c = 20$MHz center frequency and symbol rate of $R_b = 5$Mbps.

where, $S_{\nu}^{\alpha}(f)$ is the CSD of the additive noise $\nu(t)$, and $S_{s}^{\alpha}(f)$ is the CSD of the primary user signal $s(t)$. Since $\nu(t)$ is not a cyclostationary process, the CSD of ν for $\alpha \neq 0$ is zero. Based on this we can derive the test statistic for the detector in the discrete domain as,

$$\xi = \sum_{\alpha,\alpha \neq 0} \sum_{f} S_{r}^{\alpha}(f) \tilde{S}_{r}^{\alpha}(f) \tag{25}$$

where, $\tilde{S}_{r}^{\alpha}(f)$ is the conjugate of $S_{r}^{\alpha}(f)$. The detector is then given by,

$$d = \begin{cases} 0; \ \xi < \lambda \\ 1; \ \xi \geq \lambda \end{cases} \tag{26}$$

An important point to note here is that one needs sufficient number of samples (i.e. sufficient values for N and M) to get a good estimate of the CSD and hence this method is not so computationally efficient. Furthermore, when insufficient number of samples are used, the detection performance will tend to greatly degrade due to the poor estimate of the CSD.

5.4 Spectrum sensing with matched filter detection

The matched filter detection based sensing is the same as the traditional matched filter detection technique (Sklar, B.) deployed in traditional digital receivers. Obviously for match filter based spectrum sensing a complete knowledge of the primary user signal is required such as the modulation format data rate, carrier frequency etc. The matched filter detection technique is a very well treated topic in literature therefore we just present the fundamental results on matched filter detection in this section. Given a real transmit signal waveform $s(t)$ defined over $0 \leq t \leq T$ the corresponding matched filter maximizing the signal to noise ratio at the output of the filter sampler is given by,

$$h(t) = \begin{cases} s(T-t); & 0 \leq t \leq T \\ 0; & \text{elsewhere} \end{cases} \tag{27}$$

Figure-11 depicts matched filter based spectrum sensing method for primary user detection. Considering that a complete signal information of the primary user signal is required in this case the matched filter method is not really recommended by the system designers to suit our purpose here unless when the complete signal information is known to the secondary user. Then based on the test statistic $\xi(nT)$ at the output of the filter sampled every $t = nT$ seconds,

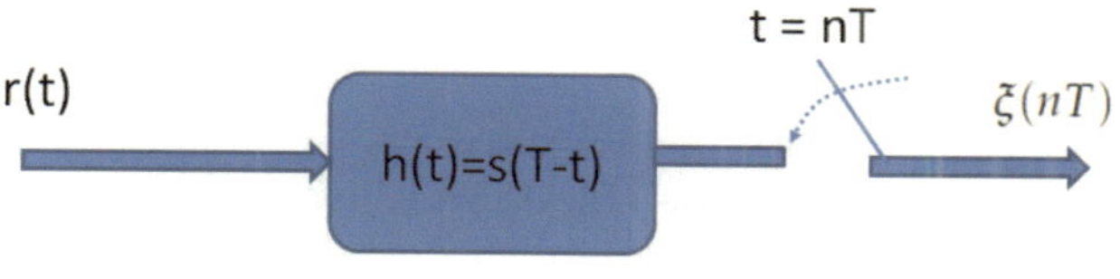

Fig. 11. Matched filter based spectrum sensing and detection of primary users.

the detector is given by,

$$d(nT) = \begin{cases} 0; & \xi(nT) < \lambda \\ 1; & \xi(nT) \geq \lambda \end{cases} \tag{28}$$

Performance wise matched filter based detector gives better detection probability compared to the previously discussed methods using the energy detector and the cyclostationary feature based detector.

5.5 Spectrum sensing in UWB radios with MB-OFDM

In many cases it is important to detect the primary users within a specific sub-band rather than in the entire band of operation. Since UWB spans a large portion of the spectrum we adopt a slightly different method to detect primary users in the particular bands of interest. Detecting a primary user in the entire band of operation can be performed using any of the methods specified for spectrum sensing in the previous sections. Here we consider detecting the primary user in a specific sub-band. Consider the FFT output of the MB-OFDM UWB receiver $y_j(i)$ corresponding to the jth time sample in the ith frequency bin with $i = 1, 2 \ldots N_{FFT}$. The primary user detection in the ith frequency bin can be performed considering the energy detector (for example) using the test statistic given by,

$$\xi(i) = \sum_{j=1}^{N} y_j(i) y_j^*(i) \tag{29}$$

followed by the threshold detection as explained before. The primary user detection for the entire band can also be obtained by extending the above test statistic to derive a new test statistic given by,

$$\xi = \sum_{i=1}^{N_{FFT}} \sum_{j=1}^{N} y_j(i) y_j^*(i) \tag{30}$$

followed by the threshold detection. A hard-combining method for the primary user detection in the entire band could also be used on the other hand. In the hard-decision based combination method we can decide that a primary user is present if at least one of the N_{FFT} frequency bin had decided on H_1. In this case the detection rule becomes,

$$d = \bigwedge(d_i) \tag{31}$$

where, $\bigwedge$ is the logical 'OR' operation in the binary format and d_i are the decision outcomes for the ith frequency bin. In this case we could also derive the detection and false alarm probabilities (for the 'OR' logic based hard decision rule) as,

$$P_D = 1 - [1 - P_D(i)]^{N_{FFT}} \tag{32}$$

$$P_{FA} = P_{FA}^{N_{FFT}}(i) \tag{33}$$

where $P_D(i)$ and $P_{FA}(i)$ are the detection and false alarm probabilities respectively for the ith frequency bin.

5.6 Collaborative spectrum sensing

In the previous section we described the hidden terminal problem with primary user nodes located at a distance from the the sensing cognitive radio node. Exploiting the spatial domain for spectrum sensing is a way to solve the hidden terminal problem which we present here. Spatial domain spectrum sensing is performed by sensing the environment at different locations and by fusing the corresponding results to make a final decision whether a primary user is present or not. This could be done by cognitive radio nodes collaboratively sensing the environment and sharing the information with each other. In such sense there are two ways collaborative sensing could be performed 1) cooperative spectrum sensing with a centralized fusion center, and 2) distributed spectrum sensing with no centralized fusion center. Below we present them in detail.

5.6.1 Cooperative spectrum sensing

In cooperative spectrum sensing (Ganesan, G.; Mishra, S.) cognitive radio nodes in the environment sense the spectrum locally and then send the decision to a centralized node termed as the cognitive base station (CBS). The cognitive base station then would fuse the data and decide upon the presence of a primary user, the decision is then reported back to the cognitive radio nodes in a secure and reliable channel. Figure-12 depicts the cooperative spectrum sensing scenario which we describe here. Error prone reporting channels between the cognitive radios and the base station can also degrade the detection performance for the cooperative sensing as presented in (Aysal, T.). In this case however we assume error free channels for reporting the sensing data back and forth.

Let us assume there are K cognitive radio nodes present in the environment, and the $k^{th}, k \in \{1, 2 \ldots K\}$ cognitive radio node performs spectrum sensing using any of the one methods described in the previous sections (energy based method for example). If the corresponding hard decision made by the respective cognitive radio node is d_k, then the cognitive base station can perform its fusion using the 'OR' rule given by $\hat{d} = \bigwedge(d_k)$, where $\bigwedge$ is the logical 'OR' operator. The corresponding detection probability after data fusion then becomes,

$$P_D = 1 - [1 - P_D(k)]^K \tag{34}$$

$$P_{FA} = P_{FA}^K(k) \tag{35}$$

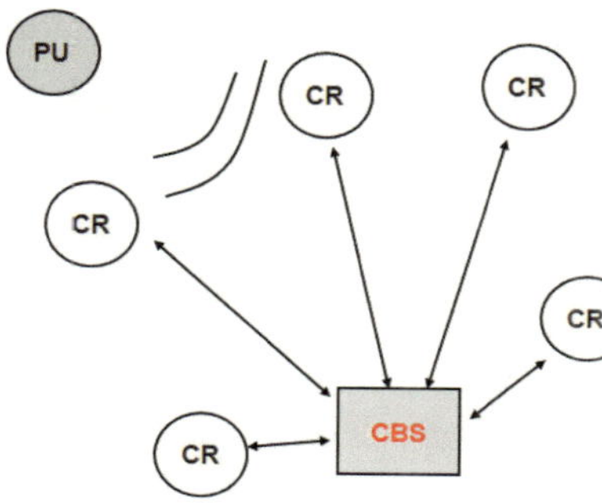

Fig. 12. Cooperative spectrum sensing with cognitive base station.

where $P_D(k)$ and $P_{FA}(k)$ are the detection and false alarm probabilities respectively for the local sensing performance at the kth cognitive radio node. The fusion rule at the cognitive base station can be varied depending on the design requirements. One could also consider the logical 'AND' rule or in general the L out-of-K rule where you decide upon the presence of the primary user if L cognitive radio nodes have detected the presence out of the K nodes. Figure-13 depicts the performance curves in terms of the complementary ROC curves for the 'OR' rule base cooperative sensing with energy based local decisions. From the figure we clearly see a great improvement in the detection performance when fusion strategy is deployed with cooperative sensing compared to the non-cooperative sensing case, especially at low signal to noise ratio levels.

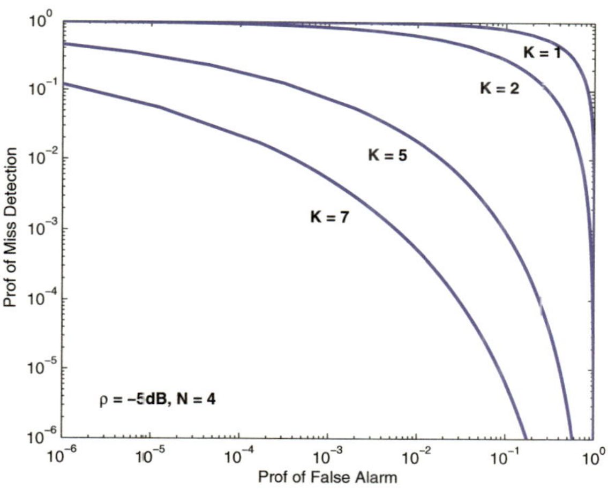

Fig. 13. C-ROC curves for the cooperative spectrum sensing with the 'OR' rule based fusion decision at the CBS, with $\rho_k = \rho = -5\text{dB}$ and $N_k = N = 4$.

The data fusion can also be performed by means of soft combination. In soft combination the cognitive radio nodes will report the soft decisions to the cognitive base station and the base station would fuse the soft decisions by appropriate methods. Some of the standard techniques considered for soft-fusion are the equal ratio combining and the maximal ratio combining. In equal ratio combining the received soft decisions are summed up at the base station and a threshold detection is performed to make the decision. In the maximal ratio combining the soft decisions from the kth cognitive radio node is weighted appropriately based on its credibility for example and then summed up before performing the threshold detection.

5.6.2 Distributed spectrum sensing

The other collaborative technique in spectrum sensing is the distributed sensing method (Bazerque, J.; Chen, Y.). In distributed sensing unlike in the cooperative sensing there is no fusion center to perform the data fusion. Instead the locally sensed data are exchanged between the cognitive radio nodes themselves in the environment and the cognitive radio nodes will perform the fusion locally with the collected information. The information exchange between the cognitive radios can be by means of broadcasting or by means one to one transmissions. Figure-14 depicts an example of the collaborative sensing strategy.
Similar to the cooperative sensing case, here too the local sensing can be performed by one of the proposed techniques for spectrum sensing in the previous sections. Instead of performing the data fusion at the base station as in the cooperative sensing strategy it is performed at the cognitive radio nodes itself in this case. The major advantage associated with distributed sensing is the non-requirement of a central fusion center and the corresponding feedback reporting channel from the base station to the cognitive radio nodes. However, distributed sensing increases the overhead at the nodal level by requiring to perform the data fusion and data management etc.

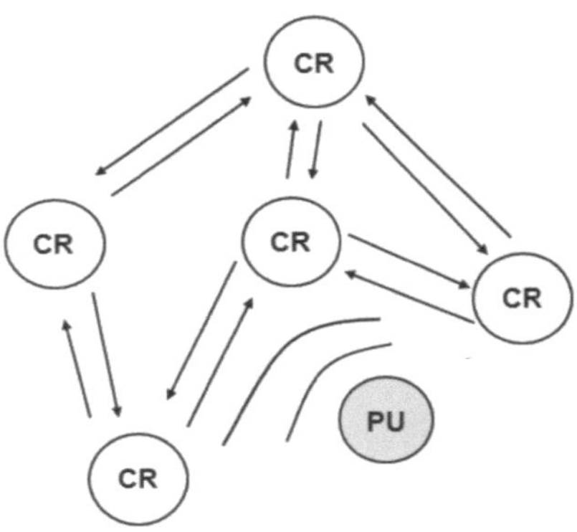

Fig. 14. Distributed spectrum sensing without a centralized fusion center.

6. Interference mitigation with detect-and-avoid techniques

The interference mitigation problem can be classified as interference caused to the cognitive radio nodes from the primary users as well as the secondary users and the interference caused by the cognitive radio nodes to the primary users and other secondary users. The interference actually depends on the geographical positioning of the radio nodes (that is the distance between the nodes), the transmit signal power from a particular node, and the channel gains of the links etc. In this section we briefly touch upon interference mitigation by means of detect-and-avoid in MB-OFDM UWB radios.
As described in the previous sections, there is a potential risk for wireless interferences of UWB technology with other wireless devices; in particular with WiMAX Customer Premise Equipment (CPE). In (Rahim, A et. al.) and (Li, Y. et. al.) the coexistence and interference issues mentioned here have been investigated to some extent. To address the risk of

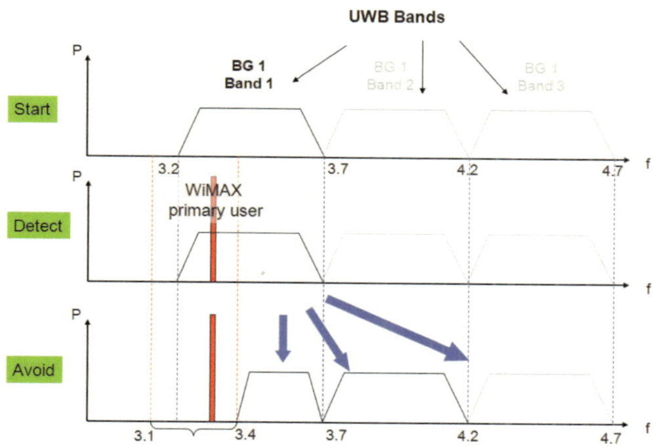

Fig. 15. Detect and Avoid of an UWB device to avoid interference to a WiMAX primary wireless service

interference of UWB on other wireless services, regulatory bodies around the world have defined stringent limits for the emission power of UWB devices. In most cases the limit is given as an Equivalent Isotropically Radiated Power (EIRP) emission mask. EIRP emission mask was defined by the FCC in 2002, the European Union in 2006, China in 2008, Japan in 2006 and Korea in 2006. The disadvantage of the EIRP mask is that UWB transmission power is limited even in the absence of WiFi or WiMAX communication. A more flexible approach is to allow higher emission power for UWB devices when no other wireless system is transmitting within the same coverage area.

In this case an opportunistic approach could be used, where secondary users (e.g., UWB devices) are required to detect the transmission of primary users in specific spectrum bands and consequently refrain from transmitting in those bands or reduce their emission power. In the case of UWB, this approach is also named Detect and Avoid (DAA) as UWB devices should *Detect* the presence of a primary user (e.g., WiMAX) in the radio frequency spectrum environment and use other frequency bands for the transmission to *Avoid* creating interference to the primary user (see Figure-15). In this context, UWB DAA can be considered a simple form of cognitive radio.

Regulations for the use of the DAA mitigation techniques for UWB are different around the world. In Europe, the regulation for generic UWB devices (i.e., not specifically DAA enabled) is composed of two ECC Decisions: the baseline Decision ECC/DEC/(06)04 (ECC Decision, 2006), which defines the European spectrum mask for generic UWB devices without the requirement for additional mitigation and Decision ECC/DEC/(06)12 (ECC Decision, 2006), recently amended by (ECC Decision, 2008), which provides supplementary mitigation techniques such as Low Duty Cycle (LDC) or DAA. The related European Commission decision is 2009/343/EC (EC Decision, 2009).

In USA, FCC (FCC Part47-15, 2007) has opened the 3.1 - 10.6 GHz frequency band for the operation of UWB devices provided that the EIRP power spectral density of the emission is lower than or equal to -41.3 dBm/MHz. FCC regulations do not specify the use of mitigation techniques for UWB devices operating in the mentioned frequency range.

In China Mainland, in the 4.2-4.8 GHz band, the maximum EIRP is restricted to -41.3dBm/MHz by the date of 31st Dec, 2010. After that, the UWB devices shall adopt an

Interference Relief Technology, such as DAA. There are no specific parameters or limit values for DAA in the current Chinese UWB regulation specification.

In Japan, in the 3.4 to 4.8 GHz frequency range, UWB devices without interference avoidance techniques such as DAA may not transmit at a level higher than -70 dBm/MHz. In the 3.4 to 4.2 GHz band, UWB devices may transmit at or below the limit of -41.3 dBm/MHz, under the condition that they are equipped with interference avoidance techniques such as DAA. In the 4.2 to 4.8 GHz band, UWB devices shall adopt an interference avoidance technique after 31st Dec, 2010.

In Korea, the UWB emission limit mask requires the implementation of an interference avoidance technique such as DAA in the 3.1 to 4.2 GHz and 4.2 to 4.8 GHz bands to provide protection for IMT Advanced systems and broadcasting services. The requirements in the 4.2 to 4.8 GHz band shall be implemented after 31st Dec, 2010.

In Hong Kong, the proposed rule is, based on the 33rd Radio Spectrum Advisory Committee (RSAC) Meeting discussion, to allow a maximum EIRP of -41.3 dBm/MHz in the 3.4 to 4.8 GHz band, provided that appropriate mitigation techniques are employed. Otherwise the maximum EIRP is restricted to -70 dBm/MHz.

In Europe, references (ECC Report 120, 2008) and (EC Decision, 2009) identify three types of victim systems to be protected by DAA mechanisms: 1) BWA Indoor terminals in the 3.4 - 4.2 GHz range, 2) Radiolocation systems in the 3.1 - 3.4 GHz range and 3) Radiolocation systems in the 8.5 - 9 GHz range.

The DAA mitigation techniques are based on the concept of *coexistence zones* which correspond to a minimum isolation distance between an UWB device and the victim system. For each DAA zone, in conjunction with the given minimum isolation distance, the detection threshold and the associated maximum UWB transmission level are defined based on the protection zone the UWB device is operating within. In the frequency range 3.4 - 4.2 GHz, three zones are defined on the basis of the detected uplink power of the victim signal: Zone 1 with a detection threshold for the uplink victim signal of -38 dBm. In this zone, the UWB device is required to reduce its emission level in the victim bands to a maximum of -80 dBm/MHz. As an alternative, the UWB device is allowed to move to a non-interfering channel. Zone 2 with an uplink detection threshold of -61 dBm. In this zone, the UWB device is required to reduce its emission level to a maximum of -65 dBm/MHz. As an alternative, the UWB device is allowed to move to a non-interfering channel. Zone 3 where the UWB device does not detect any victim signal transmitting with a power greater than -61 dBm. In this case, the UWB device is allowed to continue transmitting at maximum emission level of -41.3 dBm/MHz. Figure-16 provides a description of the different protection zones:

Reference (ECC Report 120, 2008) provides flowcharts for the implementation of the DAA algorithm as represented in Figure-17.

The flowcharts and detection algorithms are implemented on the basis of the following parameters:

- Minimum Initial Channel Availability Check Time, which is the minimum time the UWB device spends searching for victim signals after power-on.
- Signal Detection Threshold, which is the victim power level limit, employed by the UWB device in order to initiate the transition between adjacent protection zones.
- Avoidance Level, which is the maximum Tx power to which the UWB transmitter is set for the relevant protection zone.
- Default Avoidance Bandwidth, which is the minimum portion of the victim service bandwidth requiring protection.

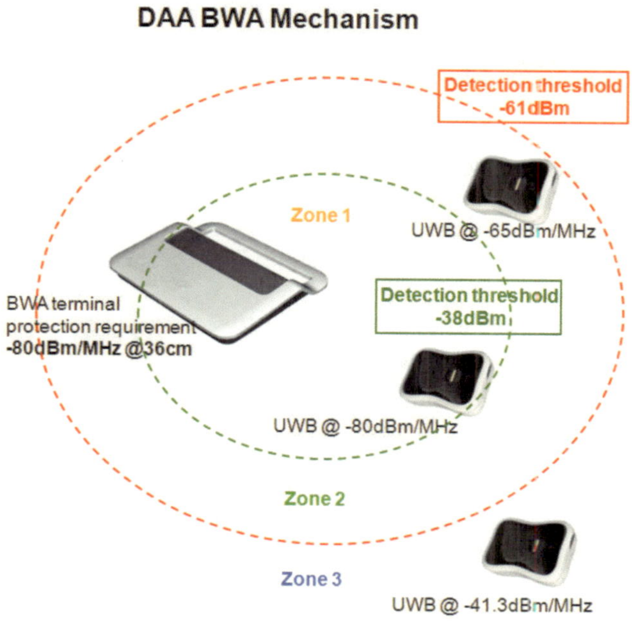

Fig. 16. Protection zones for DAA UWB devices

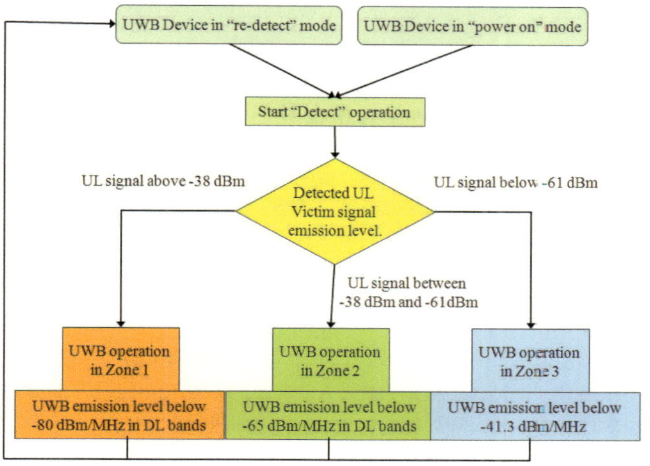

Fig. 17. Workflow of Detect and Avoid for three protection zones

- Maximum Detect and Avoid Time, which is the maximum time duration between a change of the external RF environmental conditions and adaptation of the corresponding UWB operational parameters.
- Detection Probability, which is the probability for the DAA enabled UWB device to make a correct decision either due to the presence of a victim signal before starting transmission or due to any change of the RF configuration during UWB device operation.

These parameters are also dependent on the type of communication service provided by the primary user. For example, UWB devices have different DAA times for different services (e.g., VoIP, Web surfing, Sleep mode, Multimedia broadcasting) of the primary user (e.g., Broadband Wireless Access).

In UWB networks, devices can negotiate detection capability and share detection information. For example, if one device is sending a large file to another device, it is possible for the receiving device to be the primary detecting device. DAA UWB network can implement smart detection algorithms where the most capable or powered devices can implement the detection of the primary users and distribute this information to the less capable devices.

7. Localization and radio environment mapping

For the cognitive radio nodes to perform its functionalities properly it needs to have context aware capabilities such as the spectrum sensing capability. Another context aware mechanism to support the intelligence of the cognitive radio is locating radios in the network (Giorgetti, A.). By means of localizing the radios in the network the cognitive radio node can create a map of radios which would help to perform its functionalities better. For example, knowing the location of the primary user nodes can become beneficial when considering directional transmissions for maximizing the spatial re-usage of the spectrum.

Another means getting context awareness is by means of radio environment maps. The term radio environment map or REM refers to a database of the radio environment, which can be locally maintained in a node or in a network where all the nodes could access it. A cognitive radio node in a network can get its intelligence by means of sensing or extracting information from the REM. The REM itself need to be updated periodically by means of sensing and learning operations. The advantage of maintaining a network level REM is that not all the nodes need to perform sensing on its own but rather get information from the REM and hence reducing the complexity of the cognitive radio node. A typical REM would contain information about the radio nodes in the vicinity and the related radio and network resources such as frequency channels, data rates, center frequency, location information, which network the node belongs to, what services the node offers, the regulatory and policy details of the nodes, and the nodes historical behavior etc. Getting and maintaining all the information about the nodes in the environment is not always feasible in which case the REM will contain only the information that are available. By using such REM data bases communication networks can be made much efficient especially considering wireless networks. However, many technical aspects related to the design and deployment of REM need to be addressed. For example, how often the information need to be updated in the REM, how much and what information required to be stored, what are the overheads in having such REM for maintaining and distributing the information, and finally the security and privacy requirements for the REM.

8. Scenarios and applications for UWB based CR

Finally, we present some application scenarios for the use of UWB based cognitive radios. The scenarios that we present here are derived from the two EU projects C2POWER (C2POWER, 2010) and EUWB (EUWB, 2008). The scenarios that we provide are for dynamic spectrum access (EUWB scenarios) as well as for energy efficient communications (C2POWER scenario).

Scenario-1: UWB based cognitive radios are considered for home entertainment where UWB based multimedia devices such as a hi-fi surround system with audio/video transmissions

could utilize the DAA techniques. In such an environment the UWB devices need to be aware of the 5GHz ISM band devices, WiMAX devices in 3.6GHz etc.
Scenario-2: UWB based cognitive radios are considered for airborne in-flight transmissions such as for audio/viedo delivery to the passengers. In such scenarios the UWB radios need to be aware of any custom built radios within the UWB frequency band for flighth specific applications and as well as any satellite receivers in the UWB frequency range.
Scenario-3: UWB based cognitive radios are considered for vehicular communications such between sensors and the central unit. In such situations the UWB radios need to be aware of the surrounding radios in order to avoid interference and at the same time make sure that its time critical transmissions are also not interfered with.
Scenario-4: UWB radios can also be used for energy saving in short range wireless communications. Given the favorable channel conditions a source node may opt to communicate to its destination by means of a relay node for better energy efficiency (C2POWER, 2010). In such context UWB radios with intelligence (i.e. UWB based cognitive radios) can play a prominent roll.

9. Conclusion

In this chapter we provided the concept and fundamentals of UWB based cognitive radios for having intelligence in the standard UWB radios. By having cognition in the UWB devices the transmissions could be dynamically adopted in order to improve the performance. The intelligence in the radio leads to a better usage of the radio resources such as the radio spectrum by having dynamic spectrum access capabilities in the spatio-temporal domain. The cognitive engine residing in the UWB radio learns about its surrounding and acts based on the internal and network level policies.
Even though the cognitive radio technology shows prominent advantages yet many issues are to be solved prior to its deployment, various standardization and regulatory activities are currently underway in order to regulate the dynamic spectrum access and cognitive radio technology.

10. Acknowledgement

This work was partly funded by the European Commission under the C2POWER project (EU-FP7-ICT-248577) - http://www.ict-c2power.eu, and the EUWB project (EU-FP7- ICT-215669) - http://www.euwb.eu.

11. References

Atapattu, S., Tellambura, C., and Jiang, H., (2010), 'Performance of an Energy Detector over Channels with Both Multipath Fading and Shadowing', IEEE Transactions on Wireless Communications, Vol.9, No.12, Dec 2010, pp3662-3670

Aysal, T., C., Kandeepan, S., and Piesiewicz, R., (2009) 'Cooperative spectrum sensing with noisy hard decision transmissions,' in IEEE Conf ICC, 14-18June 2009, Dresden.

Bazerque, J., A., and Giannakis, G., B., (2010), 'Distributed Spectrum Sensing for Cognitive Radio Networks by Exploiting Sparsity,' *IEEE Trans. on Signal Proc.*, vol. 58, no. 3, pp. 1847–1862, March 2010.

C2POWER Project (2010), The European Commission funded STREP Project 'C2POWER' (EU-FP7-ICT-248577) - http://www.ict-c2power.eu, 2010

Chen, Y., Zhao, Q., and Swami, A., (2009), 'Distributed Spectrum Sensing and Access in Cognitive Radio Networks With Energy Constraint,' *IEEE Trans. on Signal Proc.*, vol. 57, no. 2, pp. 783–797, Feb. 2009.

Digham, F.,F., Alouini, M., S., and Simon, M., K., (2007), 'On the Energy Detection of Unknown Signals over Fading Channels,' *IEEE Trans. on Communications.*, vol. 55, no. 1, pp. 21–24, Jan. 2007.

EC-The Commission of the European Communities (2007), Commission Decision 2007/131/EC on allowing the use of the radio spectrum for equipment using ultra-wideband technology in a harmonised manner in the Community, Official Journal of the European Union, Feb. 21, 2007. url: http://eur-lex.europa.eu/LexUriServ/site/en/oj/2007/l_055/l_05520070223en00330036.pdf

'2009/343/EC' Decision amending Decision 2007/131/EC on allowing the use of the radio spectrum for equipment using ultra-wideband technology in a harmonised manner in the Community.

'ECC/DEC/(06)04 ECC' Decision of 24 March 2006 on the harmonised conditions for devices using Ultra-Wideband (UWB) technology in bands below 10.6 GHz.

'ECC/DEC/(06)12' ECC Decision of 1 December 2006 on the harmonised conditions for devices using Ultra-Wideband (UWB) technology with Low Duty Cycle (LDC) in the frequency band 3.4-4.8 GHz.

ECC Report 120 (June 2008) on Technical requirements for UWB DAA (Detect And Avoid) devices to ensure the protection of Radiolocation in the bands 3.1 - 3.4 GHz and 8.5 - 9 GHz and BWA terminals in the band 3.4 - 4.2 GHz.

ECC Decision of 1 December 2006 amended 31 October 2008 on supplementary regulatory provisions to Decision ECC/DEC/(06)04 for UWB devices using mitigation techniques

ETSI Technical Specification ETSI TS 102 754. v.1.2.1 (2008-10). "Electromagnetic compatibility and Radio Spectrum Matters (ERM); Short Range Devices (SRD); Technical Characteristics of Detect and Avoid (DAA) mitigation techniques for SRD equipment using Ultra Wideband (UWB) technology.

EUWB (2008), The European Commission funded Integrated Project (EC: FP7-ICT-215669): http://www.euwb.eu.

FCC (2002) 'Federal communications commission: spectrum policy task force report,' Federal Communications Commission ET Docket 02-135, November 2002.

FCC-Federal Communications Commission (2003), Facilitating Opportunities for Flexible, Efficient, and Reliable Spectrum Use Employing Cognitive Radio Technologies, NPRM and Order, ET Docket no. 03-322, Dec. 2003.

FCC CFR Title 47 Part 15 Subpart F, "Ultra-Wideband Operation". Federal Communications Commission, October, 2007.

Fette, B., (Edt) (2006), 'Cognitive Radio Technology', Elsevier Publishers, 2006.

Haykin, S. (2005), 'Cognitive Radio: Brain-Empowered Wireless Communications,' IEEE Journal on Select Areas in Commun., vol. 23, no. 2, pp. 201 - 220, Feb. 2005.

Ganesan, G., and Ye Li (2007), 'Cooperative Spectrum Sensing in Cognitive Radio, Part I: Two User Networks', IEEE Trans on Wireless Communications Vol6, Iss6, June 2007 p:2204-2213

Gardner, W., A., (1991),'Exploitation of spectral redundancy in cyclostationary signals', IEEE Signal Processing Magazine,Vol8,Iss2,Apr1991,pp14-36

Giorgetti, A., Chiani, M., Dardari, D., Minutolo, R., and Montanari, M., (2007), 'Cognitive radio with ultra-wide bandwidth location capable nodes,' IEEE Military Communications Conference, Orlando, FL, Oct. 2007, pp. 1-7.

Gu, H., and Yang, C., (2008), 'Power and Admission Control for UWB Cognitive Radio Networks,' IEEE ICC, 19-23 May, 2008, Beijing.

Hossain, E., Niyato, D., and Han, Z. (2009), Dynamic Spectrum Access and Managament in Cognitive Radio Networks, Cambridge University Press, 2009.

Kandeepan, S. et. al (2009), Project Report-'D2.1.1:Spectrum Sensing and Monitoring', EUWB Integrated Project, European Commission funded project (EC: FP7-ICT-215669), May 2009, , http://www.euwb.eu.

Kandeepan, S., Baldini, G., and Piesiewicz, R., 'Experimentally Detecting IEEE 802.11n Wi-Fi Based on Cyclostationarity Features for Ultra-Wide Band Cognitive Radios', IEEE Personal Indoor and Mobile Radio Communications (PIMRC) conference, Sep 2009, Tokyo

Kay, S., (2006), 'Intuitive Probability and Random Processes using Matlab,' Springer, New York 2006.

Kyouwoong, K., Akbar, I., A., Bae, K., K., Jung-sun Urn, Spooner, C., M., Reed, J., H., (2007), 'Cyclostationary Approaches to Signal Detection and Classification in Cognitive Radio', IEEE International Symposium on New Frontiers in Dynamic Spectrum Access Networks (DySPAN), 17-20 April 2007 pp212-215, Dublin.

Li, Y., Rabaey, J., Sangiovanni-Vincentelli, A., "Analysis of Interference Effects in MB-OFDM UWB", IEEE Wireless Communications and Networking Conference 2008 (WCNC 2008), April 2008.

Mishra, S., M., Sahai, A., and Brodersen, R., (2006), 'Cooperative sensing among coginitive radios,' in IEEE Conf ICC, Istanbul, June 2006.

Mitola, J. & Maguire Jr, G. (1999), 'Cognitive Radio: Making Software Radios More Personal,' IEEE Personal Communications, vol. 6, no. 4, pp. 13- 18, Aug. 1999.

Radunovic, B., and Le Boudec, J., (2004), 'Optimal Power Control, Scheduling, and Routing in UWB Networks ,' IEEE Journal On Selected Areas in Communications, Vol. 22, No. 7, Sep 2004, pp1252-1270.

Rahim, A., Zeisberg, S., Finger, A., "Coexistence Study between UWB and WiMAX at 3.5 GHz Band," IEEE International Conference on Ultra-Wideband 2007, ICUWB 2007, pp.915-920, 24-26 Sept. 2007.

Sklar, B., (1988), 'Digital Communications, Fundamental and Applications', Prentice Hall 1988.

Urkowitz, H., (1967), 'Energy Based Detection of Unknown Deterministoc Signals,' *Proc. of the IEEE*, vol. 55, no. 4, pp. 523–531, Apr. 1967.

Wang, Z., Qu, D., Jiang, T., He, Y., (2008), 'Spectral Sculpting for OFDM Based Opportunistic Spectrum Access by Extended Active Interference Cancellation,' IEEE Globecom 30 Nov-4Dec, 2008, New Orleans.

Wimedia-PHY (2009) Wimedia Alliance standards on PHY Specifications 1.5 for UWB communications, 11-August, 2009, http://www.wimedia.org/en/index.asp

Xing, Y., Mathur, C., Haleem, M., A., Chandramouli, R., Subbalakshmi, K., P., (2007), 'Dynamic Spectrum Access with QoS and Interference Temperature Constraints,' IEEE Transactions on Mobile Computing, vol. 6, no. 4, Apr. 2007, Page(s):423-433.

Yamaguchi, H., (2004), 'Active Interference Cancellation Technique for MB-OFDM Cognitive Radio,' 34th European Microwave Conference, 13 Oct. 2004, pp1105-1108, Amsterdam.

Yucek, T. and Arslan, H., (2009), 'A Survey of Spectrum Sensing Algorithms for Cognitive Radio Applications,' IEEE Communications Surveys and Tutorials, Vol. 11, No. 1, first quarter 2009, pp116-130.

Zhang, L., Liang, Y., C., and Xin, Y., (2007), 'Joint Admission Control and Power Allocation for Cognitive Radio Networks,' in Proc. 2007 IEEE Int. Conf. on Acoustics, Speech and Signal Processing, 15-20, Apr 2007, Hawaii.

Zhao, Y., Morales, L., Gaeddert, J., Bae, K., K., Um, J., and Reed, J., H., (2007), "Applying Radio Environment Maps to Cognitive WRAN Systems," in Proc. of the Second IEEE International Symposium on Dynamic Spectrum Access Networks (DySPAN 2007), April 17-20, 2007, Dublin, Ireland.

Zhao, Y., Raymond, D., da Silva, C., Reed, J., H., and Midkiff, S., F., (2007)'Performance Evaluation Of Radio Environment Map-Enabled Cognitive Spectrum-Sharing Networks', Military Communications Conference (MILICOM 2007),29-31 Oct, 2007, Orlando.

12

Detection and Avoidance Scheme for DS-UWB System: A Step Towards Cognitive Radio

Shaoyi Xu[1] and Rumin Yang[2]
[1]*Beijing Jiaotong University*
[2]*Chongqing University of Technology*
P. R. China

1. Introduction

Cognitive radio (CR) improves spectrum efficiency to satisfy increasing demands on wireless transmission by dynamic spectrum access without interfering with legacy networks. In 2004, IEEE 802.22 Working Group was formed to develop a standard for wireless regional area networks (WRANs) based on CR technology (Hu et al et al., 2007). It is expected to obtain a broadband access to data networks on the vacant TV channels while avoiding harmful interference to licensed TV broadcasting in rural areas within a typical radius of 17km to 30km (Stevenson et al., 2006).

Ultra wideband radio (UWB), a promising technology, has found a myriad of exciting applications as well as generating a great deal of controversy, for its extremely broad bandwidth transmission as well as its revolutionary way of overlaying coexistent RF systems could cause interference on them (Lansford, 2004; Parr et al., 2003). Over the years, the co-existence problem of UWB has been all along a hot topic in the academy, industry, and regulatory bodies. After years of public debates, arguments, and comments, two important solutions to the co-existence problem are made—the policy-based power emission mask (FCC, 2002) and the device-centric cognitive radio (Lansford, 2004; Walko, 2005; Haykin, 2005). So far, several cognitive UWB schemes have been proposed, among which are soft-spectrum (Zhang & Kohno, 2003) scheme and detection-and-avoidance (DAA) scheme (Kohno & Takizawa, 2006).

Reliably detecting of weak primary signals is an essential functionality for a DAA UWB system as soon as a primary user (PU) comes back into operation on the operating channels. Two types of primary users are defined in a WRAN which are TV services and wireless microphones (WMs). Compared with TV services, it is tougher to detect WM signals for the following two reasons. Firstly, wireless microphones are low power devices and occupy a narrow bandwidth. The transmission power of a WM is as low as 50mW in a 200kHz bandwidth. When the sensor is several hundred meters away from this WM signal, the received signal-to-noise ratio (SNR) may be below -20dB (Zeng & Liang, 2007). Another, they utilize arbitrary unused TV bands and are deployed for a short time such that it is difficult for CR users to obtain much information on WM signals (De & Liang, 2007; Dhillon & Brown, 2008).

This chapter will concern two questions. Firstly, how to detect the weak primary signals. Secondly, how to avoid such interference from the primary user and how to coexist with it.

To address the first problem, we consider detecting multiple WM signals in a WRAN when UWB users want to use this spectrum and propose a singular value decomposition (SVD) based algorithm. To verify the better performance by using the suggested approach, simulation results by comparing to the traditional methods will be shown. For the second concern, a pulse-shaping scheme under the limit of a power spectrum density algorithm will be proposed. In a cognitive environment, the re-design of pulses should be agile enough and easily reconfigurable. Furthermore, to avoid interfering with the primary system, the transmission power o f UWB should be considered.

2. Detection of weak primary signals in a cognitive radio network

2.1 Basic assumptions and problem formulation

Several methods have been suggested to detect WM signals. In (Mossa & Jeoti, 2009), a cyclostationary filter is proposed to grasp the existence of WM signals and to estimate their frequency locations. Obviously, such database dependent methods can not adapt to the dynamic signal detecion. (Lei & Chin, 2008; Wu et al., 2009) proposed beacon based methods for wireless microphones but these put the onus on many already-deployed incumbent wireless microphones. (Zeng & Liang, 2009; Unnikrishnan & Shellhammer, 2007) investigate the method based on eigenvalues of received data matrix when a WM signal is present but can not solve the multiple WM signals detection in a wideband cognitive network. To the best of our knowledge, the literature of wideband spectrum sensing for multiple WM signals is very limited. Actually, it is inevitable that multiple wireless microphones appeared simultaneously. Furthermore, performing wideband spectrum sensing can improve detection efficiency and maximize the opportunistic throughput (Quan et al., 2008). (Kalke, 2005) estimated that about 25,000 licensed wireless microphones are utilized by recording studios of TV broadcasters, organizers and performers in concerts and theatres, commentators in sports events, film production crews and government agencies. To avoid interfering to each other, these WM signals must operate in different center frequencies with enough guard bandwidth. To detect multiple WM signals in a wide bandwidth, (Lim et al., 2007) suggested to use a cyclostationary filter with a filterbank to detect every sub-channel which is divided from the wide sensing spectrum in advance. If a conventional energy detector is used, the sensing process has to include two steps: coarse sensing and fine sensing. The former step determines the presence of WM signals and the latter step is required to decide which channel is occupied (IEEE 802.22 working Group for WRAN, 2006). Obviously, the system complexity and sensing periods will be greatly increased by using traditional methods to sense WM signals in a wideband spectrum.

In our work, we propose a singular value decomposition based algorithm to detect multiple WM signals in a CR network which can sense a wideband channel consisting of multiple narrowband channels. After performing SVD on the received data matrix of a wideband spectrum, the presence of WM signals is detected by comparing the singular values with a prefixed threshold and the number of WM signals can be determined at the same time. Then, the WM signals are approximated and the center frequencies of these WM signals are estimated. Consequently, guard bandwidths will be set on the two sides of the primary WM signals and CR users can still work on the other spectra within the sensing bandwith without interfering with the primary wireless microphone users. The detection threshold and probability of false alarm are derived and simulation results confirm that our method is very effective and robust to detect and estimate multiple WM signals in a wideband spectrum.

Consider a CR network with N samples utilized to perform spectrum sensing at the ith CR user. Then the received signals at this CR user have two hypotheses as

$$\begin{cases} H_0: & r_i(n) = u_i(n) \\ H_1: & r_i(n) = h_i s_i(n) + u_i(n) \end{cases}. \tag{1}$$

Here H_0 and H_1 respectively mean the primary user is inactive and the licensed user is operating. h_i is the channel gain between the PU and the ith secondary user. s_i represents the received PU signals by the ith SU and u_i is AWGN with zero mean and variance σ_u^2, respectively. The test statistic for an energy detector is given by

$$T_i = \frac{1}{N}\sum_{n=1}^{N}|r_i(n)|^2 . \tag{2}$$

Under the hypothesis H_0, it shows a Gaussian random distribution when N is large with mean σ_u^2 and variance $\frac{2}{N}\sigma_u^4$. Hence, for a given probability of false alarm P_f, the threshold η of an energy detector can be derived as

$$\eta = \sigma_u^2\left(1 + \frac{\sqrt{2}Q^{-1}(P_f)}{\sqrt{N}}\right) \tag{3}$$

where $Q(x) = (1/\sqrt{2\pi})\int_x^{\infty} e^{-t^2/2}dt$ is the normal Q-function.

In (Unnikrishnan & Shellhammer), it is pointed out that most wireless microphones use analog frequency modulation (FM) and a WM signal occupies only 200kHz. Specifically, most energy of a WM signal is contained in an only 40kHz bandwidth (Notor, 2006). However, IEEE 802.22 draft requires the sensing spectrum is at least one channel (6, 7 or 8MHz), and hence the proportion which a WM signal occupies is below 3%. Based on the above analysis, $s(t)$ can be modeled as a summation of multiple single-tone cosinoidal signals as

$$s(t) = \sum_{k=1}^{P} A_k \cos(2\pi f_k t + \theta_k) \tag{4}$$

where A_k, f_k and θ_k respectively denote the amplitude, center frequency and phase of the kth WM signal and P is the number of WM signals in the sensing spectrum. θ_k can be modeled as a uniform random variable over [0, 2π). Without loss of generality, we assume s_i and u_i are independent of each other and $SNR = P_{WM} / 2\sigma_u^2$ denotes the SNR of the primary WM signals received by the ith CR user where P_{WM} is the total power of P WM signals.

In this chapter, we consider that there are multiple WM signals in several sensing channels and each channel is a TV channel with 6MHz bandwidth. Under this assumption, we focus the detection of multiple WM signals on a wideband spectrum.

2.2 SVD based approach to detect and estimate multiple WM signals

In this section, we will present the SVD based method to detect the presence of WM signals and to estimate the number and center frequencies of these detected WM signals.

2.2.1 Technology to detect multiple WM signals

SVD plays an important role in signal processing and statistics, particularly in the area of linear systems. For a time series $r(n)$ with $n = 1,2,\ldots,N$, commonly, we can construct a Hankel matrix with $M = N - L + 1$ rows and L columns illustrated as follows:

$$\mathbf{R} = \begin{bmatrix} r(1) & r(2) & \cdots & r(L) \\ r(2) & r(3) & \cdots & r(L+1) \\ \vdots & \vdots & & \vdots \\ r(N-L+1) & r(N-L+2) & \cdots & r(N) \end{bmatrix} \tag{5}$$

then $\mathbf{R}$ is an $M \times L$ matrix. Its elements can be found by substitution of $r(n)$

$$\mathbf{R}_{ml} = r(m+l-1), \quad m = 1,2,\ldots,M \text{ and } l = 1,2,\ldots L\,. \tag{6}$$

Using the SVD, $\mathbf{R}$ can be factorized as

$$\mathbf{R} = \mathbf{U\Sigma V}^{H} \tag{7}$$

where $\mathbf{U}$ and $\mathbf{V}$ are an $M \times M$ and an $L \times L$ unitary matrix, respectively. The columns of $\mathbf{U}$ and $\mathbf{V}$ are called left and right singular vectors, respectively. $\mathbf{\Sigma} = diag(\lambda_1, \lambda_2, \cdots, \lambda_m)$ is a diagonal matrix whose nonnegative entries are the square roots of the positive eigenvalues of $\mathbf{R}^H\mathbf{R}$ or $\mathbf{RR}^H$. These nonnegative entries are called the singular values of $\mathbf{R}$ and they are arranged in a decreasing order with the largest one in the upper left-hand corner. $[\ ]^H$ denotes the complex transpose of a matrix.

When no any primary WM signal is present, the received signal $r(n)$ includes only AWGN contribution such that its singular values are similar and close to zero. When WM signals are active whose power is higher than a threshold, there will exist several dominant singular values to represent these WM signals. As a result, the WM signals can be detected by examining the presence of dominant singular values.

It is critical to determine the number of WM signals P and we will present such method in the following part. To simplify our analysis, we assume that the power values of all WM signals received in the detected spectrum are approximately same, that is to say $A_1 \approx A_2 \approx \ldots \approx A_P$. Since the SNR of primary WM signals received by the secondary detectors is usually very weak, we think this assumption is feasible. Several methods can be utilized to determine if the dominant singular values are present. It is pointed out in (Teh et al, 1995) that the relationship between the number of dominant singular values K and the number of single-tone cosinoidal signals P has the form as $K = 2P$, therefore, threshold γ can be adopted which is the ratio between the first singular value and the $(2X+1)$th singular value. That is to say, if the following equation is true, P WM signals can be declared to be present as

$$\text{If } \lambda_1 / \lambda_{2X+1} \geq \gamma, \text{ then } P = X \tag{8}$$

and the expression of γ will be derived in Section 2.3.

2.2.2 Technology to estimate the center frequencies of multiple WM signals

Once WM signals are detected to be active in the sensing channels, the center frequencies of these primary WM signals need to be estimated such that a guard bandwidth can be

retained and CR users utilize the other spectra to improve spectrum efficiency. Next, we will present the frequency estimation technique by using SVD.
After P WM signals are detected to be active, the data matrix $\mathbf{R}$ in (5) is the superposition of the WM signal space and AWGN space and $\mathbf{R}$ can be partitioned into two subspaces as follows

$$\begin{aligned}\mathbf{R} = \mathbf{U}\boldsymbol{\Sigma}\mathbf{V}^H &= [\mathbf{U}_S \quad \mathbf{U}_U]\begin{bmatrix}\boldsymbol{\Sigma}_S & 0 \\ 0 & \boldsymbol{\Sigma}_U\end{bmatrix}[\mathbf{V}_S \quad \mathbf{V}_U]^H \\ &= \mathbf{U}_S\boldsymbol{\Sigma}_S\mathbf{V}_S{}^H + \mathbf{U}_U\boldsymbol{\Sigma}_U\mathbf{V}_U{}^H = \mathbf{R}_S + \mathbf{R}_U\end{aligned} \tag{9}$$

where

$$\boldsymbol{\Sigma}_S = diag(\lambda_1, \lambda_2, \cdots, \lambda_{2P}), \tag{10}$$

and

$$\boldsymbol{\Sigma}_U = diag(\lambda_{2P+1}, \lambda_{2P+2}, \cdots, \lambda_m) \tag{11}$$

with $\lambda_1 > \lambda_2 > \cdots > \lambda_{2P} >> \lambda_{2P+1} > \lambda_{2P+2} > \cdots > \lambda_m$ corresponding to the singular values in the WM signal subspace and the noise subspace, respectively. $\lambda_1, \lambda_2, \ldots, \lambda_{2P}$ are $2P$ dominant singular values which correspond to the P WM signals. $\mathbf{R}_S = \mathbf{U}_S\boldsymbol{\Sigma}_S\mathbf{V}_S{}^H$ and $\mathbf{R}_U = \mathbf{U}_U\boldsymbol{\Sigma}_U\mathbf{V}_U{}^H$ are the WM signals subspace and the noise subspace, respectively. By rearranging $\mathbf{R_S}$ into a time serial, we can get the estimated data vector of WM signals $\boldsymbol{y} = [y_1, y_2, \cdots, y_N]^T$ which includes P WM signals. Next, we will present the algorithm to estimate P center frequencies corresponding to P WM signals.
We define $\boldsymbol{Y} = FFT(\boldsymbol{y}) = [Y_1, Y_2, \cdots, Y_N]^T$ as the N-point Fast Fourier Transform (FFT) operation so we can use the theory of the Rife and Boorstyn (Rife & Boorstyn, 1974) as the frequency estimation of the WM signal which has the maximum power

$$k_{1\_\max} = \max{}^{-1}\left[\left|\boldsymbol{Y}[k]\right|\right], \quad 1 \le k \le N \tag{12}$$

$$\hat{f}_1 = \frac{k_{1\_\max}}{N} f_s \tag{13}$$

where $|.|$ is the absolute value operator, max(.) operator means k_{1_max} is the k_1th sampling point where $|\mathbf{Y}[k]|$ obtains its maximum and f_s is the sampling frequency.
By applying equation (12) and (13), the center frequency of the WM signal which has the maximum power can be acquired. Following the similar step, we can obtain the approximate center frequency for the jth WM signal as

$$k_{j\_\max} = \max{}^{-j}\left[\left|\boldsymbol{Y}[k]\right|\right], \quad 1 \le k \le N \tag{14}$$

and

$$\hat{f}_j = \frac{k_{j\_\max}}{N} f_s \tag{15}$$

where k_{j_max} presents the k_jth sampling point corresponding to jth peak magnitude.
From the above analysis it can be concluded that this estimation algorithm is easy to implement since only FFT is required. By using FFT, the efficiency of frequency estimation can be improved greatly. Another, the inaccurate knowledge of P will not affect frequency estimation. If P is under or over estimated, then fewer or more frequencies will be estimated than the true number.
Rife and Boorstyn pointed out that when SNR is high enough, the true frequency has a high probability lies in the range (Rife & Boorstyn, 1974)

$$f \in [\hat{f} - (f_s / 2N), \hat{f} + (f_s / 2N)]. \tag{16}$$

In summary, the SVD based detection and estimation algorithm consists of the following steps:

Step 1. Pick a number L so that $k < L < N - k$ (Tufts & Kumaresan, 1982), where N is the number of sampling points and k is the number of dominant singular values. In our work, $k = 2P$.

Step 2. Arrange the received signal vector $\boldsymbol{r}$ to form a Hankel data matrix $\mathbf{R}$ as (5). Then compute the SVD of $\mathbf{R}$ and obtain all singular values of $\mathbf{R}$.

Step 3. Calculate the threshold $\gamma = \lambda_1 / \lambda_{2X+1}$ ($X = 1,2,\ldots$) and compare the ratio $\lambda_1 / \lambda_{2X+1}$ with the predefined threshold γ. If $\lambda_1 / \lambda_{2X+1} \geq \gamma$, the WM signals are determined to be present and the number of WM signals can be derived by $P = X$. Otherwise, no WM signal is declared to be active. The derivation of γ will be explained in Section 2.3.

Step 4. If P WM signals are declared to be present, compute $\mathbf{R_S}$ then arrange it into a data vector $\boldsymbol{y}$. Apply FFT on $\boldsymbol{y}$ and consecutively find the number of the point k_{j_max} at which the k_jth peak amplitude of the FFT is approached.

Step 5. Obtain the estimated center frequency of jth WM signal by using (12 –15).

2.3 Theoretical analysis and determination of threshold

In this section, we will derive the threshold γ and probability of false alarm P_f.
We denote $\mathbf{R_S}$ ($M \times L$) and $\mathbf{R_U}$ ($M \times L$) as the Hankel matrix of WM signals and an AWGN signal, respectively, such that $\mathbf{R_U} \sim N_p(0, \Sigma)$ where p is the dimension of $\mathbf{R_U}$ and Σ is the covariance matrix. Since the power of WM signals is usually very low, the distribution of $\mathbf{R_S} + \mathbf{R_U}$ can be approximated as $N_p(0, \Sigma)$. According to (Zeng & Liang, 2009; Johnstone, 2001), we have the following three theorems:

Theorem 1. Assume $M/L \geq 1$ and N is large enough, the largest singular value can be approximated as

$$\lambda_1 \approx \sqrt{\frac{\sigma_u^2}{N}\left(\sqrt{N} + \sqrt{ML}\right)^2}. \tag{17}$$

Theorem 2. Assume $M/L \geq 1$ and N is large enough, the largest singular value follows the following distribution

$$\frac{\lambda_1^2 - \mu_{ML}}{\delta_{ML}} \sim F_1 \tag{18}$$

where μ and δ are called a center constant and a scaling constant and they are defined as

$$\mu_{ML} = \left(\sqrt{M-1} + \sqrt{L}\right)^2 \tag{19}$$

and

$$\delta_{ML} = \left(\sqrt{M-1} + \sqrt{L}\right)\left(\frac{1}{\sqrt{M-1}} + \frac{1}{\sqrt{L}}\right)^{\frac{1}{3}}. \tag{20}$$

F_1 is the distribution function of Tracy-Widom distribution of order 1 which has the form as

$$F_1(s) = \exp\left\{-\frac{1}{2}\int_s^{\infty} q(x) + (x-s)q^2(x)dx\right\}, s \in \mathbb{R} \tag{21}$$

and q solves the Painlevé II differential function (Johnstone, 2001).
Theorem 3. The distribution of rth largest singular value ($r < L$) has the approximate distribution as

$$\Phi\left(\lambda_{r+1}^2 | M, L, \Sigma\right) \approx \Phi\left(\lambda_1^2 | M, L-r, I_{L-r}\right) - c_{M,L} \tag{22}$$

where $c_{M,L}$ is an empirical constant.
Based on the above three theorems, as a result, P_f can be presented as

$$\begin{aligned}
P_f &= P(\lambda_1 / \lambda_{2X+1} < \gamma) = P\left(\lambda_1^2 / \lambda_{2X+1}^2 < \gamma^2\right) \\
&= P\left(\lambda_{2X+1}^2 / \lambda_1^2 > 1/\gamma^2\right) = P\left(\lambda_{2X+1}^2 > \lambda_1^2 / \gamma^2\right) \\
&\approx P\left(\lambda_{2X+1}^2 > \frac{\sigma_u^2}{\gamma^2 N}\left(\sqrt{N} + \sqrt{ML}\right)^2\right) \\
&= 1 - \Phi\left(\lambda_{2X+1}^2 > \frac{\sigma_u^2}{\gamma^2 N}\left(\sqrt{N} + \sqrt{ML}\right)^2 | M, L, \Sigma\right) \\
&\approx 1 - \Phi\left(\lambda_1^2 > \frac{\sigma_u^2}{\gamma^2 N}\left(\sqrt{N} + \sqrt{ML}\right)^2 | M, L-2X, I_{L-2X}\right) + c_{M,L} \\
&= 1 - F_1\left(\frac{\frac{\sigma_u^2}{\gamma^2 N}\left(\sqrt{N} + \sqrt{ML}\right)^2 - \mu_{M,L-2X}}{\delta_{M,L-2X}}\right) + c_{M,L}
\end{aligned} \tag{23}$$

Hence, for a pre-determined P_f, the required threshold γ can be represented as

$$\gamma = \frac{\sigma_u\left(\sqrt{N} + \sqrt{ML}\right)}{\sqrt{N\left[\delta_{M,L-2X} F_1^{-1}\left(1 + c_{M,L} - p_f\right) + \mu_{M,L-2X}\right]}}. \tag{24}$$

2.4 Simulation results

2.4.1 Simulation parameters

Since it is difficult to derive the accurate closed form expression of γ and P_f, we need to resort to simulations for evaluating the performance of our approach.

We consider the spectrum of interest is three consecutive channels which means the sensing bandwidth is 18MHz. We assume that three WM signals are distributed on this 18MHz bandwidth and their SNRs are same. The signals are firstly down-converted into baseband and filtered by a baseband filter with bandwidth 18MHz. And then, these WM signals have the center frequency of 2.4MHz, 8MHz and 14.2MHz, respectively. The selected sampling frequency f_s must be larger than the Nyquist frequency of the WM signal which has the highest center frequency and in our simulation f_s should be larger than 28.4MHz. To find the threshold γ, we require the probability of false alarm is $P_f = 0.1$. To evaluate the performance of frequency estimation, we define the mean estimation precision for the frequency estimation as

$$\varepsilon = \frac{\sum_{j=1}^{3} \frac{\left|\widehat{f}_j - f_j\right|}{f_j}}{3}. \tag{25}$$

where $\widehat{f}_j$ and f_j are the estimated jth center frequency and the jth ($j \leq 3$) true center frequency, respectively. To investigate our proposal, we compare our simulation results with a conventional energy detector whose threshold has been given in (3).

It has been shown in (Tufts & Kumaresan, 1982) that when the column number L in a Hankel matrix satisfies the inequality $2P < L < N - 2P$, we can obtain the correct or approximately correct estimation result. However, to the best of our knowledge, it has not been seen that the optimal L theoretically, moreover, the optimal L is different in different cases. In our work, the simulation results show that satisfying $2P < L < N - 2P$, different L has no significant impact on the system performance and frequency estimation.

In our work, without specific explanation, the sampling frequency is selected as $f_s = 36$MHz and we select $L = N/5$ as the column number in our following simulations.

2.4.2 Simulation results and analysis

Fig. 1 shows simulation results of the probability of detection (P_d) vs. SNR when the proposed SVD based method and a classical energy detector are used. To investigate the effect for different WM signals, we show simulation results for the single WM signal with a center frequency of 2.4MHz and multiple WM signals, respectively. From this figure we can conclude that the detection performance can be improved greatly by using our method, especially for the single WM signal. For example, for the target P_d of 90%, a 4dB improvement can be obtained than an conventional energy detector by using the proposed approach for a single WM signal. For the multiple WM signals, an improvement of 2dB can be attained compared with the conventional energy detector. To evaluate the performance of the detector, the receiver operating characteristic (ROC) curves are illustrated in Fig. 2 when SNR is -12dB for the single WM signal and SNR = -10dB for three primary WM signals. We plot the P_d under H_1 against P_f under H_0 when P_f changes from 0.001 to the desired 0.1. We can observe that the ROC curve of our algorithm is much higher than that of the energy detector for both the single WM signal and multiple WM signals which verifies the better performance of our detector.

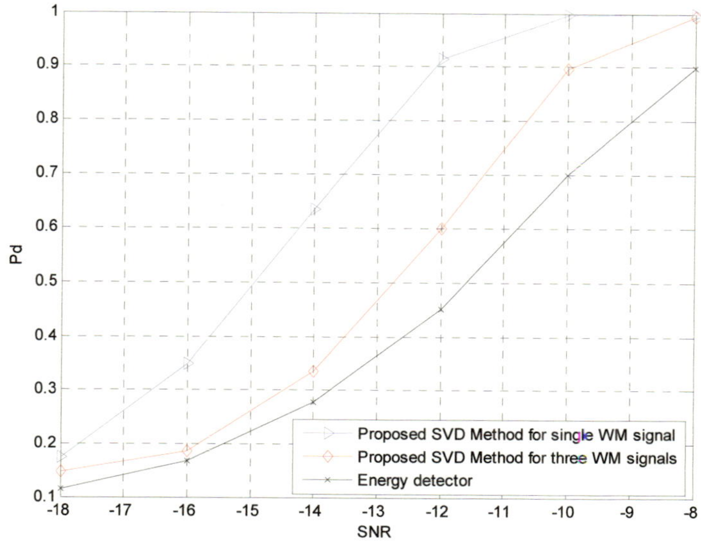

Fig. 1. Comparison of P_d between the proposed SVD-based method and a energy detector when PU is a single WM signal and three WM signals.

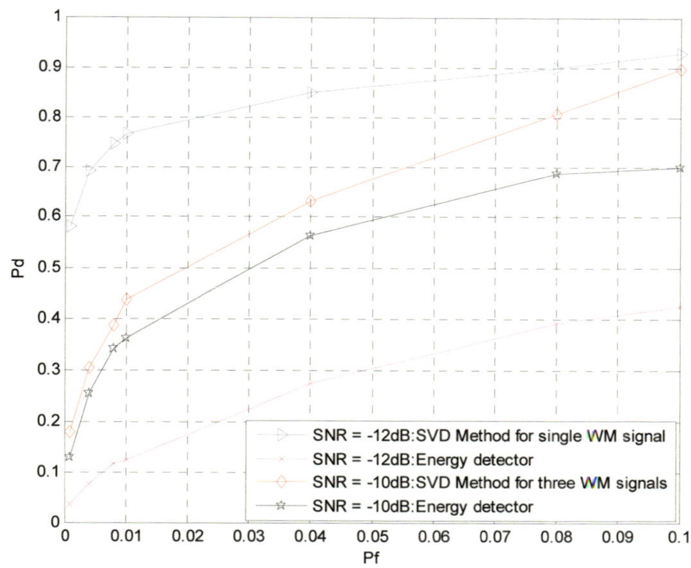

Fig. 2. Comparison of ROC curve between the SVD-based method and an energy detector when PU is a single WM signal and three WM signals.

To study the robustness of our algorithm, we first compare the P_d of our SVD based detection method under different column number L when three primary WM users operate simultaneously. Fig. 3 depicts the simulation results when $L = 3N/4$, $N/2$, $N/3$ and $N/5$, respectively. From this figure we can observe that although different L is taken, a good detection probability can be achieved with very slight difference. Then, we compare the P_d of the proposed approach under different sampling frequency f_s. The used sampling frequencies are 24MHz, 30MHz, 36MHz and 48MHz, respectively. Among these frequencies, 24MHz is lower than the Nyquist frequency of the WM signal whose center frequency is 14.2MHz. From Fig. 4 we can conclude that with the changing of f_s, the probability of detection shows very slight difference. Even for the f_s = 24MHz which is lower than the Nyquist frequency, a good P_d can be obtained which proves that our method is robust for different sampling frequency.

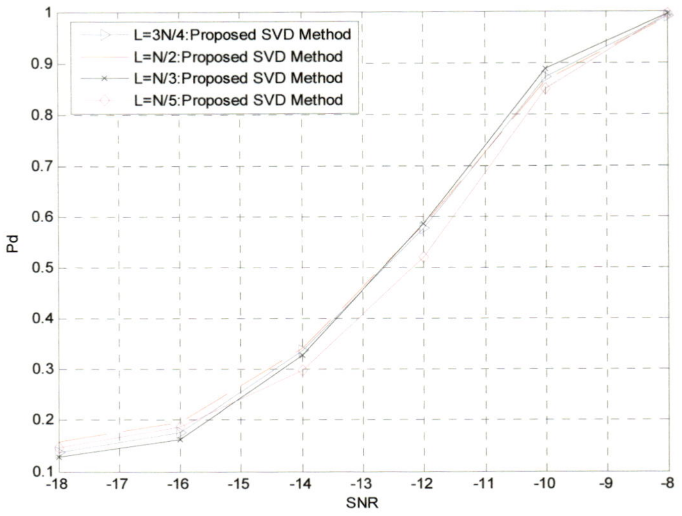

Fig. 3. Pd vs. SNR with different column number L for three WM signals.

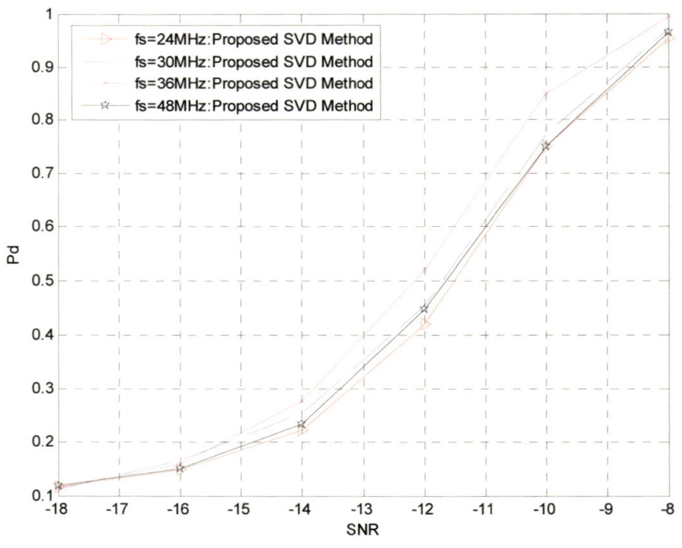

Fig. 4. Pd vs. SNR with different fs for three WM signals.

To investigate the estimation performance of the WM's center frequency, we plot the mean estimation precision ε in Fig. 5 and 6 when L and f_s change. From these two figures we can see that the proposed frequency estimation method is very effective. For example, for the f_s of 36MHz and SNR = -10dB, the absolute error of the worst estimation is within 10kHz. Whereas, a nearly perfect frequency estimation can be obtained for f_s = 36 and 48MHz when SNR = -10dB. In Fig. 5, an obvious result can be found that ε gets better with the increase of f_s. This is feasible since it is more possible to find the jth magnitude when the sampling rate is larger. However, higher f_s means a higher requirement for the system complexity. As a result, a tradeoff is needed to consider between system complexity and a satisfying ε. Figure 5 also proves that the estimation precision can be degraded severely if a f_s lower than the Nyquist frequency is used. Fig. 6 presents the mean estimation precision ε when f_s = 36MHz and L = $3N/4$, $N/2$, $N/3$ and $N/5$, respectively. From Fig. 6 we can conclude that the difference of L has no significant impact on ε, especially when SNR is higher than -12dB.

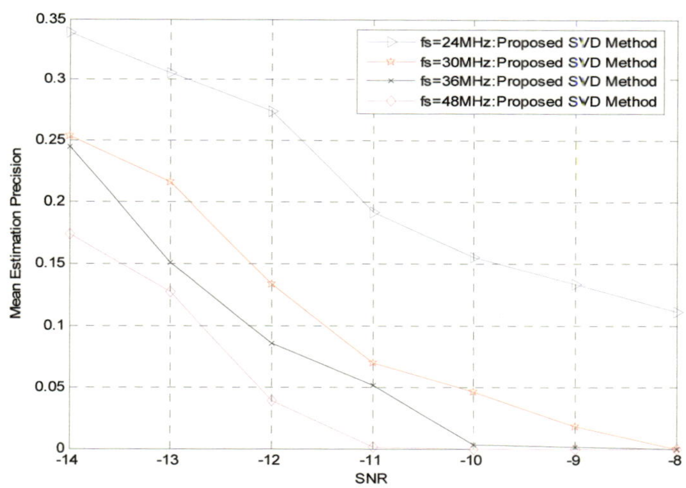

Fig. 5. Mean estimation precision of center frequency vs. SNR with different sampling frequency fs for three WM signals.

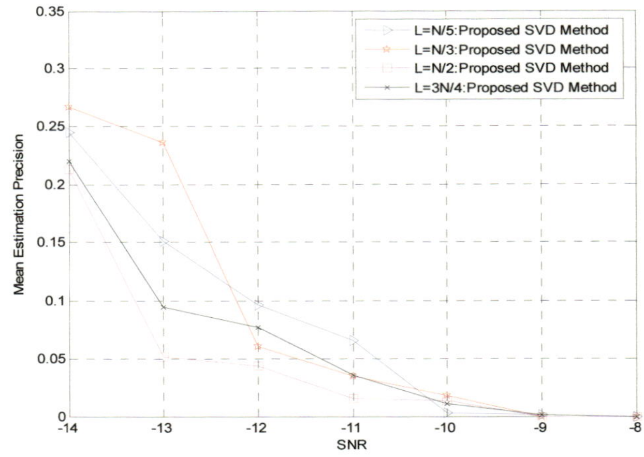

Fig. 6. Mean estimation precision of center frequency vs. SNR with different column number L for three WM signals.

2.5 Conclusions

In this part, we proposed a SVD based approach to detect and estimate multiple WM signals in a WRAN when a UWB system wants to use this spectrum. After performig SVD on the received data matrix, the presence and number of these WM signals can be detected and their center frequencies can be estimated. Consequently, guard bandwidths are retained to protect these primary users and the other detected spectra are available for the CR users. Simulation results prove that our method is simple and robust and it is especially suitable for detection and estimation of WM signals in a wideband spectrum sensing.

3. Detection and avoidance scheme based on orthonormal expansion

As well known, UWB technology is divided into two distinctive groups. The first group, known as multi-band UWB (MB-UWB), divides the entire UWB band into many sub-bands, with each sub-band being allocated a sinusoidal carrier. The DAA scheme mentioned above originates from this multi-band way of using UWB spectrum—each time a sub-band is detected being interfered, the carrier allocated for it is turned off. The second group is known as direct sequence UWB (DS-UWB), which, unlike the first, typically adopts a single-band transmission and depends entirely on varying pulse shapes to fit given spectrum masks; therefore, it is relatively difficult to turn on/off individual sub-band. A question is thus raised: Can the spectrum of the single-band DS-UWB be soft?

To answer this question, let us first investigate the currently proposed DS-UWB pulses: Rayleigh monocycle, Cubic monocycle, Gaussian monocycle, Gaussian doublet (Benedetto et al., 2006; Benedetto et al., 2004), high-order Gaussian derivatives (Win, 2000), modified Hermite polynomials (Ghavami et al., 2001) and so forth. The finding is somewhat discouraging—all of them feature fixed spectra. Used individually, they are not soft at all. Then, can the combinations of them be soft? As addressed in (Benedetto et al., 2004), a group of Gaussian derivatives have been linearly combined to generate an aggregate pulse that yields maximum spectral capacity. Such a combination adopts random-search optimization method, in the sense that a large number of combination coefficients are randomly

generated and the resulting combinations are evaluated. The combination that has minimum distance to the targeted spectrum mask is picked up as the optimal combination. This optimization method demands a huge number of iterations before finding the optimum. The converging time varies from situation to situation, so the linear combination methods are something between fixed and soft.
Moreover, cognitive UWB devices need to design and re-design the pulses on the scene of communication instead of having them preset or fixed in factories. In cognitive environment, the re-design of DS-UWB pulse must be agile enough and easily re-configurable.
To this end, we propose a soft-spectrum-based detection-and-avoidance algorithm for the single-band DS-UWB systems. The algorithm adopts a co-basis expansion method, in the sense that the well-known Hermite-Gaussian functions (HGFs) are used to constitute a common basis for both the time and frequency domains. The co-basis has twofold advantages: First, it yields the pulses directly from expanding the given soft-spectrum masks in frequency domain, so the pulses can conform to arbitrary spectrum masks. Second, the co-basis (that is, the HGFs) can be digitalized and built into matrices, such that whenever a new soft-spectrum is sensed or discovered, its expansion by the co-basis is as simple as matrix multiplications. As a result, the algorithm is really soft, low complex, always convergent, and agile enough for cognitive purpose.

3.1 The establishment of the soft-spectrum mask

The criterion for the design of DAA pulses is the ruling of the Federal Communications Commission (FCC), namely, the FCC's power emission mask (FCC, 2002), which ranges from 3.1 to 10.6GHz with power limit P_{max}=−41.3dBm/MHz. Within the allocated UWB band, other radio systems such as IEEE 802.11a or HiperLan has already been in operation. For cognitive purpose, the DS-UWB radio must be aware of the existence of such primary systems before transmission and automatically avoid the frequency bands in use by primary users.
In the design of the DAA scheme for DS-UWB radio, our emphasis is placed on the side of avoidance. In order not to digress our focus, we leave the side of detection to reference to well-established spectral estimation methods in literature, for example, the multi-taper spectral estimator that performs fast Fourier transform (FFT) and threshold inspection (Haykin, 2005; Welch, 1967). Before transmission, the DS-UWB radio senses the ambient radio environment with a detecting unit. Upon detecting an in-use sub-band, it calculates the 10dB-bandwidth of the sub-band and marks the sub-band as forbidden. In a recursive manner, DS-UWB radio sweeps the entire UWB band and records all the forbidden sub-bands. After the sensing process is over, the UWB radio establishes a soft-spectrum model that conforms not only to the FCC mask but also to the real-time radio environment. The soft-spectrum model so-established can be expressed as

$$R_{ss}(f)=\begin{cases} P_{\max} & f \in I - I_s \\ 0 & f \in I_s \end{cases} \tag{26}$$

where P_{max}=−41.3dBm/MHz, I=[3.1GHz, 10.6GHz], and I_s represents the union of the forbidden sub-bands.

3.2 The relationship between the soft-spectrum and the frequency response

The DS-UWB radio is by nature a spread spectrum system, whose transmitted waveforms can be characterized as follows (Ye et al., 2004),

$$s(t)=\sum_{k=-\infty}^{\infty}\sum_{j=1}^{N_c}p(t-kT_b-jT_c)b_k c_p^j \tag{27}$$

where b_k is the kth data bit with duration T_b; T_c is the chip duration; N_c is the spreading factor (that is, T_b=N_cT_c); c_p^j is the jth chip of the pseudorandom code; $p(t)$ is the pulse waveform. Through substitution of variables Eq. (27) can be simplified as:

$$s(t)=\sum_{i=-\infty}^{\infty}d_i p(t-iT_c) \tag{28}$$

where

$$i=kN_c+j \text{ and } d_i=d_{k,j}=c_p^j b_k \tag{29}$$

The autocorrelation function of $s(t)$ is given by (Proakis, 2003)

$$r_{ss}(\tau)=\frac{1}{T_c}\sum_{l=-\infty}^{\infty}r_{dd}(l)r_{pp}(\tau-lT_c) \tag{30}$$

where $r_{dd}(\bullet)$ represents the autocorrelation function of the information sequence $\{d_i \in\{\pm1\}\}$; $r_{pp}(\bullet)$, the autocorrelation function of the pulse. Correspondingly, the PSD of $s(t)$ is given by

$$R_{ss}(f)=\frac{1}{T_c}R_{dd}(f)|P(f)|^2 \tag{31}$$

which indicates that the PSD of the transmitted waveforms depends not only on the frequency response of the pulse, $P(f)$, but also on the PSD of the information sequence, $R_{dd}(f)$, and on the chip duration T_c as well. However, since the sequence $\{d_i \in\{\pm1\}\}$ can be viewed as an uncorrelated random process with zero mean and unitary variance (Benedetto et al., 2004; Ye et al., 2004), that is, $r_{dd}(l)=\delta(l)$, and R_{dd}(f)=1, the autocorrelation function defined by Eq. (30) and the PSD defined by Eq. (31) can be further simplified respectively as follows,

$$r_{ss}(\tau)=\frac{1}{T_c}r_{pp}(\tau)=\frac{1}{T_c}\int_{-\infty}^{\infty}p(t)p^*(t+\tau)dt \tag{32}$$

and

$$R_{ss}(f)=\frac{1}{T_c}|P(f)|^2 \tag{33}$$

By substituting Eq. (26) into Eq. (33), we obtain the frequency response $P(f)$ of the transmitted DS-UWB waveforms that conforms both to the FCC mask and to the ambient RF environment, such $P(f)$ we refer to as soft-spectrum mask, namely

$$P(f)=\sqrt{T_c R_{ss}(f)}=\begin{cases} A & f\in I-I_s \\ 0 & f\in I_s \end{cases} \tag{34}$$

where

$$A = \sqrt{10^{P_{\max}/10} \times 10^{-9} \times T_c} \quad (\text{V/Hz}) \tag{35}$$

3.3 The establishment of a co-basis for both the time and frequency domain

The frequency response given by Eq. (34) is inherently an energy signal and can be uniquely expanded by orthonormal functions that span the signal base. But, an ordinary expansion does not suffice here. In the design of soft-spectrum-based DAA pulse, we need a common basis for both the time and frequency domains, that is, a co-basis. With this co-basis, the well-known orthonormal expansion method will do wonders for the design of pulses, yielding the waveform of the pulses (time domain) by expanding the soft-spectrum in the frequency domain.

Hermite-Gaussian functions (HGFs) constitute ideally such a co-basis.

The HGFs are combinations of Hermite polynomials with a Gaussian function, as written as follows (Ozaktas et al., 2000),

$$\varphi_l(u) = a_l H_l(\sqrt{2\pi}\, u) e^{-\pi u^2}, \quad a_l = 2^{1/4} / \sqrt{2^l l\,!} \tag{36}$$

where $H_l(\bullet)$ denotes the lth order Hermite polynomial. The generation function for Eq. (36) is

$$\varphi_l(u) = \frac{(-1)^l a_l}{(\sqrt{2\pi})^l} e^{\pi u^2} \frac{d^l}{du^l} e^{-2\pi u^2}, \quad l = 0,1,2\ldots \tag{37}$$

Note that the HGFs defined by Eqs. (36) and (37) are slightly different from those defined in classical mathematical textbooks—here, they are $\sqrt{2\pi}$-scaled, so that they turn out to be the eigenfunctions of fractional Fourier transform (Ozaktas et al., 2000). In other words, because the HGFs are $\sqrt{2\pi}$-scaled, they are shape-invariant to fractional Fourier transform (Ozaktas et al., 2000), that is,

$$F^{\alpha}\{\varphi_l(u)\} = e^{-i\alpha\, l\pi/2} \varphi_l(\mu) \tag{38}$$

where $\mathcal{F}^{a}$denotes fractional Fourier transform (FRFT) operator; $e^{-il\alpha\pi/2}$, the corresponding eigenvalues; α, the order of the FRFT.

The FRFT is a generalization of the ordinary Fourier transform with an order parameter α. The α-th order fractional Fourier transform is the α-th power of the ordinary Fourier transform operation. When α=−1, the corresponding FRFT operation is exactly the ordinary inverse Fourier transform. Under such circumstance, Eq. (38) becomes

$$F^{-1}\{\varphi_l(u)\} = i^{\,l} \varphi_l(\mu) \tag{39}$$

which indicates that the HGFs are shape-invariant to the inverse Fourier transform except for a phase shift. This nice property makes the HGFs constitute a common basis for both the frequency and time domain. To emphasize this, we introduce two normalized variables u and μ in place of the natural frequency f and time t. The relationship among them will be addressed later on.

3.4 The dimension of the co-basis

Before proceeding, we need to determine the dimension of the co-basis, M, or equivalently, the number of HGFs involved in the expansion. Mathematically, the orthonormal expansion is least-mean-square approximation to the signal being expanded (Proakis, 2003), as stated as follows,

$$Q = \int_{-\infty}^{\infty}[P(f) - \sum_{l=0}^{M-1} c_l \varphi_l(f)]^2 df = \int_{-\infty}^{\infty} P^2(f)df - \sum_{l=0}^{M-1} c_l^2 \tag{40}$$

where the square error Q manifests itself as in-band and out-of-band ripples in the expansion of the signal. In the case that the magnitude of the ripples is specified, Q can be computed according to the specifications and then M is obtained by solving Eq. (40). In other cases that the ripples are not of major concern, as in the current DAA case, M can be determined by rule of thumbs. For example, it has been verified by computer simulations that a medium M between 30 and 100 is sufficient for performing DAA functionalities that avoid two to four sub-bands.

3.5 The representation of the soft-spectrum mask by the co-basis

After determining the number of HGFs participating in the expansion, we need next to normalize the natural frequency argument of the soft-spectrum mask as given by Eq. (34). The purpose of such normalization is to let the support of the soft-spectrum mask match the support of the co-basis. The support of the co-basis is not compact, but its energy is mostly (say 90%) concentrated on a finite interval (Ozaktas et al., 2000) $[-u_H, u_H]$, where

$$u_H = \left\lceil \sqrt{\frac{(M+0.5)}{\pi}} \right\rceil \tag{41}$$

where $\lceil \cdot \rceil$ means rounding up to the nearest integer. u_H is closely related to the dimension of the co-basis and is used to determine a scaling factor τ,

$$\tau = \frac{u_H}{f_H} = \left\lceil \sqrt{(M+0.5)/\pi} \right\rceil / f_H \tag{42}$$

which has the dimension of time and helps to normalize the frequency and time arguments as follows:

$$u = \tau f \tag{43}$$

$$\mu = t / \tau \tag{44}$$

With the frequency and time thus-normalized, the soft-spectrum mask can be expanded by the co-basis as

$$P(u) = \sum_{l=0}^{M-1} c_l \varphi_l(u) \qquad -u_H \le u \le u_H \tag{45}$$

where c_l denotes the lth-order expansion coefficient given by

$$\begin{aligned} c_l &= \int_{-\infty}^{\infty} \varphi_l(u)P(u)du \\ &= A\int_{u\in\tau\times I} \varphi_l(u)du - A\int_{u\in\tau\times I_s} \varphi_l(u)du \end{aligned} \tag{46}$$

where $\tau\times I$ =[3.1τ, 10.6τ], representing the τ-scaled UWB band, and $\tau\times I_s$ represents the τ-scaled union of the forbidden sub-bands. Put in matrix form, Eq. (46) means

$$\mathbf{C} = \frac{Au_H}{N}(\mathbf{H}\mathbf{U}^T - \mathbf{H}\mathbf{S}^T) \tag{47}$$

where **C**=$\{c_l, 0\le l\le M-1\}$ represents a column vector; N is an integer denoting the number of sample points required for numerically calculating the integration, and by rule of thumb, is chosen as N=2048 for general DAA functionalities; **U** and **S** are two N-sampled and unitary-valued row vectors representing the entire UWB band and the forbidden sub-bands respectively; the superscript T denotes the transpose operation; **H** is an M by N matrix with its lth row denoting the (l-1)th-order HGFs, which are N-point-sampled on the interval [$-u_H$, u_H]. The waveform of the transmission pulse with normalized time argument is then obtained by applying FRFT to both sides of Eq. (45), namely,

$$p(\mu) = F^{-1}\{P(u)\} = \sum_{l=0}^{M-1} i^l c_l \varphi_l(\mu) \tag{48}$$

Finally, the waveform of the transmission pulse $p(t)$ is obtained by de-normalizing the argument μ, that is, by substituting t/τ for μ in Eq. (48), namely

$$p(t) = p(\mu)\big|_{\mu=t/\tau} = \sum_{l=0}^{M-1} i^l c_l \varphi_l(t/\tau) \tag{49}$$

3.6 Implementation of the pulse by Software Defined Radio technique

The pulse given by Eq. (49) involves linear combinations of M τ-scaled HGFs, which are continuous functions. The analogue generation of the pulse is very difficult, whereas the digital generation is much easier. Through modern SDR technique (Reed, 2002), the analogue $p(t)$ can be sampled into a discrete time sequence as

$$p[n] := p(t)\big|_{t=nT_0} = \sum_{l=0}^{M-1} i^l c_l \varphi_l(nT_0/\tau) \tag{50}$$

where := means sampling. The period of T_0 is determined as follows,

$$T_0 \le \frac{1}{2f_H} \tag{51}$$

In the case of Nyquist sampling, the equality can be used, namely T_0=1/(2f_H); thus, Eq. (50) becomes

$$p[n] = \sum_{l=0}^{M-1} i^l c_l \varphi_l\left(\frac{n}{2u_H}\right) \tag{52}$$

where n is defined on the interval equating to $2u_H$ times the support of the co-basis, namely,

$$-2u_H^2 \le n \le 2u_H^2 \tag{53}$$

Let $K=4u_H^2$. The length of the resulting sequence $\{p[n], -K/2 \le n \le K/2\}$ is then K or $(K+1)$ depending on K being even or odd. Let us take it as K for easy discussion. Putting Eq. (52) in matrix form and denoting $\mathbf{C}=\{c_l, 0 \le l \le M-1\}$ where $\mathbf{C}$ is given by Eq. (47), we obtain

$$\mathbf{P} = \frac{Au_H}{N}(\mathbf{U}-\mathbf{S})\mathbf{H}^T\mathbf{E}\boldsymbol{\Phi} \tag{54}$$

where $\mathbf{P}=\{p[n], -K/2 \le n \le K/2\}$ represents the pulse; $\mathbf{E}$ is an M by M diagonal matrix with the diagonal elements $\{e_{ll}\}$ equating to the unitary complex values $\{i^{l-1}, l=1,2,\ldots,M\}$; **H**, **U** and **S** have been defined previously; $\boldsymbol{\Phi}$ is an M by K matrix with the l-th row representing the (l-1)th-order scaled-and-discrete HGFs $\phi_l(n/2u_H)$.
We comment that the resultant **P** from Eq. (54) is a complex-valued sequence: Its real part is even, known as the inphase part; its imaginary part is odd, known as the quadrature part. In other words, **P** is a complex baseband signal, and consequently its generation needs two signal branches (one for the inphase and the other for the quadrature) (Reed, 2002).

3.7 Detection and avoidance scheme

In essence, the pulse **P** given by Eq. (54) is a sum of two quantities, each involving a series of matrix multiplications. The matrices **H**, **E**, and $\boldsymbol{\Phi}$ are known *a priori*, keeping unchanged throughout the DAA operation. The row vector **U** represents the entire UWB band, keeping also unchanged. Therefore, the minuend in Eq. (54) is a constant sequence, which only needs to be computed once and then stored in the read-only memory (ROM), whereas the row vector **S** represents the sub-bands in use by primary users, and it may keep changing under a cognitive environment. Therefore, the subtrahend in Eq. (54) requires to be updated corresponding to the changing radio environment. Obviously, matrix multiplication is a dominant operation. To reduce computation burden, the product $\mathbf{H}^T\mathbf{E}\boldsymbol{\Phi}$ is pre-computed and stored in ROM, so **P** only needs to be updated by computing the product of the row vector **S** (1 by N) with a matrix $\mathbf{H}^T\mathbf{E}\boldsymbol{\Phi}$ (N by K). Accordingly, the DAA scheme can be devised as follows:
DAA scheme (as illustrated in Fig. 7)

Initialization:
Input A, u_H, N, H, E, Φ, and U.
Compute product $\mathbf{P} = \frac{Au_H}{N}\mathbf{U}\mathbf{H}^T\mathbf{E}\boldsymbol{\Phi}$ and store it in ROM .
Compute and store in ROM the intermediate matrix $\mathbf{I} = \frac{Au_H}{N}\mathbf{H}^T\mathbf{E}\boldsymbol{\Phi}$.
Detecting:
Sense the ambient channel and take samples during some *quiet* period.
Perform FFT computation and multi-taper spectral estimation (Haykin, 2005; Welch, 1967).
Inspect the power emission level and make decision according to a prescribed soft spectrum policy.
If a subband is in use by a primary user(s), mark it as forbidden.
Repeat detecting until the entire UWB band is swept.

Avoiding:
Build row vector S according to the following criteria:
Let the amplitude of the forbidden sub-bands be equal to one.
Let the amplitude of the other sub-bands be equal to zero.
Insert smoothing curves at transitions from one to zero to prevent Gibbs phenomenon.
Compute the instantly changing part $\mathbf{P}' = \mathbf{S} \times \mathbf{I}$.
Update $\mathbf{P} = \mathbf{P} - \mathbf{P}'$.
Output the updated sequence P.
Stop.

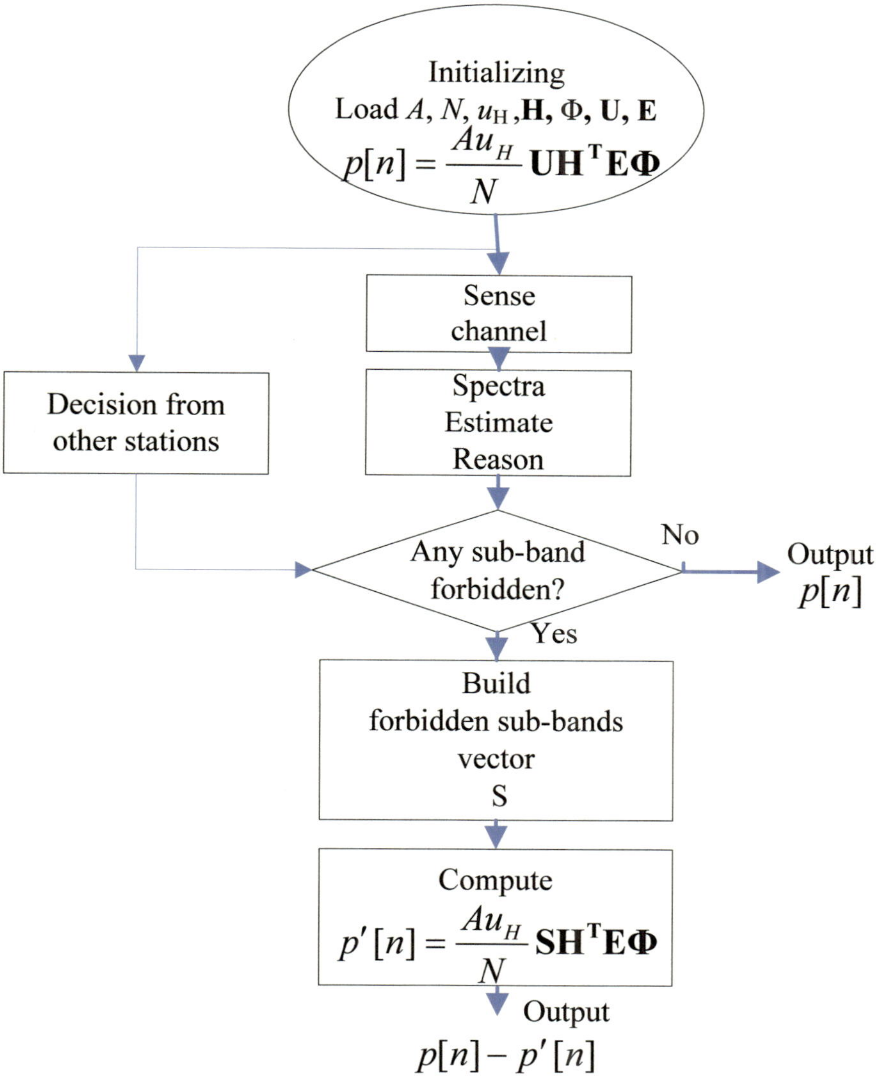

Fig. 7. Flowchart of the DAA Scheme

3.8 Computer simulations

3.8.1 Examples

As examples, we investigate the following two scenarios:

1. No sub-band is in use by primary users; the entire DS-UWB band is accessable.
2. Two sub-bands are in use by primary users: one at central frequency f_1=4.5 GHz and the other at f_2=7.5GHz. The bandwidth are both 500 MHz.

The parameters involved are listed in Table 1.

Name	Notation	Typical Values
Lower frequency bound	f_L	3.1 GHz and Below
Upper frequency bound	f_H	10.6 GHz
Chip duration	T_c	10 ns
Amplitude limit	A	2.7×10^{-11} V/Hz
Nyquist sampling period	T_0	0.0472ns
Dimension of the co-basis	M	48
Number of spectrum samples	N	2048
Length of the time sequence	K	64
Normalization factor	τ	0.3774 *ns*
Interfered central frequencies	f_1 f_2	4.5GHz 7.5GHz
Bandwidth of the interfered sub-band	Δf	500MHz

Table 1. Basic Parameters involved

For the first scenario, the PSD of the resulting pulse is illustrated in Fig. 8. As expected, the pulse occupies the entire regulated DS-UWB band. Raised-cosine curves have been inserted to smooth the transitional edges of the spectrum vector **U** and **S**, so the out-of-band ripples are ideally suppressed.

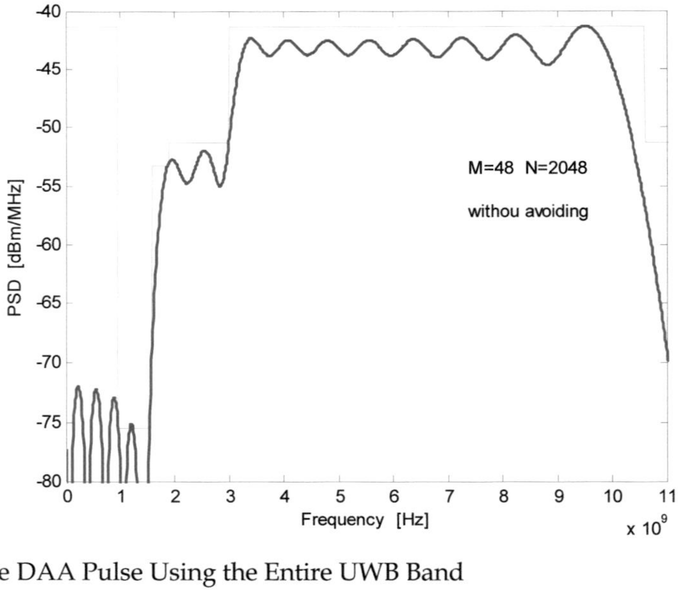

Fig. 8. PSD of the DAA Pulse Using the Entire UWB Band

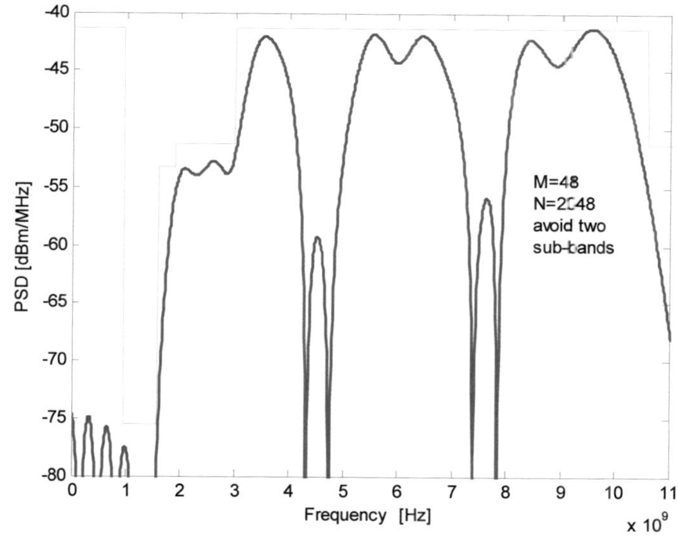

Fig. 9. PSD of the DAA pulse avoiding two sub-bands on which primary users are operating

Fig. 9 illustrates the PSD of the resulting pulse for the second scenario. As expected, the DAA pulse forms two 15dB deep valleys around the two sub-bands in use by the assumed two primary users, effectively avoid interfering the primary users.

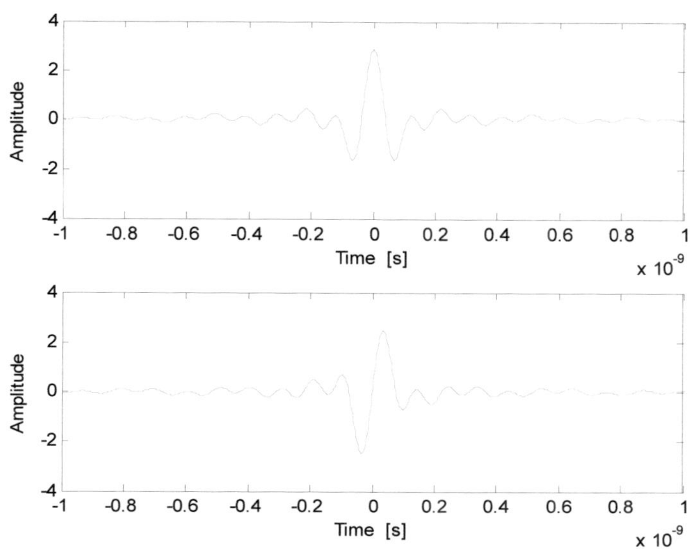

Fig. 10. Waveforms of the DAA Pulse Avoiding Two Sub-bands

The pulse waveforms for the second scenario (for simplicity, the waveforms of the first, which is similar to the second, is left out) is shown in Fig. 10. As seen, the pulse consists of two parts: The real part (on the top) is even, and the imaginary (on the bottom) is odd.

The autocorrelation function, as given by Eq. (32), of the DAA pulse is illustrated in Fig. 11, in which the narrow main-peak suggests that the DAA pulse is sensitive to time jitter, possibly more sensitive than an ordinary pulse, this is the price to pay for DAA.

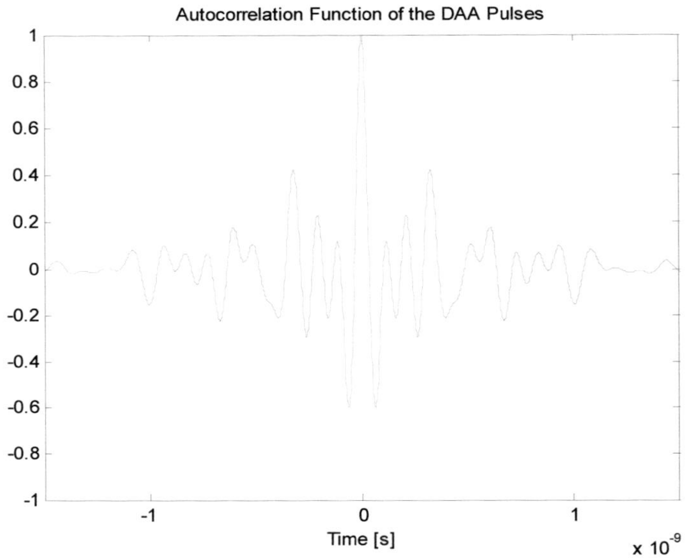

Fig. 11. Autocorrelation Function of the DAA Pulses Avoiding Two sub-bands.

In multi-user or multi-access situations, the DAA pulse works in a similar manner as in a general DS-UWB spread-spectrum scheme. So the performance for multi-user or multi-access under DAA operation is guaranteed by the performance of the pseudorandom or pseudo-noise (PN) sequence assigned for differentiating multi-users. Therefore, the cross-correlation properties of the DAA pulses are out of concern here.

3.8.2 Complexity

The suggested DAA algorithm involves mainly matrix multiplications, for which the dominating operation is complex multiplication. Three factors dictate the total number of multiplications: first, N--the number of sampling points for performing numerical integration; second, M-- the dimension of the co-basis; finally, K--the length of the resulting sequence representing the transmission pulse.

The unchanging part $\mathbf{P}$ requires roughly $K(MN+M+N)$ complex-value multiplications and thus consumes most of the computer time. Given N=2048, M=48, and the resultant K=64, the complex-value multiplications totals 6,425,600 in the simulations. However, the $\mathbf{P}$ only needs to be calculated once, so it does not represent the real computational complexity. On the other hand, the changing part $\mathbf{P}'$ requires to be updated frequently, but its computational time is reduced to NK because the intermediate matrix ($\mathbf{I} = \frac{Au_H}{N}\mathbf{H}^T\mathbf{E}\mathbf{\Phi}$) has already existed; therefore, the real computational complexity is $\mathcal{O}(NK)$, totaling roughly 131,072, roughly equivalent to 0.1 second if the digital signal processor embedded in the UWB radio operates at one million instructions per second. The amount of time does not vary regardless of the central frequencies and bandwidths of the sub-bands in use by primary users—as opposed to the changeable computational time in the linear combination method addressed in (Benedetto et al., 2004). Therefore, the DAA algorithm has predictable and managable processing delay, and is robust in real-time communications.

3.9 Conclusion

Detection and avoidance, as a cognitive radio scheme, has been proven effective for multi-band UWB group. The basic idea underlying the DAA is turning off individual carrier-tone on the interfered sub-band. However, coming to direct-sequence UWB, a competing technology group with the multi-band UWB, this idea of turning off tones ceases to be true because shutting off any sub-band would mean to re-design the pulses all over again. In a cognitive environment, the re-design should be agile enough and easily reconfigurable. To this end, we devise a DS-UWB-oriented DAA scheme by emphasizing the side of avoidance (that is, the re-design of the pulse) while de-emphasizing the side of detection by referencing the well-established spectral estimation methods in existing literatures. We propose a domain-less co-basis expansion method in the sense that Hermite-Gaussian functions are used to constitute a common basis (co-basis) for the time and frequency domains. One advantage of the co-basis is that the transmission pulses are directly obtained from the expansion of given soft-spectrum masks, so the resulting pulses fit into arbitrary spectrum masks. Another advantage is that the co-basis functions (that is, the HGFs) are discretized, built as matrices, and stored in ROM, such that whenever a soft spectrum is sensed or discovered, the DAA-enabled pulse is generated by merely matrix multiplying. The amount of computational time is thus trivial, and the re-design of the pulse can respond quickly to a rapidly-changing soft spectrum. The algorithm can be implemented through software defined radio (SDR) techniques.

Computer simulation verifies that the DAA algorithm is low complex, easily configurable, robust, and agile enough to avoid the intended subbands.

4. References

Brent Parr et al., "A novel Ultra-Wideband Pulse Design Algorithm," *IEEE Communications letters*, Vol. 7, No. 5, May 2003.

De, P. & Liang, Y. C. Blind Sensing Algorithms for Cognitive Radio, *Proceedings of 2007 IEEE Radio and Wireless Symposium*, pp. 201–204, Long Beach, USA, January 9-11, 2007

Dhillon, R. S. & Brown, T. X. Models for Analyzing Cognitive Radio Interference to Wireless Microphones in TV Bands, *IEEE DySPAN 2008*, pp. 1 – 10, Chicago, USA, October 14-17, 2008

FCC, "Revision of Part 15 of the Commission's Rules Regarding Ultra-wideband Transmission Systems: First Report and Order," *Technical Report* FCC 02-48, 2002.

Haldun M. Ozaktas et al., *The Fractional Fourier Transform with Applications in Optics and Signal Processing*, John Wiley & Sons, LTD, Chichester, New York, 2000.

Honggang Zhang & Ryuji Kohno, "Soft-Spectrum Adaptation in UWB Impulse Radio," *the 14th IEEE 2003 International Symposium on Personal, Indoor and Mobile Radio Communication Proceeding.*

Hu, W. et al. Dynamic Frequency Hopping Communities for Efficient IEEE 802.22 Operation. *IEEE Communications Magazine*, Vol. 45, No. 5, (May 2007), pp. 80–87, ISSN 0163-6804

IEEE P802.22 working Group for WRAN, Cognitive Wireless RAN Medium Access Control (MAC) and Physical Layer (PHY) specifications: Policies and procedures for operation in the TV Bands, IEEE P802.22/WDv0.4.7 Draft Standard for WRAN Part 22, 2006.

Jeffrey H. Reed, *Software Radio: A modern Approach to Radio Engineering*, Prentice Hall PTR, 2002.

Jim Lansford, "UWB Coexistence and Cognitive Radio," *Ultra Wideband Systems 2004*, Joint with Conference on Ultra wideband Systems and Technologies

John G. Proakis, *Digital Communications, Fourth Edition*, McGraw-Hill Companies, 2003.

John Walko, "Cognitive Radio," *IEE Review*, Http://www.iee.org, May 2005.

Johnstone, I. M. On the distribution of the largest eigenvalue in principle components analysis, *The Annals of Statistics*, Vol. 29, No. 2, (2001), pp. 295–327.

Kalke, J. TV Band Low Power Devices the need for Coexistence with IEEE 802.22, *IEEE Plenary Tutorial*, pp. 1-10, Vancouver, Canada, November 14-18, 2005

Lei, Z. D. & Chin, F. A Reliable and Power Efficient Beacon Structure for Cognitive Radio Systems, *IEEE Transactions on Broadcasting*, Vol. 54, No. 2, (June 2008), pp. 182 - 187, ISSN 0018-9316

Lim, S.; Kim, S.; Park, C. & Song, M. The detection and classification of the Wireless Microphone signal in the IEEE 802.22 WRAN system, *Asia-Pacific Microwave Conference 2007*, pp.1–4, Bangkok, Thailand, December 11-14, 2007

Maria-Gabriella Di Benedetto et al. (Ed), *UWB Communication Systems: A Comprehensive Overview*, Hindawi Publishing Corporation, 2006.

Maria-Gabriella Di Benedetto et al., *Understanding Ultra Wide Band Radio Fundamentals*, Prentice Hall PTR, 2004.

Moe Z. Win. "Ultra-Wide Bandwidth Time-Hopping Spread-Spectrum Impulse Radio for Wireless Multiple-Access Communications," *IEEE Transaction on Communications*, Vol. 48, No. 4, April 2000.

Mohammad Ghavami et al., "Hermite Function Based Orthogonal Pulses for Ultra Wideband Communications," *Proc. IEEE Wireless Personal Multimedia Conference (WPMC'01)*, Aalborg, Denmark, September 2001.

Mossa, A. M. & Jeoti, V. Cognitive Radio: Cyclostationarity-Based Classification Approach for Analog TV and Wireless Microphone Signals, *IEEE Innovative Technologies in Intelligent Systems and Industrial Applications 2009*, pp. 107 - 111, Kuala Lumpur, Malaysia, July 25-26, 2009

Notor, J. The Evolution of Spectrum Sharing in the IEEE 802.22 WRAN Standards Process, http://www.eecs.berkeley.edu/~dtse/3r_notor.ppt , Feb. 2006.

P. D. Welch, "The Use of Fast Fourier Transform for the Estimation of Power Spectra: A Method base on Time-Averaging over Short, Modified Periodograms," *IEEE Trans. Audio Electroacoustics*, Vol. AU-15, pp. 70-73, 1967.

Quan, Z.; Cui, S. G.; Poor, H. & Sayed, A. Collaborative wideband sensing for cognitive radios, *IEEE Signal Processing Magazine*, Vol. 25, No. 6, (November 2008), pp. 60 - 73, ISSN 1053-5888

R. Kohno & K. Takizawa, "Detection and Avoidance Based on Soft-Spectrum Adaptation of UWB Interference to Existing Radio Systems," *IEEE Ninth International Symposium on Spread Spectrum Techniques and Applications*, 2006.

Rife, D. C. & Boorstyn, R. R. Single Tone Parameter Estimation from Discrete-Time Observations, *IEEE Trans. Inform. Theory*, Vol. IT-20, (September 1974), pp. 591–598, ISSN 1089-7771

Simon Haykin, "Cognitive Radio: Brain-Empowered Wireless Communications," *IEEE Journal on Selected Areas in Communications*, Vol. 23, No. 2, Feb. 2005.

Stevenson, C. R. et al. Functional Requirements for the 802.22 WRAN Standard r47, (January 2006)

Teh, K. C.; Teng, C. C.; Kot, A. C. & Li, K. H. Jammer Suppression in Spread Spectrum, *IEEE Conf. on Information Engineering*, pp. 220–224, 1995.

Tufts, D. & Kumaresan, R. Singular Value Decomposition and Improved Frequency Estimation Using Linear Prediction, *IEEE Trans. on Acoustics, Speech, and Signal Processing*, Vol. 30, No. 4, (August 1982), pp. 671–675

Unnikrishnan, J. & Shellhammer, S. Simulation of Eigenvalue based sensing of wireless mics, IEEE 802.22-07/0357r0, (July 2007)

Wu, Y. C; Wang, H. G. & Zhang, P. Protection of Wireless Microphones in IEEE 802.22 Cognitive Radio Networks, *IEEE International Conference on Communications Workshops 2009*, pp. 1 - 5, Dresden, Germany, June 14-18, 2009

Zeng, Y. H. & Liang, Y. C. Covariance Based Signal Detections for Cognitive Radio, *Proceedings of 2nd IEEE DySPAN 2007*, pp. 202– 207, Dublin, Ireland, April 17-20, 2007

Zeng, Y. H. & Liang, Y. C. Eigenvalue-Based Spectrum Sensing Algorithms for Cognitive Radio, *IEEE Transactions on Communications*, Vol. 57, No. 6, (June 2009), pp. 1784 - 1793, ISSN 0090-677*8*

Zhenzhen Ye, Madhukumar A. S & Francois Chin, "Power Spectral Density and In-Band Interference Power of UWB Signals at Narrowband Systems," *IEEE International Conference on Communications 2004*, Volume 6, pp. 3561-3565, June 2004.

13

Performance Analysis of Spectrum Management Technique by Using Cognitive Radio

Keisuke Sodeyama and Ryuji Kohno
Yokohama National University
Japan

1. Introduction

The usage of the radio spectrum and the regulation of radio emissions are coordinated by national regulatory bodies. As part of radio regulation, the radio spectrum is divided into frequency bands, and licenses for the usage of frequency bands are provided to operators, typically for a long time such as one or two decades. With licensed frequency bands, operators have often the exclusive right to use the radio resources of the assigned bands for providing radio services. Depending on the type of radio service and on the efficiency of the radio systems, frequency bands may be used inefficiently.

Therefore, many national regulatory and standards bodies such as the Federal Communications Commission (FCC)[2], IEEE 802.22 WG [3], and the Ministry of Internal Affairs and communications in Japan have paid attention to the dynamic spectrum access (DSA) technology. Using DSA technology, radio systems can dynamically use and release radio spectrum wherever and whenever they are available. Moreover, DSA technology helps to minimize unused radio spectrum band [9]. This technology is also referred to as *cognitive radio technology*. Cognitive radio is defined as an intelligent wireless communication system, which may be aware of its environment and adapt to statistical variations in the input stimuli [8].

On the other hand, wireless communications systems such as wireless local area network (WLAN) and Bluetooth are becoming pervasive throughout the world. Especially, the main application of WLAN is wireless connection of PC's to a network but even includes such uses as wireless transmission of moving pictures at indoor environment. Thus, WLAN has dramatically grown in popularity. Bluetooth is also a promising wireless interface solution in mobile ubiquitous environments and is expected to predominate among such applications soon. Meanwhile, some new technologies such as ultra wideband (UWB) radio systems have been proposed for short-range wireless applications [7]. They are expected to spread as a complement to developed technologies such as WLAN and Bluetooth or to be merged with such established technologies.

UWB radio may inherently degrade the performance of the primary systems since the radio band of the UWB radio systems overlaps that of primary systems such as worldwide interoperability for microwave access (WiMAX), 4th generation mobile cellular systems (4G) and field pickup unit (FPU). The technical conditions on the usage of UWB radio system were set up by the Ministry of Internal Affairs and Communications on March 2006, in Japan. In the conditions, it is essential for UWB radio to equip interference mitigation technique, *detect and avoid* (DAA) [11][12].

In the environment for the usage of UWB, coexistence of heterogeneous wireless communications systems are enabled by using the concepts and techniques of the cognitive radio. Cognitive radio is a radio system that can sense the surrounding radio wave environment and use the radio resources efficiently by flexible reconfiguration of the system as a function of the environment changes [4].

Although UWB radio systems with DAA are allowed to transmit with power level of -41.3 dB/MHz, those without DAA technique must limit their emission level by -70 dBm/MHz, which is lower than the noise level. Therefore, DAA is essential for UWB radio systems in order to allow them to transmit with the maximum allowed power level.

The question that may arise at this point is how to design the MAC layer of cognitive radio systems such as UWB radio with DAA. Therefore, in this paper, this coexistence environment is analyzed by introducing two important benchmarks and the design issue is discussed based on these results. Moreover, we discuss the detection technique of primary system signals for UWB system with DAA and the effect of UWB system performance using DAA.

The rest of this paper is organized as follows. In Section 2, the cognitive radio system design issue is analyzed. The performance analysis of UWB radio system with DAA technique is presented in Section 3. Finally, conclusions are drawn in section 4.

2. Analysis of cognitive radio system design issue

2.1 System model

2.1.1 Channel and traffic model

We omit the effect of channel errors in order to make the analysis tractable. Hence, the channel is either busy or idle. The offered traffic is modelled with two random processes per radio systems [10], offered traffic and departure rate.

2.1.2 Radio spectrum usage model

Without loss of generality, radio spectrum usage model having two different radio systems is considered to analyze this coexistence environment. As shown in Fig. 1, radio system A operates on one frequency channel (center frequency f_2) and radio system B operates on three frequency channels (center frequencies f_1, f_2, f_3).

Radio system A can be considered as an UWB system with DAA technique and radio system B as a primary system. Radio system B access the channel based on the scheduling algorithm such as a time-division multiple access (TDMA). Radio system A can occupy a wideband radio resource if and only if all of the channels of radio system B are idle. Moreover, radio system A can recognize available channels without
sensing error and delay.

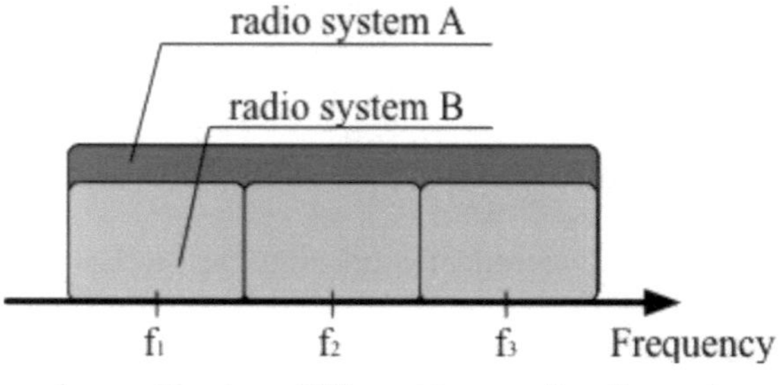

Fig. 1. Frequency channels used by two different types of radio system

2.2 Definitions of benchmarks

In this chapter, we employ "*air time*" and "*interference time*" as benchmarks. The "*air time*" means the ratio of allocation time per radio system to the reference time (say one hour) [10]. Namely,

$$air\ time_{type} = \frac{1}{N_{type}} \sum_{i=1}^{N_{type}} \frac{allocation\ time(i)}{reference\ time} \tag{1}$$

where N_{type} is the number of channels belonging to $type \in A, B$ and $allocation\ time(i)$ is the total time of radio resources allocated to $type$. It characterizes the share of resources each radio system can allocate.

The "*interference time*" refers to the ratio of *interfere time* to the *reference time*. Hence,

$$inerference\ time = \frac{1}{N_B} \sum_{i=0}^{N_B} \frac{inerference\ time}{reference\ time}, \tag{2}$$

where *interference time(i)* is the total time when radio system A and B use channels simultaneously. Note that *allocation time (i)* does not include *interference time(i)*.

The radio systems with different channel bandwidths have different requirements for the throughput performance so that the fairness of the network should be considered. However, the mutual interference of the radio systems is significantly important for the design of this coexistence network compared to throughputs and the fairness since the radio system B is a licensed system, which must be protected from the interference from the radio system A such as the UWB system. Thus, in this paper, the throughput performance is not investigated and our interest is restricted to analyze the mutual influence and the actual channel usage rate.

2.3 Numerical results

In this section, computer simulation and the theoretical analysis are presented. We reported the theoretical analysis in [6]. Fig. 2 and Fig. 3 show *airtime* and *interference time* versus offered traffic of radio system A or B, respectively. Also, Fig. 4 shows *airtime* and *interference time* versus the departure rate of radio system A.

From Fig. 2, *interference time* is approximately zero over wide range of offered traffic of radio system B because of DAA function of system A. The *airtime* of system B can achieve about 0.65 without increasing *interference time*. However, *airtime* of system A is decreased by increasing offered traffic of B. Therefore, a trade-off between *airtime* of system A and that of system B can be found.

From Fig. 3, *airtime* of system A may be increased by increasing its offered traffic. However, maximal *airtime* of system A cannot exceed 0.1. On the other hand, offered traffic of system A also increases *interference time*, of which maximal value is about 0.2. Therefore, if the system A requires more offered traffic, then that of system A should be increased at the cost of increasing *interference time*.

From the Fig. 4, while *interference time* is decreased by increasing the departure rate of radio system A, *airtime* of radio system B becomes longer. However, *airtime* of system A is decreased since the occupancy time of channels becomes shorter by increasing the departure rate of system A.

In order to minimize the *interference time*, the offered traffic of radio system A should be small and the departure rate large. The *airtime* of system B is 0.3 and *interference time* is 0.5 even if offered traffic of system A is one. On the other hand, the *airtime* of system B becomes zero and *interference time* becomes 0.5 if the departure rate of system A is zero. Therefore, the occupancy time of channels should be shortened for system A rather than decreasing offered traffic since the departure rate is inversely proportional to the occupancy time.

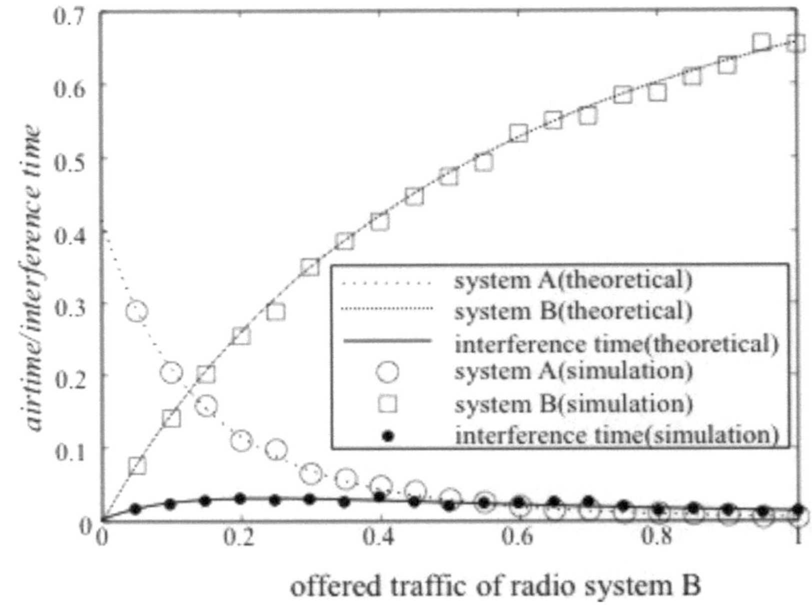

Fig. 2. *Airtime* of each system and *interference time* vs. offered traffic of radio system B.

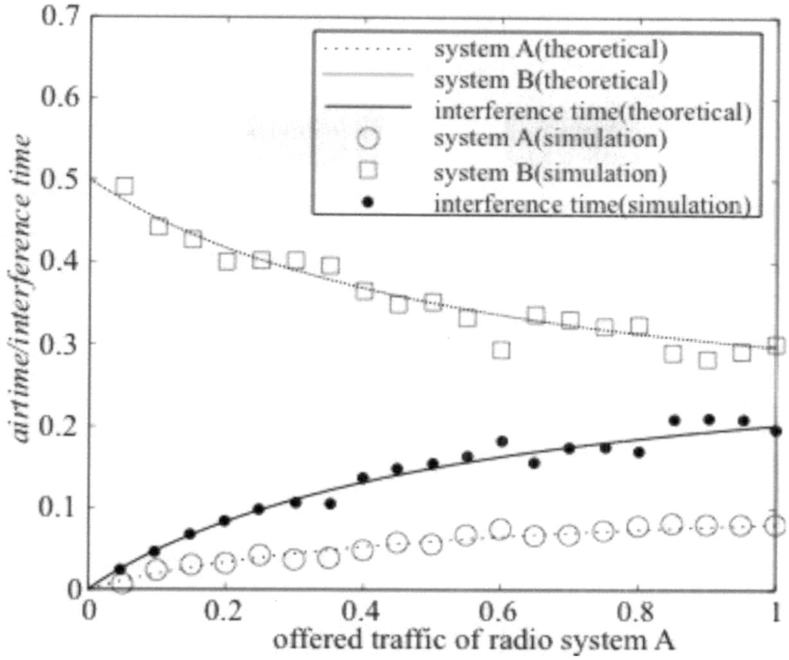

Fig. 3. *Airtime* of each system and *interference time* vs. offered traffic of radio system A.

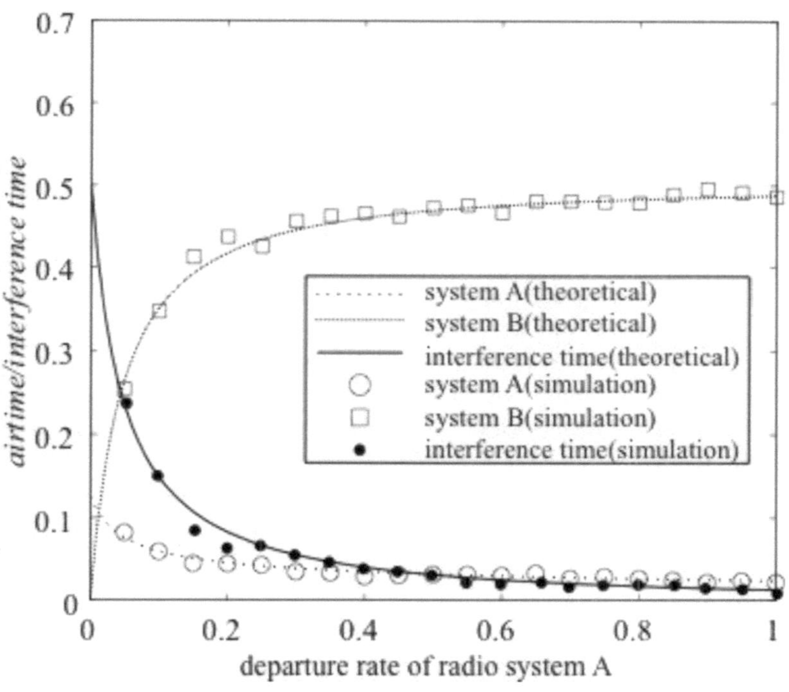

Fig. 4. *Airtime* of each system and *interference time* vs. departure rate of radio system A.

3. The performance analysis of UWB radio system with DAA

In Section 3, we show that the mutual interference is inherently occurred even with the ideal cognitive radio technology. However, in practical situations, the cognitive radio technologies cannot detect the primary systems ideally and this effect may degrade the performance of the primary systems. Therefore, in this section, the performance of the interference mitigation technique for UWB radio communications is investigated under a practical scenario. Detection technique of primary system for UWB system is discussed and also the avoidance technique of interference to primary system for UWB system is presented [5].

3.1 Detection technique primary system

We consider coexistence environment between UWB system and primary system such as 4G and WiMAX under the indoor environment. The detection scheme of primary system signal using MB-OFDM receiver is studied in order to realize DAA operation. By comparing the estimated average power of primary system signal obtained from the fast fourie transform (FFT) outputs corresponding to the primary band with the background noise average power, detection is performed. For the comparison, it is necessary to determine the appropriate threshold value so that the miss-detection probability can be made smaller. In order to perform value for detection this, theoretical detection and miss-detection probabilities are obtained and they are used to determine the threshold value for detection.

The MB-OFDM system has N sub-carriers and the continuous M sub-carriers are interfered from the primary system signals within the limits of the band of primary system. Therefore, N sub-carriers the observation signal of MB-OFDM receiver is represented as component interference and component noise.

If the average power of N sub-carriers is larger than the average power of (N--M) sub-carriers, then the detector assumes that the primary system exists.

3.2 Avoidance technique of interference for primary system

Among the interference avoidance techniques to primary systems, active interference cancellation (AIC) is the simplest one in the frequency domain. In this technique the MB-OFDM receiver detects the primary system signals and the part of sub-carriers which overlaps the frequency band of primary system are not transmitted therefore forming a notch. Moreover, this technique reduces the effect of side lobe from neighbor sub-carriers. Therefore, in this paper, we assume MB-OFDM system can avoid the interference to primary system by using arbitrary avoidance techniques. Therefore, we consider the performance of MB-OFDM system with interference mitigation technique for primary systems. Also, we show significant performance improvement by introducing convolutional codes with bitwise interleaving.

3.3 Numerical results

3.3.1 Analysis of detection technique

The detection and miss-detection probabilities of primary system in MB-OFDM receiver is shown in Fig. 5.

The detection probability depends on signal to noise ratio (SNR) and threshold, and the miss-detection probability depends on only the threshold. Here, SNR is defined the MB-OFDM signal to noise ratio. Although the low threshold should be chosen to obtain the high detection probabilities, the low threshold also increase the miss-detection probabilities. Hence, the threshold value should be changed dynamically according to the SNR to keep the constant detection probability. Fig. 6 indicates the relationship between threshold and SNR which is satisfied the arbitrary detection probability such as 60, 70, 90%. The miss-detection probability of the dynamic threshold is shown in Fig. 7.

In Fig. 6, a constant detection probability is obtained since threshold is dynamically changed by following SNR. The high threshold is required to obtain the high data rate. However, the high threshold inherently increase the miss-detection probability and thus the throughput of the MB-OFDM system is decreased.

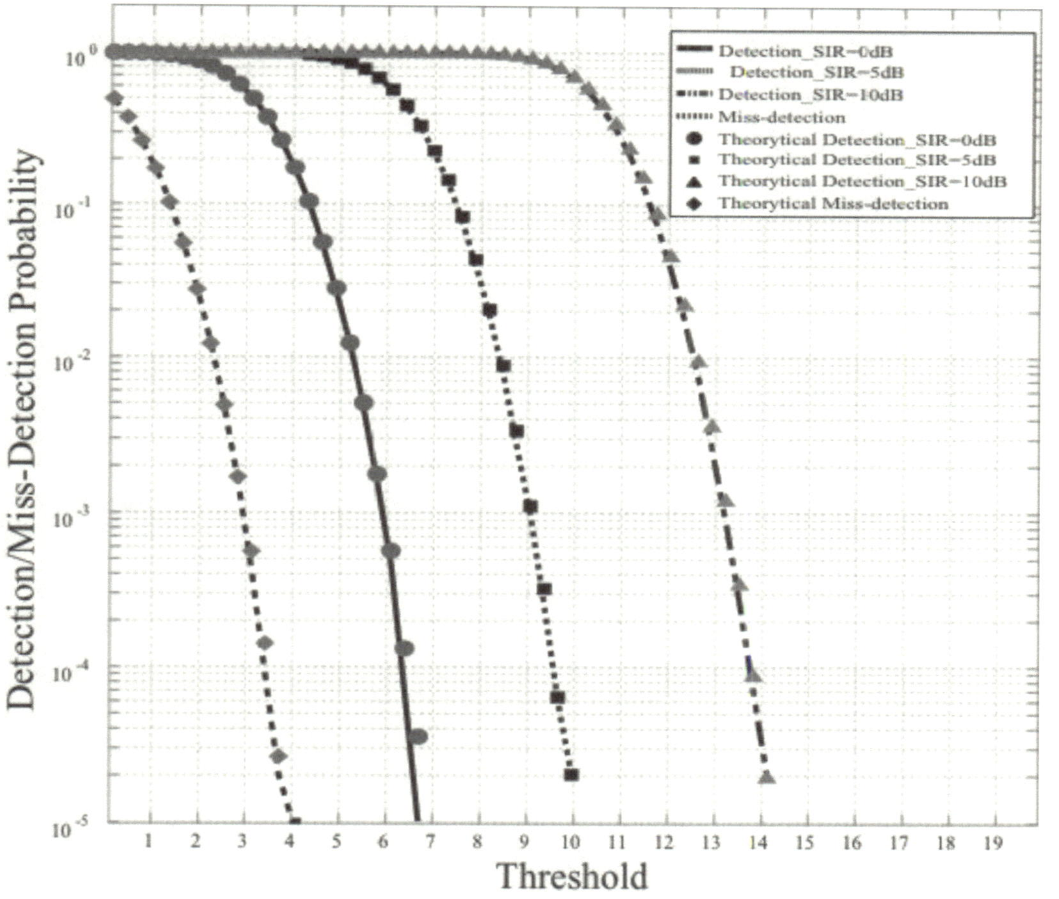

Fig. 5. The detection/miss-detection probability of primary system

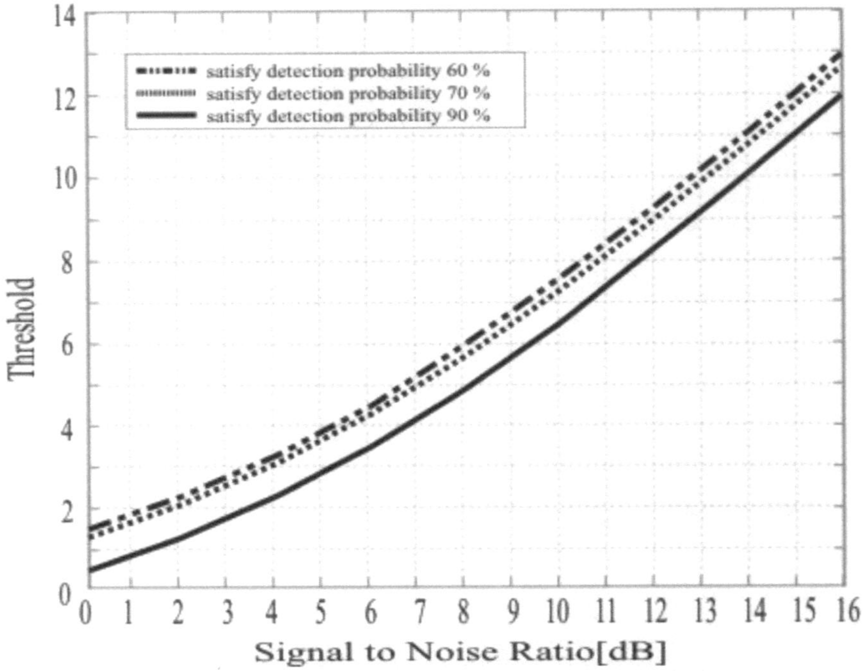

Fig. 6. The relationship between threshold and SNR which is satisfied the arbitrary detection probability

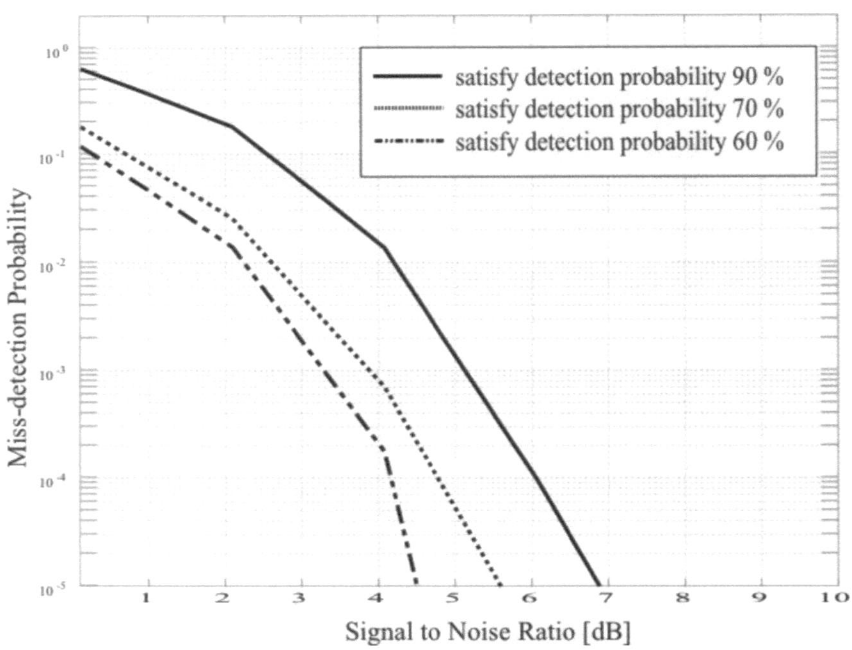

Fig. 7. The miss-detection probability of dynamic threshold following SNR

3.3.2 Analysis of interference avoidance technique

The BER of MB-OFDM system with error correction code and interleave is illustrated in Fig. 8.

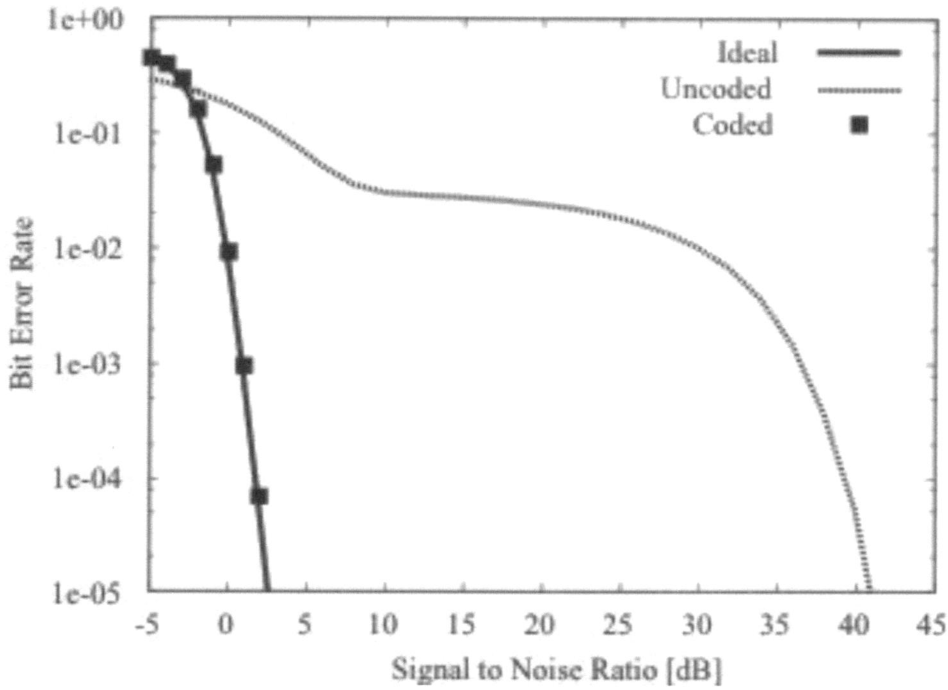

Fig. 8. The performance of MB-OFDM system with DAA

In Fig. 8, the device with DAA technique has a high bit error rate (BER). The error floor observed in moderate SNR region is due to the power control, which makes the transmission power level as -70 dB/MHz in the band overlapping with the primary system. The BER is decreased exponentially again in the high SNR region more than 35dB. Moreover, combining the bit-interleaved convolution code with DAA, the performance of MB-OFDM system can be further improved and is asymptotically identical with the ideal performance.

4. Conclusion

In this chapter, we studied dynamic spectrum access technology in the coexistence environment of primary systems and cognitive radio systems. We can conclude that the occupancy time of channels should be shortened for cognitive radio systems rather than decreasing offered traffic (i.e., arrival rate) since the departure rate is inverse proportion to the occupancy time.

Moreover, we showed the performance of UWB radio system with DAA in the coexistence environment between UWB systems and primary systems. DAA technique should be chosen in consideration to the required performance quality of UWB applications. The realtime applications such as verbal communication and high-quality video across the wireless communication are essential to high data rate. Therefore, the unused frequency band by the primary systems needs to allocated. In this case, the detector of UWB needs a high detection probability. On the other hand, in the application that allows time delay, transmitter power control to avoid the interference to primary systems is applied. Interestingly, the BER performance of bit-interleaved convolutional coded MB-OFDM with the transmit power control is almost identical with the ideal performance. Hence, this fact leads that the miss-detection of the primary system may not affect the performance of the UWB systems. Thus, DAA technology is the effective interference mitigation techniques of high data rate UWB system.

5. Acknowledgment

This work is partly supported by Research Fellowships of the Japan Society for Promotion of Science.

6. References

[1] A. Batra, "Time Multi-Band OFDM Physical Layer Proposal for IEEE 802.15 Task Group 3a," *IEEE P802.15-04/493r1-TG3a*, Sep. 2004.

[2] FCC, ET Docket No 03-222 Notice of proposed rule making and order, Dec. 2003.

[3] Functional Requirements for the 802.22 WRAN Standard IEEE.

[4] John Polson, "Cognitive Radio Applications in software Defined Radio," *Software Defined Radio Technical Conference*, Phoenic, Arizona, Nov. 2004.

[5] K. Sodeyama, M. Itami, R. Kohno, "Performance Analysis of Interference Mitigation Method for 4G Cellular Systems with MB-OFDM UWB Systems", *The 29th Symposium on Information Theory and Its Applications*, Nov. 2006.

[6] K. Watanabe, K. Ishibashi, R. Kohno, "An Analysis of Spectrum Management for Coexistence of Fixed and Cognitive Radio Systems," *2006 Software Defined Radio Technical Conference*, Nov. 2006.

[7] M. Z. Win, R. A. scholtz, "Ultra-Wide Bandwidth Time-Hopping Spread-Spectrum Impulse Radio for Wireless Multiple-Access Communications," *IEEE Trans. Commun.*, vol. 48, no. 4, pp. 679?691, April 2000.

[8] Q. Zhao. B. Sadler, A Survey of Dynamic Spectrum Access," *IEEE Signal Processing Mag.*, vol. 24, no. 3, pp. 79-89, May 2007.

[9] S. Haykin, "Cognitive Radio: Brain-Empowered Wireless Communications," *IEEE J-SAC*, vol. 23, no. 2, pp. 201-220, Feb. 2005.

[10] S. Mangold, K. Challapali, "Coexistence of Wireless Networks in Unlicensed Frequency Bands," *WWWRF#9 meeting*, Jul. 2003.

[11] T. Zasowski, A. Wittneben, "Performance of UWB Systems using a Temporal Detect-and-Avoid Mechanism," *The International Conference on Ultra-Wideband*, Sep. 2006.

[12] V. S. Somayazulu, J. R. Foerster, R. D. Roberts, "Detect and Avoid (DAA) Me chanisms for UWB Interference Mitigation," *The International Conference on Ultra-Wideband*, Sep. 2006.

Part 4

Novel UWB Applications in Medicine

14

The Future of Ultra Wideband Systems in Medicine: Orthopedic Surgical Navigation

Mohamed Mahfouz, Michael Kuhn and Gary To
University of Tennessee,
United State of America

1. Introduction

Ultra-wideband (UWB) technology has been utilized in low probability of detection radar and communications systems for decades since its inception from time domain electromagnetics in the 1960s (Fontana, 2004). Interest in UWB for unique indoor communications and positioning applications has skyrocketed since the FCC released its notice of inquiry in 1998 and then opened up the 3.1-10.6 GHz and 22-29 GHz frequency bands for UWB use in 2002 (FCC, 2002).

1.1 General overview of ultra-wideband technology for indoor positioning systems

A depiction of a typical indoor positioning system is shown in Figure 1 where the base stations are connected to a master processing unit, and a reference tag is needed to bring the mobile tag into the 3-D global coordinate frame. The use of time difference of arrival for 3-D triangulation combined with leading-edge detection at the UWB receiver help mitigate the stringent requirements needed in terms of base station synchronization and ranging sensitivity to dense indoor multipath interference. Although the system architecture shown in Figure 1 is well known and has been implemented in other wireless positioning systems including GPS, realizing this architecture for high accuracy indoor 3-D positioning has proven to be deceptively difficult.

Central difficulties in achieving high 3-D real-time accuracy for indoor localization systems include indoor multipath interference, sampling-rate limitations, local oscillator phase noise, phase center effects, system clock jitter and drift, etc. Many techniques have been proposed for ranging in UWB positioning systems which includes (see Figure 2a): leading-edge detection (Fontana, 2004), matched filter (Low et al., 2005), received signal strength (Ekahau Inc, 2008), first peak detect (Alavi & Pahlavan, 2006), and peak search and subtract (Falsi et al., 2006). The main limitation of peak detection algorithms (e.g. received signal strength, first peak detect, and peak search and subtract) is assuming *a priori* information concerning the received pulse shape. The matched filter is useful in low signal-to-noise ratio (SNR) environments in conjunction with leading-edge detection. In realistic indoor environments which contain significant multipath interference, this introduces large 1-D ranging errors resulting in poor overall 3-D accuracy (Figure 2b). In many current systems, a tunnel (or square law) diode is used at the receiver for energy detection which integrates a nanosecond of the incoming signal, putting an upper bound of roughly 5 cm on the 1-D ranging accuracy even when using leading-edge detection (Fontana, 2004). The ranging limitations due to multipath interference are compounded with sampling rate limitations. The use of conventional analog-to-digital converters (ADC), even if 5-10 GSPS, places an upper bound on the 1-D ranging accuracy of 5 cm.

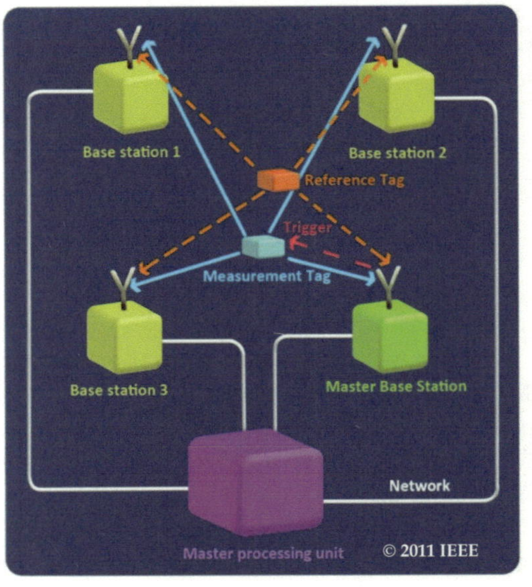

Fig. 1. A typical indoor positioning system where four or more base stations triangulate the 3-D position of a mobile tag, and a reference tag in a known location serves as a reference point for the global coordinate frame (Mahfouz, Kuhn, Wang et al., 2011).

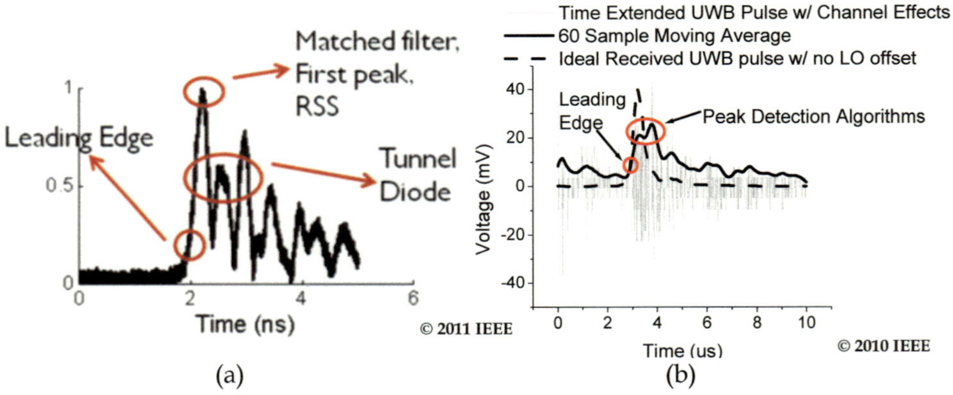

Fig. 2. (a) Comparison of peak and leading-edge detection methods utilized at the UWB receiver where the leading-edge of the multipath signal provides the highest accuracy and does not assume *a priori* pulse shape information (Mahfouz, Kuhn, Wang et al., 2011), (b) received signal where significant pulse distortion causes noticeable error in locating the correct peak position while the leading-edge of the UWB pulse results in mm 1-D ranging (Kuhn et al., 2010).

1.1.1 Commercial systems

As shown in Table 1, current commercial UWB systems can achieve 3-D localization accuracy in the range of 15 - 30 cm. A comparison of the DART Ultra-Wideband system from Zebra Enterprise Solutions and the real-time location system (RTLS) from Ubisense is given in Table 1 (Zebra Enterprise Solutions, 2011, Ubisense, 2011). As shown in Table 1, the two systems share many commonalities including frequency range, operating range, compact tag size, and overall 3-D accuracy. One major difference is the method of digital communication: the Zebra Enterprise Solutions system uses UWB pulse-based modulation whereas the Ubisense system uses a narrowband 2.4 GHz digital communication scheme. Although 2.4 GHz communication is a more mature technology than UWB pulse-based modulation, it has a distinct disadvantage in that it is much more susceptible to multipath interference and may experience severe degradation depending on the operating environment of the system. Figure 3 shows the two commercial systems. The Zebra Enterprise Solutions system is shown in Figure 3a including the main controller and four hubs while Figure 3b shows two different sized tags from the Ubisense RTLS.

Company	Frequency Range (GHz)	Operating Range (m)	Tag Size (cm^3)	Number of Tags	Refresh Rate (Hz)	Localization Method	Accuracy (cm)
Zebra Enterprise Solutions	5.94 - 7.12	> 50	1.12x 4.01x 2.11	10,000	<1-100	TDOA	< 30
Ubisense	5.8 – 7.2	> 50	3.8x 3.9x 1.65	> 1000	<1-34	TDOA and AOA	< 15

Table 1. Comparison of commercial UWB localization systems with specifications of their compact tags.

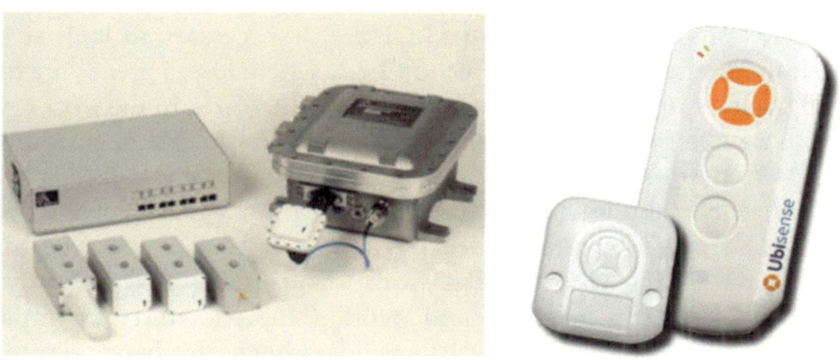

(a) (b)

Fig. 3. Commercial indoor UWB localization systems (a) Zebra Enterprise Solutions, (b) Ubisense.

1.1.2 Research systems

Competing technologies for high accuracy indoor positioning include frequency modulated continuous wave (FMCW), impulse-based (i.e. carrier-free) UWB, and carrier-based UWB.

Table 2 provides a summary of the various research groups utilizing these three approaches for high accuracy indoor positioning. Similar accuracy levels (0.5-20 cm) have been achieved for both carrier-based (Meier, 2007; Mahfouz, 2008; Zhang, 2010; Waldmann, 2008; McEwan, 2004) and impulse-based (Low, 2005; Zetik, 2004; Ossberger, 2004; Fuiji, 2007) UWB positioning systems, although carrier-based systems have shown the potential for mm and sub-mm range accuracy even for 3-D indoor environments (Mahfouz, 2008; Zhang, 2010; Meier, 2007; McEwan, 2004). FMCW has proven to be a successful competing technology for high accuracy positioning systems (Stelzer, 1999, 2004; Ellinger, 2007; Mosshammer, 2007; Tragas, 2007; Feger, 2009). In the 5.8 GHz band for industrial, medical, and scientific (ISM) applications, documented accuracy of 5-20 cm for 2-D has been achieved (Stelzer, 2004; Ellinger, 2007; Mosshammer, 2007; Tragas, 2007). FMCW systems operating at higher frequencies including 35 GHz and 77 GHz have achieved accuracy levels of 0.1 mm (Stelzer, 1999, Feger, 2009) with the system described by Feger et al. working at ranges of up to 10 m (Feger, 2009). The most recent FMCW trend is a European-wide push to create low power wireless sensor networks built on 5.8 GHz FMCW technology (Ellinger, 2007; Mosshammer, 2007; Tragas, 2007; Feger, 2009).

1.1.3 Available frequency bands

Wireless medical devices, for both positioning and communication, have stringent requirements on the frequency bands in which they can operate. Table 3 highlights the different bands both in the United States and Europe which can be used for indoor medical applications for both narrowband and UWB applications. UWB has available frequency bands from 3.1 - 10.6 GHz and 22 - 29 GHz in the U.S. Only portions of that 3.1 - 10.6 GHz band are currently available in Europe. A number of telemetry bands exist in the U.S., and both the U.S. and Europe have instrumentation, scientific, and medical (ISM) bands available, mainly in the 300 MHz to 3 GHz range. As shown in Table 3, *in vivo* telemetry applications typically use bands in the RF and lower microwave frequency range for operation (i.e. 315 MHz, 402 - 405 MHz, and up to 1427 - 1432 MHz in the United States while Europe uses 433.05 - 434.79 MHz and 868 - 870 MHz). In order to understand why telemetry bands are not allocated above 1.43 GHz, it is necessary to look at the complex permittivity of human tissues as done in Section 1.2.1, where losses greatly increase above 1 GHz. The UWB positioning system operates from 5.4 - 10.6 GHz in the upper region of the 3.1–10.6 GHz band while most wireless telemetry systems for *in vivo* operate at 433.92 MHz in the 433.05 - 434.79 European ISM band and at 315 MHz for the U.S. ISM band. As shown in Figure 4, the allocated band in the United States goes from 3-10.6 GHz at a power level of -41.3 dBm/MHz. Compared to Europe, where the power level is required to be at -71.3 dBm/MHz from 4.8-6 GHz, -65 dBm/MHz from 8.5-10.6 GHz, and can only be at -41.3 dBm/MHz from 3.4-4.8 GHz if detect and avoid (DAA) circuitry is implemented to minimize interference with other wireless systems operating in this band. Figure 5 compares the UWB bands in Japan versus the United States. Similar to Europe, the Japanese regulations also require DAA in the 3.4-4.8 GHz band. The main difference between the European band and the Japanese band is in the no DAA band: in Europe, this exists from 6-8.5 GHz while in Japan, this band goes from 8.5-10.6 GHz. From looking at Figure 4 and Figure 5, it is clear that the various restrictions imposed around the world make designing one system for worldwide operation difficult to achieve. Multiple variations of a system may be needed to meet the various worldwide regulations.

Research Group/ Company	System Architecture	Frequency (GHz)	Reported Error	Operating Range
Mahfouz, 2009	Carrier-Based UWB	5.4-10.6	2-5 mm (3-D)	5 m/ Indoor
Waldmann, 2008	Carrier-Based UWB	7-8	1.7 cm (1-D)	10 m/ Indoor
Meier, 2007	Carrier-Based UWB	22.58-25.7	0.1-2 mm (1-D)	8 m/ Indoor
McEwan, 2004	Carrier-Based UWB	5.8	< 2 mm (2-D)	10 cm x 10 cm
Ossberger, 2004	Impulse-Based UWB	~2-7	5-10 mm (1-D)	5 m/ Indoor
Fujii, 2007	Impulse-Based UWB	3.7-5	20 cm (2-D)	8 m/ Indoor
Low, 2005	Impulse-Based UWB	3.2-5.2	1 cm (1-D)	8 m/ Indoor
Zetik, 2004	Impulse-Based UWB	0.01-5	1.5 cm (2-D)	2 m/ Indoor
Stelzer, 1999	FMCW and Interferometry	35	0.1 mm (1-D)	< 1 m/ Indoor
Stelzer, 2004	FMCW	5.8	10 cm (2-D)	500 m/ Outdoor
Ellinger, 2007 Mosshammer, 2007	FMCW	5.8	18 cm (2-D)	40 m/ Indoor
Feger, 2009	FMCW	77	0.1 mm (1-D)	1.5 m/Chamber

Table 2. Comparison of current research high accuracy positioning systems.

Location	Frequency Band	Frequency (MHz)
U.S.	Medical Implant Communications Service	402 - 405
U.S.	Wireless Medical Telemetry Service	608 – 614; 1395 – 1400; 1427 - 1432
U.S.	Instrumentation, Scientific, and Medical (ISM)	315; 902 – 928; 2400 – 2483.5; 5150 – 5875
Europe	ISM	433.05 - 434.79; 868 - 870 (short-range) 2400 - 2483.5
U.S.	UWB	3.1 – 10.6 GHz; 22 – 29 GHz, center freq > 24.075 GHz
Europe	UWB	3.4 – 4.8 GHz; 6 – 8.5 GHz
Japan	UWB	3.4 – 4.8 GHz; 8.5 – 10.6 GHz

Table 3. Summary of licensed medical wireless frequency bands.

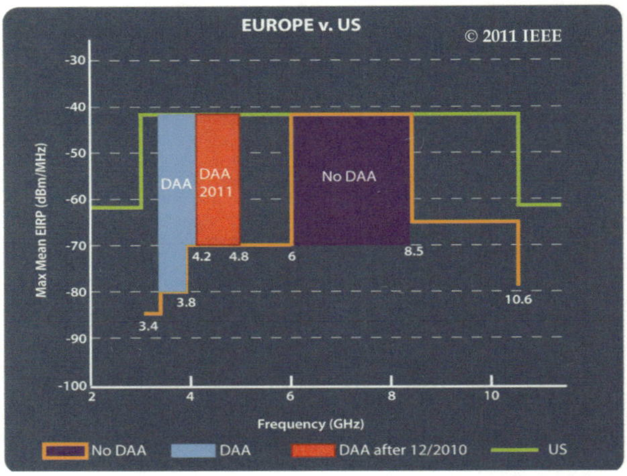

Fig. 4. Comparison of allocated UWB bands between 3-11 GHz in the U.S. versus Europe (Mahfouz & Kuhn, 2011).

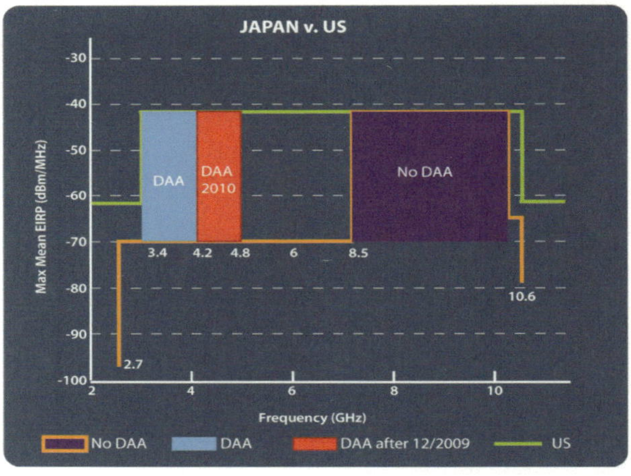

Fig. 5. Comparison of allocated UWB bands between 3-11 GHz in the U.S. versus Japan.

1.2 Microwave interaction with biological tissues

The electrical characteristics of biological tissues change dramatically from DC through higher frequencies such as X-ray and gamma radiation. At very low frequencies (the kHz range), the primary means through which electrical current travels through the body is conduction via the extracellular matrix. At the visible frequency range and even higher in ultraviolet (UV) and X-ray frequency ranges, most electromagnetic waves are able to pass through biological tissues, with differing amounts of energy being absorbed by different tissues. Between these two extremes lie the RF/Microwave frequency bands. Electrical properties of biological tissues change dramatically over this frequency range. There are specific techniques, such as coaxial probe dielectric measurements, which can be followed to apply a uniform method for electrical characterization of biological tissues (or other lossy media) over this frequency range.

RF/Microwave radiation is considered from 100 kHz - 300 GHz. Although the interaction of radiation with biological tissues changes significantly over this range, there are some similar properties that provide coherence for grouping RF/Microwave frequencies together. First, over this whole frequency range radiation can be considered non-ionizing. Ionizing radiation includes UV and X-ray, and these frequency ranges are characterized by disruption of atomic structures. Second, it is convenient and useful to use RF/Microwave frequencies for the purpose of radiating electromagnetic (EM) energy. At low frequencies (LF) and extremely low frequencies (ELF), the wavelength of the transmitting signal is much larger than the structures used in transporting it. Radiation at LF and ELF is not typically seen. Conversely, the wavelengths in the RF/Microwave range are comparable to their corresponding transmission lines (millimeter to meter in size), which makes them ideal for radiation applications. Central to designing a microwave system which utilizes biological tissues as a transmission medium is a thorough understanding of the electrical properties of the tissues that will be exposed to the EM energy. An enormous amount of research has been done in this area. This has included use of a wide range of measurement techniques, testing on many different types of animals as well as humans, and testing of biological tissues in different environments (e.g. *in vivo, in vitro,* over a range of temperatures, etc.).

1.2.1 Complex permittivity

The complex permittivity is a common method used to characterize the electrical properties of an arbitrary medium. Combined with other electrical parameters, such as conductivity and skin depth, it can be used to provide complete electrical characterization. (1) shows the basic definition of complex permittivity

$$\varepsilon^* = \varepsilon' - j\varepsilon'' \tag{1}$$

where the first term, ε', represents the capacitive nature of the tissue (amount of charge stored in it) while ε'' characterizes the lossy nature of the medium. Using these two terms, it is possible to calculate α and β, which can then be used to characterize how an EM wave behaves inside the medium (Vorst, 2006). (2) defines the loss tangent $tan\delta$ where ω is the angular frequency, σ is the conductivity, and $\varepsilon', \varepsilon''$ are defined in (1). (3) describes how to calculate α and β using the wavelength in free space λ_0 , $tan\delta$ is the loss tangent, and ε' is the relative permittivity. (4) shows how the time and distance varying electrical field $E(z,t)$ is calculated using α and β.

$$\tan\delta = \frac{\omega\varepsilon'' + \sigma}{\omega\varepsilon'} \tag{2}$$

$$\alpha = \frac{2\pi}{\lambda_o}\sqrt{\frac{\varepsilon'}{2}(\sqrt{1+\tan^2\delta} - 1}$$
$$\beta = \frac{2\pi}{\lambda_o}\sqrt{\frac{\varepsilon'}{2}(\sqrt{1+\tan^2\delta} + 1} \tag{3}$$

$$E(z,t) = E_0 e^{j\omega t-(\alpha+j\beta)z} \tag{4}$$

Finally, the power dissipated through the region can be calculated with the electric field

$$P = |E(z)|^2 \sigma \tag{5}$$

where $E(z)$ only depends on spatial variation and σ is the conductivity of the medium. Estimating the power dissipated provides a key metric in designing microwave systems which utilize biological tissues as the propagation medium. The electrical losses can be attributed to both ionic conduction and dipole rotation (Nelson, 1991)

$$\varepsilon'' = \varepsilon''_d + \varepsilon''_c \tag{6}$$

where ε''_d is loss from dipole rotation and ε''_c is the loss from ionic conduction. The standard way to model the complex permittivity for an arbitrary material is through the Cole-Cole equation

$$\varepsilon(w) = \varepsilon_\infty + \frac{\Delta\varepsilon}{1 + (j\omega\tau)^{1-\alpha}} \tag{7}$$

where there is a dispersion peak in the loss factor whose size and place depends on the relaxation time, τ, and dispersion factor α. Figure 6 shows typical curves for the complex permittivity of human tissues as modeled through the Cole-Cole equation (7) (Gabriel et al., 1996). A multi-peak Cole-Cole equation (8) is used by Gabriel et al. and allows a parametric model for complex permittivity to be established where multiple dispersion peaks and relaxation times are taken into account (Gabriel et al., 1996).

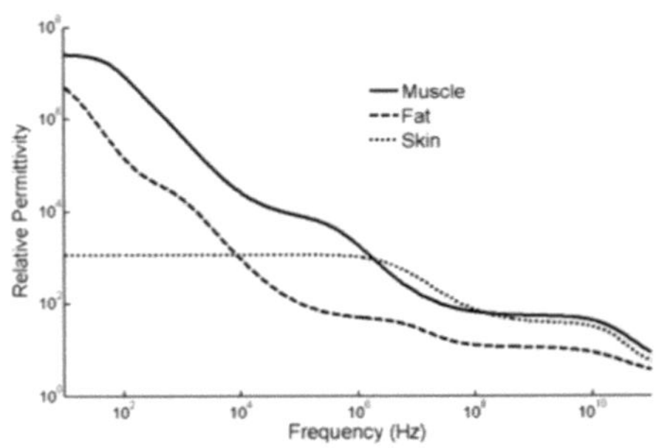

(a)

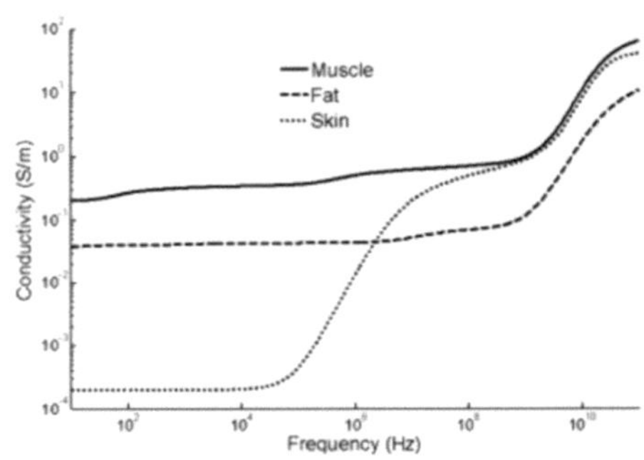

(b)

Fig. 6. Electrical properties of human muscle, fat, and skin at 37°C. (a) relative permittivity, (b) conductivity.

$$\varepsilon(w) = \varepsilon_{\infty} + \sum_{i=1}^{4} \frac{\Delta\varepsilon_i}{1+(j\omega\tau_i)^{1-\alpha_i}} + \frac{\sigma_{ionic}}{j\omega\varepsilon_0} \tag{8}$$

The loss factor in an arbitrary medium (i.e. ε") is due to both ionic conductivity and dipole rotation, and dipole rotation is the primary means through which losses occur over the RF and microwave frequency range. The various losses experienced through the microwave frequency range are shown in Figure 7. The temperature increases the ionic conductivity, which is the main reason the loss factor tends to increase with increases in temperature over lower microwave frequency range (DC – 1 GHz) (Komarov, Wang, & Tang, 2005). Above 1 GHz, the effects due to ionic conductivity lessen, and the loss factor is primarily due to dipole rotation (or free-water dispersion). This has the effect of decreasing the loss factor of water-based tissues, at least over the range of 3 – 17 GHz, with increase in temperature. Figure 8 shows how changes in temperature affect both ε′ and ε″ for water over a range of frequencies and temperatures (Komarov, Wang, & Tang, 2005). As Figure 8 shows in a qualitative sense, over the microwave frequency range of interest (3-17 GHz), increases in temperature tend to decrease ε′ by an almost consistent amount (decrease of ~8 for an increase in temperature of 25°C) and also tend to decrease ε″.

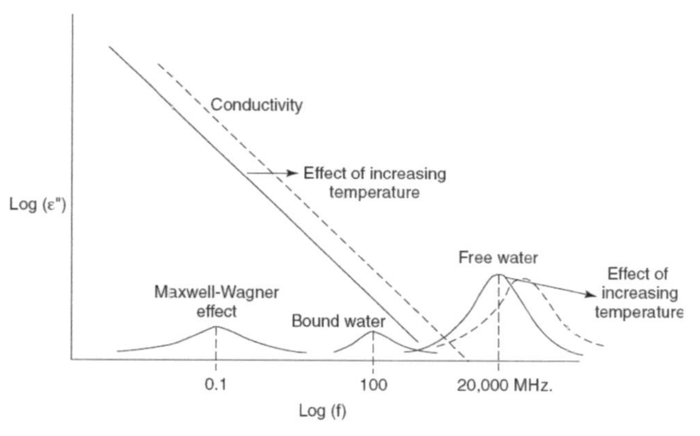

Fig. 7. Water and temperature effects on the loss factor (Komarov, Wang, & Tang, 2005)

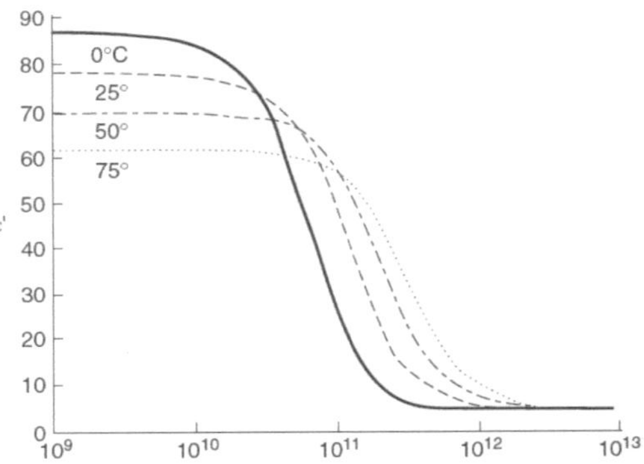

(a)

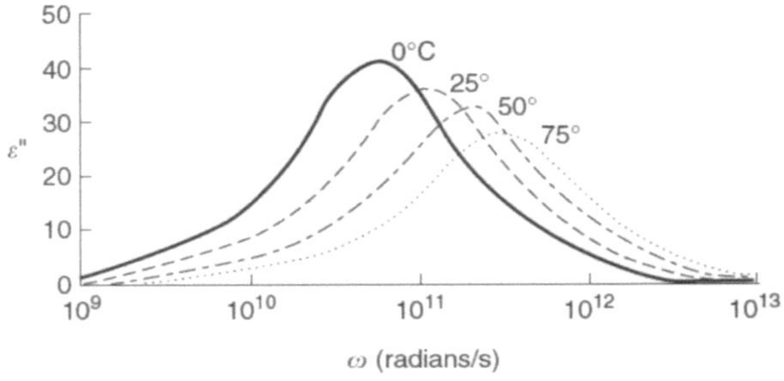

(b)

Fig. 8. Effects of temperature on complex permittivity of water: (a) ε′, (b) ε″ (Komarov, Wang, & Tang, 2005)

Since most biological tissues are highly water-based, especially muscle and liver, they tend to follow similar trends for variation of complex permittivity over a range of temperatures. The change in complex permittivity of water is mostly due to the change in viscosity that occurs in water for varying temperatures. For example, a 32.4% decrease in viscosity of water is observed when going from 20°C to 37°C. This causes a decrease in the relaxation time τ of approximately 38.2%.

2. State-of-the-art in wireless medical systems

The introduction of reliable low cost wireless technologies has significantly changed the medical industry over the past decade. There is a great demand for wireless medical equipment, most noticeably in the patient monitoring and personal health care. The primary reason for migrating from wired tools to wireless solutions is to eliminate the clumsy and restrictive wirings. It improves the overall efficiency by allowing multiple medical equipments to be controlled and monitored wirelessly by one single terminal device. The following sections illustrated a few of the commercially available wireless medical devices, as well as a number of researches being conducted.

2.1 Commerical devices

Vital signs monitoring devices are the most common medical systems being implemented with wireless technologies. Wireless blood pressure monitoring systems shown in Figure 9 can be used either in doctor office or for personal health monitoring. The wireless blood pressure cuff, as shown in Figure 10, was developed by SunTech and the Harvard Sensor Network Labs. They can be used by paramedics while responding to emergency. Wireless Oximeter (Figure 11) can measure the blood oxygen level of the patient up to 120 hours continuous monitoring from a fully charged battery. Figure 12 shows a wireless ECG monitoring device. It supports up to 12 channels and can be used continuously for 24 hours on a fully charged battery. This particular device demonstrates one of the great improvements for wireless medical system, which is enhancing the mobility of the patients. It allows nonstop monitoring of the patients' ECG and eliminates the need to attach and removing the ECG electrodes while being transferred around the hospital. A wireless EEG headset is shown in Figure 13. The current system support 14 sensors and can be used up to 12 hours on a full battery.

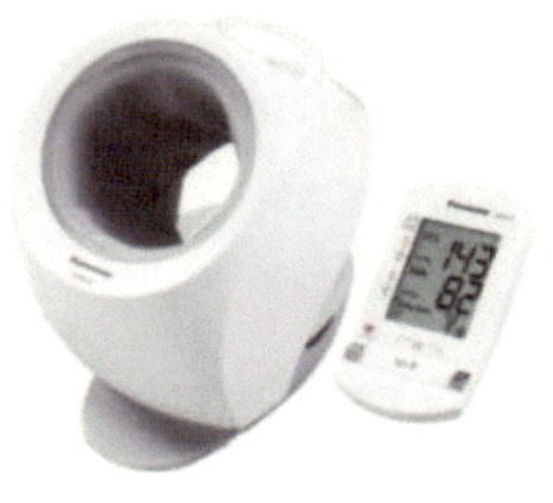

Fig. 9. Wireless blood pressure measuring device (Panasonic (n.d)).

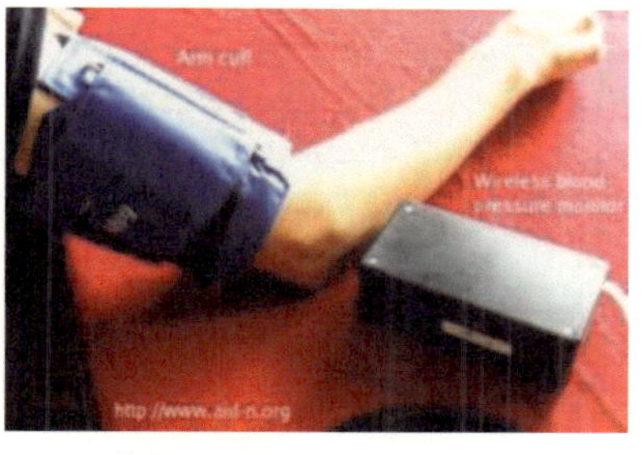

Fig. 10. Wireless blood pressure cuff (SunTech (n.d)).

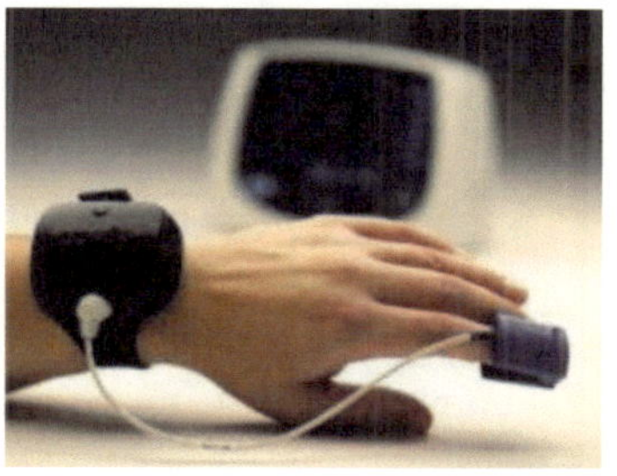

Fig. 11. Wireless Oximeter (Nonin (n.d)).

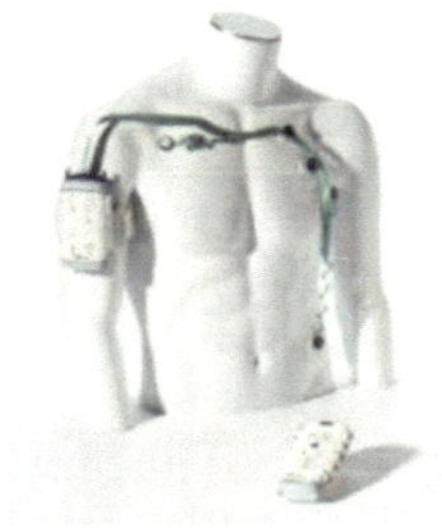

Fig. 12. Wireless ECG monitoring system (Life Science Corp (n.d)).

Fig. 13. Wireless EEG headset (Emotiv (n.d)).

While the devices mentioned above are used primary in hospital, there is also a great effort to develop personal healthcare devices to improve the quality of life of the patients. Devices such as glucometer, insulin pump and alcoholmeter have also been incorporated wirelessly with personal mobile devices (Figure 14 - Figure 16). In the recent years, a lot of focus has been given to minimally invasive methods to collect data for diagnosis. One of the most popular solutions to collect information within the gastrointestinal (GI) track is to use capsule camera as shown in Figure 17. The images taken are transmitted wirelessly to a computer to process. The capsule has also demonstrated the capability of producing images to help with diagnosis that cannot be accomplished with ordinary endoscopy and radiological methods.

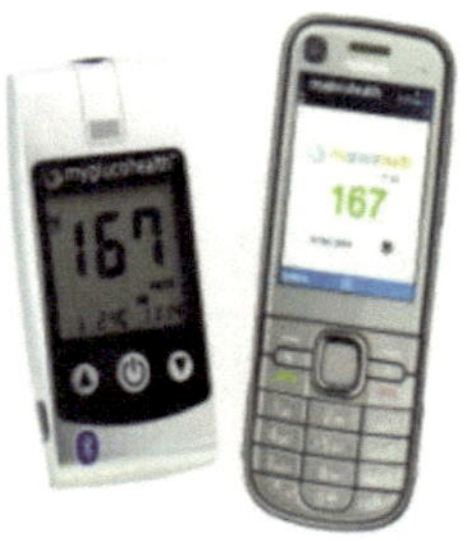

Fig. 14. Wireless glucometer (Myglucometer (n.d)).

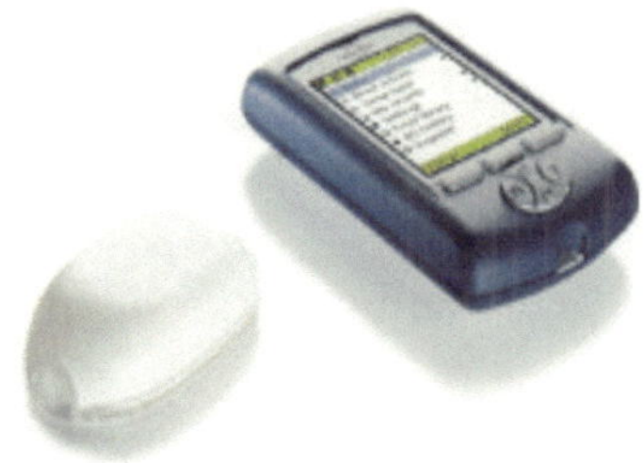

Fig. 15. Wireless insulin pump manager (Omnipod (n.d)).

Fig. 16. Wireless alcoholmeter (Alcosystem (n.d)).

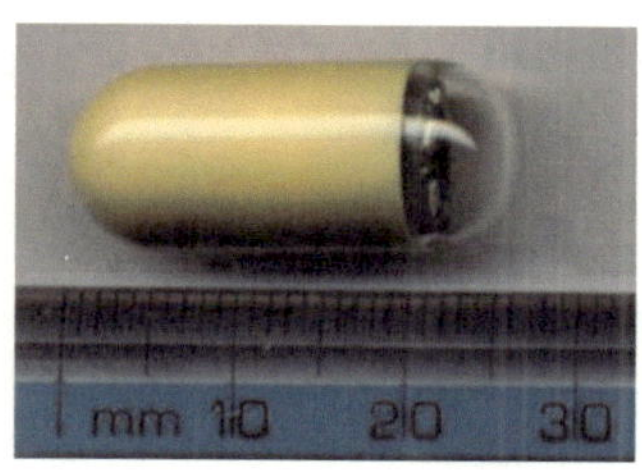

Fig. 17. Capsule Endoscopy (Public Domain (n.d)).

Apart from ambulatory and personal medical devices, wireless surgical tracking devices have also been developed to improve the accuracy and efficiency of diagnosis and surgery. Image guidance surgical navigation system uses optical and electromagnetic trackers to track the surgical instruments in the attempt to minimize the human error during surgery. Optical system (Figure 18), uses two infrared cameras to triangulate the position of the target instrument. Figure 19 shows an electromagnetic tracking device developed by Ascension and GE healthcare. The system provides real time feedback of the current position of the biopsy needle, as well as the needle path projection.

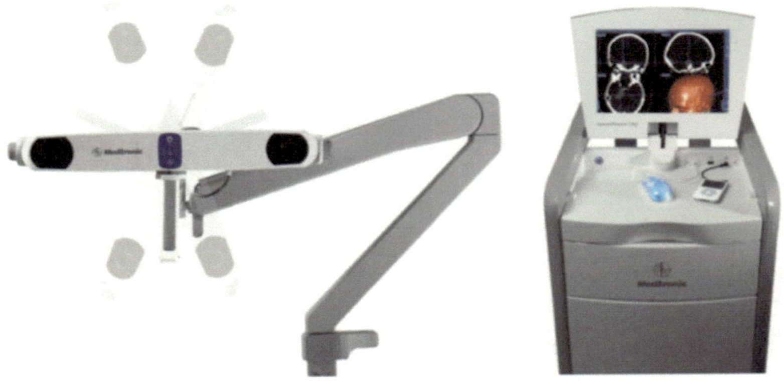

Fig. 18. Optical tracking devices for surgical navigation (Metronics).

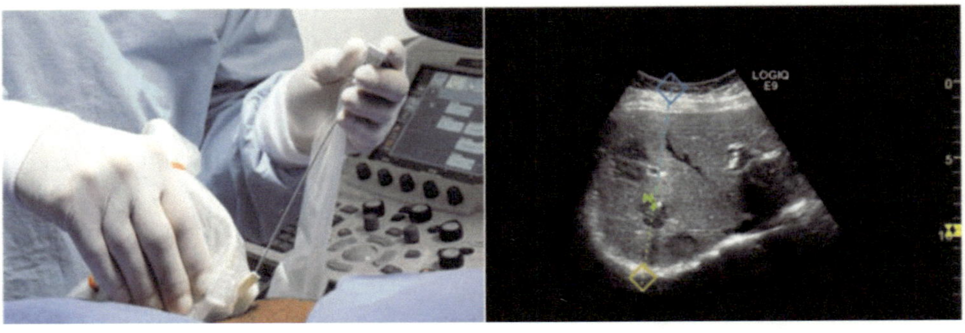

Fig. 19. The biopsy needle is coupled with electromagnetic tracking device to provide feedback of the needle positions (Ascension), (G.E. Healthcare).

2.2 Current research

The commercially available devices mentioned in previous section have undergone many years of research and development. The following section is going to look at some of the current researches being done with wireless medical device.

While there are many wireless ambulatory monitoring systems mentioned above, most of them operate in a standalone mode with its own receiver. It would be more beneficial to the physicians and health care professional to centralize all the information into one single device. Tia Gao et al. introduced a wireless sensor network (WSN) system for medical devices. (Gao, et al., 2008) The information from the sensors is wirelessly transmitted to the server, and it can be accessed through handheld devices and computers (Figure 20). The authors tested the system along with medical professions in a mock emergency situation with satisfying results. Another focus of the research is to develop applications from the sensor technologies. Pekka Iso-Ketola et al. developed a wireless medical device using an accelerometer to monitor patient's posture after total hip replacement (THR) surgery (Figure 21). (Iso-Ketola et al., 2008) The devices are also given to the patient such that they can monitor and follow the precautions given by the surgeons.

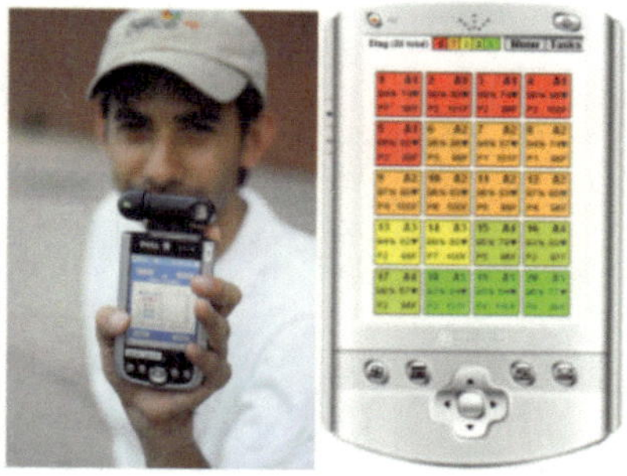

Fig. 20. Patients' conditions are being monitored through a hand held device (Gao, et al., 2008).

Fig. 21. Wireless hip posture monitoring system (Iso-Ketola, Karinsalo, & Vanhala, 2008).

Shyamal Patel et al. developed a network of wireless acceleration sensing nodes that are attached to different sections of the patient's body as shown in Figure 22 (Patel, et al., 2009). The data collected were analyzed. The calculated parameter can help with the diagnosis of the severity of Parkinson's disease. Stacy Bamberg et al. developed a wireless gait analysis system. A force measuring system is placed within a shoe, and a triaxial accelerometers and gyroscopes attached on the outside of the shoes as shown in Figure 23. (Morris & Paradiso, 2002) The sensors measure the forces and motion on the foot during gait.

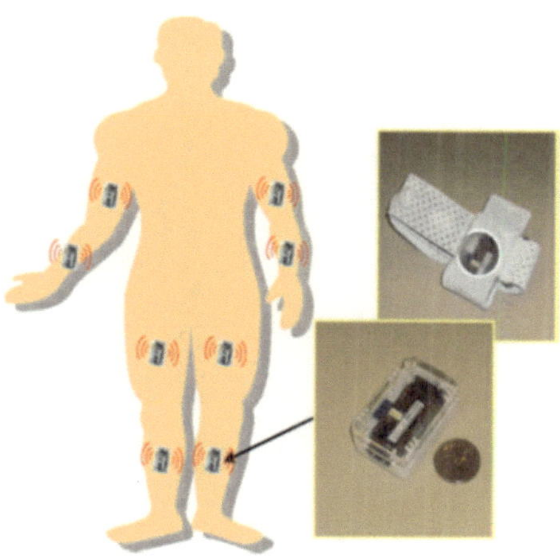

Fig. 22. A network of wireless sensing nodes consists of accelerometers (Patel, et al., 2009).

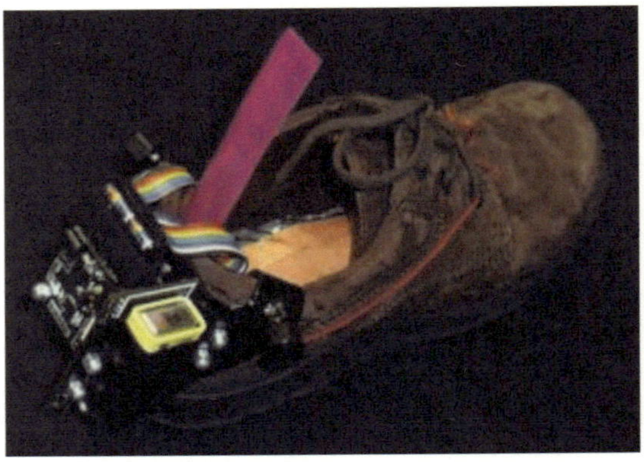

Fig. 23. Wireless gait analysis system (Morris & Paradiso, 2002).

Aside from the patient monitoring and diagnostic tool, several research groups have been concentrated on implantable medical devices. The technology to design and fabricate micro-electromechanical system (MEMS) sensors and application specific integrated circuit (ASIC) enables embedded measuring systems to be made in an extremely compact fashion. It is now possible to measure *in-vivo* condition that was once impossible. Graichen Friedmar et al. developed a complete embedded system to measure strain within a Humerus implant (Figure 24) (Graichen et al., 2007). Antonius Rohlmann et al. also completed an embedded system to measure the post operative load of spiral implants wirelessly as shown in Figure 25 (Rohlmann et al., 2007). D'Lima and Colwell modified existing knee implants with four load sensors to measure the *in-vivo* stress on the implant after the total knee arthoplasty (Figure 26) (D'Lima et al., 2005). Chun-Hao Chen et al. designed a wireless Bio-MEMS system to measure the C-reactive proteins as shown in Figure 27 (Chen, et al., 2009).

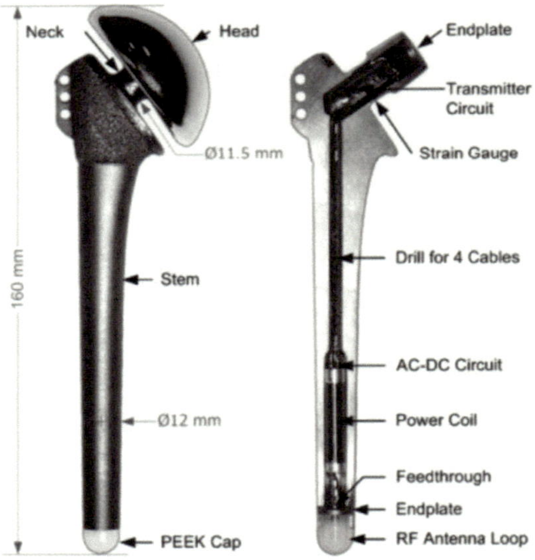

Fig. 24. Telemetry strain measuring Humerus implant (Graichen et al., 2007).

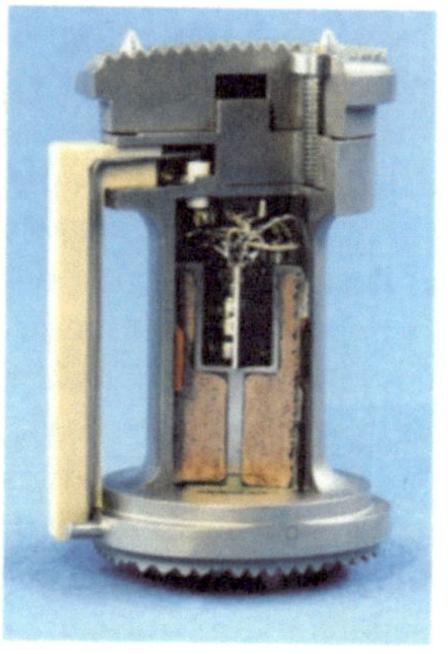

Fig. 25. Wireless load measuring system for vertebral body replacement (Rohlmann et al., 2007)

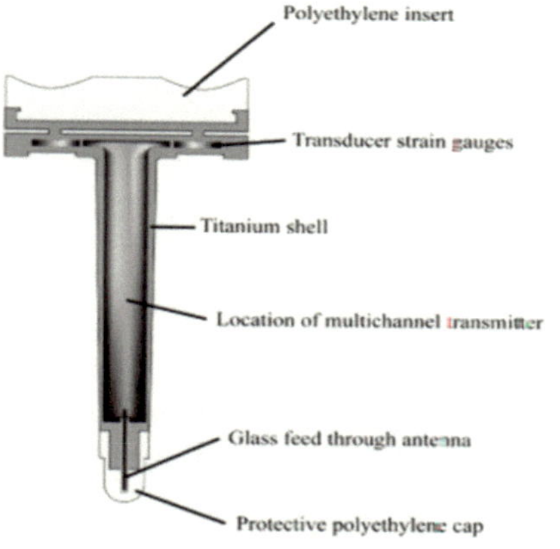

Fig. 26. Telemetry stress measuring knee implants (D'Lima et al., 2005).

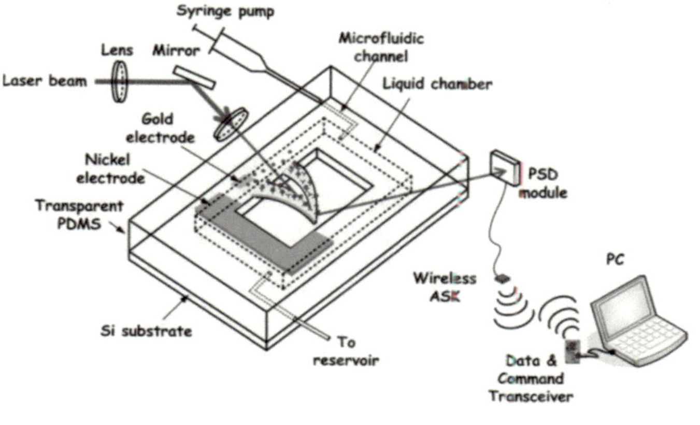

Fig. 27. Wireless Protein detection with BioMEMS (Chen, et al., 2009)

Measuring the forces and contact areas in vivo is extremely valuable to researchers, implant designers, clinicians, and patients. Measuring these values post operatively allows for evaluation of the performance of current designs and prediction of future design performance. Data on the in vivo load state of joint replacement components is required to understand the structural environment and wear characteristics of that component. Normal loads, load center, contact area, and the rate of loading need to be measured in order to fully understand the kinematics and kinetics of the orthopedic implant. This data can be used to help patients by allowing clinician to monitor implant kinematics, wear, and function. In the cases of predicted premature wear, preventative measures such as orthotics, bracing, or physical therapy could be used to avert the need for revision procedures. Additionally, one of the major postoperative concerns was inflection. Currently, there is no effective way to prevent it until symptoms are developed. Biosensing devices that react to disease related protein can monitor and alert physicians to administrate antibiotic during early stage of the infection.

3. Wireless signal propagation in hospital environments

The main concern with using wireless tracking and communication technology in the operating room (OR) and other hospital environments is the high level of scatterers and corresponding multipath interference experienced when transmitting wireless signals. While the experiment from Clarke et al. provides quantitative data on how wireless real-time positioning systems perform in the OR, it is also useful to look into narrowband and UWB channels and their effect on narrowband and UWB signals for communication and positioning applications (Clarke & Park, 2006). There are two typical approaches used when modeling wireless channels: the first is statistical models used to model generic environments (e.g. industrial, residential, commercial, etc.), which incorporate LOS or non-line-of-sight (NLOS) measurements taken in the time and frequency domains, which are then used in setting the parameters of these statistical models. The second method uses ray tracing techniques to model specific geometrical layouts (e.g. buildings, cities) and can provide a more accurate depiction of which obstacles and structures will have the greatest effect on wireless propagation. The drawback with ray tracing is the static nature of the results (i.e. results are only valid for a certain scenario of objects placed in the scene). Even if the wireless systems in the operating room are static, other objects will not be including people, patients, the operating table, and medical equipment.

3.1 Channel modelling in the operating room

A useful technique for modeling the operating room channel is to take time domain and frequency domain measurements in the operating room. This can be done both during surgery (live) and not during surgery (non-live) with variable Tx-Rx distances (e.g. 0.5 m to 4 m). Figure 28 and Figure 29 show the time domain and frequency domain setups to collect data in the OR. Figure 30 and Figure 31 show the live and non-live setups where the layout of the dual OR is shown to highlight the Tx and Rx locations for both the live and non-live experiments. Note that both monopole and single element Vivaldi antennas are used for transmission and reception. The basic strategy in the time domain is to send out a narrow UWB pulse, either baseband or modulated by a carrier signal, in the 3.1-10.6 GHz band approved by the FCC. Indoor measurements can also be measured at bands higher than the standard 3.1- 10.6 GHz (e.g. 22-29 GHz) with the understanding that the effective isotropic radiated power (EIRP) is limited to -51.3 dBm/MHz rather than the -41.3 dBm/MHz available in the lower band (FCC, 2002). Figure 32 shows the experimental setup during the non-live case (Figure 31) for obtaining both time and frequency domain data while Figure 33 shows the experimental setup during an orthopedic surgery. When performing measurements in the frequency domain, the typical approach is to use a vector network analyzer to sweep across the UWB frequency range (e.g. 3.1 - 10.6 GHz) and measure the S-parameter response of the channel where a UWB signal is passed between a transmitting and receiving antenna. The inverse Fourier transform can then be used to convert the signal from a frequency response into an impulse response in the time domain. This allows frequency dependent fading and path loss as well as the RMS delay spread and power delay profile measurements to be obtained. In Figure 29, a vector network analyzer is used to collect data for frequency domain measurements.

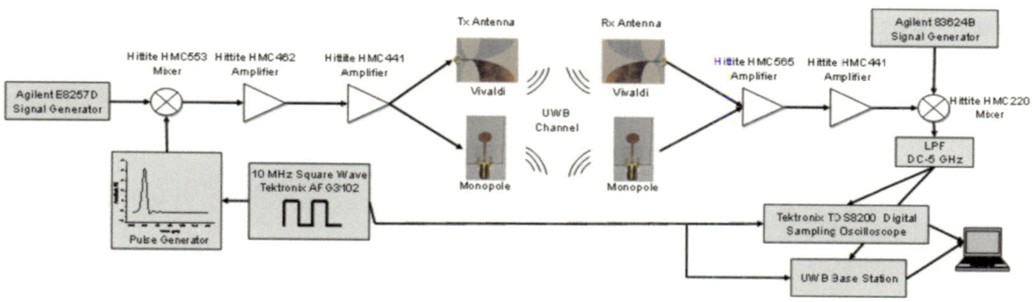

Fig. 28. Experimental setup to collect time domain data in the operating room with the UWB localization system (Mahfouz & Kuhn, 2011).

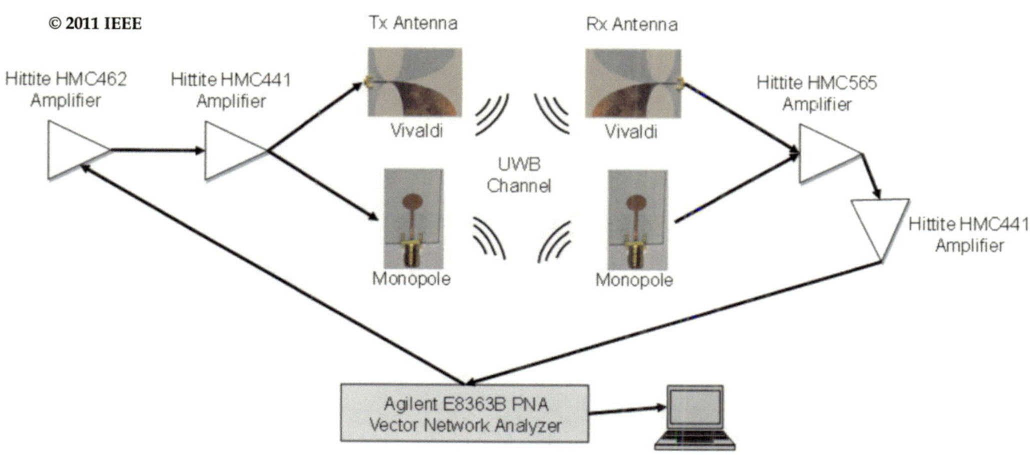

Fig. 29. Experimental setup to collect frequency domain data in the operating room for characterization of the 3.1-10.6 GHz UWB band.

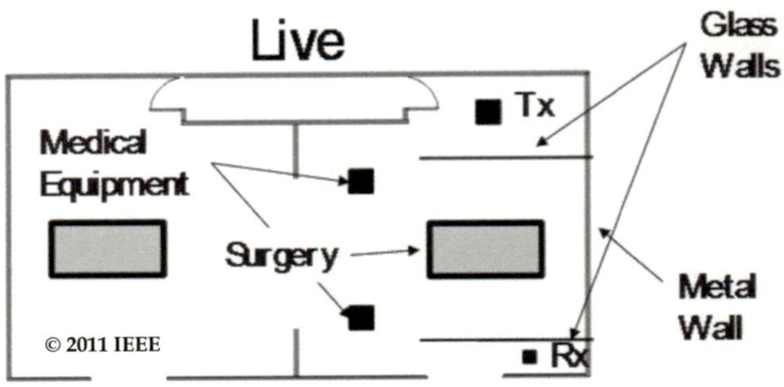

Fig. 30. Layout of dual operating room during surgery outlining the patient table, glass walls, medical equipment, doors, and walls. The Tx and Rx were positioned 4 m apart across the surgery (Mahfouz & Kuhn, 2011).

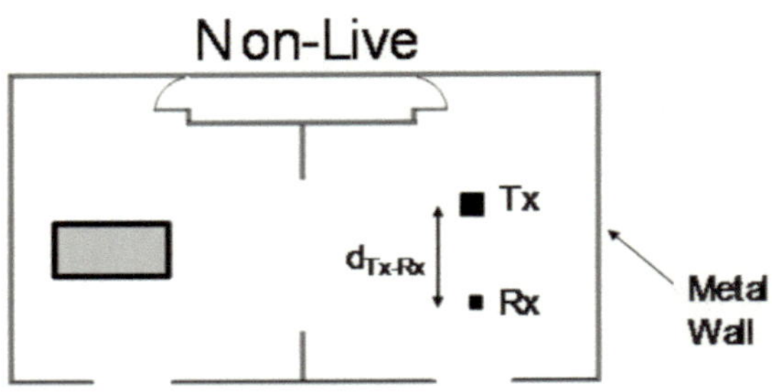

Fig. 31. Layout of dual operating room without surgery taking place where medical equipment, glass walls, and the patient table have been removed. The Tx and Rx were placed in the surgical area and moved from 0.5-4 m apart.

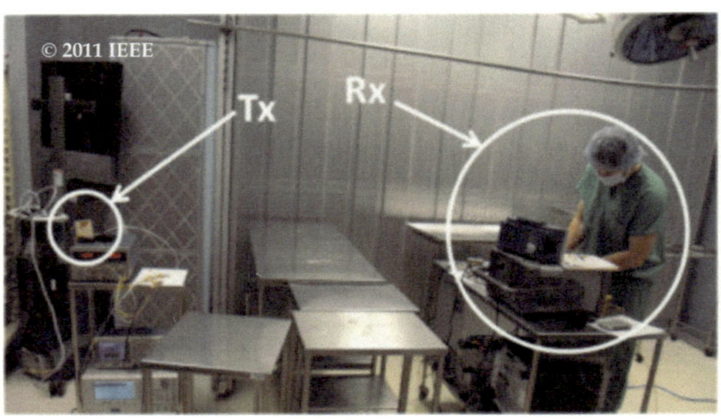

Fig. 32. Experimental setup in the operating room during non-live scenario (Mahfouz & Kuhn, 2011).

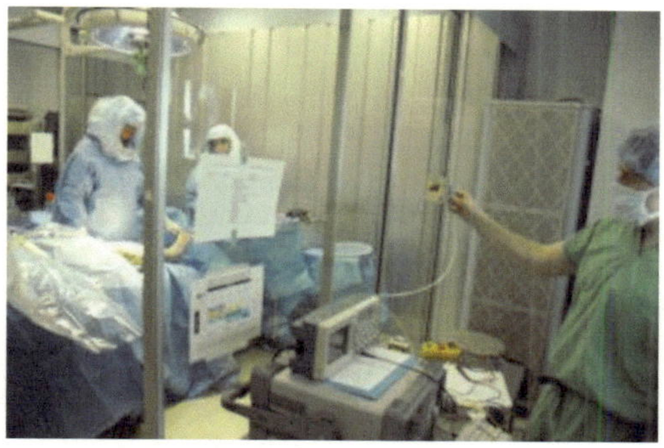

Fig. 33. Experimental setup in the operating room during an orthopedic surgery.

3.2 Experimental results

Table 4 shows a truncated list of parameters for the LOS operating room environment fit to the IEEE 802.15.4a channel model which were obtained with time domain and frequency domain experimental data. Figure 34 shows the pathloss for the OR environment obtained by fitting experimental data and compared to residential LOS, commercial LOS, and industrial LOS. The pathloss in the OR is most similar to residential LOS, although this can change depending on which instruments are placed near the transmitter and receiver or the locations of the UWB tags and base stations in the room. Figure 35 shows pathloss obtained for a Tx-Rx distance of 0.49 m where the transmitting (monopole) and receiving (Vivaldi) antenna effects have been removed. Small scale fading effects can be seen as well as frequency dependent pathloss, which is captured in the parameter κ in Table 4.

Figure 36 shows an example time domain signal where significant multipath interference is caused by reflections from metal tables and walls. Figure 37 shows an example time domain received signal for a Tx-Rx distance of 1.49 m using the monopole antenna for transmitting and single element Vivaldi antenna for receiving. A decaying exponential is overlayed on the received signal to highlight the intra-cluster decay, defined by γ_0 = 1.33 in Table 4. The pathloss of the LOS OR channel is most like a residential LOS environment whereas the power delay profile (PDP) is closer to an industrial LOS environment (γ_0 = 0.651) where dense clusters of multipath quickly decay (rather than the residential LOS environment where γ_0 = 12.53). The mean number of clusters ($\bar{L}$ = 4) is in between the residential and industrial LOS environments ($\bar{L}$ = 3 and $\bar{L}$ = 4.75). The inter-cluster decay constant and inter-cluster arrival rate (Λ and Γ) for the operating room channel are more similar to the industrial LOS channel rather than the commercial or residential LOS channels. The operating room LOS channel is similar to the industrial LOS channel in its time domain characteristics (i.e. multipath interference and decay) while it is similar to the residential LOS channel in its frequency domain characteristics.

Operating Room	LOS
PL_0 [dB]	-47.5
n	1.33
κ	0.95
$\bar{L}$	4
Λ [1/ns]	0.095
λ [1/ns]	n/a
γ_0 [ns]	1.33
k_γ	0.217
Γ [ns]	10.8

Table 4. Summary of parameters fit to IEEE 802.15.4a channel model with experimental UWB data taken in the operating room (Mahfouz et al., 2009).

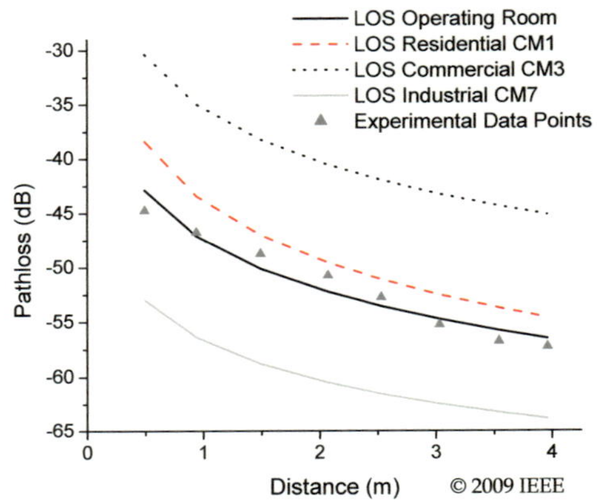

Fig. 34. Comparison of pathloss for IEEE 802.15.4a LOS channels. The pathloss for the OR environment is most similar to residential LOS (Mahfouz et al., 2009).

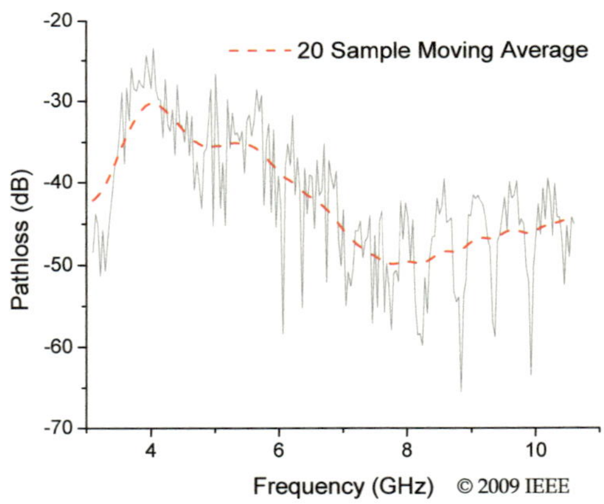

Fig. 35. Pathloss obtained with the Tx and Rx placed 0.49 m apart where effects from the transmitting (monopole) and receiving (Vivaldi) antennas have been removed. The frequency dependence, κ, can clearly be seen as well as small scale fading effects (Mahfouz et al., 2009).

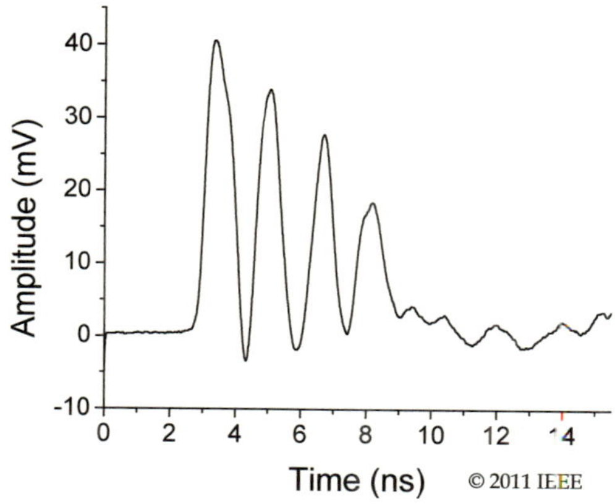

Fig. 36. Experimental received time domain signal with noticeable multipath interference caused by metal tables and walls in the operating room (Mahfouz & Kuhn, 2011).

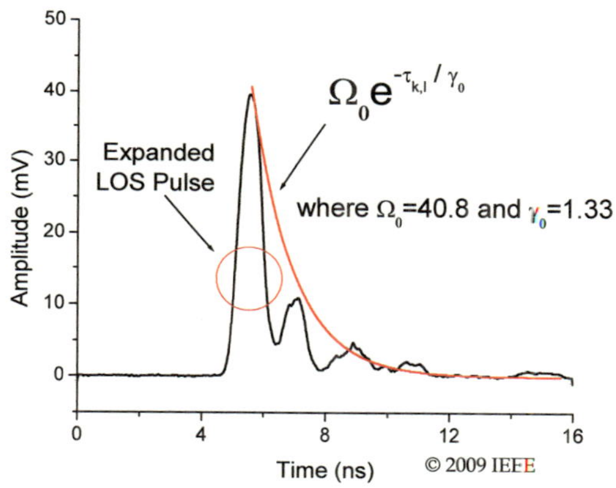

Fig. 37. Example received signal in the time domain for a Tx-Rx distance of 1.49 m highlighting the distortion (seen as expansion) in the LOS pulse due to a dense cluster of multipath rays. The overlayed exponential is fit using γ_0 as outlined in Table 4 to show the intra-cluster decay of the LOS cluster (Mahfouz et al., 2009).

3.3 Electromagnetic interference in the operating room

Electromagnetic interference (EMI) in the OR was measured across a wide frequency range in the context of comparing the interference present in useable frequency bands for narrowband and UWB communication and localization systems (for available bands see Table 3).

3.3.1 OR indoor environment

EMI was measured over a large frequency band (200 MHz – 26 GHz) in the OR during four separate orthopedic surgeries. Figure 38 shows the experimental setup in the OR. Besides the operating table, numerous other pieces of medical equipment were present during the surgery including an anesthesia machine, ventilator, surgical lamps, various monitoring equipment, visualization screens, carts containing necessary orthopedic surgical tools, drills, etc. Also, numerous people were present including the surgical team, orthopedic company representatives, and spectators observing the surgery. The combination of people and medical equipment closely packed into the OR creates a dense multipath indoor environment that can greatly disrupt standard RFID tracking systems. UWB systems have inherent advantages that make them a strong candidate for use in dense multipath environments such as the OR.

3.3.1 Experimental setup

Various hardware was needed to get accurate measurements across the wide band of 200 MHz - 26 GHz. It should be noted that all reported gain and noise figure values are averages across the frequency range of operation. Figure 39 shows the four antennas used to cover the entire frequency range. The standard setup for each of the frequency bands measured included an antenna, two stages of amplification, and a spectrum analyzer for visualization. Commercial off-the-shelf components were used whenever possible. Table 3 lists the major medical, scientific, and UWB frequency bands in the US and Europe. A majority of the scientific and medical bands in both Europe and the US fall between the frequencies of 200 MHz - 3 GHz. Also, most RFID systems operate in the MHz range up to 3 GHz. Even though RFID systems can operate at 5.8 GHz or 24.125 GHz, limitations still exist on how well a system with small bandwidth can handle the dense multipath environment of the OR at these high frequencies. When looking at different wireless bands currently in use, whether WLAN, cellular phones, GPS, or medical, the advantages of operating in the higher frequency bands of 3.1 - 10.6 GHz and 22 - 29 GHz useable for UWB become clear.

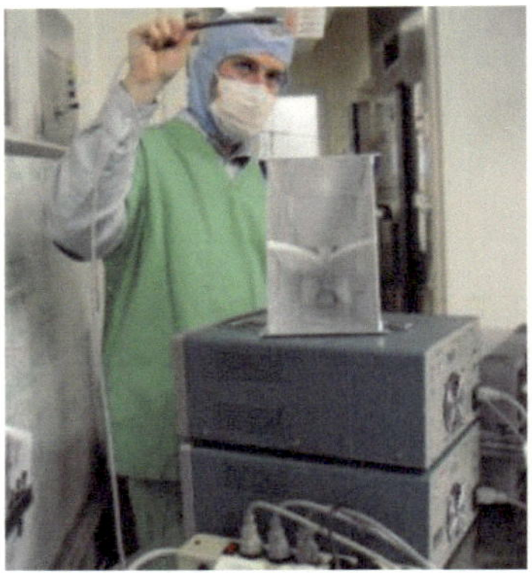

Fig. 38. Experimental setup in the OR.

3.3.2 Experimental results

Electromagnetic interference was measured over the frequency range of 200 MHz - 26 GHz. The results from these measurements can be seen in Figure 40-42. A number of signals were detected in the lower frequency range of 400 MHz - 2.5 GHz. As shown in Figure 40, no appreciable signals were picked up between 200 - 800 MHz. Although there is a small spike near 470 MHz, it is only 6dB above the noise floor and is considered noise. Also, there are no licensed frequency bands in the US that could correspond to the 470 MHz peak. Figure 41 shows the frequency band from 800 MHz - 3 GHz. A number of different signals were found in this frequency range. The two strongest signals, which were found at 872 MHz and 928 MHz, correspond to CDMA2000 uplinks and downlinks. The peak at 1.95 GHz also corresponds to a US cellular band. Finally, the peak at 2.4 GHz is caused by WLAN and Bluetooth components. Figure 42 shows the frequency band from 3 - 26 GHz. No noticeable signals were picked up across this entire band. This is somewhat unexpected since there are ISM and WLAN bands between 5 - 6 GHz, which could be the major culprit causing interference that could affect UWB systems.

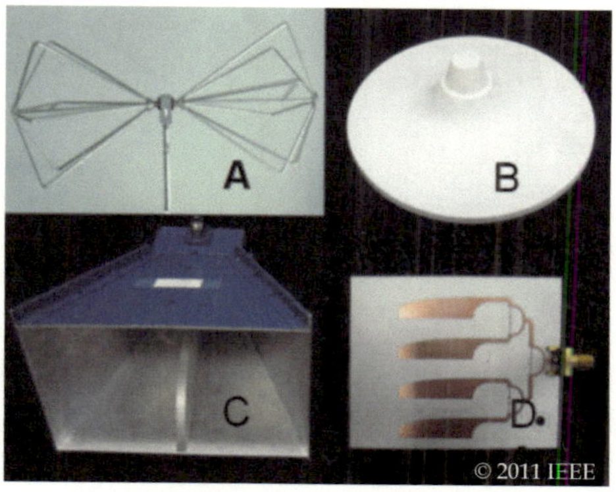

Fig. 39. Antennas used in OR measurements: a) biconical, b) multiband disc, c) broadband TEM horn, d) 4-element Vivaldi array (Mahfouz & Kuhn, 2011).

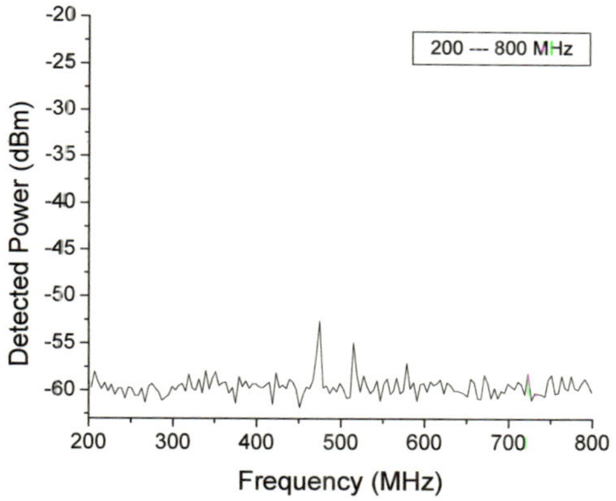

Fig. 40. Measured EMI over frequency range of 200 - 800 MHz (Mahfouz & Kuhn, 2011).

The frequency bands containing noticeable EMI correspond to widespread technologies that will likely be seen in the average OR. One surprise was the almost complete absence of US scientific and medical bands. Many medical devices do conduct wireless operations at the frequency bands summarized in Table 3, but besides the WLAN signal at 2.4 GHz seen in Figure 41, no significant EMI corresponding to these frequency bands was detected in the OR. As outlined in Table 3, there is another UWB frequency band from 22 - 29 GHz that can be used for localization systems. As seen from Figure 42, there is no EMI in the band from 22 - 26 GHz. One reason for having no EMI is that very few licensed bands exist between 22 - 29 GHz that would affect an OR. Also, signals in this frequency band tend to be attenuated more by the atmosphere and are typically used for short range applications. Using UWB for localization in the OR holds a distinct advantage over other technologies because of both the large bandwidth used as well as the higher frequencies available for operation.

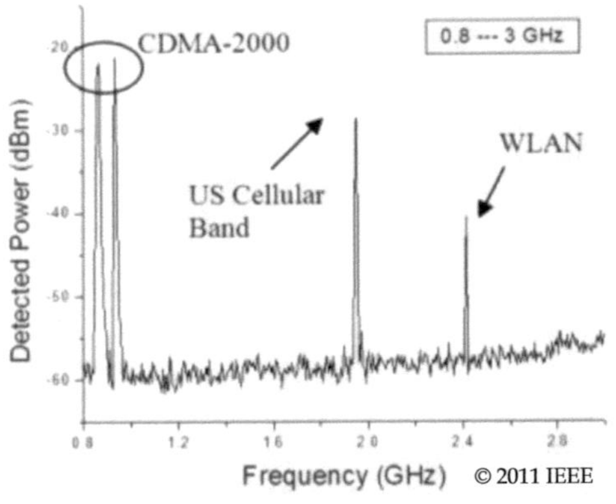

Fig. 41. Measured EMI over frequency range of 800 MHz – 3 GHz (Mahfouz & Kuhn, 2011).

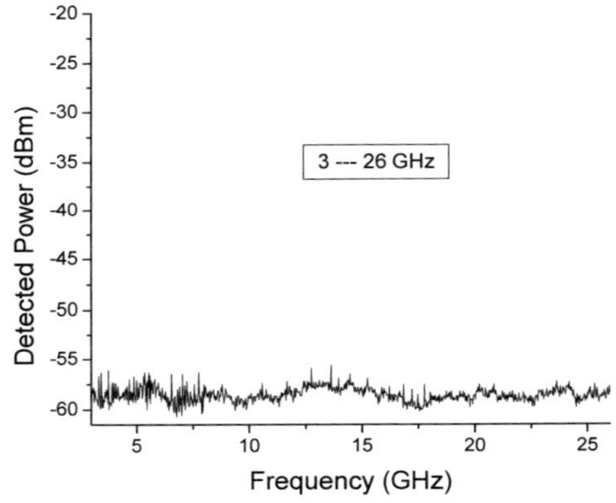

Fig. 42. Measured EMI over frequency range of 3 – 26 GHz.

4. High accuracy positioning systems for indoor environment

Although UWB positioning systems are well established in their use for indoor applications requiring 3-D real-time accuracy on the level of 10-15 cm, current commercial systems have not been able to meet the stringent accuracy specifications (e.g. 1-2 mm or sub-mm 3-D) of the next level of applications including smart medical instruments, surgical navigation, and tracking in wireless body-area-networks.

4.1 Development of a high accuracy ultra-wideband positioning system

The challenges in developing a millimeter range accuracy real-time non-coherent UWB positioning system include: generating ultra-wideband pulses, pulse dispersion due to antennas, modeling of complex propagation channels with severe multipath effects, need for extremely high sampling rates for digital processing, noise and sensitivity of the UWB receiver, local oscillator phase noise (in the case of a carrier-based system), antenna phase center variation, time scaling, jitter, and degradation due to overall system calibration. For such a high precision system with mm or even sub-mm accuracy, all these effects should be accounted for and minimized. The complete setup of the non-coherent UWB positioning system is shown in Figure 43. The source of the non-coherent UWB positioning system is a step-recovery diode (SRD) based pulse generator with a pulse width of 300 ps and bandwidth of greater than 3 GHz. The Gaussian pulse is up-converted with an 8 GHz carrier and then transmitted through an omni-directional monopole UWB antenna. Multiple base stations are located at distinct positions to receive the modulated pulse signal. The received modulated Gaussian pulse at each base station first goes through a directional Vivaldi receiving antenna and then is amplified through a low noise amplifier (LNA) and demodulated to obtain the I signal. Only one channel rather than I/Q is required since energy detection and carrier offsets are also applied at the UWB receiver. After going through a low pass filter (LPF), the I channel is sub-sampled using an UWB sub-sampling mixer, extending the signal to a larger time scale while maintaining the same pulse shape (Zhang et al., 2007). The PRF clocks are set to be 10 MHz with an offset frequency of 1-2 kHz between the tag and base stations which corresponds to an equivalent sampling rate of 50-100 GS/s. Finally, the extended I channel is processed by a conventional analog to digital converter (ADC) and standard FPGA unit. Leading-edge detection is performed on the FPGA. The time sample indices are sent to a computer where additional filtering and the final time-difference-of-arrival (TDOA) steps are performed to localize the 3-D position of the UWB tag.

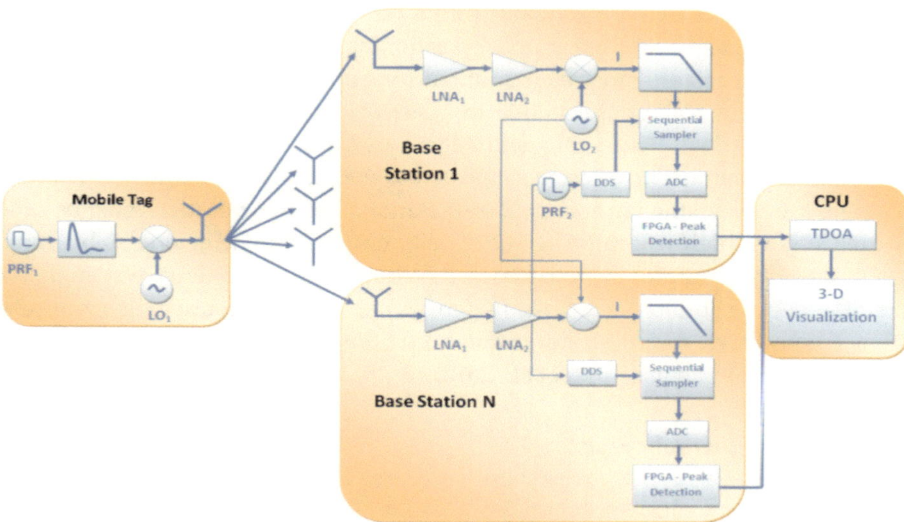

Fig. 43. System architecture of non-coherent UWB positioning system which includes a carrier-based transmitted signal at the tag and a combination of downconversion and energy detection at the UWB receiver.

To detect narrow pulses on the order of a few hundred picoseconds (i.e. 300 ps or 3 GHz bandwidth in our system), analog to digital converters with at least 6 GS/s are needed to satisfy the Nyquist criterion. However, such high performance ADC units are currently either not commercially available or too expensive for most applications. A realistic alternative approach to real-time sampling is to sub-sample the UWB pulses while maintaining the initial pulse shape through extended time techniques. The extended UWB signals can then be handled by readily available commercial ADCs, reducing overall system cost (Zhang et al., 2007). The sampler utilizes a simple broadband balun structure and a balanced topology.

The non-coherent architecture of the current UWB positioning system places stringent requiremens on phase noise specifications of the local oscillators at the transmitter and receiver. The use of a reference tag partially mitigates the local oscillator phase noise and temperature effects at the UWB receivers. Even with a reference tag, the phase noise presents a formidable challenge to achieving millimeter 3-D real-time accuracy. High phase noise carriers (e.g. free running voltage controlled oscillators) cause up to an order of magnitude (e.g. cm) greater error than low phase noise carriers. When attempting to achieve millimeter and sub-mm accuracy, phase center variation of the antennas at the Tx/Rx is an important source of error which needs to be taken into account. The transmitter employs a UWB monopole antenna which provides an omni-directional radiation pattern with minimal phase center variation while the receiver utilizes a single element Vivaldi antenna for a radiation pattern directed at the view volume of interest. Noticeable variation of the phase center is observed in both the E and H cuts especially for angles greater than ±30°. High accuracy positioning systems must employ calibration techniques to remove the phase center effects. For example, antennas used for GPS systems go through an advanced automated calibration process which uses high precision robots to move the antennas to 6000-8000 distinct points in calibrating out phase center effects. More challenges appear in achieving high accuracy real-time indoor positioning at the system-level. Cable length effects at the UWB receivers must be accounted for and statically calibrated and removed from the system. Time scaling effects due to system clock drift must be characterized and calibrated out of the final TDOA calculations in a dynamic manner when moving around the view volume. Time scaling effects change across the view volume due to the differences in LOS ranges r_i between the tag and each base station. The 3-D variation must be calibrated out in order to get a highly accurate indoor positioning system achieving stable millimeter range accuracy. Future improvements for this UWB indoor positioniong system include the addition of real-time, multi-tag access (Kuhn et al., 2011) and utilizing comprehensive simulation frameworks for accurate simulation of advanced mixed signal systems in realistic indoor environments (Kuhn et al., 2010).

4.2 Real-time experimental results

Two 3-D experiments with unsynchronized LOs and PRF clock sources were carried out, where a minimum of four base stations are needed for the 3-D measurements.

4.2.1 3-D dynamic free motion

Figure 44 shows a four base station setup where the 3-D positions were measured for each base station utilizing the Optotrak 3020 system, which also serves as a reference for comparing the 3-D real-time accuracy of our UWB localization system. The Optotrak 3020 has 3-D real-time accuracy of better than 0.3 mm. It should be noted that the spatial spread of the base stations along the z-axis is the largest (2498 mm), while the x-axis is the smallest

(1375 mm). In the dynamic mode, the tag is moving randomly inside the 3-D space as shown in Figure 44. The 3-D motion of the tag is then plotted and UWB measurements are compared with Optotrak measurements. RMSE is used to report the error since it is the true unbiased error when data values fluctuate above and below zero. Figure 45 plots the UWB trace and Optotrak trace in the 3-D dynamic mode.

Figure 46 shows the 3-D dynamic errors in the *x, y,* and *z* axes over 1000 measured points. The overall 3-D RMSE is 6.37 mm. The error along the *x-axis* contributed most to the overall distance error, which can be explained by the limited spatial spread of base stations along the *x-axis* and can be calculated using the PDOP definitions in (Mahfouz et al., 2008). Such error can be mitigated through better arrangement of the base stations along the *x-axis*.

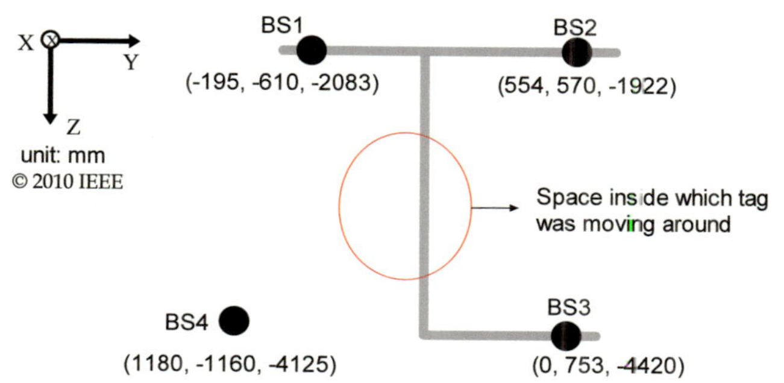

Fig. 44. 3-D unsynchronized localization experiments, 4 base station distribution with locations for each base station (Zhang et al., 2010).

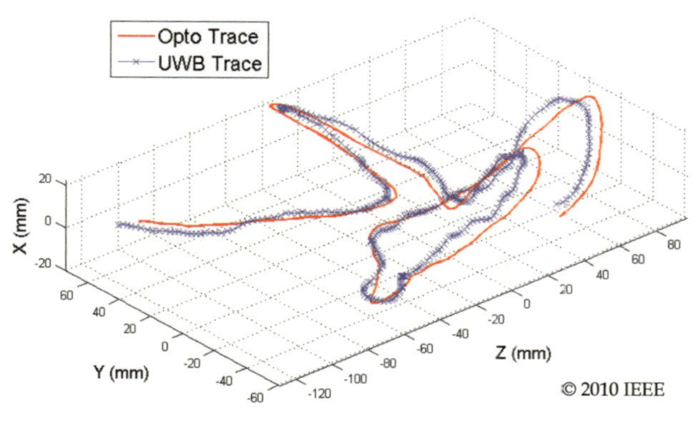

Fig. 45. 3-D dynamic random mode with energy detection. UWB trace is compared to Optotrak trace (Zhang et al., 2010).

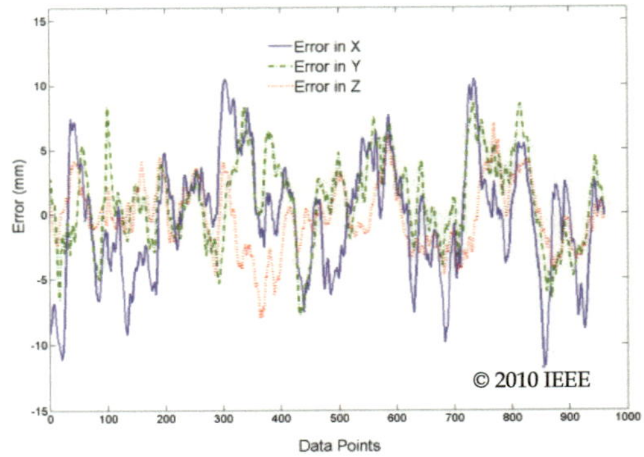

Fig. 46. 3-D dynamic mode with energy detection. x, y and z axes error compared to Optotrak measurements (Zhang et al., 2010).

4.2.2 3-D robot tracking

The next non-coherent 3-D experiment is to dynamically track the robot position. The monopole antenna and the reference Optotrak probe are tied together and fixed to the arm of the CRS A465 robot. The robot arm set up is shown in Figure 47. Finally, the base stations can be seen in Figure 48. The robot was pre-programmed to specifically cover 20 distinct static positions in a 3-D volume, stopping for three seconds at each position and then moving to the next position and so on. The measured traces by the UWB system are compared to the Optotrak reference system as shown in Figure 49. Figure 50 shows the 20 distinct static positions taken by both the UWB and the Optotrak systems. The overall dynamic 3-D robot tracking RMSE is 5.24 mm. In Table 5 the real-time non-coherent 3-D experimental results are summarized under various scenarios. The reported RMSE are based on 1000 continuous data points recorded and compared to the Optotrak 3020 system, which served as the real-time reference of our UWB localization system and provides a 3-D accuracy of better than 0.3 mm.

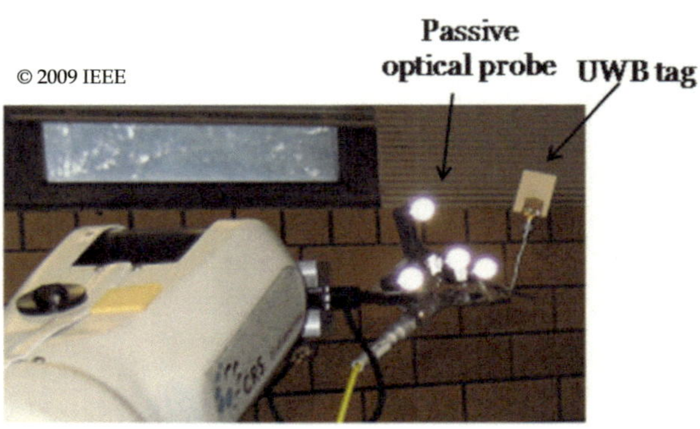

Fig. 47. Robot arm with UWB monopole and optical tracker attached (Mahfouz et al., 2009).

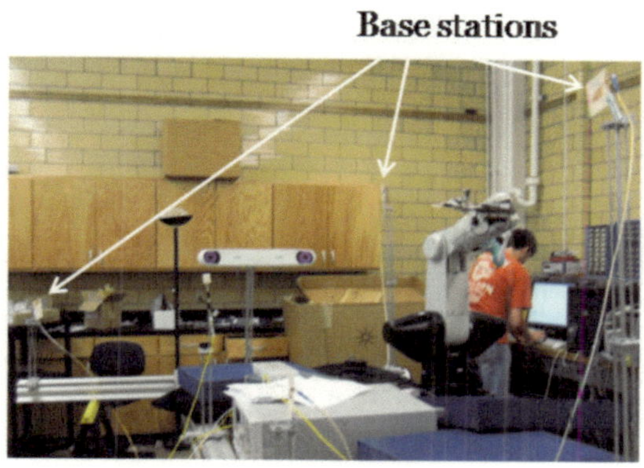

Fig. 48. Experimental setup outlining base station positions.

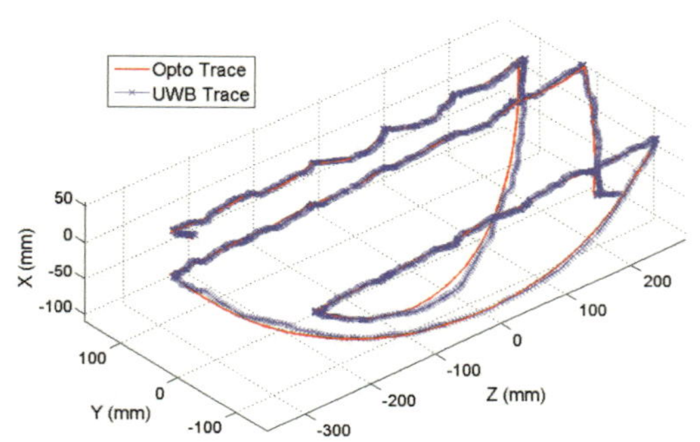

(a) 3-D view

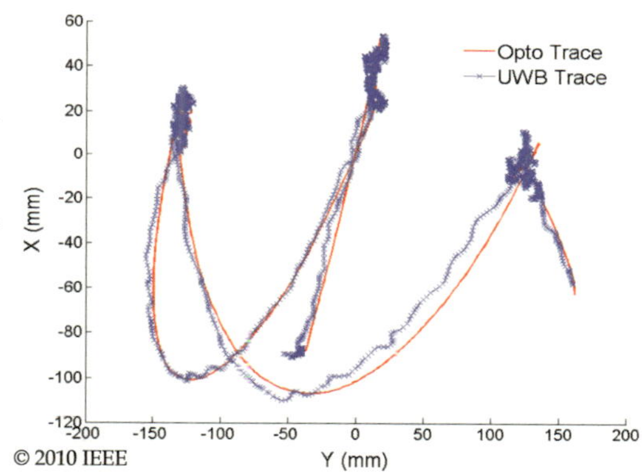

(b) XY plane

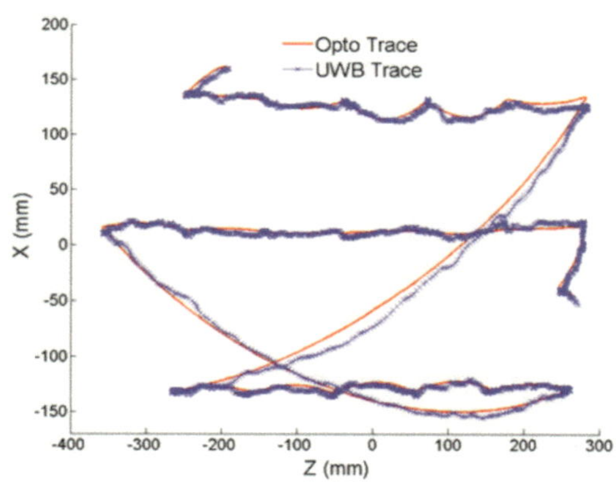

(c) XZ plane

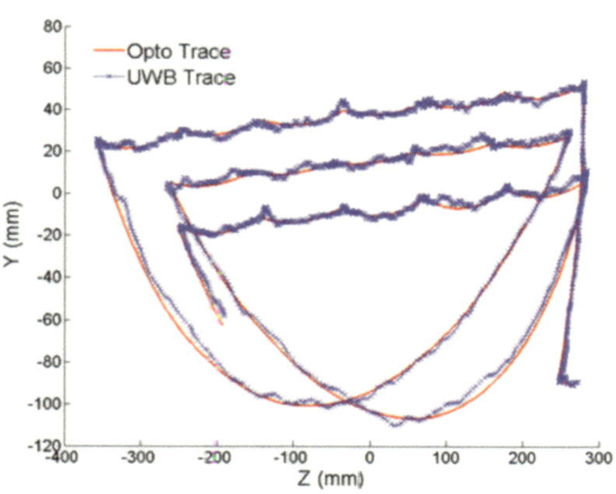

(d) YZ plane

Fig. 49. 3-D dynamic robot tracking. UWB trace compared to Optotrak trace: (a) 3-D view; (b) XY plane; (c) XZ plane; (d) YZ plane (Zhang et al., 2010).

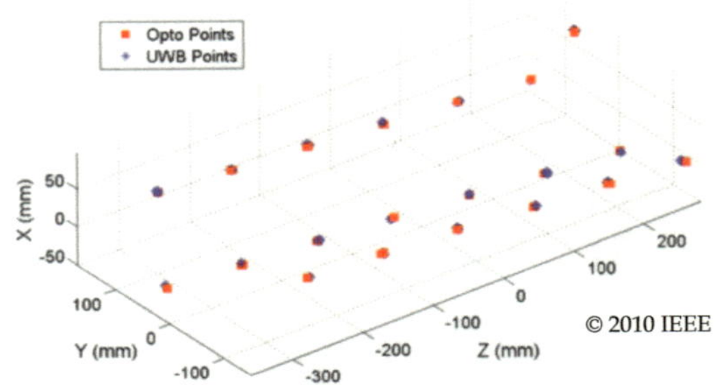

Fig. 50. 3-D robot tracking at static positions. UWB points compared to Optotrak points (Zhang et al. 2010).

3-D Experiments	RMSE (mm)
Tag free random motion	6.37
Robot dynamic tracking	5.24
Robot static positions (20 distinct locations)	4.67
Static position w/ 106 times of average	1.98

Table 5. Error Summary – 3-D unsynchronized localization experiments (Zhang et al. 2010).

5. Wireless MEMS sensors used as feedback control in an orthopedic surgical navigation system

Over the past decade, orthopedic companies have been trying different methods and protocols to eliminate one of the primary causes of implant failure in total knee arthroplasty (TKA), which is the malalignment of the implants to the biomechanical axis of the patient. To properly place the implant, the gaps after the resections between the femur and tibia during extension and 90 degrees flexion have to be parallel to each other and the gap size have to be the same (Figure 51). However, the surgeons are usually working with a small incision with limited access to the joint. Moreover, the knee joint are stabilized by the medial and lateral collateral ligaments. The laxity of the ligaments can affect the gap balance.

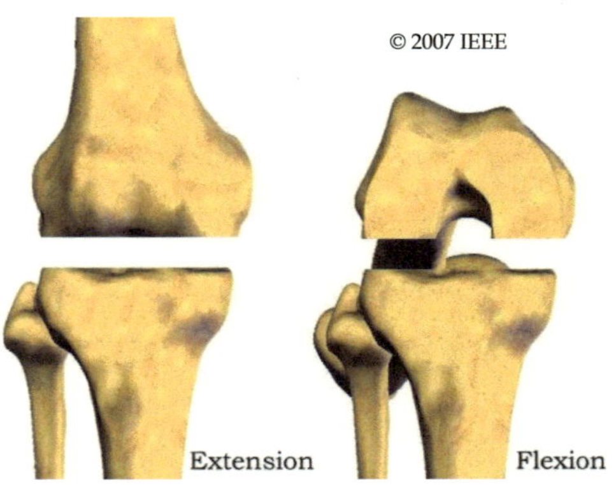

Fig. 51. Flexion and Extension gap between the femur and tibia (To, 2007)

In order to help the surgeons to assess the tightness of the joint after resection, an instrument was designed to provide quantitative feedback to the users. A wireless strain measuring device was designed. The high level system design is shown in Figure 52. Two types of sensors were investigated in the design of this instrument. The first type of sensor is piezo-resistive based microcantilever as shown in Figure 53. When a piezo-resistive element undergoes stress, the resulting strain causes changes in the resistance of the material. Hence, it is possible to use to measure strain by monitoring the resistance of the material.

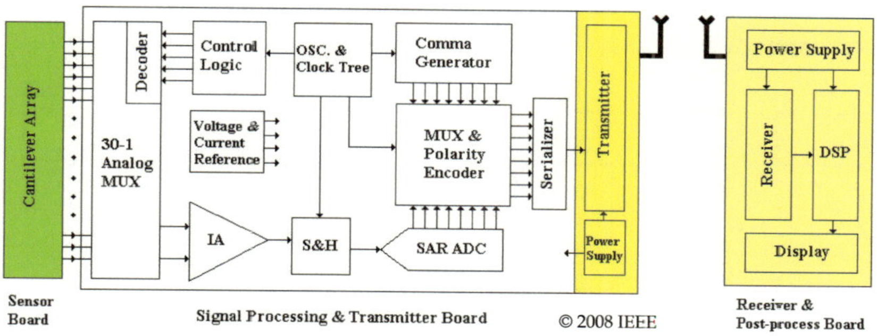

Fig. 52. High level design of a wireless strain measuring system (Qu et al., 2008)

The piezo-resistive microcantilevers here are used to measure a macro pressure that causes a deflection in the microcantilever beam. The microcantilevers are tiny and extremely fragile. In addition, silicon is not a FDA-approved biocompatible material unless specifically doped. This specific application to measure macro forces requires a protective layer with a material that damps the applied stress and provides a biocompatible interface for bodily contact. Medical grade epoxy was used as a protective material for the sensors as well as providing a bio-compatible surface to interface with the soft tissues. The epoxy was cured over the microcantilevers to protect and to give a desirable force readout range. Curing procedures and epoxy homogeneity were investigated to create the most reliable, non-interfering encapsulation. Parameters investigated included viscosity, cure time, working time, heat cure, and minimization of bubbles and microbubbles. EP30MED (Masterbond, Inc.) was chosen as the most favorable epoxy for encapsulation. A microcantilever that was encapculated with a 2mm thick epoxy was used for mechanical testing as shown in Figure 54. An Instron 5544 testing machine was used. The properties of the encapsulated sensor are shown in Table 6.

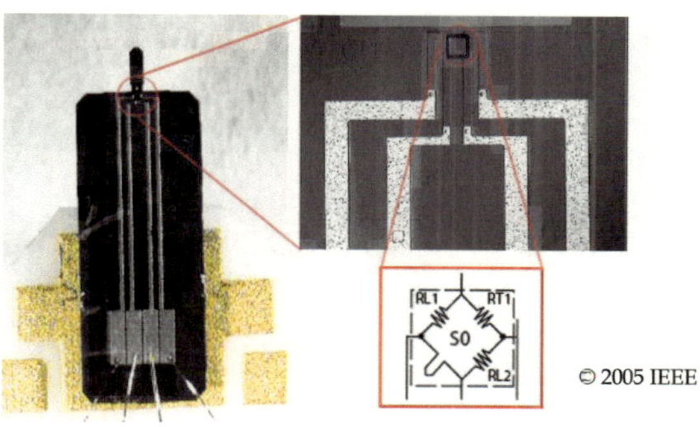

Fig. 53. Piezo resistive microcantilever (Nascatec, Stuttgart, Germany) [To & Mahfouz, 2005]

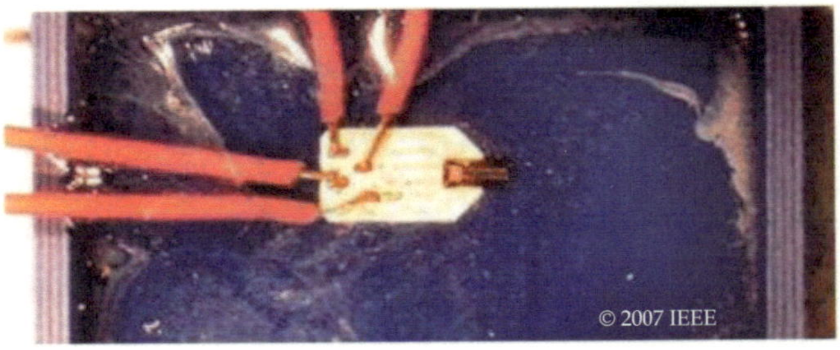

Fig. 54. Microcantilever encapsulated in EP30MED epoxy.

Parameter	Value
Range	0 - 300 kPa
Input	0 - 3.3V +/- 1%
Linearity	0.625mV/kPa (over range)
Repeatability	0.6444mV/kPa (over range)
Sensitivity	0.35455mV/kPa (over range)

Table 6. Properties of microcantilever encapsulated in 2mm of EP30MED (Qu et al., 2010)

The readout circuit for the microcantilever system was tested with off-the-shelf components using an MSP430 (Texas Instrument) as microcontroller, ADG726 (Analog Device) as multiplexer, INA331A2 (Texas instrument) as instrumental amplifier, and MAX1472/1473 as transmitter and receiver. The readout circuit is too bulky to be fitted inside a surgical instrument. As a result, an application specific integrated circuit (ASIC) is designed specifically for the reading of the microcantilever sensors. The ASIC includes the multiplexer, signal conditioning circuit, analog to digital converter (ADC), and a buffer interfacing with the transmitter. The footprint of the ASIC is shown in Figure 55. The specification of the ASIC is shown in Table **7**. The gain of the amplifier can be adjusted via an external resistor. After examining the outputs of the microcantilever, the gain was configured to 72. The overall system RSS error with microcantilever embedded within 2mm thick of EP30MED is approximately +/- 1.79kPa.

Fig. 55. ASIC designed for microcantilever readout (Left: ASIC footprint, Right: ASIC with testing package) (Qu et al., 2010)

Parameters	Values
Analog input channels	16
Analog MUX switching frequency	Oscillator dependent
A/D Converter input range	~ 200mV - 1589mV
A/D Converter resolution	8bit
A/D Converter rate	772 kHz
Band gap reference	1.249V
INA gain	Gain resistor dependent
INA phase margin	65°
INA Unit gain bandwidth	~ 2.4 GHz
A/D ENOB	7.24 bit
A/D SNDR	45.4 dB
A/D SFDR	56.4 dB
DNL	+0.57/-0.42 LSB
INL	+1.3/-0.2 LSB
Power supply	2.6 - 4.4V

Table 7. ASIC specification (Qu et al., 2010)

The final design of the instrument is designed to fit within a spacer block (Figure 56). The spacer block is placed within the resection gap to identify the tightness of the joint. Moreover, identifying the location of the high strain area can help the surgeons in balancing the joint with appropriate ligaments release. The system design is separated into 3 layers. An array of 30 microcantilever are arranged and wirebonded onto the circuit board. The bottom most circuit board is the ASIC and the battery layer as shown in Figure 57. Two switches are used to connect the poly Li^+ batteries to the electronics and sensors. Traditional coin cell batteries are not suitable for this design as they are too large in size and they are incapable of powering all 30 microcantilevers, which is about 70mA. The poly Li^+ batteries can be made in customable shape and they are rechargeable. For the prototype, a USB socket is used to recharge the batteries. High density sockets are used to connect the ASIC layer to the sensors layer.

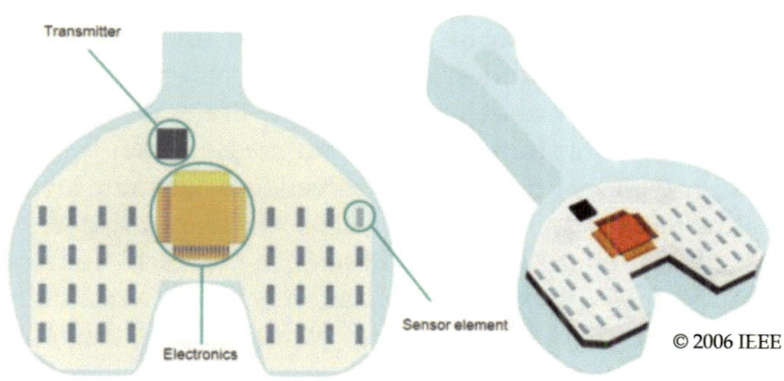

Fig. 56. Instrumented Spacer Block [To et al., 2006].

The middle layer is the TX PCB. The transmitter is using MAX1473 and configured the carrier frequency to 433MHz. The material for the circuit board was changed to 0.0020" rogers 4350 for better performance. A chipped antenna is used to further reduce the volume required from traditional whipped antenna. The assembled PCB is shown in Figure 57. Each side of the PCBs has 15 active sensing microcantilevers and 1 additional microcantilever for reference on the left side of the PCB.

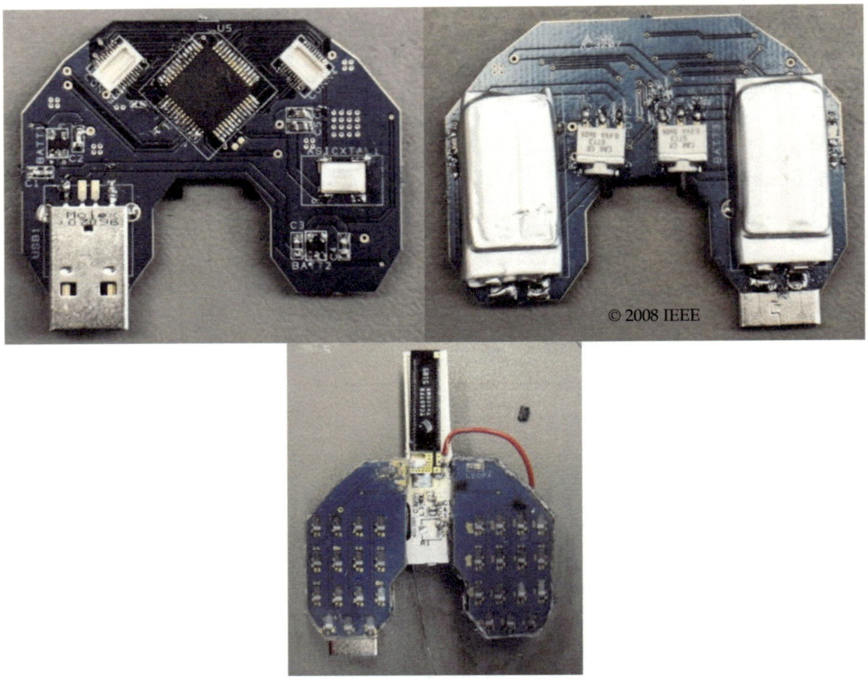

Fig. 57. Left: Top view of the signal processing layer; Center: Bottow view showing the batteries; Right: Top view of assembled PCB (Right) with 15 microcantilevers arrayed on each condyle. (Qu et al., 2008)

The second type of sensor being investigated was capacitive based MEMS device. Strain sensing is accomplished by embedding pairs of electrodes with specific geometries in a biocompatible material. Deformation of the embedding materials causes changes in the configuration of the capacitor electrodes. However fabricating MEMS devices on polymeric materials is not as straight forward as with silicon substrate. Researchers have shown that polyimide can be used as a substrate material, and parylene can be used as the dielectric material (1.5 μm). It is noted that parylene has served as a substrate layer in early capacitve fabrication when the sensor was left on the silicon wafer, but it poses a problem due to adhesion and mechanical strength of the thin film during removal from the silicon substrate. A negative-resist based photolithography fabrication was implemented to reduce time and number of steps for fabrication. The electrodes consist of a 10 nanometer (nm) titanium adhesion layer and 300 nm of gold deposited on the substrate via physical vapor deposition. Array design is multi-faceted to understand the behavior of the sensors at a small scale and to optimize design to boost readout speed, increase nominal capacitance, and decrease crosstalk and parasitic effects specific to the configuration of this array. Increasing nominal capacitance is most easily achieved through larger electrode size and thinner dielectric layers, thus presenting a tradeoff between keeping sensor size to a minimum and nominal capacitance at an appropriate level for accurate measurement. Similarly, the spacing needs to be optimized between closeness (providing high spatial resolution across the array), and crosstalk (sensors too close to one another affecting readout). A uniaxial and triaxial strain measuring device was fabricated as shown in Figure 58.

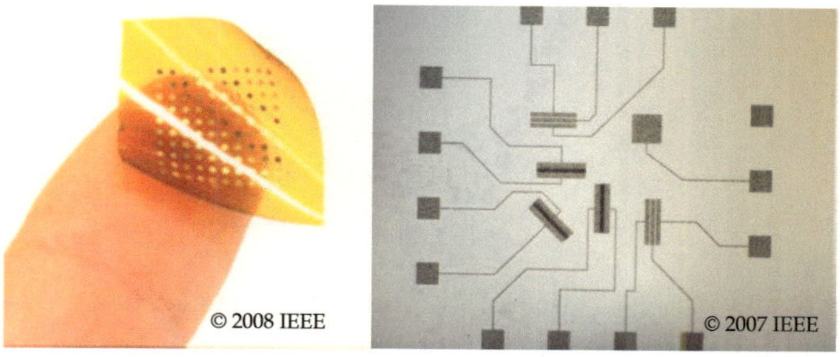

Fig. 58. Capacitive based MEMS strain measuring device (Left: Uniaxial(Pritchard et al., 2008), Right: Triaxial (Evans III, 2007)).

An array of sensors was tested using an MTS (Eden Prairie, MN) 858 Table Top System mechanical testing machine with a 2.5 kN load cell. The load profile is shown in Figure 59. A protective polyimide layer was placed over the electrodes and a second protective layer over the entire assembly. Unlike the microcantilever sensors, no protective epoxy layer was required. Similar to the piezoresistive microcantilever, a transition was made from using off-the-shelf IC to ASIC electronics for the capacitive MEMS sensors. An ASIC consisting of diode array, matched capacitor capacitance to voltage converter and a custom designed instrumental amplifier as shown in Figure 60.

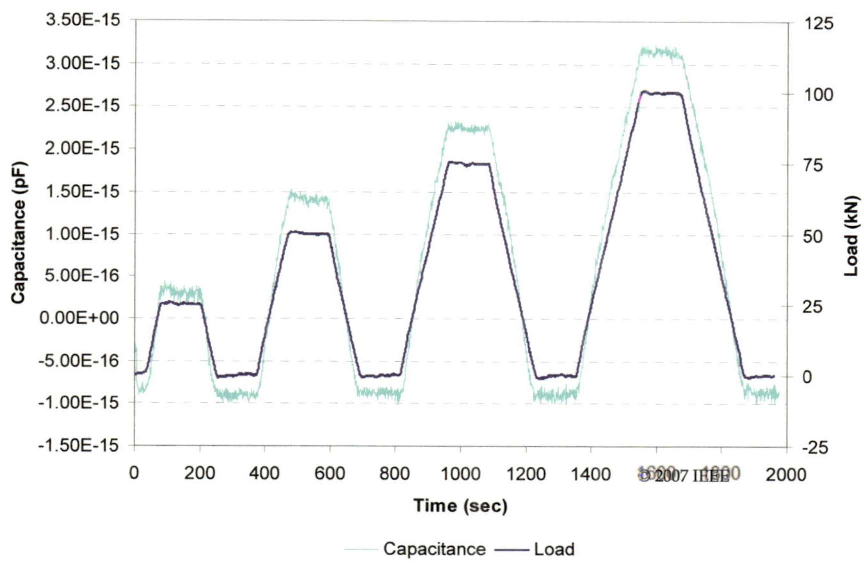

Fig. 59. Load profile for capacitance array test. Test is from 5 pF capacitor array (Evans III, 2007).

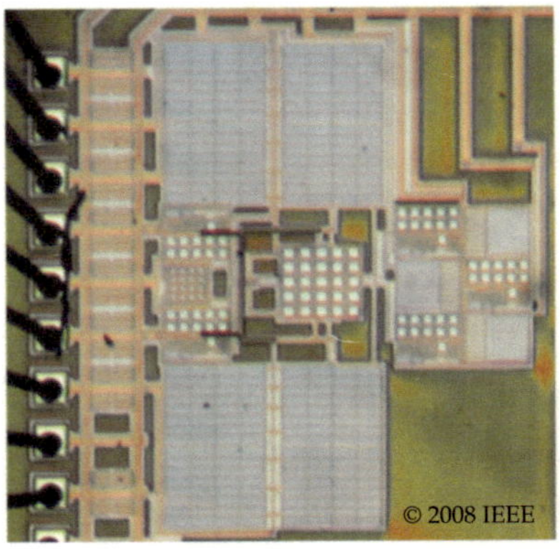

Fig. 60. Microscopic view of ASIC after fabrication (Haider et al., 2008)

The capacitive MEMS sensors can be fabricated at high density such that hundreds of sensors can be placed within the same area as the microcantilever system. Addressing and processing these sensors, as well as transmitting these data at a high speed manner are some of the major challenges in this system.

6. Conclusion

Wireless medical devices can greatly improve efficiency and productivity for the healthcare sector. Current wireless technologies enable compact design and are inexpensive to manufacture. With the increasing amount of personal information being monitored and transmitted wirelessly, security measures are one future trend in wireless technology research. Protecting patient privacy is a vital aspect in remote patient monitoring systems.

Despite the increasing number of commercial wireless medical products, many of them use different wireless protocols, and data are transferred at various data rates. Hence, one of the focuses will be increasing the degree of integration for wireless medical healthcare monitoring systems. There will also be an increasing use of implantable and ingestible wireless medical devices, as it provides means to access and gather data and information that were previously impossible.

As research advances, there will be a demand for higher resolution to improve the accuracy of indoor wireless positioning systems. Highly accurate wireless positioning has many applications in medicine including computer assisted surgery, smart surgical instruments, and accurately locating sensors in body area networks. As the amount of integration and the complexity of the system increases, the current wireless protocols will not be sufficient to provide a large enough bandwidth for real time applications. Thus, high speed, high data rate design and protocols will be needed in future wireless technologies.

While it is important to research and develop more advanced wireless medical devices, it is also important to focus on application development for these devices. The ultimate goal for biomedical devices is to provide tools to assist the physicians, and to improve the quality of life of the patients. Application development plays a crucial role in developing these instruments. It is also worth mentioning that some of the devices measured in this article are very similar to each other, in which they share the same basic components. However, they are fine tuned to be used in different applications, and this is critical in developing a device robust enough to meet the stringent requirements of the medical field.

7. Reference

Alavi, B., & Pahlavan, K. (2006). Modeling of the TOA-based distance measurement error using UWB indoor radio measurements. *IEEE Comm. Let.*, *10* (4), 275-277.

Alcosystem (n.d). Retrieved from http://www.alcosystems.se/iBAC_1sidig.pdf

Ascension (n.d). Retrieved from http://www.ascension-tech.com

Chen, C.-H., Hwang, R.-Z., Huang, L.-S., Lin, S.-M., Chen, H.-C., Yang, Y.-C., et al. (2009). A Wireless Bio-MEMS Sensor for C-Reactive Protein Detection Based on Nanomechanics. *IEEE Transactions on Biomedical Engineering*, *56*(2), 462-470.

Clarke, D., & Park, A. (2006). Active-RFID system accuracy and its implications for clinical applications. *IEEE Symp. on Computer-Based Med. Sys*, (pp. 21-26). Salt Lake City, USA.

Lifescience Corp, L. (n.d.). Retrieved from http://www.lifesynccorp.com/products/wireless-system.html

D'Lima, D., Townsend, C., Arms, S., Morris, B., & CW, C. (2005). An implantable telemetry device to measure intra-articular tibial forces. *Journal of Biomechanics*, *38*, 299-304.

Ekahau Real-Time Location System Brochure. (2008). (Ekahau, Inc) Retrieved from http://www.ekahau.com/images/stories/documents/ekahau_rtls_brochure_en_122007_lo.pdf

Ellinger, F., Eickhoff, R., Gierlich, R., & Hüttner, J. (2007). Local positioning for wireless sensor networks. *IEEE Globecom Workshops*, (pp. 1-6). Washington DC, USA.

Emotiv. (n.d.). Retrieved from http://www.emotiv.com

Evans III, B. M. (2007). *MEMS Capacitive Strain Sensing Elements for Integrated Total Knee Arthroplasty Prosthesis Monitoring*. Knoxville, TN, USA: University of Tennessee.

Falsi, C., Dardari, D., Mucchi, L., & Win, M. (2006). Time of arrival estimation for UWB localizers in realistic environments. *EURASIP Journal on Applied Signal Processing*.

FCC. (2002). The first report and order regarding ultra-wideband transmission systems. *FCC 02-48, ET Docket*, 98-153.

Feger, R., Wagner, C., Schuster, S., Jager, H., & Stelzer, A. (2009). 77-GHz FMCW MIMO radar based on an SiGe single-chip transceiver. *IEEE Trans. Microwave Theory and Tech*, *57* (5), 1020-2035.

Fontana, R. (2004). Recent system applications of short-pulse ultra-wideband (UWB) technology. *IEEE Trans. Microwave Theory and Tech*, *52* (9), 2087-2104.

Fuiji, A., Sekiguchi, H., Asai, M., Kurashima, S., Ochiai, H., & Kohno, R. (2007). Impulse radio UWB positioning system. *IEEE Radio and Wireless Symposium*, (pp. 55-58).

Gabriel, S., Lau, R., & Gabriel, C. (1996). The dielectric properties of biological tissues: III parametric models for the dielectric spectrum of tissue. *Phys. Med. Biol.*, *41*, 2271-2293.

Gao, T., Pesto, C., Selavo, L., Chen, Y., Ko, J., Lim, J., et al. (2008). Wireless medical sensor networks in emergency response: implementation and pilot results. *IEEE International Conference on Technologies for Homeland Security*. Waltham, MA.

Graichen, F., Arnold, R., Rohlmann, A., & Bergmann, G. (2007). Implantable 9-Channel Telemetry System for In Vivo Load Measurements With Orthopedic Implants. *IEEE transactions on biomedical engineering*, *54* (2), 253-261.

Haider, M., Mahfouz, M., Islam, S., Eliza, S., Qu, W., & Pritchard, E. (2008). A low-power capacitance measurement circuit with high resolution and high degree of linearity. *51st Midwest Symp on Circuits and Systems*, (p. 261). Knoxville, TN.

Healthcare, G. (n.d.). Retrieved from http://www.gehealthcare.com

Iso-Ketola, P., Karinsalo, T., & Vanhala, J. (2008). HipGuard: A wearable measurement system for patients recovering from a hip operation. *2nd Intl Conf. on Pervasive Computing Tech. for Healthcare.* Tampere, Finland: PervasiveHealth 2008.

Komarov, V., Wang, S., & Tang, J. (2005). Permittivity and measurements. In K. Chang (Ed.), *Wiley Encyclopedia RF Microwave Engineering* (pp. 3693-3711). New York: John Wiley & Sons.

Kuhn, M., Mahfouz, M., Turnmire, J., Wang, Y., & Fathy, A. (Jan, 2011). A multi-tag access scheme for indoor UWB localization systems used in medical environments. *Proceedings of IEEE Topical Conference on Biomedical Wireless Technologies, Networks, and Sensing Systems (BioWireleSS),* Phoenix, AZ, 75-78.

Kuhn, M., Mahfouz, M., Zhang, C., Merkl, B., & Fathy, A. (2010). A system level simulation framework for UWB localization. *IEEE Transactions on Microwave Theory and Techniques , 58* (12), Part 1, 3527-3537.

Low, Z., Cheong, C., Ng, W., & Lee, Y. (2005). Pulse detection algorithm for line-of-sight (LOS) UWB ranging applications. *IEEE Ant. and Wireless Prop. Letters , 4*, 63-67.

Mahfouz, M., Zhang, C., Merkl, B., Kuhn, M., & Fathy, A. (2008). Investigation of high accuracy indoor 3-D positioning using UWB technology. *IEEE Transactions on Microwave Theory and Techniques , 56* (6), 1316-1330.

Mahfouz, M., Kuhn, M., To, G., & Fathy, A. (2009). Integration of UWB and wireless pressure mapping in surgical navigation. *IEEE Transactions on Microwave Theory and Techniques , 57* (10), Part 2, 2550-2564.

Mahfouz, M., Kuhn, M. (Jan, 2011). UWB channel measurements and modeling for positioning and communications systems in the operating room. *Proceedings of IEEE Topical Conference on Biomedical Wireless Technologies, Networks, and Sensing Systems (BioWireleSS),* Phoenix, AZ, 47-50.

Mahfouz, M., Kuhn, M., Wang, Y., Turnmire, J., & Fathy, A. (Jan, 2011). Towards sub-millimeter accuracy in UWB positioning for indoor medical environments. *Proceedings of IEEE Topical Conference on Biomedical Wireless Technologies, Networks, and Sensing Systems (BioWireleSS),* Phoenix, AZ, 47-50.

McEwan, T. E. (2004). *Patent No. 6,747,599.* United States of America.

Meier, C., Terzis, A., & Lindenmeier, S. (2007). A robust 3D high precision radio location system. *IEEE International Microwave Symp.,* (pp. 397-400).

Meier, C., Terzis, A., & Lindenmeier, S. (2007). Investigation and suppression of multipath influence on indoor radio location in the millimeter wave range. *Conf. Wave Prop. Comm., Microwave Sys. Nav,* (pp. 21-24). Chemnitz, Germany.

Metronics. (n.d.). Retrieved from http://www.medtronic.com/

Morris, S. J., & Paradiso, J. A. (2002). Shoe-integrated sensor system for wireless gait analysis and real-time feedback. *Proceedings of the 2nd Joint IEEE EMBS (Engineering in Medicine and Biology Society) and BMES (the Biomedical Engineering Society) Conference.*

Mosshammer, R., Huener, M., Szumny, R., Kurekt, K., Hittner, J., & Gierlichli, R. (2007). A 5.8 GHz local positioning and communication system. *IEEE MTT-S International Microwave Symposium,* (pp. 1237-1240). Honolulu, USA.

Myglucometer. (n.d.). Retrieved from http://www.myglucometer.com/

Nelson, S. (1991). Dielectric properties of agricultural products - measurements and applications. *IEEE Trans. Elec. Insul , 26* (5), 845-869.

Nonin. (n.d.). Retrieved from http://www.nonin.com/documents/M-5507%204000%20Cleveland%20Whitepaper.pdf

Omnipod. (n.d.). Retrieved from http://www.myomnipod.com

Ossberger, G., Buchegger, T., Schimback, E., Stelzer, A., & Weigel, R. (2004). Non-invasive respiratory movement detection and monitoring of hidden humans using ultra wideband pulse radar. *IEEE International Conf UWB Sys. Tech*, (pp. 395-399). Kyoto, Japan.

Panasonic. (n.d.). Retrieved from http://service.us.panasonic.com/OPERMANPDF/EW3153.PDF

Patel, S., Lorincz, K., Hughes, R., Huggins, N., Growden, J., Standaert, D., et al. (2009). Monitoring Motor Fluctuations in Patients With Parkinson's Disease Using Wearable Sensors. *IEEE Transactions on Information Technology in Biomedicine*, *13* (6).

Pritchard, E., Mahfouz, M., Evans, B., Eliza, S., & Haider, M. (2008, Oct). Flexible capacitive sensors for high resolution pressure measurement. *IEEE sensors*, 1484.

Public Domain Retrieved from http://en.wikipedia.org/wiki/File:CapsuleEndoscope.jpg

Qu, W., Islam, S., Mahfouz, M., To, G., & Mofasta, S. (2010). Micro-cantilever Array Pressure Measurement System for Biomedical Instrumentation. *IEEE Sensors Journal*, *10* (2), 321-330.

Qu, W., Islam, S., To, G., & Mahfouz, M. (2008). Design of analog signal processing integrated circuit for multi-channel biomedical strain measurement instrument. *International Conference on Biomedical Electronics and Devices*. Funchal, Portugal.

Rohlmann, A., Gabel, U., Graichen, F., Bender, A., & Bergmann, G. (2007). An instrumented implant for vertebral body replacement that measures loads in the anterior spinal column. *Medical Engineering & Physics*, *29* (5), 580-585.

Stelzer, A., Diskus, C., & Thim, H. (1999). A microwave position sensor with sub-millimeter accuracy. *47* (12), 2621-2624.

Stelzer, A., Pourvoyeur, K., & Fischer, A. (2004). Concept and application of LPM-a novel 3-D local position measurement system. *IEEE Trans. Microwave Theory and Tech*, *52* (12), 2664-2669.

SunTech. (n.d.). Retrieved from www.suntechmed.com

To, G. (2007). *Development of the Telemetrical Intraoperative Soft Tissue Tension Monitoring System in Total Knee Replacement with MEMS and ASIC Technologies.* Knoxville, TN, USA: University of Tennessee.

To, G., & Mahfouz, M. (2005). Development for Intra/Post Operative Surgical Instrument using MEMS/IC. *International Conference on Biomedical Engineering*. Singapore.

To, G., Qu, W., & Mahfouz, M. (2006). ASIC Design for Wireless Surgical MEMS Device and Instrumentation. *IEEE EMBS*. New York, USA.

Tragas, P., Kalis, A., Papadias, C., Ellinger, F., & Eickhoff, R. (2007). RESOLUTION: reconfigurable systems for mobile local communication and positioning. *Mob. and Wireless Comm. Summit*, (pp. 1-5). Budapest, Hungary.

Ubisense. (2011). Retrieved from www.ubisense.net.

Vorst, A.V., Rosen, A., & Kotsuka, Y. (2006). *RF/Microwave Interaction with Biological Tissues*, John Wiley & Sons Inc., ISBN: 9780471732778, Hoboken,NJ.

Waldmann, R., Weigel, P., & Gulden, P. (2008). Method for high precision local positioning radar using an ultra wideband technique. *International Microwave Symp,* (pp. 117-120). Atlanta, USA.

Zebra Enterprise Solutions (2011). Retrieved from http://www.zebra.com/id/zebra/na/en/index/products/location/ultra_wideband.html.

Zetik, J., Sachs, J., & Thomä, R. (2004). UWB localization - active and passive approach. *Proceedings of the 21st IEEE IMTC, 2,* pp. 1005-1009.

Zhang, C., Fathy, A., & Mahfouz, M. (2007). Performance enhancement of a sub-sampling circuit for ultra-wideband signal processing. *IEEE Micro. and Wireless Comp. Lett , 17* (2), 873-875.

Zhang, C., Kuhn, M., Merkl, B., Fathy, A., & Mahfouz, M. (2010). Real-time non-coherent UWB positioning radar with millimeter range accuracy: theory and experiment. *IEEE Transactions on Microwave Theory and Techniques , 58* (1), 9-20.

15

Ultra-Wideband Pulse-Based Microwave Imaging for Breast Cancer Detection: Experimental Issues and Compensations

Joshua C. Y. Lai, Cheong Boon Soh, Kay Soon Low and Erry Gunawan
Nanyang Technological University
Singapore

1. Introduction

Recent research based on numerical modeling (Bond, 2003; Xu, 2001) that ignored hardware characteristics and simple experiments using homogenous breast phantoms (Sill, 2005; Xu, 2004) have shown the potential of ultra-wideband to detect early stage breast cancer. However, clutter interference from heterogeneous breast tissues and hardware characteristics like pulse jitter, finite dynamic range and precision for signal acquisition can severely degrade the detectability of breast tumors. This chapter discusses the experimental issues encountered and compensation methods used to improve the detectability of tumor.

In order to bridge the gap between numerical simulations and experiments, it is important to identify the experimental issues before conducting experiments with more realistic breast phantoms so that the source of imaging artifacts can be identified and compensated. An ideal imaging scenario is first created where the simple sum-and-delay algorithm (Xu, 2001) is working perfectly. In this ideal scenario, the breast phantom is round and symmetrical such that the averaging method can perfectly remove the skin reflection. The breast medium is homogeneous such that propagation of signal in the medium is constant with accurate delay time estimation. Skin is approximated by a single interface (air to breast phantom) because its thickness is small compared to UWB pulse width in space.

In this chapter, several important experimental issues are discussed.

- Impulse Generator – Pulse Jitter Artifact
- Real Time Oscilloscope – Limited Dynamic Range
- Breast Phantom Positioning Error – Ring Artifact
- Signal Loss Compensation – Noise Amplification
- Filtering and Correlation – Noise Reduction
- Averaging and Antenna Number – Signal SNR vs Image SNR

To facilitate the discussion, the experimental setup will first be described in the following subsection 1.1. Experiments discussed in this chapter are conducted in time domain using an impulse generator and a real-time oscilloscope.

1.1 Experimental setup

A breast phantom is fabricated using soy bean oil (ε_r = 2.6, σ = 0.05 S/m), contained in a cylindrical polypropylene container (diameter 16 cm, height 12 cm). Tumor is simulated

with a small cylindrical shape jelly ($\varepsilon_r = 8$, $\sigma = 0.4$ S/m) with 4 mm diameter and 1 cm length made of tissue-mimicking phantom material (Lazebnik, 2005).
The excitation signal is generated using the Picosecond Pulse Labs 3500D impulse generator, which produces gaussian pulses with full width at half maximum (FWHM) of 80 picoseconds. Agilent DSO81204B real-time oscilloscope with 40 GHz sampling rate is used for recording the backscattered signals from the breast phantom.
Thales UWB antennas (Chua, 2005) are used as the transmitter and receiver of the UWB signals. The antennas dimension is 3 cm width and 4 cm height. The antennas gain is 11 dB with azimuth beamwidth of 60 degrees and elevation beamwidth of 40 degrees. The antennas return loss measured with Agilent N5230A vector network analyzer is lower than -10 dB from 2.4 to 12 GHz.
Breast phantom is placed on a rotary stage with antennas scanning at the side to simulate the human breast in prone position. Breast phantom is rotated for 360 degrees relative to the stationary antennas to simulate a circular antenna array around the breast circumference. The overall experimental setup is shown in Figure 1.
The collected signals are processed with averaging method (Xu, 2001) to remove the signal artifacts, which includes incident pulse, boundary reflection and multipath. The processing is also known as calibration in the literature. Delay-and-sum beamforming (Xu, 2001) algorithm is used to generate the image as in confocal imaging technique. Breast image is formed by synthetically focusing the signals received from the antenna array to every point within the region of interest.

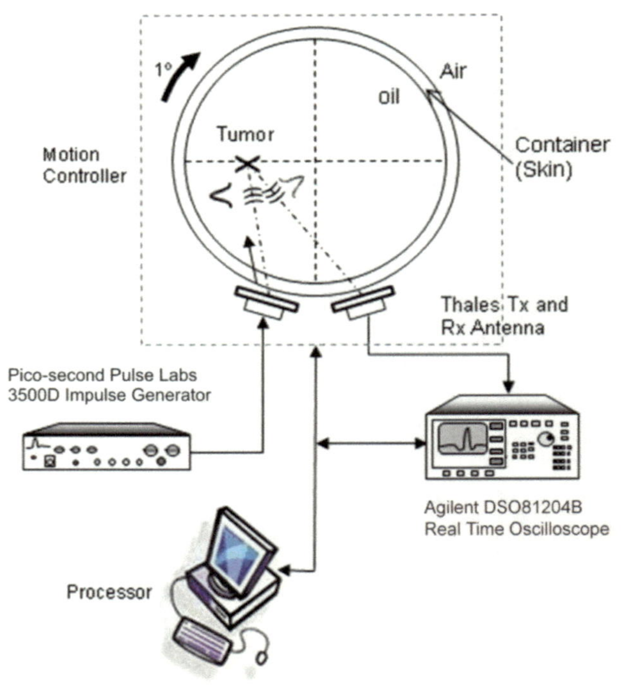

Fig. 1. Overall experimental setup.

2. Pulse jitter artifact and compensation

As mentioned in the introduction, averaging method is applied to remove artifacts in the received signals before delay-and-sum beamforming. An average of all received signals from different antennas is calculated. The averaged signal is used as a template artifact and is subtracted from individual received signals. Clean tumor responses can be obtained if the system is free of noise. The processing is also known as calibration in the literature.

However, averaging method will not work well if the received signals are not aligned perfectly and have unequal amplitudes. Pulse delay jitter is caused by the impulse generator being unable to maintain a constant delay time between trigger signal and the output UWB pulse. The maximum delay timing jitter measured with oscilloscope in the experiment is 31 ps. Pulse amplitude instability is caused by the impulse generator which is unable to maintain constant amplitude of the output UWB pulse.

Figures 2 and 3 show that pulse instability causes phase shift of 31 ps or 20 sample points in the signals. The resultant artifact would be large if the signals are not first compensated. For phase jitter compensation, all received signals are aligned by finding the zero-crossing point between maximum and minimum peaks and phase shifting the signals to the same zero-crossing point. For amplitude instability, compensation is done by normalizing all the signals peak-to-peak amplitude to one unit.

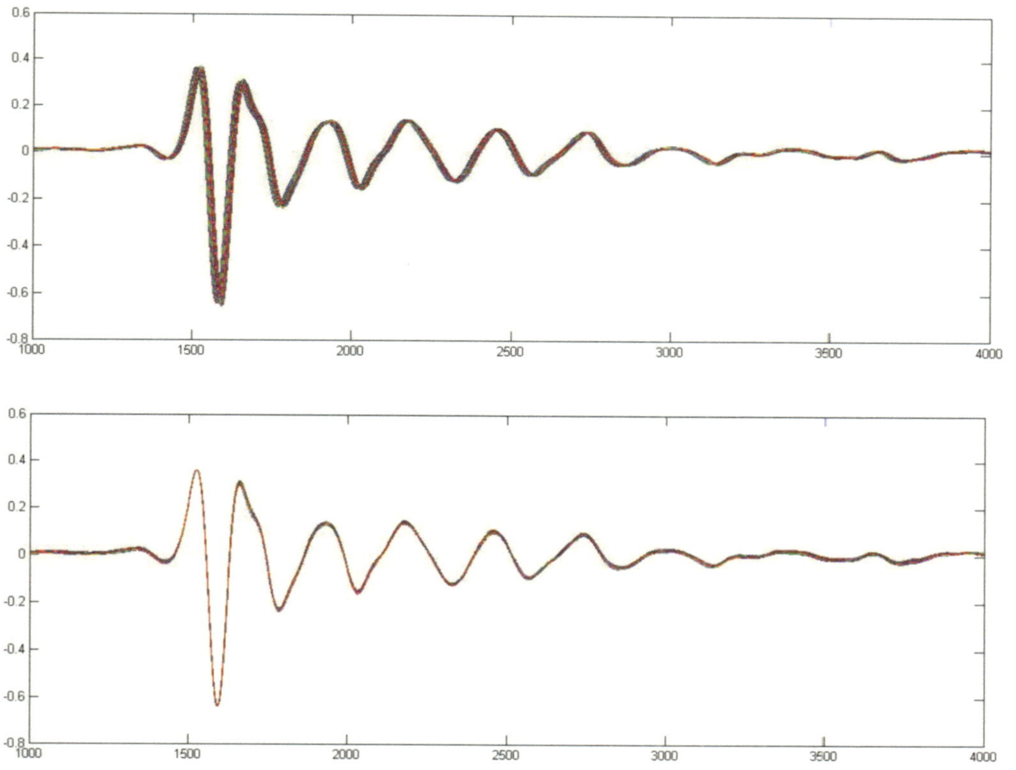

Fig. 2. Signals before and after pulse jitter compensation.

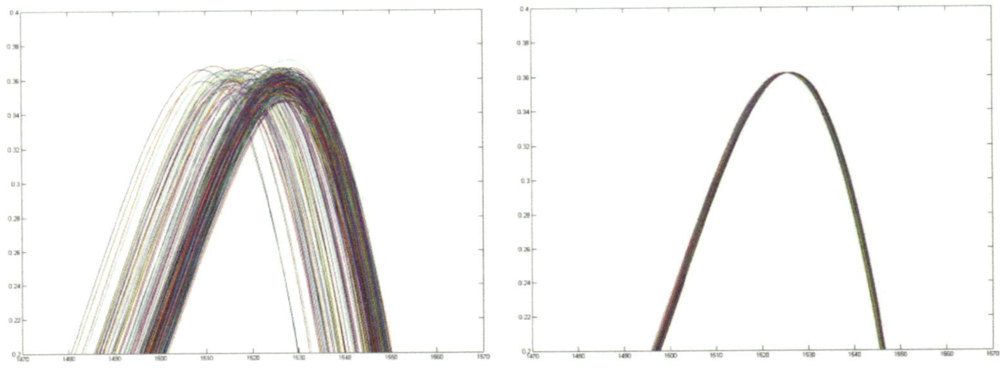

Fig. 3. Zoom in view of signals given in Figure 2.

To further minimize the phase error, the received signals are extrapolated to higher sampling rates. Table 1 shows the phase error for received signals and resultant signal artifact after pulse instability compensation at different extrapolated sampling rates. Measured noise amplitude is obtained after applying averaging method to a set of data collected from a tumor-free breast phantom. Simulated noise amplitude is calculated with Matlab by subtracting two identical signals, one signal is phase shifted by one sample time from another signal.

The noise amplitude shown in Table 1 is normalized to the incident pulse amplitude. The measured noise amplitude does not further decrease with higher sampling rates greater than 1 THz because the other contributors of noise such as environmental noise becomes significant, whereas the simulated noise amplitude continues to decrease with higher sampling rate as expected.

Sampling Rate	Worst Phase Error	Simulated Noise Amplitude	Measured Noise Amplitude
No compensation	31.25 ps	-	0.7866
40 GHz	25.00 ps	0.5437	0.5217
80 GHz	12.50 ps	0.2816	0.2781
160 GHz	6.25 ps	0.1684	0.1734
320 GHz	3.13 ps	0.0815	0.0942
640 GHz	1.56 ps	0.0533	0.0565
1 THz	1.00 ps	0.0344	0.0436
2 THz	0.50 ps	0.0173	0.0435

Table 1. Calculated and measured noise amplitude after pulse instability compensation at different extrapolated sampling rates.

Figure 4 shows the effects of pulse instability compensation on images formed by delay-and-sum beamforming. The tumor is located at 3 cm to the right from the center, or at coordinate (130, 100). The pulse jitter artifact located at (150,130) is significantly suppressed.

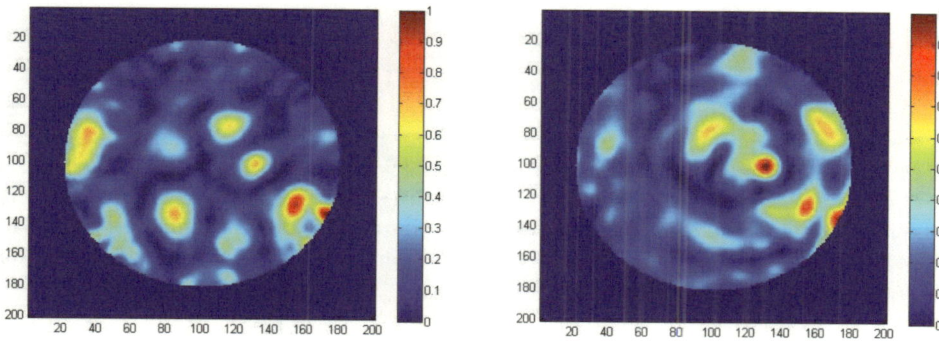

Fig. 4. The effects of pulse instability compensation on breast phantom images. Left: without compensation. Right: with compensation.

3. Real-time oscilloscope

This section discusses the dynamic range of Agilent DSO-081204B real-time oscilloscope used in the experiment. The information of the dynamic range will determine whether the pulse reflected from a tumor is too small to be detected. In this discussion, dynamic range is defined as the ratio of the oscilloscope vertical-scale range and the amplitude of the smallest possible digitized pulse.

Agilent DSO-081204B real time oscilloscope has analog-to-digital converters (ADC) with 8-bit resolution. After averaging and interpolation, the oscilloscope is able to increase the vertical resolution and store the data in 16-bit resolution. Due to quantization of the signal, at least 4-bit resolution is needed to construct the pulse shape of tumor response as shown in Figure 5. So the remaining 12-bit range is the maximum dynamic range of the oscilloscope.

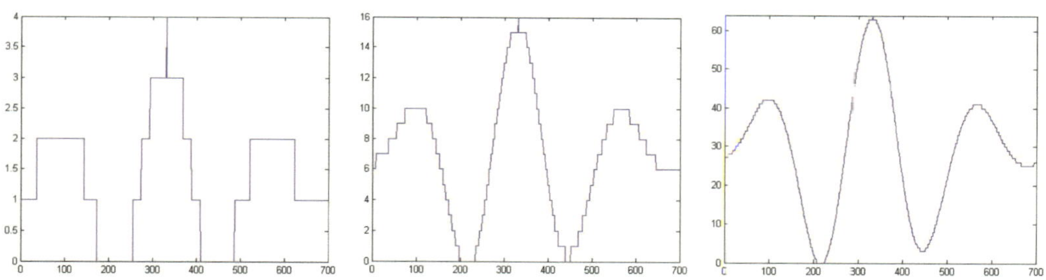

Fig. 5. Ideal tumor response constructed with 2-bit, 4-bit, and 6-bit vertical resolutions showing that the minimum required resolution is 4 bits.

An experiment is conducted to determine the dynamic range of the oscilloscope by investigating its ability to construct a pulse. Due to the presence of noise, a pulse cannot be observed from a single signal. Instead, 360 sets of signals are collected and an averaged signal is obtained. The random noise will be averaged out, and the pulse can be observed if it is within the dynamic range.

The incident pulse is set very small so that its amplitude is equal to the expected smallest peak for different possible dynamic range as shown in the Table 2. The recorded signals are averaged to determine at which dynamic range the pulse can still be constructed.

$$Expected\,peak = \frac{Full\,range\,voltage}{Dynamic\,range} = \frac{8V}{2^{12}\,bits} = 0.002V \tag{1}$$

From the experiment, , the pulse cannot be seen from the averaged signals if its amplitude is set to 2 mV, while the maximum noise amplitude is 5.9 mV. Whereas for a pulse amplitude of 4mV, the pulse is merely noticeable at sample time 1600 as shown in Figure 6. From this experiment which considers the system noise, the dynamic range is estimated to be 11-bits, which has 2048 values available for the whole range. Thus, the maximum detectable ratio between incident pulse and tumor response is 2048.

Dynamic Range 2^	Smallest peak can be detected	Recorded peak
8bits	31mV	33.2mV
9bits	16mV	14.9mV
10bits	8mV	7.8mV
11bits	4mV	6.1mV
12bits	2mV	5.9mV

Table 2. Expected and recorded peak of incident pulse for different dynamic range.

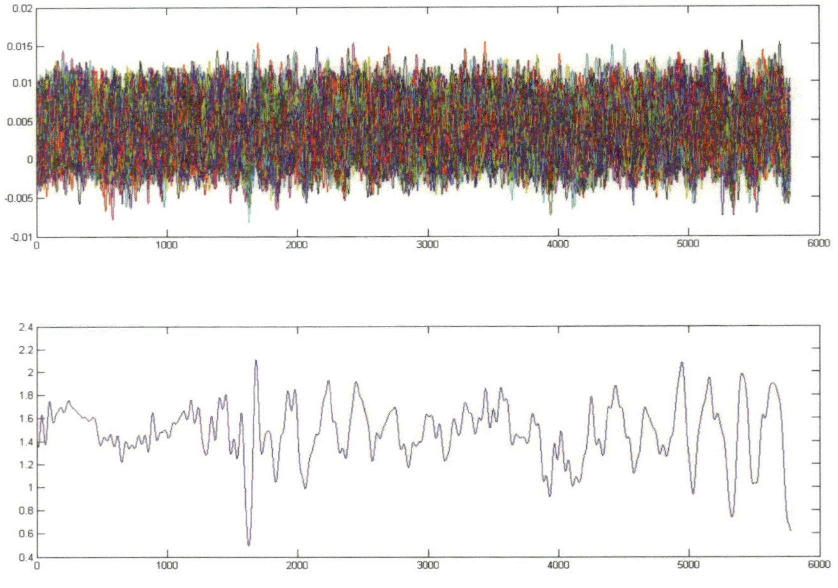

Fig. 6. Constructed pulse to test dynamic range of 2^{11} bits. The pulse is just noticeable at sample 1600. Upper trace shows the 360 signals and lower trace shows the averaged signal.

4. Ring artifact

If the experiments use a breast phantom with higher permittivity, ring artifacts will appear as shown in Figure 7. The phantom is fabricated using tissue mimicking phantom material (Lazebnik, 2005) with 80% oil, contained in a polypropylene cylindrical container (diameter 16cm, height 12cm). The material is able to closely simulate the dielectric properties of human tissues.

Tumor simulant is an 8 mm cube made of phantom material with 10% oil buried at 25 mm right from center of the phantom. The measured dielectric constant at 5 GHz for phantom medium is 9 whereas for tumor is 50 which is representative of normal and malignant human breast tissues. Ring artifacts have not been reported previously by other researchers because their breast phantoms use only lower dielectric constant materials.

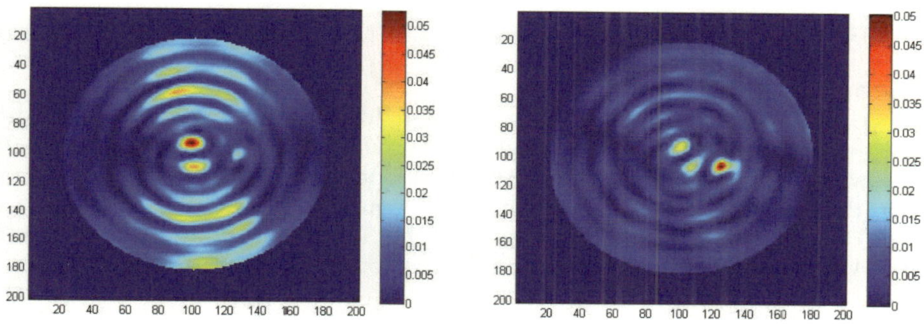

Fig. 7a. Images of breast phantom with 1 mm off-center positioning error (left) and breast phantom less than 0.5 mm off-center error (right).

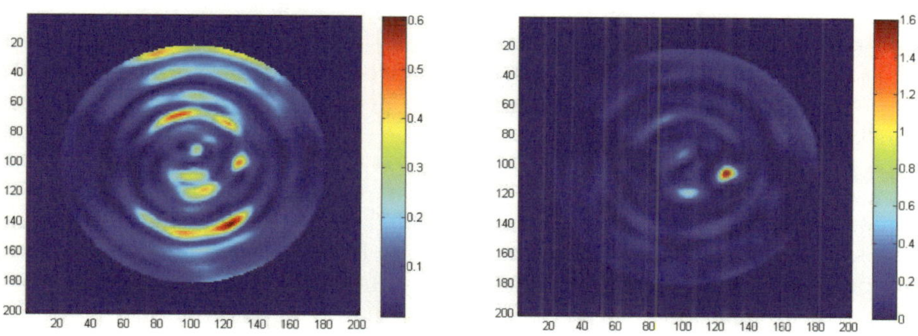

Fig. 7b. Same Images from Figure 7a with correlation applied.

Ring artifact arises from positioning error (off-center) of the breast phantom. Ideally, averaging method works perfectly with round phantom. Ring artifact appears when the phantom is not positioned perfectly on the rotary axis of the experimental stage. This causes small displacement error of the phantom boundary relative to the antennas.

Ring artifact is caused by the coherent adding of the residue boundary reflections after delay-and-sum beamforming. The ring-to-ring distance is proportional to the wavelength of

the incident signal. Ring artifact also indicates the direction of phantom off-center displacement. For instance, when the phantom is displaced to right side, the ring artifact will appear on right indicating positive x-axis off-center displacement and left indicating negative x-axis off-center displacement.
In experiments with oil medium which has lower dielectric constant, ring artifact is not noticeable because the tumor response is large enough to dominate the ring artifacts in the image. The correlation method is not able to improve the image quality when ring artifacts arise as illustrated in Figure 7b. This is due to the high similarity between tumor response and the residue incident pulse ringing.
Adjusting the phantom to the best position using visual inspection will result in a positioning error of 1 mm to 3 mm. Better placement can be achieved by placing a reference object on the antenna to measure the antenna to phantom boundary distance and adjusting the phantom position until the error is smaller than 0.5 mm. The resulting ring artifact is reduced significantly.

4.1 Experiment on ring artifact

An experiment is conducted to test the amplitude of the ring artifact for different displacement errors of a breast phantom without tumor. The phantom is adjusted to the best position with error less than 0.5 mm. Measurements are taken for the phantom at best position, then with off-center displacements of 1 mm, 2 mm, and 3 mm from the best position. The resultant signal artifact shown in Table 3 is computed after applying averaging method. When the breast phantom is perfectly positioned, the signal artifact should be zero.

Position error	Artifact RMS Amplitude ($x10^{-3}$)	Artifact P-P Amplitude ($x10^{-3}$)
<0.5 mm	1.0	2.3
1 mm	3.3	5.7
2 mm	5.5	9.9
3 mm	8.7	15.0

Table 3. Averaged RMS and peak-to-peak amplitude of signal artifact after averaging method for different off-center positioning errors.

5. Loss compensation

This section discusses the power loss during propagation of UWB pulse in breast medium, and the loss models used for compensation. Loss can be contributed by the radial spreading of UWB pulse originating from the antenna as well as attenuation caused by the breast medium. Loss compensation is a signal processing step to equalize all the received signals originating from different locations such that the whole scanning region has unity gain.

5.1 Radial spreading loss

Most studies approximate the propagating signal as a uniform cylindrical wave and thus the radial spreading loss equal to $1/r$, where r is distance from antenna to the particular scanning point. Considering both transmit and receive paths make the loss proportional to distance square. Compensation is done by multiplying the signals by r^2. Figure 8 shows the decrease of reflected signal amplitude from a tumor considering only radial spreading loss. The tumor is located nearest to the antenna at 0 degree and furthest at 180 degrees.

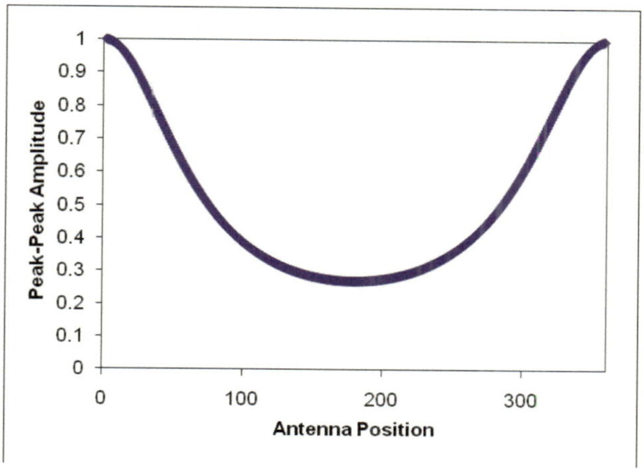

Fig. 8. Simulated received pulse amplitude considering only radial spreading loss.

5.2 Effects of loss compensation

To see the effects of loss compensation, compensation is applied on experiment data to compare the results obtained without compensation applied.

Imaging results in Figure 9 show that loss compensation tends to amplify noise near the phantom boundary. The compensation applied here is only considering radial spreading loss. Worse results will be expected if other loss factors are incorporated, since the signals are multiplied by larger factors.

Radial spreading compensation is a commonly used signal processing step in breast cancer detection algorithms in numerical noise-free modeling. In view of the deteriorating effects of radial spreading compensation on image quality, it is recommended not to apply the compensation.

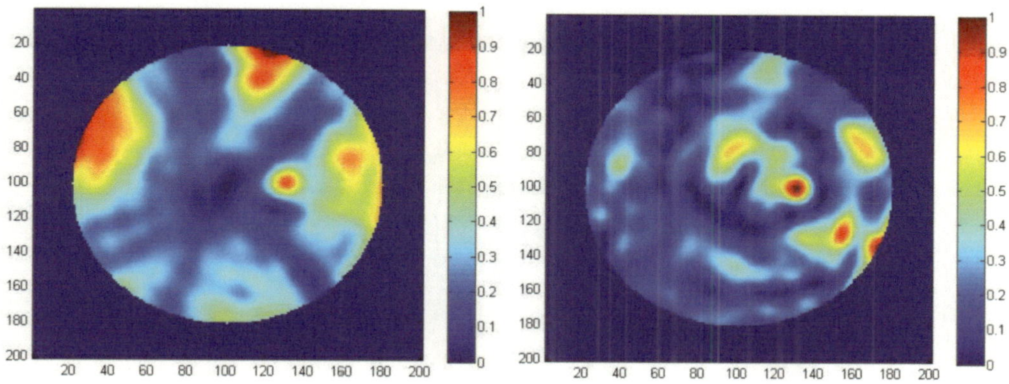

Fig. 9. Image with radial spreading compensation (left) and without radial spreading compensation (right).

6. Filtering and correlation

This section describes two signal processing methods applied to improve the signal-to-noise ratio (SNR) of the breast phantom images.

6.1 Filtering

The ultra-wideband (UWB) antenna used in the experiments has a bandwidth of 1.8 to 6.3 GHz. Two significant narrowband interferences for the experiments conducted are cell phone noise at 1.8 GHz and wireless local area network (LAN) noise at 2.4 GHz. Thus, digital notch filters at 1.8 GHz and 2.4 GHz are applied to reduce the interferences. The signals and power spectra before and after filtering applied are given in Figure 10. The image quality has been improved with filtering as shown in Figure 11.

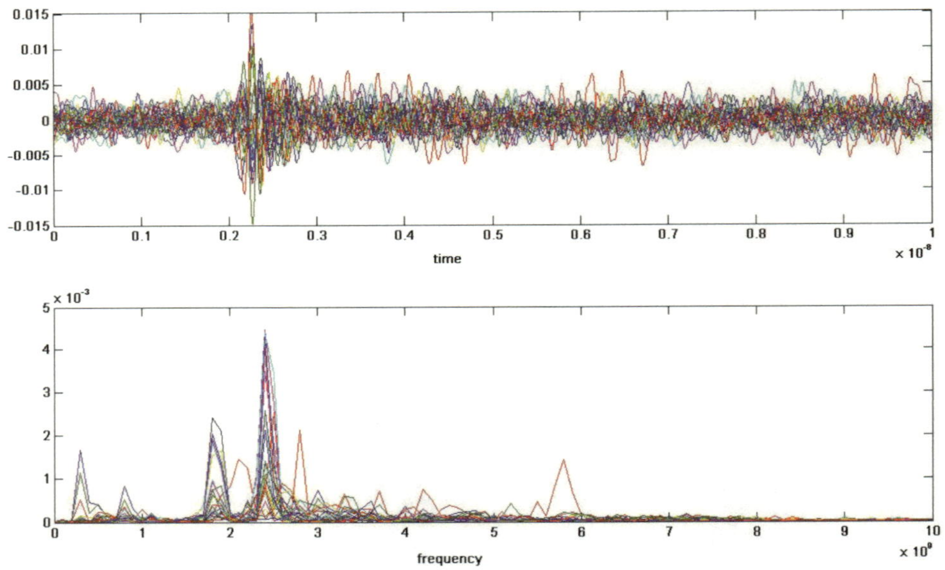

Fig. 10a. Signals (upper trace) and power spectra (lower trace) before filtering.

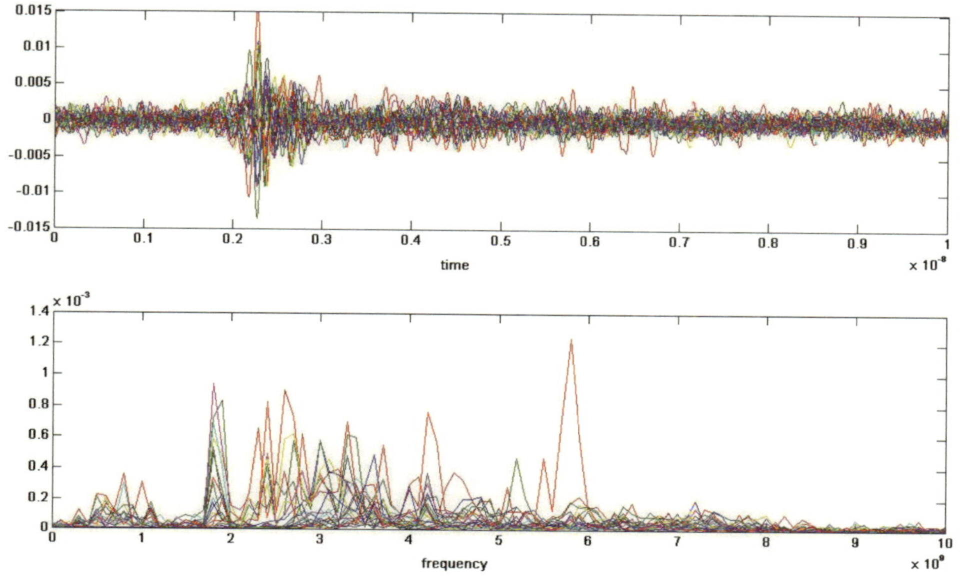

Fig. 10b. Signals (upper trace) and power spectra (lower trace) after filtering.

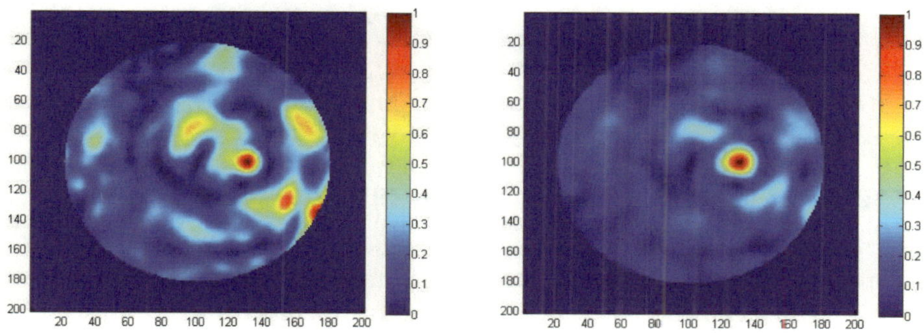

Fig. 11. Images formed without filtering (left) and with filtering applied (right). The tumor is located at 3 cm to the right of the center.

6.2 Correlation

A tumor response template is created in Matlab as shown in Figure 12. Correlation is applied by multiplying the tumor response with the filtered signals after the delay-and-sum operation at each pixel. Then the signals are windowed and summed to give a value for every pixel in the breast phantom image given in Figure 13. The image quality has been enhanced with correlation.

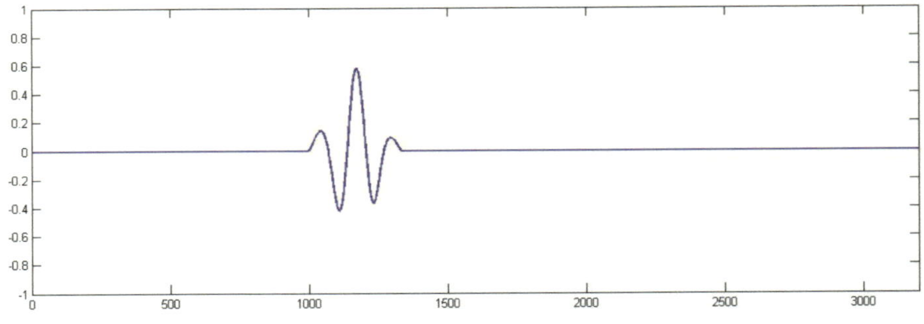

Fig. 12. Tumor response template.

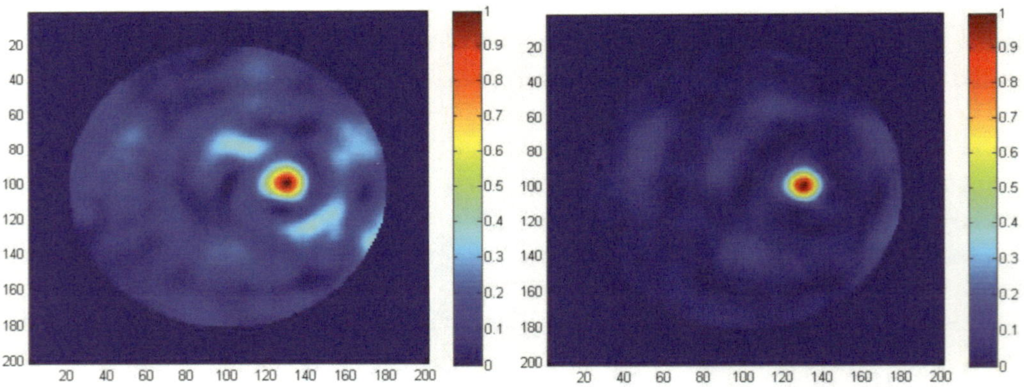

Fig. 13. Comparison for filtered images without correlation (left) and with correlation (right).

The discussion above focuses on only a single experiment data. The result may not be representative as the occurrences of artifacts in images are random due to coherence alignment of noise at certain points. Thus large scale experiments are performed to study the effect of applying filtering and correlation on breast phantom images. Ten experiments are conducted with the same experimental setup and breast phantom. Tumor simulant is at position (130, 100) for all experiments. Experiments are repeated with different antennas array of 6, 12, 24, and 36 elements in one phantom rotation. Forty sets of data are collected and processed. A total of 80 images are formed for SNR analysis. Figure 14 show 20 images from experiments conducted with 24 antennas array.

SNR is calculated from the ratio of the pixel value at tumor location (130, 100) over the highest value of noisy pixels 6 mm radius outside the tumor location. Since the SNR value from individual experiment is highly variable, due to random occurrences of artifacts, an averaged SNR value is calculated using ten SNR values from ten experiments.

Detectability is the ability to observe the tumor in the images although the tumor may not appear as the strongest pixel. Detectability is defined as one when the tumor pixel value is above half of the maximum scale and above twice of the adjacent region pixel values, otherwise zero is given for that image. Table 4 shows that detectability is at least 80% when SNR is positive.

Figure 15 show the SNR versus number of antennas per phantom. There is a significant improvement on SNR with correlation applied. Increasing the number of antennas improves the SNR but cost and complexity of implementation increases. To archive positive SNR, the minimum number of antennas needed without correlation is 12 compared to 10 for correlation.

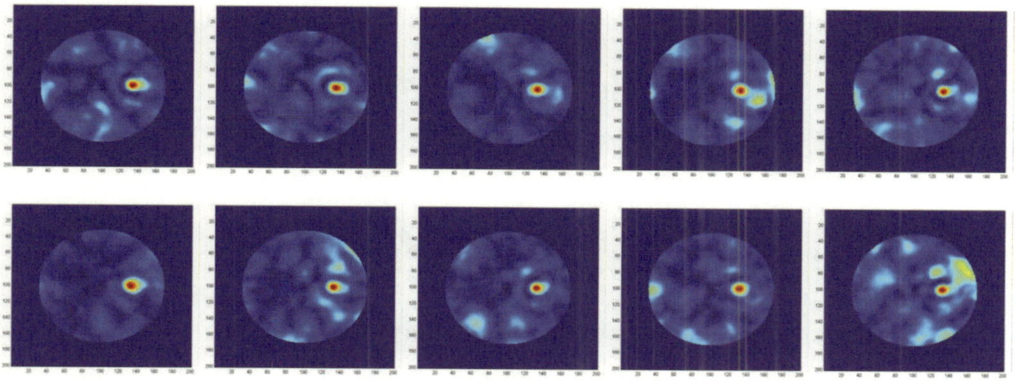

Fig. 14a. Ten filtered images without correlation applied.

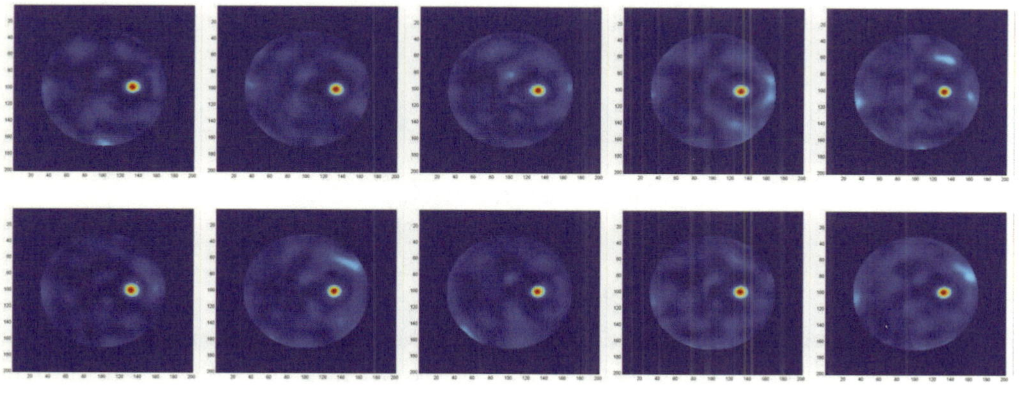

Fig. 14b. Ten filtered images with correlation applied.

SNR	Array of 6	Array of 12	Array of 24	Array of 36
Delay & Sum	-8.7 dB	-3.6 dB	-0.3 dB	4.3 dB
Filtering	-6.3 dB	0.5 dB	6.1 dB	8.2 dB
Correlation	-5.7 dB	3.5 dB	11 dB	13 dB
Detectability	**Array of 6**	**Array of 12**	**Array of 24**	**Array of 36**
Delay & Sum	0/10	5/10	8/10	10/10
Filtering	1/10	8/10	10/10	10/10
Correlation	2/10	10/10	10/10	10/10

Table 4. SNR and detectability for different antennas arrays.

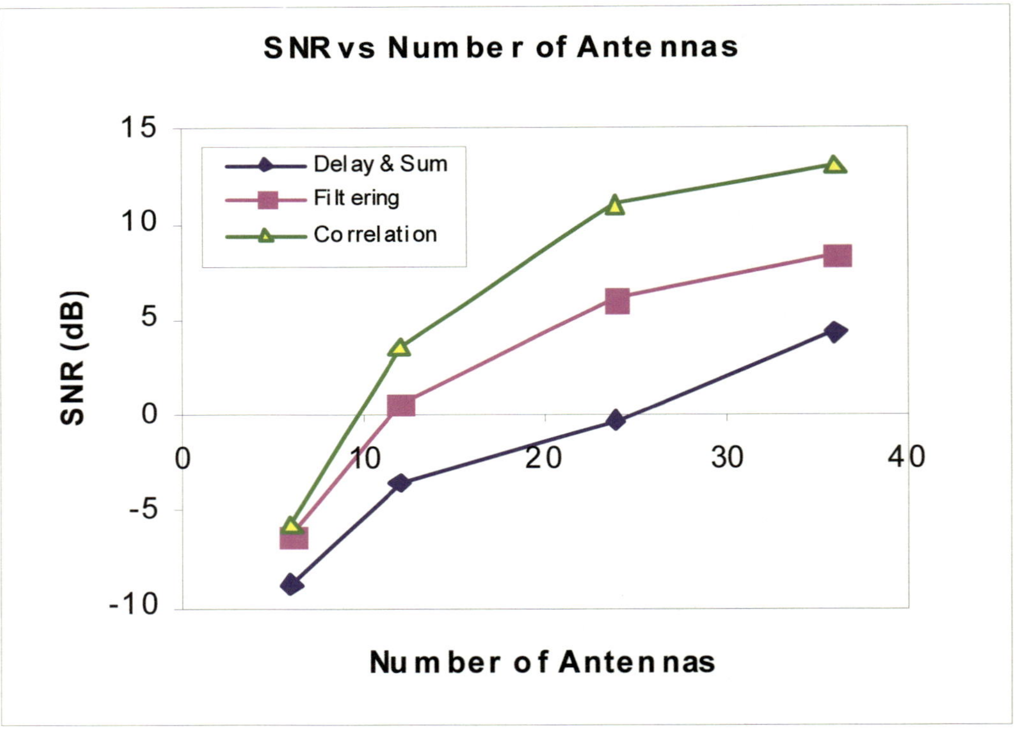

Fig. 15. SNR versus number of antennas.

To study the robustness of the correlation method, the tumor stimulant is positioned at different distances, 1 cm to 6 cm, from the center of the breast phantom with the experimental results given in Figure 16. Images in Figure 16 show that correlation improves the image quality.

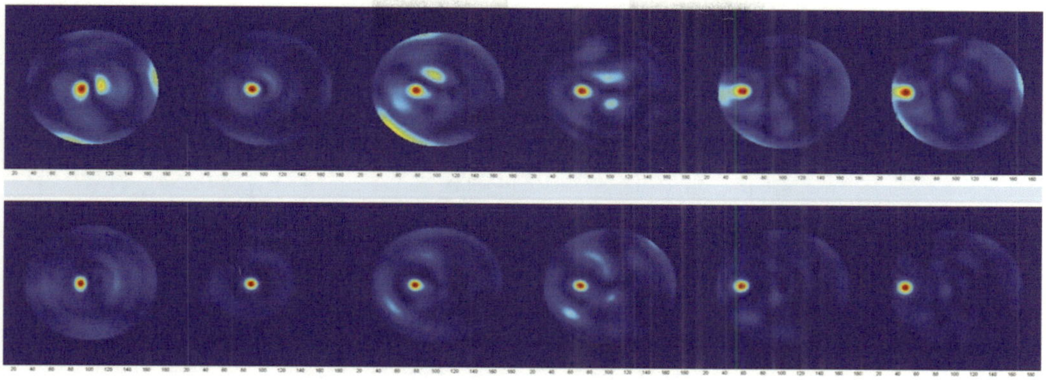

Fig. 16. Images formed without correlation (top) and with correlation (bottom), with tumor located 1 cm to 6 cm from center.

7. Signal and image SNR

This section discusses the effects of improving the image SNR by increasing the averaging number and antenna number. The tradeoff of increasing these two factors is the increase of acquisition time.

In most other studies, an array of only a few antennas are considered as more antennas do not improve the image resolution but at higher simulation time. However, the advantage of more antennas is noise reduction. Better image SNR can be obtained by increasing the averaging number and antenna number.

Some terms used in the discussion are defined as follows:

- Averaging number is the number of waveforms acquired by the oscilloscope to produce an averaged waveform for each acquisition at one antenna position.
- Antenna number is the number of antennas in the synthetic array around the circumference of the phantom. It is also the number of steps for a complete phantom rotation relative to a stationary antenna.
- Signal SNR is the ratio of the root mean square (RMS) value of averaged tumor response to the RMS value of the averaged noise.
- Image SNR is the ratio of the tumor pixel intensity to the highest-value artifact pixel intensity. The definition differs from SCR which is Signal-to-Clutter Ratio, since there is no clutter considered in this discussion. Artifact pixels are caused by coherent summation of noise and occur randomly whereas clutters remain at their positions and averaging is unable to remove them.

7.1 SNR improvement with larger averaging and antenna number

Signal SNR can be improved by using larger averaging number, whereas image SNR can be improved by using larger antenna number.

To measure the noise level, antenna is placed stationary without the presence of breast phantom or any moving object. Incident pulse amplitude is set to different attenuation setting from 0 dB (8 V) to 24 dB (0.5 V). A total of 360 measurements with only incident pulse are taken as in breast scanning. Noise is calculated by subtracting the individual measurement with the average measurements. The RMS and peak-to-peak noise is shown in

Table 5 and Figure 17, which is the average of 360 noise amplitudes relative to the incident pulse amplitude.

Averaging	1	4	16	64	256	1024	4096
RMS noise	4.8 mV	1.7mV	0.83 mV	0.45 mV	0.24 mV	0.15 mV	0.10 mV

Table 5. RMS amplitude of noise for different averaging number with maximum pulse amplitude of 8V (attenuation setting 0 dB).

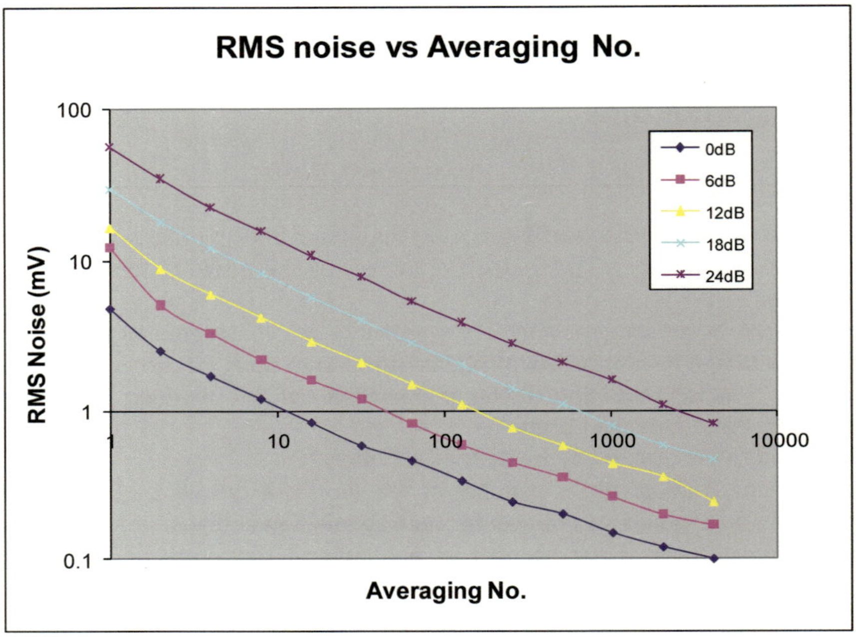

Fig. 17. RMS noise versus averaging number for different pulse attenuation settings.

Matlab simulations are conducted to determine the image SNR. Ten sets of UWB noise are created by applying a bandpass filter to white noise with all the resultant noise RMS amplitudes set to 0.5 mV. The noise is scaled to the noise RMS amplitude of different averaging numbers as shown in Table 5 and added to the ideal tumor response as shown in Figure 12. The tumor response has peak-to-peak amplitude of 1 mV and RMS of 0.2567 mV. The signals are delayed as though they are received from the 360 antennas spaced regularly around the breast phantom. The same delay factors are used in creating the signals and in subsequent beamforming. Thus there is no error caused by delay estimation. A total of 120 sets of data are created for confocal imaging for each antenna number of 12, 24, 45, 90, 180, and 360.

Since the image SNR is random due to random occurrences of artifacts, ten SNR values are obtained and averaged based on simulation with ten sets of data for each averaging number and antenna number. A total of 720 images are produced from the simulations. The results are summarized in Table 6 and Figure 18.

Antenna No.		12	24	45	90	180	360
Averaging No.	**Signal SNR**	**Image SNR**					
2	**-18.66**	-2.8	-2.8	-2.2	-2.4	-2.0	-1.3
4	**-15.89**	-2.8	-2.8	-2.1	-2.1	-1.4	0.0
8	**-13.4**	-2.7	-2.7	-1.9	-1.7	-0.2	1.7
16	**-10.09**	-2.6	-2.2	-1.3	-0.2	1.9	4.0
32	**-7.08**	-2.3	-1.3	-0.2	1.9	4.0	5.8
64	**-4.68**	-1.9	-0.2	1.3	3.8	5.5	6.8
128	**-2.69**	-1.2	1.1	2.6	5.1	6.5	7.5
256	**-0.11**	0.3	2.8	4.4	6.4	7.4	8.1
512	**2.17**	1.9	4.3	5.7	7.2	8.0	8.6
1024	**3.58**	2.9	5.2	6.4	7.7	8.3	8.8
2048	**6.6**	4.8	6.5	7.5	8.4	8.8	9.1
4096	**8.19**	5.6	7.1	8.0	8.7	9.0	9.2

Table 6. Image SNR for different averaging number and antenna number. Signal SNR is calculated with tumor response RMS of 0.2567 mV.

Images from the simulations are shown in Figures 19, 20, and 21 for different sets of noise, noise amplitudes, and antenna numbers. As observed from Figure 18, the cost to increase the image SNR is getting higher when the signal SNR is positive. When the signal SNR is negative, the cost to increase the image SNR is lower. In other word, it is worth using array of more antennas when the tumor response received is weaker than the noise.
In this simulation, the amplitude of the tumor response is fixed at 1 mV, with RMS value of 0.2567 mV. The tumor response is 1000 times smaller than the 1V incident pulse. In this case, we could use antenna number of 12 and averaging number of 1024 to produce reasonably good results.

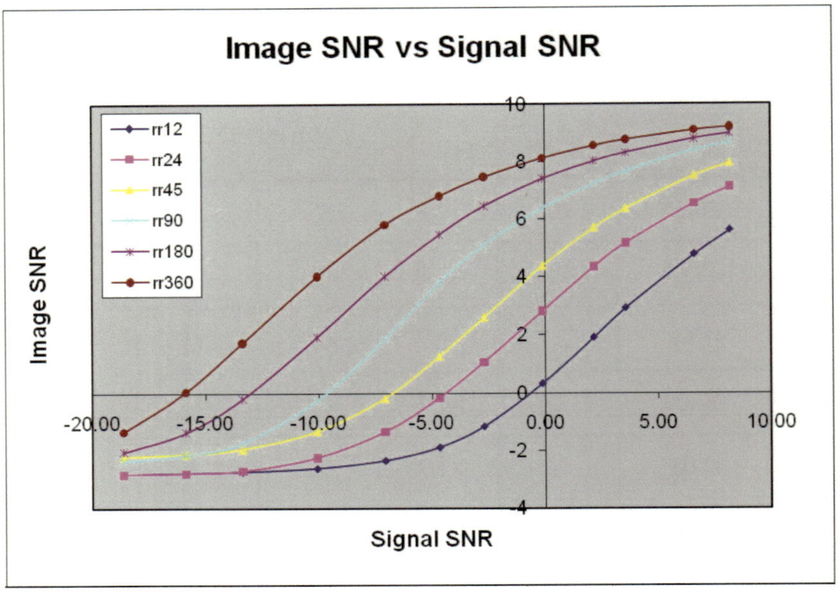

Fig. 18. Image SNR versus Signal SNR for different antenna numbers.

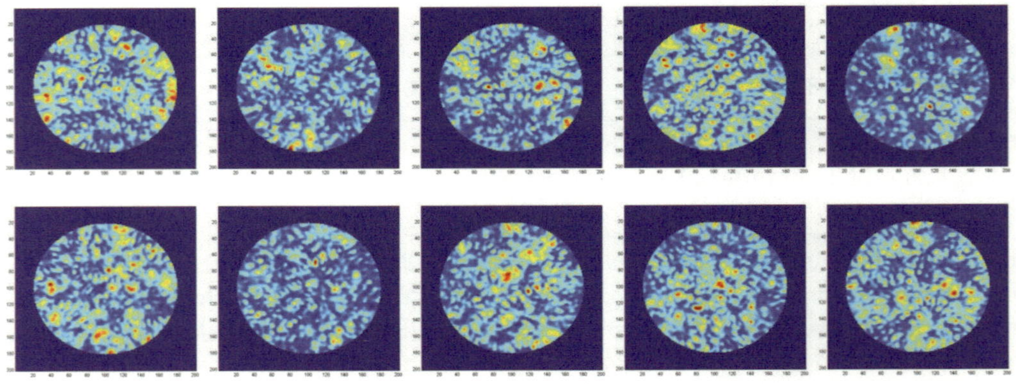

Fig. 19. Images obtained with ten sets of noise processed with antenna number of 360 and RMS noise level of 2.2 mV.

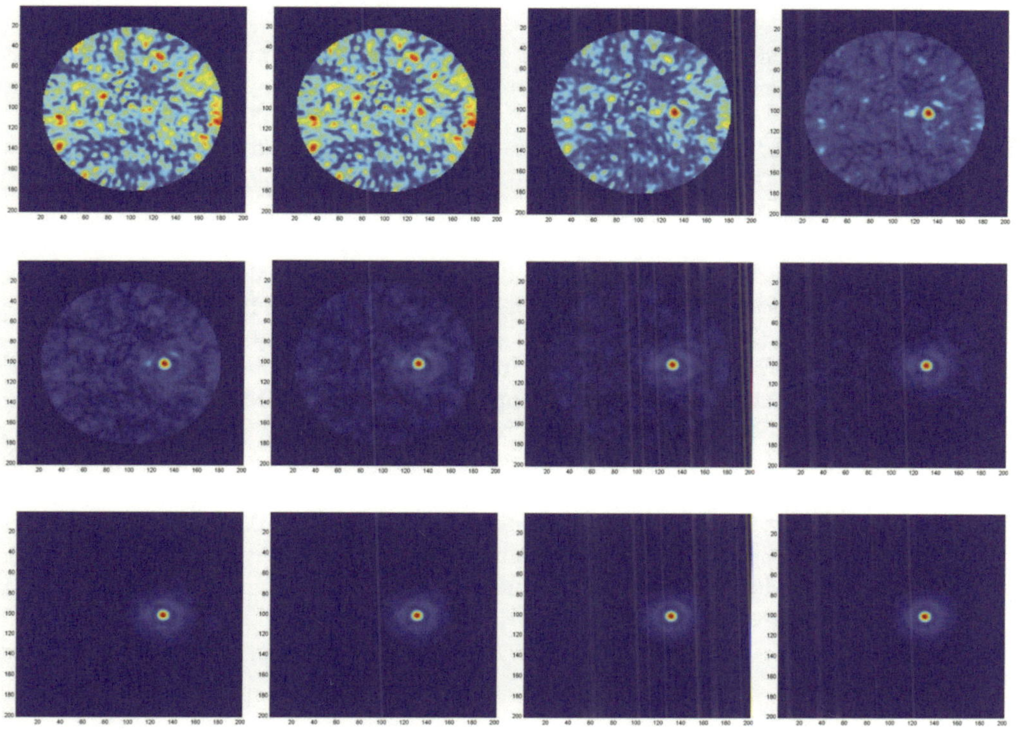

Fig. 20. Images obtained with the same set of noise processed with antenna number of 360 and RMS noise level from 2.2 mV (top left) to 0.10 mV (bottom right).

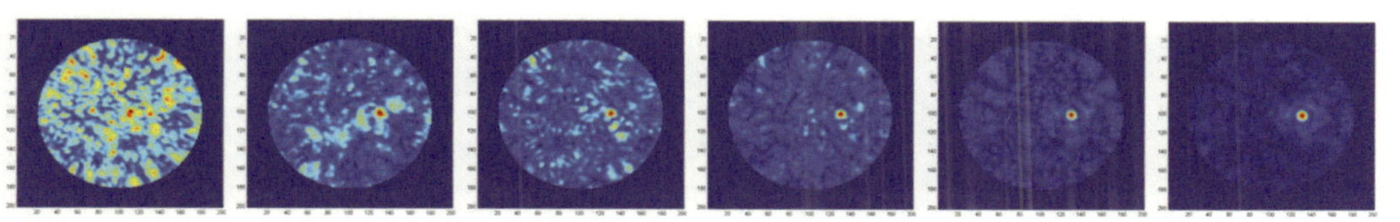

Fig. 21. Images obtained with the same set of noise processed with antenna number of 12 (top left), 24, 45, 90, 180, 360 (bottom right) and RMS noise level of 0.45 mV.

8. Conclusion

In this chapter, experimental study of UWB breast cancer detection in time domain and several important experimental issues are discussed. These include pulse jitter artifact, dynamic range of oscilloscope, ring artifact caused by positioning error, noise amplification caused by radial spreading loss compensation, noise reduction after applying filtering and correlation, as well as the improvement on signal SNR and image SNR by using larger averaging and antenna number. The identified issues and compensation methods will facilitate the future experiments in UWB breast cancer detection with more realistic breast phantoms.

9. References

Bond E. J., Li X., Hagness S. C., & Van Veen B. D. (2003). Microwave Imaging via Space-Time Beamforming for Early Detection of Breast Cancer, *IEEE Transactions on Antennas and Propagation*, vol. 51, no. 8, pp. 1690-1705, ISSN 0018-926X

Chua L. W. (2005). A new UWB antenna with excellent time domain characteristics, *Proceedings of The European Conference on Wireless Technology*, ISBN: 2-9600551-1-X, Paris, October 2005.

Lazebnik M., Madsen E. L., Frank G. R., and Hagness S. C. (2005). Tissue-mimicking phantom materials for narrowband and ultrawideband microwave applications, *Physics in Medicine and Biology*, vol. 50, no. 18, pp. 4245-4258, ISSN 0031-9155

Sill J. M. & Fear E. C. (2005). Tissue sensing adaptive radar for breast cancer detection - experimental investigation of simple tumor models, *IEEE Transactions on Microwave Theory and Techniques*, vol. 53, no. 11, pp. 3312-3319, ISSN 0018-9480

Xu Li & Hagness S. C. (2001). A Confocal Microwave Imaging Algorithm for Breast Cancer Detection, *IEEE Microwave and Wireless Components Letters*, vol. 11, no. 3, pp. 130-132, ISSN 1531-1309

Xu Li, Davis S. K., Hagness S. C., van der Weide D. W., & Van Veen B. D. (2004). Microwave Imaging via Space-Time Beamforming: Experimental Investigation of Tumor Detection in Multi-Layer Breast Phantoms, *IEEE Transactions on Microwave Theory and Techniques*, vol. 52, no. 8, pp. 1856-1865, ISSN 0018-9480

16

Frequency Domain Skin Artifact Removal Method for Ultra-Wideband Breast Cancer Detection

Arash Maskooki, Cheong Boon Soh, Erry Gunawan and Kay Soon Low
Nanyang Technological University
Singapore

1. Introduction

According to the Center for Disease Control and Prevention (CDC) (Center for Disease Control and Prevention (CDC), 2007) and American Cancer Society (ACS) (American Cancer Society (ACS), 2007), breast cancer is the most prevalent type of cancer among women only after skin cancer and the number two cause of cancer death in women. Among every 8 women in U.S. one has breast cancer. In addition, about 2.5 million survivors of breast cancer are living in the United States and it is estimated that 40,170 lives perished due to this mortal disease in 2009. Many risk factors have been associated with breast cancer including, but not limited to, gender, age, family history, race and lifestyle (American Cancer Society (ACS), 2007). Some of these risk factors could not be controlled such as age or family history. In contrast, factors such as life style could be changed to prevent or reduce the chance of breast cancer. Various techniques are practiced today for treating breast cancer. Nevertheless, in all treatment methods early detection is a crucial factor. Detecting breast cancer in early stages can lead to more efficient treatment and comfort for the patient. Hence, an accurate diagnosis modality for breast cancer detection can save many lives annually.
Currently, X-Ray mammography is the gold standard method in detecting breast cancer and is used widely due to relatively easy access to the equipment. In this method, a low dose of X-Ray is used to obtain a mapping of the distribution of the tissue density inside the compressed breast. Mammography can be used for both screening and diagnosis of breast cancer. In screening mammography, X-Ray imaging of the breast is used in women with no signs of the breast cancer for the purpose of investigation of the breast tissue changes and examining the lesions inside the breast. As a diagnostic tool, X-Ray mammography could be used to assess the malignancy of the lesions that has already been found in the breast. It has been shown that mammography can reduce the mortality rate due to breast cancer (J. Elwood and B. Cox and A. Richardson, 1993) because detecting breast cancer in the early stages can lead to a more efficient treatment. Although X-Ray mammography has benefits such as ease of use, availability and being reasonably accurate, it is not a perfect method for screening and diagnosing breast cancer. A major drawback of X-Ray mammography is the danger of exposing the patient to a low dose of ionizing radiation which can increase the risk of cancer. This makes X-ray mammography a less suitable tool for screening and more suitable for diagnosis. In addition, to avoid a blurred image and preserve the uniformity of the tissue, the breast is compressed before X-Ray imaging which can be uncomfortable for the patients (Fear et al., 2002).

Another modality based on Ultra-Wide Band (UWB) pulse confocal imaging radar for screening and possibly diagnosis of breast cancer has been suggested and investigated in recent years (Li & Hagness, 2001). UWB imaging of the breast employs malignant tissue's significant dielectric contrast to detect the location of the tumor relative to an antenna array. To obtain a 2-D or 3-D (depending on the application) map of dielectric contrast inside the breast medium, an array of antenna around the breast transmit a UWB pulse sequentially and record the backscattered signal. In the next step, the backscattered signals are synthetically focused at a synthetic focal point. To focus the beam at the synthetic focal point, the travel time between the antenna and the synthetic focal point is calculated and the energy received from that point in all channels in the antenna array are added together. Existence of a strong scattering point (high contrast in dielectric value) at the synthetic focal point results in a coherent summation of pulse energy and hence a high total energy value assigned to that point. On the other hand, if no scattering point is present at the synthetic focal point, energy values will add incoherently and the total energy value assigned to that location won't be so high. To produce a 2D or 3D image of the breast region, the synthetic focal point is scanned through all the breast region and the summation of the acquired energy values are mapped as the pixel value at that location.

In contrast to the X-ray mammography, UWB confocal imaging does not expose the patient to any ionizing radiation. This means that it can be used frequently as an screening tool specially for younger women. In addition, the uncomfortable breast compression is eliminated is this modality.

Although it has been shown that the cancerous tumors can be located by UWB breast imaging, there are still issues that must be addressed. One of these issues is the problem of the surface skin backscatter. Human breast skin has a high dielectric contrast with both air and skin tissue. This will produce a strong backscatter from the skin surface in the received signal. Although skin backscatter has its major components in the early time of the received signal, it continues to affect parts of the late time response. The problem is that the late time effects of the skin backscatter are still strong enough to mask the weak tumor reflection completely. Hence, the skin effects should be removed from the response prior to the imaging process. A simple method to remove the skin artifact is to mask out the early time part of the signal. However, as mentioned before, skin artifact continues its effect to the late time response and masking out the early time part of the signal would not remove the late time effects. Another problem of the simple masking method is that prior to the imaging process the location of the tumor is not known. Hence, masking the early time part of the signal might remove the tumor reflection. A more practical method has been suggested by researchers based on averaging. In averaging method, the backscattered signal is averaged over all signals from different antennas in the array. The average is then subtracted from the individual signals. The idea is that the skin reflection would appear in the average because it is similar in all channels and will add coherently, but the tumor response has different locations in different channels and would not add up coherently and hence would not appear in the average. Although averaging can significantly suppress the early time portion of the skin reflection, it is not able to remove the late time parts of the artifact. In addition, averaging removes parts of the reflection from the tumor especially in cases where the tumor is equidistant from different antennas. As an improvement over averaging, an averaging based filtering process is proposed in literature to remove the skin artifact. Although filtering greatly improves the skin backscatter removal, it still deteriorates the tumor response. Furthermore, the same as averaging method, it will remove the tumor reflection together with the skin artifact when the tumor is equidistant to several antennas in the array.

This chapter introduces a frequency domain approach to the skin artifact removal problem in UWB breast cancer detection. In this approach, the received signal is first converted into the frequency domain. Then, the skin related information is identified and removed from the frequency domain response. The signal is then converted back to the time domain and used for the imaging process. This approach does not affect the tumor response as it processes each signal individually and does not incorporate any subtraction or addition to the signals. Besides, the skin reflection is eliminated from both early time and late time responses. In addition, it is shown that the frequency domain approach can preserve the tumor response when the tumor is located equidistant to all antennas in the array, such as central axis of the array. Using computer simulations, the frequency domain approach is compared with the existing averaging based methods in different scenarios and is shown to be able to outperform the current existing methods.

This chapter is organized as follows. The next section provides a brief overview of the confocal imaging method in UWB breast cancer detection. In Section 3, the skin artifact problem and existing methods to overcome this problem are described. Section 4 introduces the frequency domain approach to remove the skin artifact. In Section 5, performance of the frequency domain approach is compared with the existing methods through computer simulations in different cases and each case is discussed in detail. Finally, Section 6 will conclude this chapter.

2. Confocal Microwave Imaging (CMI) in UWB breast imaging

To obtain the tumor signature, breast is illuminated by a UWB microwave pulse from a number of antennas arranged in a planar configuration over the breast or in a circular configuration around the breast (Fear et al., 2002). Due to the contrast between dielectric properties of the tumor and the normal breast tissue, malignant tumors have a significant scattering cross section compared to the normal tissue and benign tumor types (Hagness et al., 1998). To obtain a high resolution in the imaging process, the pulse bandwidth has to be as high as possible. However, frequencies higher than 10 GHz will face a dramatic attenuation inside the human body and the breast medium. Hence, the upper limit for frequency content of the signal is about 10 GHz. This limit ensures enough penetration for human breast imaging (Fear & Stuchly, 2000).

To obtain enough penetration for the UWB imaging of the breast, the antenna is excited by a differentiated Gaussian pulse (Fear et al., 2002):

$$V(t) = (t - t_0)e^{-(\frac{t-t_0}{\tau})^2} \tag{1}$$

where $t_0 = 4\tau$ and $\tau = 0.0625ns$. The resulting signal has temporal duration (full-width half-maximum) of 0.17 ns and approximately 6 GHz bandwidth with maximum power near to 4GHz. Figure 1 shows the temporal pulse shape and its frequency spectrum (Fear et al., 2002). The heterogeneous nature of the breast medium produces clutter in the UWB signature of the tumor which makes detecting the tumor difficult or in some cases impossible. To overcome this problem, a modified version of UWB Ground Penetrating Radar (GPR) is suggested for breast medium to amplify the tumor signature and eliminate the clutter (Hagness et al., 1998). In this method, clutter suppression is carried out by coherent addition of the tumor responses from all channels. To spatially focus the signal at the candidate location, the backscattered

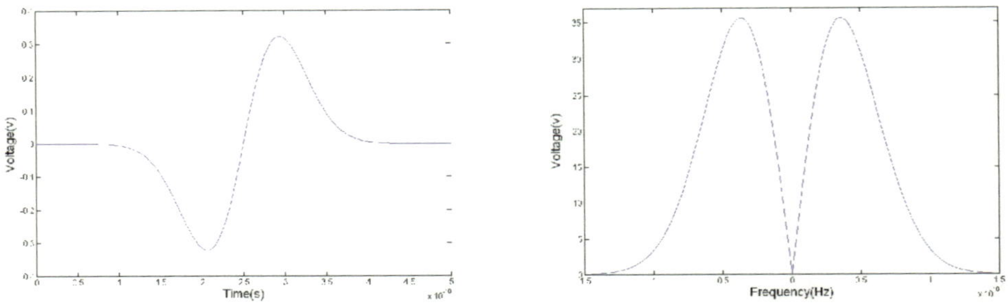

(a) Differentiated Gaussian Pulse

(b) Frequency Spectrum of Differentiated Gaussian Pulse

Fig. 1. Pulse Waveform

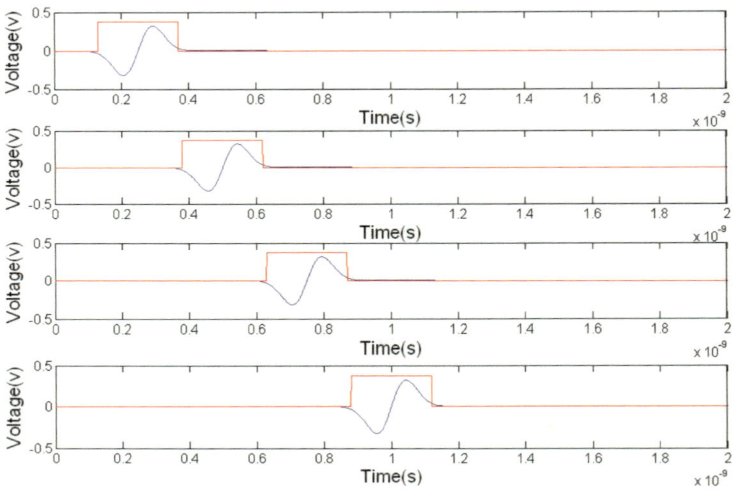

Fig. 2. A time window is applied on different channel responses in Confocal Imaging Process

signal is time shifted and summed based on the pulse travel time to the synthetic focal point. The energy value of the windowed portion of the signals from all the channels are calculated and summed up to form a pixel value for the image of the breast. The synthetic focal point is then scanned through the entire breast medium by appropriately changing the time shifts and windowing to achieve a 2D or 3D image of the breast. As mentioned before, existence of a strong scatterer at the synthetic focal point would produce a replica of the pulse at the corresponding time instance in the backscattered signal in all channels. Hence, summation of this portion of the response from all channels yields a high energy value which points to the existence of a scatterer. However, if only clutter and noise are the dominant components at the focal point, the energy levels do not add coherently and the clutter will be suppressed (Fear & Stuchly, 2000). Figure 2 depicts the windowing and summation process. The window length is chosen equal to the temporal duration of the original pulse to include the whole backscattered energy.

3. Skin backscatter removal

The signal received by the antenna contains contributions from lesions in the breast, clutter due to heterogeneity of the tissue and skin backscatter. The skin backscatter artifact is in the orders of magnitude larger than the other parts of the signal and can swamp the tumor response. Thus, to detect the tumor, skin response should be removed from the signal. This section describes and reviews the existing skin artifact removal algorithms.

3.1 Time window

The simplest way to get rid of the skin artifact is to mask out the early time part of the response. This method could be considered as multiplying a time window which is a step function of zero amplitude when the skin response exists and amplitude one in the rest of the time interval (Zhi & Chin, 2006). However, this method could not be used in practice. The problem is that there is no clear border between early time and late time parts of the backscattered signal. In addition, skin reflection can be present in the late time response and mask the tumor backscatter. Hence masking the early time part of the signal may remove the tumor response as well. This is because the location of the tumor reflection is not a priori information.

3.2 Averaging

A more practical method to remove the skin reflection from the backscattered signal is to use the average of all channels to reconstruct and remove the skin reflection (Li & Hagness, 2001). In this method, the average of the backscattered responses from several antenna elements is subtracted from the response of each element. The idea in averaging method is that the skin response is similar in different channels. This is because the skin structure is assumed symmetric in the view of all elements of the antenna array. To compensate for the variations in each element's distance from skin surface, all responses are aligned to the local maximum of the early time response (which corresponds to the skin backscatter). When the signals in different channels are averaged, the similar parts such as skin backscatter will add up coherently and hence appear in the averaged signal. However, other reflections such as tumor and clutter backscatters differ in amplitude and location from channel to channel. The tumor reflection is at different time instants in different channels. This is because of the difference in the relative distance of the tumor to each antenna element. Hence the tumor signature will add incoherently in the averaging process and is suppressed in the average signal. Thus, removing the average from the signal removes mostly the skin artifact.

Averaging best works when the skin reflection is similar in all channels. In such case, it can totally remove skin artifact from the data. Nevertheless, skin structure is not the same all over the breast and this affects the similarity of the skin reflections at different receivers. One of the factors that changes the skin response in different locations is the variation in skin thickness in different parts of the breast (Ulger et al., 2003). Another contributing factor is the heterogeneity of the breast tissue under the skin surface which can cause variations in the skin backscatter.

As mentioned above, skin response does vary from channel to channel and this variation affects the ability of the averaging in removing the skin artifact thoroughly. Another inherent problem of the averaging method is the distortion to the possible tumor reflection. This is because during the averaging process the tumor and clutter responses from all channels are added and averaged incoherently. This average is then subtracted from each channel and consequently an unwanted averaged version of the clutter and tumor response is added to each channel which deteriorates the tumor signature and is undesirable.

3.3 Weighted average (Space-Time filter)

An improvement over averaging method is the Space-Time filter proposed in (Bond et al., 2003). In this method, a weighted average algorithm makes use of the similarity of the early time responses in antenna elements to estimate and remove the artifacts in this portion of the received signal. To do this, a coefficient is assigned to each channel response and then averaging is performed through all the channels using the filter coefficients. The filter weights assigned to each channel is chosen to minimize the mean-square error over the early time response, where only the skin reflection exists. The following paragraphs discuss this method in detail (Bond et al., 2003).

To remove the skin artifact from the first channel out of N channel responses, the filter weights could be derived as follows. Suppose that the skin artifact at n^{th} sample of the first channel is to be removed. A window of length $2J+1$ is used to select part of each channel response (except for the first channel) centered at the sample number n. Define $\mathbf{b}_i[n]$, a $(2J+1)\times 1$ vector of these samples from the ith channel as

$$\mathbf{b}_i(n) = [b_i(n-J), \cdots, b_i(n), \cdots, b_i(n+J)]^T, \; 2 \le i \le N \tag{2}$$

Vector $\mathbf{b}_{2N}[n]$ is formed by concatenation of $\mathbf{b}_i$ vectors of channels 2 through N as

$$\mathbf{b}_{2N}(n) = [\mathbf{b}_2^T(n), \cdots, \mathbf{b}_N^T(n)]^T \tag{3}$$

Let $\mathbf{q}_i$ be the filter coefficients in the ith channel; similar to (3), $\mathbf{q}$ is formed as

$$\mathbf{q} = [\mathbf{q}_2^T, \cdots, \mathbf{q}_N^T]^T \tag{4}$$

The error function for this filter is:

$$e = \sum_{n=n_0}^{n_0+m-1} |b_1(n) - \mathbf{q}^T\mathbf{b}_{2N}(n)|^2 \tag{5}$$

In (5) the time interval $n_0 \le n \le (n_0+m-1)$ is part of the signal that contains no backscatter from lesions. The solution to these filter coefficient values are chosen to minimize the error function defined in the (5) which is (Haykin, 1996):

$$\mathbf{q} = \mathbf{R}^{-1}\mathbf{p} \tag{6}$$

where,

$$\mathbf{R} = \frac{1}{m}\sum_{n=n_0}^{n_0+m-1} \mathbf{b}_{2N}(n)\mathbf{b}_{2N}(n)^T \tag{7}$$

and,

$$\mathbf{p} = \frac{1}{m}\sum_{n=n_0}^{n_0+m-1} \mathbf{b}_{2N}(n)b_1(n). \tag{8}$$

High degree of similarity in early time response among the channels, results in an ill-conditioned $\mathbf{R}$ matrix. To prevent error in calculation of $\mathbf{R}^{-1}$ in (6), $\mathbf{R}$ is approximated by reduced rank $\mathbf{R}_p$ (Bond et al., 2003)

$$\mathbf{R}_p = \sum_{i=1}^{p} \lambda_i \mathbf{u}_i \mathbf{u}_i^T \tag{9}$$

where λ_i is the ith of the first p significant eigenvalues of $\mathbf{R}$ and $\mathbf{u}_i$ is the corresponding eigenvector. $\mathbf{R}_p^{-1}$ is determined by

$$\mathbf{R}_p^{-1} = \sum_{i=1}^{p} \frac{1}{\lambda_i} \mathbf{u}_i \mathbf{u}_i^T. \tag{10}$$

Substituting (10) in (6) yields the filter coefficients. These coefficients could later be used to remove skin response in channel 1 by subtracting the weighted average of $N-1$ other channels as follows

$$b(n)_1 = \sum_{n} |b_1(n) - \mathbf{q}^T \mathbf{b}_{2N}(n)|^2 \tag{11}$$

The same concept can be used for the other channels.

This algorithm improves the former averaging method in removing the skin response effectively (Zhi & Chin, 2006). Although weighted average can remove the skin artifact more effectively than the simple averaging, it still adds unwanted averaged data from clutter and tumor responses from all signals to each channel response and hence deteriorates the tumor signature (Bond et al., 2003). Another major drawback of this method is the need for a criterion to decide on the moment when the early time response ends. This is because the tumor response should not be included in the MSE minimization used to define filter coefficients otherwise it will be removed from the output. However, the tumor response location is determined in the detection process which is performed after skin removal step and is not a priori knowledge.

4. Frequency domain approach for skin backscatter removal

4.1 Introduction

According to the Geometrical Theory of Diffraction (GTD) (Keller, 1958; Naishadham & Piou, 2004), backscattered signal from a number of scattering points has a pattern in frequency domain. As described in (Cuomo et al., 1999) each scatterer in the view of the antenna will create a harmonic term (pole) in frequency domain response of a UWB pulse radar. This means that the frequency response of the backscattered signal is a combination of some complex harmonic terms with different attenuation coefficient and frequencies. As predicted by GTD, the attenuation coefficient is a function of the shape of the object and its frequency is determined by the distance between the scatterer and the receiver. This representation has interesting applications in UWB microwave radars. Fitting an appropriate model to this pattern helps to reconstruct the signal beyond the bandwidth of the system which leads to a higher image resolution without additional hardware cost. Cumo et. al (Cuomo et al., 1999) have used this bandwidth extrapolation method to estimate the frequency response of the backscattered signal in the gap between two different incoherent subbands.

In this section, we exploit this frequency domain representation to identify and remove the harmonic terms produced by the skin, based on the backscattered energy of the object. After removing the skin related information, the frequency response is reconstructed using the mathematical model and subsequently the time domain response is reproduced by the inverse Fourier transform. The reconstructed time domain signal is then used to construct an image of the breast medium.

4.2 Model parameters estimation

As stated before, the frequency response of the signal reflected from a number of scattering points could be represented as the sum of a number of complex sinusoids. The number of these terms equals to the number of the scattering points in the view of the antenna and the multiple scattering effect (Moore et al., 1997). Mathematical model of the frequency domain signal is (Piou, 2005),

$$y(f) = \sum_{i=1}^{N} A_i(f) e^{(j\frac{4\pi}{c}R_i)f} \tag{12}$$

where, f is the frequency, c is the speed of light, N is the number of scattering points and R_i is the range of the ith scattering point. $A_i(f)$ is the frequency dependence function corresponding to the ith scattering point. This frequency dependence function is of the form f^{α} and the exponent α is known for some common scattering mechanisms. For example, a flat plate has $\alpha = 1$ or a sphere will have $\alpha = 0$ (Moore et al., 1997). As stated by Cumo et. al (Cuomo et al., 1999), f^{α} scattering behavior can accurately be estimated by exponential functions over a finite bandwidth interval. Hence, the following discrete model can represent the frequency behavior of the reflected signal given in (12) above.

$$y(k) = \sum_{i=1}^{N} a_i e^{-(\alpha_i + j\frac{4\pi}{c}R_i)k\Delta f} \tag{13}$$

where a_i are constant coefficients of the sinusoids, α_i and R_i refer to the frequency decay/growth factor and the range of the ith scatterer, respectively, and Δf is the sampling frequency. In the rest of this section we provide the formulation to estimate model parameters in (13). More details and derivations are available at (Piou, 2005).

4.3 State - space representation of the signal

From the system theory we know that the following state-space equations hold for input-output relation in a linear system.

$$\begin{aligned} \mathbf{x}(k+1) &= \mathbf{A}\mathbf{x}(k) + \mathbf{B}\mathbf{w}(k) \\ \mathbf{y}(k) &= \mathbf{C}\mathbf{x}(k) + \mathbf{w}(k) \end{aligned} \tag{14}$$

where $\mathbf{x}(k)$ is the state vector, $\mathbf{w}(k)$ is the input vector and $\mathbf{y}(k)$ is the output of the system. $\mathbf{A}$, $\mathbf{B}$ and $\mathbf{C}$ are matrices characterizing the system and define its state-space behavior. The transfer function of the system described in (14) is given in (15).

$$\mathbf{T}(z) = \frac{\mathbf{Y}(z)}{\mathbf{X}(z)} = \mathbf{C}(z\mathbf{I} - \mathbf{A})^{-1}\mathbf{B} + 1 \tag{15}$$

The impulse response of such system in general comprises a number of complex sinusoids or poles of the system which are the roots of the denominator or as seen in (15) are the eigenvalues of **A**, the open-loop matrix of the system. Hence, the output signal of the system $y(k)$ can be written as

$$y(k) = \sum_{i=1}^{M} a_i e^{-(\alpha_i + j\beta_i)k\Delta t} \tag{16}$$

In (16) M is the number of the poles of the system or the eigenvalues of **A**, a_i are the constant coefficients of each complex sinusoid and α_i and β_i are the damping factor and frequency of the ith harmonic, respectively. Δt is the sampling time interval.
Comparing (16) and (13) reveals that the frequency response of the backscattered signal and the impulse response of a linear system have a similar mathematical structure. Thus, we can use the mathematics of linear system identification to estimate the parameters of the frequency model for the backscattered data. In the rest of this section the formulation to obtain model parameters is presented.
Suppose that the frequency response of the backscattered signal is the impulse response of a hypothetical linear system. Here we try to extract the system matrices and consequently a model for the impulse response based on the eigenvalues or poles of this hypothetical system. As mentioned before, we can derive the desired frequency model parameters from this impulse response model.
The process of finding the hypothetical system matrices involves forming forward prediction or Hankel matrix from the sample data of the frequency response of the backscattered signal and deriving **A** through singular value decomposition of **H**, the Hankel matrix which is defined as follows,

$$\mathbf{H} = \begin{pmatrix} y(1) & \cdots & y(L) \\ \vdots & \ddots & \vdots \\ y(N-L+1) & \cdots & y(N) \end{pmatrix} \tag{17}$$

where, $y(i)$ are the samples of frequency domain response of the backscattered data, N is the number of data samples and L is chosen as $N/3$ (Naishadham & Piou, 2004). By singular value decomposition, **H** is decomposed into three matrices,

$$\mathbf{H} = \mathbf{U\Sigma V^*} \tag{18}$$

In which **U** is the left unitary matrix, $\mathbf{V}^*$ is the right unitary matrix, and $\mathbf{\Sigma}$ is a diagonal matrix containing singular values of **H** in a descending order. $(*)$ denotes the complex conjugate and transpose.
Singular values of **H** could be separated into two subspaces, the signal plus noise subspace and the only noise subspace. If the SNR value is high enough there would be a sharp transition between singular values of the signal and those of noise. The criterion for separating the two parts is described in (Naishadham & Piou, 2005). Hence, **U**, $\mathbf{\Sigma}$, $\mathbf{V}^*$ could be divided into two different subspaces as follows,

$$\mathbf{H} = \begin{bmatrix} \mathbf{U_{sn}} & \mathbf{U_n} \end{bmatrix} \begin{bmatrix} \mathbf{\Sigma_{sn}} & 0 \\ 0 & \mathbf{\Sigma_n} \end{bmatrix} \begin{bmatrix} \mathbf{V^*_{sn}} \\ \mathbf{V^*_n} \end{bmatrix} \tag{19}$$

which in (19) the subscripts *'sn'* and *'n'* refer to the signal-noise and noise subspaces respectively. Removing the noise part, $\tilde{\mathbf{H}}$ can be formed as follows,

$$\tilde{\mathbf{H}} = \mathbf{U_{sn}\Sigma_{sn}V^*_{sn}} \tag{20}$$

Using the balanced coordinate method (Piou, 2005), $\tilde{\mathbf{H}}$ could further be factorized as

$$\tilde{\mathbf{H}} = \mathbf{\Omega\Gamma} \tag{21}$$

where

$$\mathbf{\Omega} = \mathbf{U_{sn}}\mathbf{\Sigma_{sn}}^{1/2} \text{ and } \mathbf{\Gamma} = \mathbf{\Sigma_{sn}}^{1/2}\mathbf{V}^*_{\mathbf{sn}} \tag{22}$$

where $\mathbf{\Omega}$ and $\mathbf{\Gamma}$ are the observability and controllability matrices respectively. **A** could be derived from both the observability or the controllability matrices, here $\mathbf{\Omega}$ is used to derive **A**.

$$\mathbf{A} = (\mathbf{\Omega}^*_{-\mathbf{rl}}\mathbf{\Omega}_{-\mathbf{rl}})^{-1}(\mathbf{\Omega}_{-\mathbf{rl}}\mathbf{\Omega}_{-\mathbf{rf}}) \tag{23}$$

where $\mathbf{\Omega}_{-\mathbf{rl}}$ and $\mathbf{\Omega}_{-\mathbf{rf}}$ are obtained by removing the last and first rows of $\mathbf{\Omega}$, respectively. Now α_i and R_i are related to the eigenvalues of **A** by:

$$\alpha_i = \frac{-log|\lambda_i|}{\Delta f} \text{ and } R_i = -c\frac{\Phi_i}{4\pi\Delta f} \tag{24}$$

where R_i is the range and α_i is the damping factor of the sinusoid related to the ith scattering point. Φ_i is the phase of λ_i, the ith eigenvalue of **A**. To find the constant coefficients a_i we use the following equation,

$$a_i = \frac{(\mathbf{Cm_i})(\mathbf{v_i B})}{\lambda_i^{f_1/\Delta f}} \tag{25}$$

where $\mathbf{m_i}$ are eigenvectors of **A**, and $\mathbf{v_i}$ are defined as

$$\mathbf{V} = [\mathbf{m_1} \cdots \mathbf{m_p}]^{-1} = \begin{bmatrix} \mathbf{v_1} \\ \vdots \\ \mathbf{v_p} \end{bmatrix} \tag{26}$$

In (25), **C** is the first row of $\mathbf{\Omega}$, f_1 is the carrier frequency of the pulse and the kth element of frequency vector is related to carrier frequency by

$$f_k = f_1 + (k-1)\Delta f \tag{27}$$

To derive **B**, $\mathbf{\Omega_N}$ is defined,

$$\mathbf{\Omega_N} = \begin{bmatrix} \mathbf{C} \\ \mathbf{CA} \\ \vdots \\ \mathbf{CA}^{N-1} \end{bmatrix} \tag{28}$$

and **B** is obtained by,

$$\mathbf{B} = (\mathbf{\Omega_N^*}\mathbf{\Omega_N})^{-1}(\mathbf{\Omega_N^*}\mathbf{y^T}) \tag{29}$$

where **y** is the vector of the frequency samples of the backscattered data. Now a_i could be derived from (25). Now that all the parameters of the model in (13) are derived the frequency response of the system could be reconstructed.

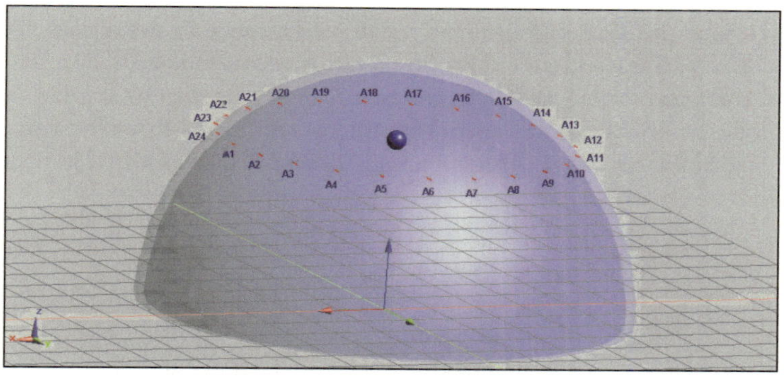

Fig. 3. Antenna array and tumor configuration

5. Evaluation of the performance of the frequency domain skin removal in comparison with other methods

The present section applies the frequency domain skin removal method described in section 4 in different scenarios and compares its performance with other methods. The focus in the first part is more on details of applying the formulation provided in Section 4 on a simplified breast model. The second part will apply the method in a more realistic scenario and compares the results with the other methods.

5.1 Simplified Breast Model

As discussed in Section 4, the backscattered signal of a UWB pulse is the summation of some harmonic terms. The number of these terms depend on the number of scattering points and the multiple scattering effect. Each harmonic term consists of a complex exponential and a coefficient. The argument of this complex exponential is the pole of the hypothetical system mentioned in Section 4. By removing the poles corresponding to the skin reflection from the frequency domain signal, all the skin related information will be removed from time domain. The process is as follows.

The received signals are first converted into frequency domain using Fast Fourier Transform (FFT) algorithm. The frequency domain signals are then processed to extract the model parameters stated in the previous section. Among these parameters, a_is are directly related to the amplitudes of each of the backscattered pulses. This can be explained as follows. In Equation (13), a_i is a complex coefficient which can be written as $|a_i|e^{j\theta_i}$ where θ_i is the phase of a_i. Taking the inverse Fourier transform of Equation (13) yields

$$x(t) = \sum_{i=1}^{N} |a_i| \frac{2\alpha_i cos(\theta_i)}{\alpha_i^2 + (t - 4\pi R_i/c)^2} \tag{30}$$

As seen in Equation (30), $|a_i|$ is proportional to the amplitude of the pulse in the time domain. However, the amplitude of the pulse backscattered from the tumor is much smaller than the skin backscatter. Hence, a threshold could be defined to remove poles with dominant a_i values from the frequency response of the signal. Removing the poles over the stated threshold ensures that only the poles corresponding to the skin will be removed from the signal. This will remove the skin effect both in the early time and the late time responses as the elimination in frequency domain will affect the whole time domain signal. Hence, the tumor reflection will be preserved without the skin late time response interference in the signal. After removing the skin related poles, the frequency domain signal is reconstructed using the mathematical model (13) and then converted back into the time domain using the inverse-FFT algorithm. Hence, the reconstructed signal will only contain contributions from the tumor and clutter. Clutter will be rejected later using confocal imaging algorithm described in Section 1. We will

Ant. No.	x	y	z	Ant. No.	x	y	z
1	35.71	0	35	13	-35.71	0	35
2	34.50	10.24	35	14	-34.50	-10.24	35
3	30.93	17.86	35	15	-30.93	-17.86	35
4	25.26	26.26	35	16	-25.26	-26.26	35
5	17.86	30.93	35	17	-17.86	-30.93	35
6	9.24	35.50	35	18	-9.24	-35.50	35
7	0	35.71	35	19	0	-35.71	35
8	-9.24	35.50	35	20	9.24	-35.50	35
9	-17.86	30.93	35	21	17.86	-30.93	35
10	-25.50	26.26	35	22	25.26	-26.26	35
11	-30.93	17.86	35	23	30.93	-17.86	35
12	-34.50	10.24	35	24	34.50	-10.24	35

Table 1. Antenna Arrangement

first describe the idea in detail using a simplified simulated breast model using SEMCAD X (version 13) software package for an antenna array with 24 elements in a circular configuration around the breast in order to show the ability of the method to remove the skin reflection from the backscattered signal. The breast medium is modeled by a hemisphere with a radius of 50mm and thickness of 2mm as the skin layer. A spherical tumor with a radius of 2mm is placed on the central axis of the hemisphere and at a height of 35mm from the center of the hemisphere (x = 0mm, y = 0mm, z = 35mm). The model and the antenna locations are shown in Figure 3 and Table 1 respectively. The relative permittivities of the skin and breast tissues are set to the values given by (Fear et al., 2002) ($\epsilon_r(skin) = 36$, $\epsilon_r(tissue) = 9$). The dielectric value assigned to the tumor is the measured dielectric value of the malignant tumor $\epsilon_r = 50$ (Fear et al., 2002). Figure 4 shows the signal received in channel 1 and its spectrum. As the

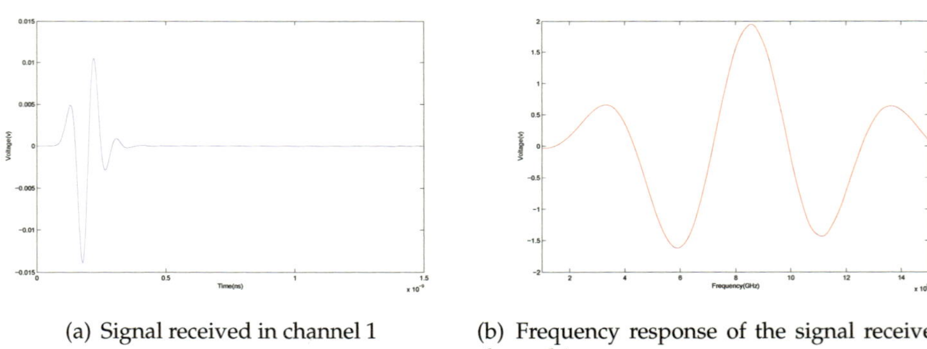

(a) Signal received in channel 1

(b) Frequency response of the signal received in channel 1

Fig. 4. Signal received in channel 1 and its frequency response skin reflects the largest energy among the reflectors in the breast medium, the high energy dominant poles in the frequency domain will correspond to the skin backscatter. Hence a threshold may be used to remove these dominant poles. The threshold is defined based on the ratio of the backscattered energies of the skin to the tumor and is obtained as follows. We fix the threshold value a little higher than the ratio of the largest possible peak tumor to the skin response times the maximum reflection coefficient value a_i. The maximum reflection coefficient corresponds to the largest scatterer which is the skin surface. Hence, by removing all the poles with a_i values larger than this threshold from early time response we make sure that only reflections larger than the tumor reflection is removed from the signal. Many factors can affect the skin to tumor response ratio and more study is needed to consider all the factors affecting this ratio and obtain an optimized threshold value. Here, to show the basic idea of the current method, we consider three factors, tumor size, skin thickness and tumor location to determine the highest possible ratio. To experimentally estimate the highest possible skin to tumor response ratio, the tumor reflection is isolated from the other reflections by performing two different simulations. One simulation is done without the tumor and the second one is with the tumor. Subtracting the results of these simulations yields the tumor signature. According to (Ulger et al., 2003) breast skin thickness varies in the range of 0.5-3.1mm; hence two extreme cases (0.5 and 3.1mm) are simulated in the experiments. The tumor size is set 2mm and 5mm which is well within the range of the early breast cancer. Then the tumor location is varied on the line connecting the center to the antenna location from the center of the breast hemisphere to 5mm below the inner layer of the skin as the tumors so close to the skin can be detected by examining the surface of the breast.

Tables 2 and 3 show the tumor to skin peak response ratio for the skin thickness of 0.5mm and 3.1mm respectively.

Location \ Tumor size	2mm	5mm
Center	9.04E-05	1.81E-04
Under The Skin	3.80E-03	5.20E-03

Table 2. Skin to Tumor Ratio (Skin Thickness: 0.5mm)

Location \ Tumor size	2mm	5mm
Center	7.72E-05	1.69E-04
Under The Skin	2.10E-03	4.00E-03

Table 3. Skin to Tumor Ratio (Skin Thickness: 3.1mm)

As expected, the tumor to skin response ratio increases as the tumor size increases. As seen in the tables, the maximum ratio is obtained when the tumor radius is 5mm and is located 5mm below the skin, the highest tumor to skin response ratio is 0.0021, i.e. the skin reflection is about 476 times stronger than the largest tumor reflection. Hence, by setting the threshold a little larger than 0.21% of the largest reflection coefficient (a_{max}) and removing all the poles with a_i values larger than this threshold from early time response we ensure that all the reflections larger than the tumor reflection is removed from the signal. This would be true in all other cases as we chose the largest possible tumor response to define the threshold.

Here, we chose 0.0025 $\times$ the largest reflection coefficient as the threshold value. The poles extracted from the signal in channel 1 are shown in Table 4; Eliminated poles are indicated by a '*'.

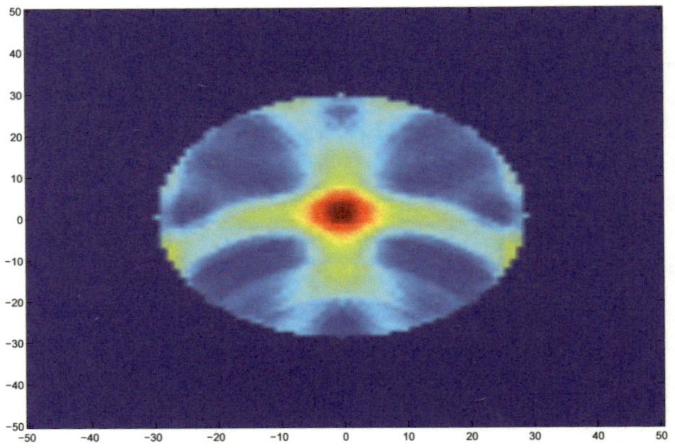

Fig. 5. Confocal imaging of the breast after removing the skin reflection

Figure 6 shows the backscattered signal after removing the skin reflection. In the figure, solid line represents the reconstructed signal super-imposed with the original signal represented by the dotted line. As seen in the figure the skin backscatter is removed from the signal. Figure 7

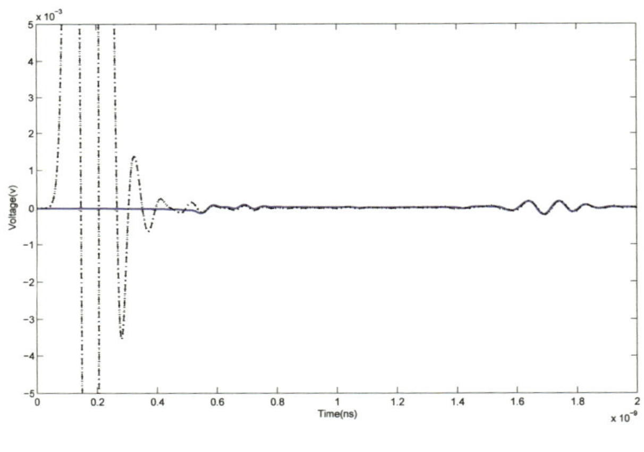

Fig. 6. The dotted line shows the original signal and the solid line is the signal after skin backscatter removal.

shows a larger view of the late time part of the response. As shown in the figure, the late time part of the signal, where the tumor response exists, has not been affected significantly.

5.2 Comparison with the Averaging and Weighted Average Methods

In this section, the performance of the frequency domain method is compared with the averaging (Li & Hagness, 2001) and weighted average filter (Bond et al., 2003). To make the breast model more realistic, the mapping of the dielectric values inside the breast medium is obtained from an MRI image of a real breast as shown in Figure 8. The clutter produced due to the heterogeneity of the breast tissue has significant effect on the effectiveness of the skin subtraction methods. In the averaging based methods, the averaged clutter from all other

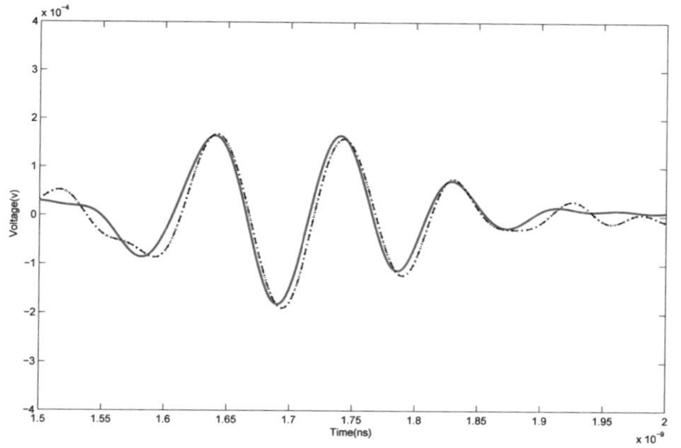

Fig. 7. The late time response of the reconstructed signal (line with dots) vs. the original signal (solid)

Pole No.	Reflection Coefficient	Pole No.	Reflection Coefficient
1	0.006030024	16	0.000576253
2	*0.330168871	17	0.000817322
3	*2.769236715	18	0.001800899
4	*8.978906551	19	0.003023339
5	*17.21322261 (MAX)	20	0.002570893
6	*16.81465757	21	0.000965909
7	*7.610199166	22	0.00044818
8	*1.504612836	23	0.000653675
9	*0.255119890	24	0.001118491
10	*0.052006092	25	0.000111348
11	0.005144400	26	0.00232989
12	0.005526546	27	0.000489255
13	0.004168027	28	0.010678232
14	0.001581127	29	0.022283477
15	0.001108609	30	0.022266937

Table 4. Reflection Coefficients(Eliminated poles are identified by *)

channels is added to each channel and makes the tumor detection even more difficult. As seen in Figure 8, different dielectric constants of the breast internal regions appear as varying intensities in the gray scale image. The scale for this mapping is given beneath the image. Here, the regions with different dielectric values are approximated by spheres. The radius of the sphere is chosen as the circumference of the region divided by 2π. The center of the spheres are located at the same height and distance as the center of the corresponding region from the center of the breast. Assume that the vertical axis in Figure 8 is z and the horizontal axis is x in the Cartesian coordinates. In this configuration, y would have an inward direction perpendicular to the xz plane. To make the model 3D, the angle ϕ_i (between the position vectors of ith sphere center and x axis) are chosen randomly in the interval $[-\pi, \pi]$. In this experiment, the tumor coordinates are x=0, y=0, z=35 (mm). Figure 9 shows the model obtained. The locations of the sphere centers are given in Table 5.

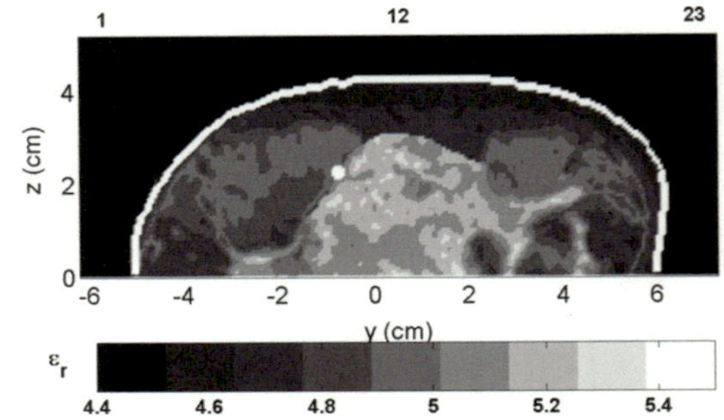

Fig. 8. 2D mapping of the dielectric values of the different regions of the breast tissue(source:(Kosmas & Rappaport, 2005))

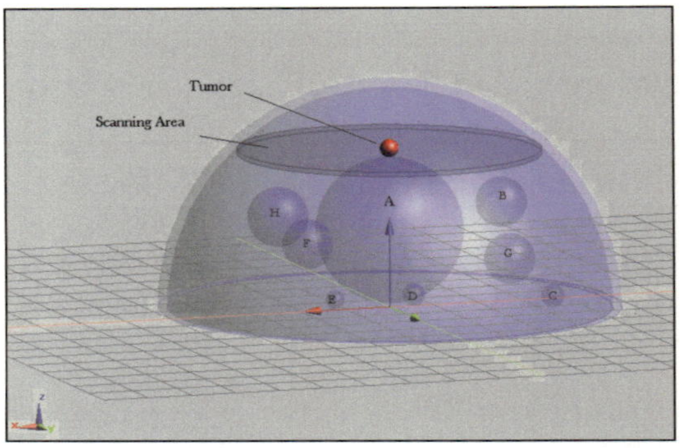

Fig. 9. 3D Model Constructed based on MRI image, shaded region shows the scanning area

Region	x	y	z	ϵ_r	Region	x	y	z	ϵ_r
A	0	0	17	5.3	E	16	10	4	4.8
B	-25	0.9	22	5.2	F	11	-9	11	5
C	-28	26	5	4.8	G	-36	-25	5.5	5
D	4	27	7	4.8	H	27	4.7	22	5.2

Table 5. Dielectric region centers (mm)

In this model, the skin layer thickness is set as 2mm. The antenna placement, physical parameters of the normal breast tissue and tumor are set as described in the previous section. As for the clutter regions, dielectric values are obtained from the MRI image as stated above. These values are given in Table 5.

The skin reflection is removed from the simulated backscattered signals using all three methods: frequency domain approach, averaging and weighted average filter to compare the performance of these methods. A 2D image of the breast is formed by applying confocal imaging process on the processed signals. The resulting images from the three methods are shown in Figure 10. Due to the symmetry of the tumor location to the antenna elements in the array, the tumor response is totally eliminated from the image processed by averaging and filtering methods. This is because, the tumor response will add coherently in the averaging process (due to the symmetry) and hence will appear in the average signal. Hence, subtracting the average removes the tumor backscatter as well as the skin backscatter. However, in frequency domain approach, each signal is processed separately and no other data is added

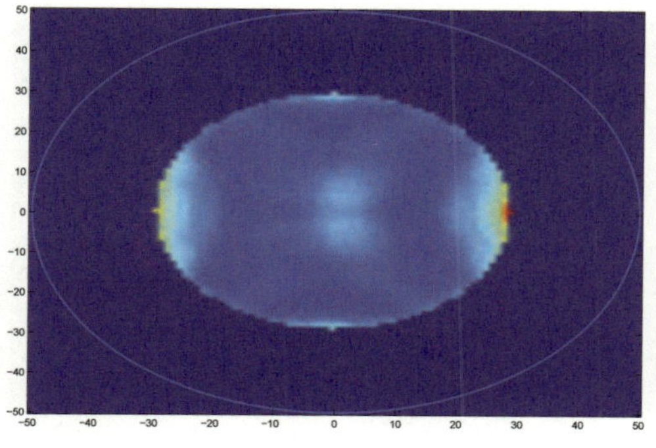

(a) Averaging

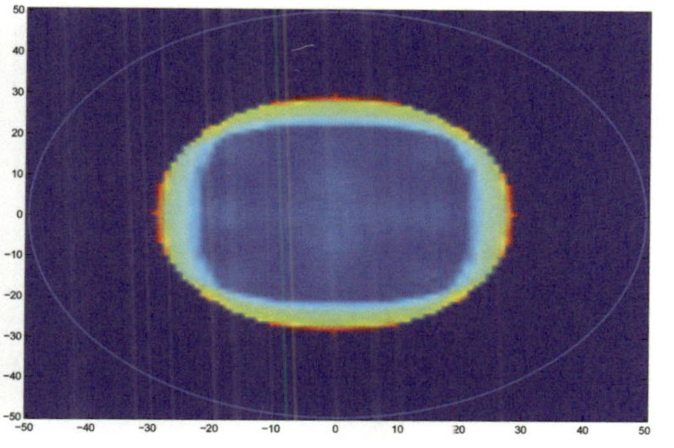

(b) Weighted Average Filter

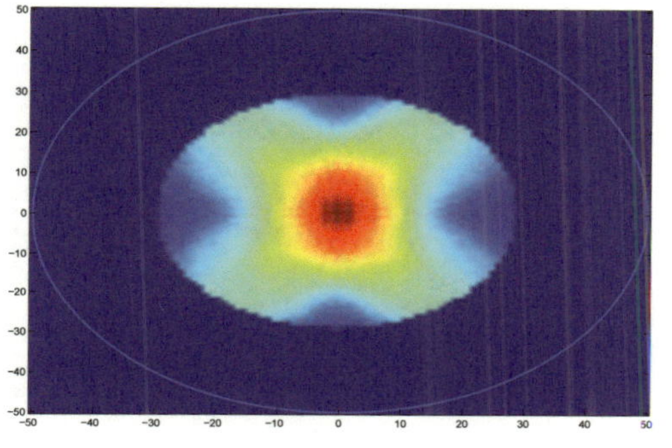

(c) Pole Removal

Fig. 10. Breast images using three skin subtraction methods: Averaging(a), Weighted Average(b), Pole removal(c)

to or subtracted from the signal, the tumor signature remains intact. This is confirmed in Figure 10. As seen in the figure, the tumor is detected at the central axis of the breast.

To compare the performance of the three methods in general case, the tumor is located in off center coordinates (x = 35 , y = 0, z = 15). The other parameters of the model are the same as the previous model.

Again, the skin reflection is removed using the three mentioned methods. The results are shown in Figure 11.

As the figure reveals, all three methods have eliminated the skin effect and the tumor is detected in the resulting image. To further evaluate the performance of the skin removal methods, the peak Tumor to Clutter Ratio (TCR) for the three methods is compared in Table 6. As seen in the table, the tumor to clutter ratio is the highest for frequency domain approach

Skin-Removal Method	TCR
Pole-Removal	3.831
Weighted Average	2.082
Averaging	1.837

Table 6. Tumor to Clutter Ratio (TCR)

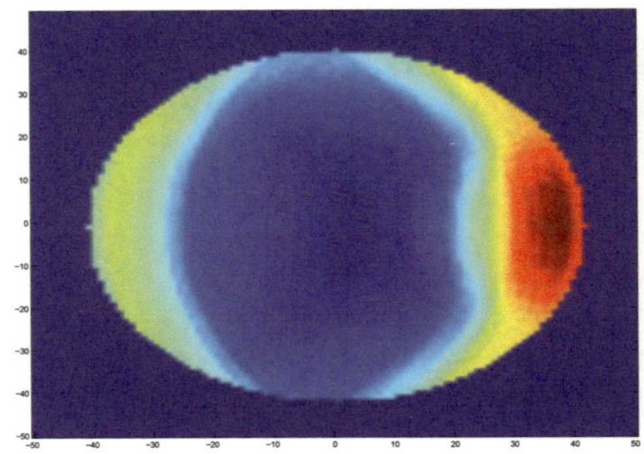

(a) Averaging

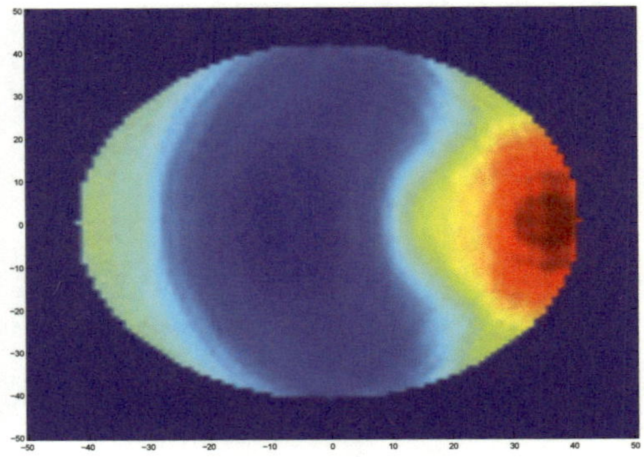

(b) Weighted Average Filter

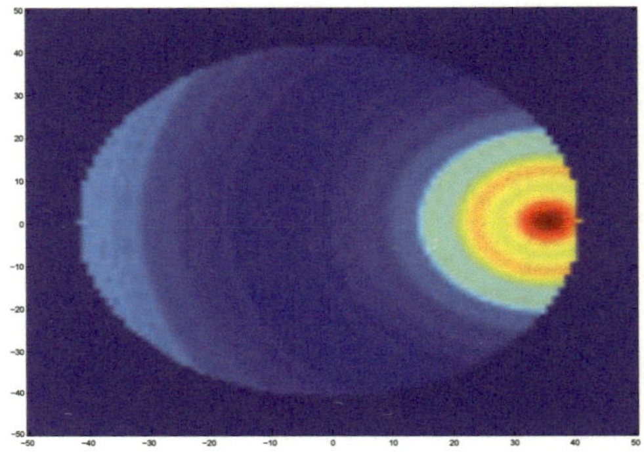

(c) Pole Removal

Fig. 11. Tumor at: x = 35, y = 0, z = 15 (mm), Averaging(a), Weighted Average(b), Pole removal(c)

and is the lowest for simple averaging. This is expected since pole removal method processes each signal individually unlike the other two methods which add clutter from other signals and degrade the tumor reflection.

6. Conclusion

The high contrast in the dielectric value of the skin relative to the normal breast tissue and air produces a strong backscatter in UWB breast cancer detection method. Such strong backscatter can totally mask the tumor reflection and hence has to be removed from the signal. Currently, two methods are used in practice to remove skin reflection. Both methods exploit the similarity of the skin reflection in the signals collected by an array of antennas to reconstruct and remove the skin reflection. Although these methods can significantly reduce the skin contribution in the backscattered signal, they have some shortcomings. Both methods use averaging to estimate the skin backscatter from the signals collected in different elements of the antenna array. As a result, if the tumor is approximately equidistant to some of the elements of the array, its reflection will suffer a high attenuation in the processed signals. This will make the tumor detection very difficult or even impossible. Another problem of the averaging based methods is that they add the averaged version of the noise and clutter from all channels to each individual channel which makes the tumor detection even more difficult. In addition, as the tumor reflection should not be included in the skin reflection estimation process, these methods need to determine the early time part of the signal where only the skin reflection exists. However, the location of the tumor in the signal is not known prior to the detection process.

This work introduces a new approach in removing the skin reflection from the backscattered signal in UWB breast cancer detection. In this approach, the backscattered signals are analyzed in frequency domain to identify and remove the skin related information from the frequency response. Based on Geometrical Theory of Diffraction (GTD), a mathematical model is applied on the frequency response of the signal. Then, the terms corresponding to the skin are removed from the model and the signal is reconstructed. Performance of this method is compared with the other existing methods in Section 5. As shown in Section 5, the frequency domain approach can detect the tumor even when it is equidistant to all the elements of the array. Besides, no extra noise and clutter is added to the signal as each signal is processed individually. Thus, the frequency domain approach shows higher tumor to clutter ratio in comparison with the other two methods. However, more investigations is needed to determine some parameters of the process such as the threshold used to remove the skin related terms from the frequency response. To optimize parameters such as the number of the antenna elements needed in the array, type of the antenna, pulse shape, etc. the method has to be applied on more realistic scenarios similar to the human breast.

7. References

Bond, E., Li, X., Hagness, S. & Van Veen, B. (2003). Microwave imaging via space-time beamforming for early detection of breast cancer, *IEEE Transactions On Antennas and Propagation* 51(8): 1690–1705.

Cuomo, K., Piou, J. & Mayhan, J. (1999). Ultrawide-band coherent processing, *IEEE Microwave Magazine* 47(6): 1094–1107.

Fear, E., Li, X., Hagness, S. & Stuchly, M. (2002). Confocal microwave imaging for breast cancer detection: Localization of tumors in three dimensions, *IEEE Transactions on Biomedical Engineering* 49: 812–821.

Fear, E. & Stuchly, M. (2000). Microwave detection of breast cancer, *IEEE Transactions On Microwave Theory And Techniques* 48(11): 1854 – 1863.

Hagness, S., Taflove, A. & Bridges, J. (1998). Two dimensional FDTD analysis of a pulsed microwave confocal system for breast cancer detection: Fixed-focus and antenna-array sensors, *IEEE Trans. Biomed. Eng.* 45: 1470–1479.

Haykin, S. (1996). *Adaptive Filter Theory*, 3rd edn, Prentice-Hall.

J. Elwood and B. Cox and A. Richardson (1993). The effectiveness of breast cancer screening by mammography in younger women, *The Online journal of current clinical trials* 32. URL: *http://www.ncbi.nlm.nih.gov/pubmed/8305999*

Keller, J. (1958). *A geometrical theory of diffraction*, Courant Institute of Mathematical Sciences, New York University.

Kosmas, P. & Rappaport, C. (2005). Time reversal with the FDTD method for microwave breast cancer detection, *IEEE Transactions on microwave theory and techniques* 53(7): 2317–2322.

Li, X. & Hagness, S. (2001). A confocal microwave imaging algorithm for breast cancer detection, *IEEE Microwave and Wireless Components Letters* 11(3): 130–132.

Moore, T., Zuerndorfer, B. & Burt, E. (1997). Enhanced imagery using spectral-estimation-based techniques, *Lincoln Laboratory Journal* 10(2): 171–186.

Naishadham, K. & Piou, J. (2004). A super-resolution method for extraction of modal responses in wideband data, *IEEE Antennas and Propagation Society International Symposium* 4: 4168–4171.

Naishadham, K. & Piou, J. (2005). State-space spectral estimation of characteristic electromagnetic responses in wideband data, *IEEE Antennas and Wireless Propagation Letters* 4: 406–409.

Piou, J. (2005). A state identification method for 1-d measurements with gaps, *Proc. American Institute of Aeronautics and Astronautics Guidance Navigation and Control Conf.* .

American Cancer Society (ACS) (2007). What are the key statistics for breast cancer? URL: *http://www.cancer.org/docroot/CRI/content/CRI_2_4_1X_What_are_the_key_statistics_for_breast_cancer_5.asp*

Center for Disease Control and Prevention (CDC) (2007). Statistics. URL: *http://www.cdc.gov/cancer/breast/statistics/*

Ulger, H., Erdogan, N., Kumanlioglu, S. & Unur, E. (2003). Effect of age, breast size, menopausal and hormonal status on mammographic skin thickness, *Skin Research and Technology* 9: 284–289.

Zhi, W. & Chin, F. (2006). Entropy-based time window for artifact removal in uwb imaging of breast cancer detection, *IEEE Signal Processing Letters* 13(10): 585–588.

Part 5

Novel UWB Application in Radars and Localization Systems

17

Full-Wave Modelling of Ground-Penetrating Radars: Antenna Mutual Coupling Phenomena and Sub-Surface Scattering Processes

Diego Caratelli and Alexander Yarovoy
Delft University of Technology
The Netherlands

1. Introduction

Ground-penetrating radar (GPR) technology finds applications in many areas such as geophysical prospecting, archaeology, civil engineering, environmental engineering, and defence applications as a non-invasive sensing tool [3], [6], [18]. One key component in any GPR system is the receiver/transmitter antenna. Desirable features for GPR antennas include efficient radiation of ultra-wideband pulses into the ground, good impedance matching over the operational frequency band, and small size. As the attenuation of radio waves in geophysical media increases with frequency [9], [13], ground-penetrating radars typically operate at frequencies below $1GHz$ [4]. For either impulse [13] or stepped-frequency continuous-wave applications [17], the wider the frequency range, the better the range resolution of the radar. Continuous wave multi-frequency radars are advantageous over impulse radars in coping with dispersion of the medium, the noise level at the receiver end, and the controllability of working frequency. They require, however, mutual coupling between the transmit (Tx) and receive (Rx) antennas, which determines the dynamic range of the system, to be kept as small as possible [12].

In this book chapter, the full-wave analysis of electromagnetic coupling mechanisms between resistively loaded wideband dipole antennas operating in realistic GPR scenarios is carried out. To this end, a locally conformal finite-difference time-domain (FDTD) technique, useful to model electromagnetic structures having complex geometry, is adopted [1], [2]. Such a scheme, necessary to improve the numerical accuracy of the conventional FDTD algorithm [19], [21], by avoiding staircase approximation, is based on the definition of effective material parameters [14], suitable to describe the geometrical and electrical characteristics of the structure under analysis. By doing so, the losses in the soil, as well as the presence of ground-embedded inhomogeneities with arbitrary shape and electrical properties, are properly taken into account. Emphasis is devoted to the investigation of the antenna pair performance for different Tx–Rx separations and elevations over the ground, as well as on scattering from dielectric and metallic pipes buried at different depths and having different geometrical and electrical characteristics. Novelty of the analysis lies in the fact that at the lowest operational frequency both the receive antenna and a pipe are situated in the near-field, whilst at the highest operational frequency only the far field is playing the role. The obtained numerical results provide a physical insight into the underlying mechanisms of subsurface diffraction and antenna mutual coupling processes. This information in turn can be usefully employed to optimize the performance of detection algorithms in terms of clutter rejection.

Finally, a frequency-independent equivalent circuit model of antenna pairs is provided in order to facilitate the design of the RF front-end of ground-penetrating radars by means of suitable software CAD tools. The procedure employed to extract the equivalent circuit is based on a heuristic modification of the Cauer's network synthesis technique [10] useful to model ohmic and radiation losses. In this way, one can obtain a meaningful description of the natural resonant modes describing the electromagnetic behaviour of antenna pairs for GPR systems.

2. Locally conformal finite-difference time-domain technique

The analysis and design of complex radiating structures requires accurate electromagnetic field prediction models. One such widely used technique is the FDTD algorithm. However, in the conventional formulation proposed by Yee [19], [21], each cell of the computational grid is implicitly supposed to be filled by a homogeneous material. For this reason, the adoption of Cartesian meshes could result in reduced numerical accuracy when structures having curved boundaries have to be modelled. In this case, locally conformal FDTD schemes [1], [2] provide clear advantages over the use of the stair-casing approach or unstructured and stretched space lattices, potentially suffering from significant numerical dispersion and/or instability [19]. Such schemes, necessary to improve the numerical accuracy of the conventional algorithm, are based on the definition of effective material parameters suitable to describe the geometrical and electrical characteristics of the structure under analysis.

In this section, a computationally enhanced formulation of the locally conformal FDTD scheme proposed in [1] is described. To this end, let us consider a three-dimensional domain D filled by a linear, isotropic, non dispersive material, having permittivity $\varepsilon(\boldsymbol{r})$, magnetic permeability $\mu(\boldsymbol{r})$, and electrical conductivity $\sigma(\boldsymbol{r})$. In such a domain, a dual-space, non-uniform lattice formed by a primary and secondary mesh is introduced. The primary mesh M_D is composed of space-filling hexahedrons, whose vertices are defined by the Cartesian coordinates:

$$\left\{(x_i, y_j, z_k) \mid i = 0,\ldots,N_x;\, j = 0,\ldots,N_y;\, k = 0,\ldots,N_z\right\}. \tag{1}$$

As a consequence, the edge lengths between adjacent vertices in M_D result to be expressed as:

$$\begin{cases} \Delta x_i = x_{i+1} - x_i, & i = 0,1,\ldots,N_x - 1, \\ \Delta y_j = y_{j+1} - y_j, & j = 0,1,\ldots,N_y - 1, \\ \Delta z_k = z_{k+1} - z_k, & k = 0,1,\ldots,N_z - 1. \end{cases} \tag{2}$$

The secondary or dual mesh $\tilde{M}_D$ (see Fig. 1) is composed of the closed hexahedrons whose edges penetrate the shared faces of the primary cells and connect the relevant centroids, having coordinates $x_{i+1/2} = x_i + \Delta x_i / 2$, $y_{j+1/2} = y_j + \Delta y_j / 2$, $z_{k+1/2} = z_k + \Delta z_k / 2$. A set of dual edge lengths is then introduced in $\tilde{M}_D$ as follows:

$$\begin{cases} \ell_{x_i} = \left(\Delta x_{i-1} + \Delta x_i\right)/2, & i = 1,2,\ldots,N_x - 1, \\ \ell_{y_j} = \left(\Delta y_{j-1} + \Delta y_j\right)/2, & j = 1,2,\ldots,N_y - 1, \\ \ell_{z_k} = \left(\Delta z_{k-1} + \Delta z_k\right)/2, & k = 1,2,\ldots,N_z - 1. \end{cases} \tag{3}$$

As usual, the electric field components are defined along each edge of a primary lattice cell, whereas the magnetic field components are assumed to be located along the edges of the secondary lattice cells. In this formulation, the relationship between $E-$ and $H-$field components is given by Maxwell's equations expressed in integral form, specifically using Faraday-Neumann's law and Ampere's law, respectively. In particular, the enforcement of the Ampere's law on the generic dual-mesh cell surface $\tilde{S}_x\big|_{i+1/2,j,k}$ having boundary $\partial\tilde{S}_x\big|_{i+1/2,j,k} = \tilde{C}_x\big|_{i+1/2,j,k}$ (see Fig. 1) results in the following integral equation:

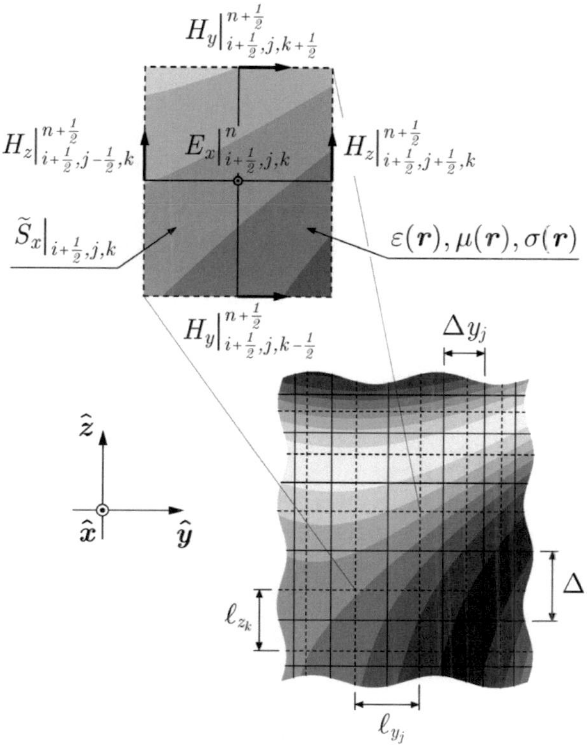

Fig. 1. Cross-sectional view of the FDTD computational grid in presence of curved boundaries between different dielectric materials.

$$\oint_{\tilde{C}_x|_{i+1/2,j,k}} \boldsymbol{H}(\boldsymbol{r},t)\cdot d\boldsymbol{l} = \iint_{\tilde{S}_x|_{i+1/2,j,k}} \sigma(\boldsymbol{r})E_x(\boldsymbol{r},t)dS + \frac{\partial}{\partial t}\iint_{\tilde{S}_x|_{i+1/2,j,k}} \varepsilon(\boldsymbol{r})E_x(\boldsymbol{r},t)dS, \tag{4}$$

where:

$$\begin{aligned}\oint_{\tilde{C}_x|_{i+1/2,j,k}} \boldsymbol{H}(\boldsymbol{r},t)\cdot d\boldsymbol{l} = \ell_{z_k}\left[H_z\left(x_{i+\frac{1}{2}},y_{j+\frac{1}{2}},z_k,t\right) - H_z\left(x_{i+\frac{1}{2}},y_{j-\frac{1}{2}},z_k,t\right)\right] + \\ -\ell_{y_j}\left[H_y\left(x_{i+\frac{1}{2}},y_j,z_{k+\frac{1}{2}},t\right) - H_y\left(x_{i+\frac{1}{2}},y_j,z_{k-\frac{1}{2}},t\right)\right] + o\left(\ell_{y_j}\right) + o\left(\ell_{z_k}\right)\end{aligned} \tag{5}$$

as ℓ_{y_j} and ℓ_{z_k} tend to zero. Under the assumption that the spatial increments Δx_i, Δy_j, Δz_k of the computational grid are small compared to the minimum working wavelength, the infinitesimal terms of higher order appearing in (5) can be neglected. Furthermore, it should be noticed that the $x-$component of the electric field is continuous along the interfaces crossing $\tilde{S}_x|_{i+1/2,j,k}$ so that, under the mentioned hypothesis, the following approximation can be made:

$$\iint_{\tilde{S}_x|_{i+1/2,j,k}} \begin{Bmatrix}\varepsilon\\ \sigma\end{Bmatrix}(\boldsymbol{r})E_x(\boldsymbol{r},t)dS \simeq E_x\left(x_{i+\frac{1}{2}},y_j,z_k,t\right)\iint_{\tilde{S}_x|_{i+1/2,j,k}} \begin{Bmatrix}\varepsilon\\ \sigma\end{Bmatrix}(\boldsymbol{r})dS. \tag{6}$$

Hence, combining the equations above yields:

$$\begin{aligned}\overline{\varepsilon}_x^{eff}\Big|_{i+\frac{1}{2},j,k}\frac{\partial}{\partial t}\mathcal{E}_x(t)\Big|_{i+\frac{1}{2},j,k} + \overline{\sigma}_x^{eff}\Big|_{i+\frac{1}{2},j,k}\mathcal{E}_x(t)\Big|_{i+\frac{1}{2},j,k} = \\ = \mathcal{H}_z(t)\Big|_{i+\frac{1}{2},j+\frac{1}{2},k} - \mathcal{H}_z(t)\Big|_{i+\frac{1}{2},j-\frac{1}{2},k} - \mathcal{H}_y(t)\Big|_{i+\frac{1}{2},j,k+\frac{1}{2}} + \mathcal{H}_y(t)\Big|_{i+\frac{1}{2},j,k-\frac{1}{2}},\end{aligned} \tag{7}$$

where we have introduced the normalized field quantities:

$$\mathcal{E}_x(t)\Big|_{i+\frac{1}{2},j,k} = \Delta x_i\, E_x\left(x_{i+\frac{1}{2}},y_j,z_k,t\right), \tag{8}$$

$$\mathcal{H}_z(t)\Big|_{i+\frac{1}{2},j+\frac{1}{2},k} = \ell_{z_k}\, H_z\left(x_{i+\frac{1}{2}},y_{j+\frac{1}{2}},z_k,t\right), \tag{9}$$

$$\mathcal{H}_y(t)\Big|_{i+\frac{1}{2},j,k+\frac{1}{2}} = \ell_{y_j}\, H_y\left(x_{i+\frac{1}{2}},y_j,z_{k-\frac{1}{2}},t\right), \tag{10}$$

and the averaged effective permittivity $\overline{\varepsilon}_x^{eff}\Big|_{i+1/2,j,k}$, and conductivity $\overline{\sigma}_x^{eff}\Big|_{i+1/2,j,k}$, defined as follows:

$$\begin{Bmatrix}\overline{\varepsilon}\\ \overline{\sigma}\end{Bmatrix}_x^{eff}\Bigg|_{i+\frac{1}{2},j,k} = \frac{1}{\Delta x_i}\int_{z_{k-1/2}}^{z_{k+1/2}}\int_{y_{j-1/2}}^{y_{j+1/2}} \begin{Bmatrix}\varepsilon\\ \sigma\end{Bmatrix}\left(x_{i+\frac{1}{2}},y,z\right)dydz. \tag{11}$$

The time derivative in (7) is then evaluated using a central-difference approximation that is second order-accurate if $E-$ and $H-$ field components are staggered in time domain [19]. This results in the following explicit time-stepping relation:

$$\mathcal{E}_x\Big|_{i+\frac{1}{2},j,k}^{n+1} = \bar{\alpha}_x^{(E)}\Big|_{i+\frac{1}{2},j,k} \mathcal{E}_x\Big|_{i+\frac{1}{2},j,k}^{n-1} + \bar{\beta}_x^{(E)}\Big|_{i+\frac{1}{2},j,k} (\nabla\times\mathcal{H})_x\Big|_{i+\frac{1}{2},j,k}^{n+\frac{1}{2}}, \tag{12}$$

where:

$$(\nabla\times\mathcal{H})_x\Big|_{i+\frac{1}{2},j,k}^{n+\frac{1}{2}} = \mathcal{H}_z\Big|_{i+\frac{1}{2},j+\frac{1}{2},k}^{n+\frac{1}{2}} - \mathcal{H}_z\Big|_{i+\frac{1}{2},j-\frac{1}{2},k}^{n+\frac{1}{2}} - \mathcal{H}_y\Big|_{i+\frac{1}{2},j,k+\frac{1}{2}}^{n+\frac{1}{2}} + \mathcal{H}_y\Big|_{i+\frac{1}{2},j,k-\frac{1}{2}}^{n+\frac{1}{2}} \tag{13}$$

denotes the finite-difference expression of the normalized $x-$component of the magnetic field curl at the time step $n+1/2$. In (12), the information regarding the local physical and geometrical properties of the electromagnetic structure under analysis is transferred to the position-dependent coefficients:

$$\bar{\alpha}_x^{(E)}\Big|_{i+\frac{1}{2},j,k} = \frac{1-\bar{Q}_x^{eff}\Big|_{i+\frac{1}{2},j,k}}{1+\bar{Q}_x^{eff}\Big|_{i+\frac{1}{2},j,k}}, \tag{14}$$

$$\bar{\beta}_x^{(E)}\Big|_{i+\frac{1}{2},j,k} = \frac{\Delta t/\bar{\varepsilon}_x^{eff}\Big|_{i+\frac{1}{2},j,k}}{1+\bar{Q}_x^{eff}\Big|_{i+\frac{1}{2},j,k}}, \tag{15}$$

with:

$$\bar{Q}_x^{eff}\Big|_{i+\frac{1}{2},j,k} = \frac{\bar{\sigma}_x^{eff}\Big|_{i+\frac{1}{2},j,k}\Delta t}{2\bar{\varepsilon}_x^{eff}\Big|_{i+\frac{1}{2},j,k}}. \tag{16}$$

The update equations of the remaining components of the electric and magnetic field can be easily derived by permuting the spatial indices i, j, k and applying the duality principle in the discrete space.

As it can be readily noticed, the computation of position-dependent coefficients (14)-(16) can be carried out before the FDTD-method time marching starts. As a consequence, unlike in conformal techniques based on stretched space lattices, no additional correction is required in the core of the numerical algorithm. Furthermore, the resulting FDTD update equations (12)-(13) have a very convenient structure, leading to a 14% reduction of the number of floating-point operations needed to determine the unknown field quantities in the generic mesh cell compared to the Yee algorithm [19], [21]. It is also to be pointed out that the proposed scheme has the same numerical stability properties as the conventional FDTD formulation, although it introduces a significant improvement in accuracy over the stair-casing approximation as well as alternative weighted averaging FDTD approaches [7]. In order to assess the effectiveness of the developed technique, several test cases have been considered. For the sake of brevity, only the computation of the fundamental resonant frequency of a metallic cavity loaded with a cylindrical dielectric resonator is presented in the Appendix. The obtained results clearly demonstrate the suitability of the proposed scheme to efficiently handle the problem of modelling antennas for ground-penetrating radar applications, where the accurate characterization of complex metal-dielectric objects having irregular geometry (think about the shape of buried targets and ground-embedded inhomogeneities) is required.

3. The full-wave antenna modelling

It is our intention to focus the attention on the full-wave analysis of a resistively loaded dipole antenna pair located above a ground modelled as a lossy homogeneous half-space having relative permittivity $\varepsilon_r = 6$ and electrical conductivity $\sigma = 15 mS/m$. The geometry of the structure is depicted in Fig. 2.

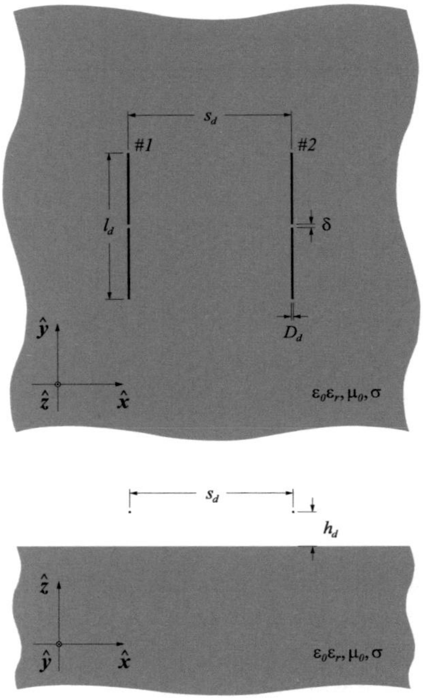

Fig. 2. Top and cross-sectional view of a resistively loaded dipole antenna pair located above a lossy homogeneous half space. Structure characteristics: $l_d = 40cm$, $D_d = 5mm$, $\delta = 2.5mm$, $\varepsilon_r = 6$, $\sigma = 15mS/m$. The reference system used to express the field quantities is also shown.

The dipoles are denoted as dipole #1 and dipole #2, respectively. In the considered antenna configuration, dipole #1 is driven by a delta-gap voltage source with internal resistance $R_G = R_1 = 300\Omega$, whereas dipole #2 is closed on a matched load having resistance $R_L = R_2 = R_G$. To properly enlarge the antenna bandwidth, thus reducing late-time ringing phenomena, a continuous resistive loading, having Wu-King-like profile [3], [15], [16]:

$$\sigma_d(y) = \sigma_0\left(1 - \frac{2|y|}{l_d}\right), \tag{17}$$

has been applied to the flairs of the considered radiators. In (17), σ_0 denotes the electrical conductivity value at the input terminals of the antennas ($y = 0$), and $l_d = 40cm$ is the length of each dipole, assumed to have diameter $D_d = 5mm$. In particular, σ_0 has been determined by means of a dedicated parametric analysis. In this way, the optimal value $\sigma_0 = \sigma_{opt} \simeq 200S/m$, resulting in a fractional bandwidth $FBW \simeq 51\%$ centred on the fundamental resonant frequency $f_r \simeq 450MHz$, has been found (see Fig. 3).

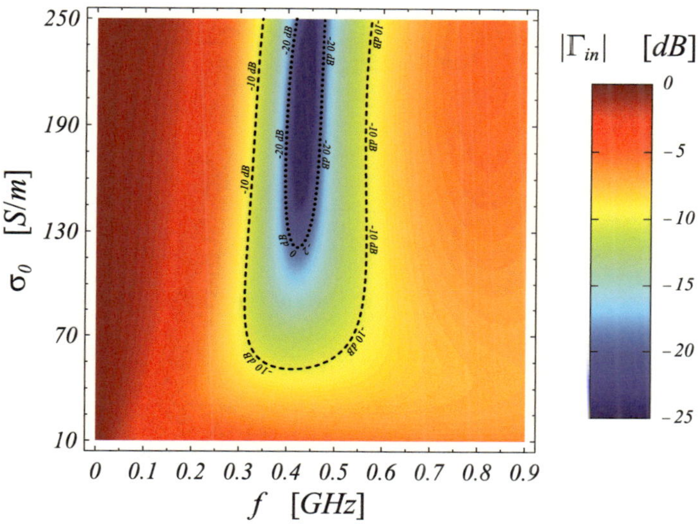

Fig. 3. Frequency behaviour of the individual antenna input reflection coefficient for different loading profiles. The antenna is elevated $h_d = 3cm$ over the ground.

The FDTD characterization of the structure has been carried out by using a non-uniform computational grid with maximum cell size $\Delta h = \lambda_{\min}/24 = 2.5mm$, where $\lambda_{\min} \simeq 6cm$ is the wavelength in the ground at the upper $-10dB$ cut-off frequency $f_{\max} = 1GHz$ of the excitation voltage signal, which is the Gaussian voltage pulse described by the equation:

$$v_G(t) = \hat{V}_G \exp\left[-\left(\frac{t - T_0}{T_G}\right)^2\right]u(t), \tag{18}$$

where $\hat{V}_G = 1\,mV$, $T_0 = 10T_G$, and:

$$T_G = \frac{\sqrt{\ln 10}}{\pi f_{\max}} \simeq 0.48\,ns. \tag{19}$$

In (18), $u(t)$ denotes the usual Heaviside function. The source pulse is coupled into the finite-difference equations used to update the time-domain electric field distribution within the feed point of dipole $\#1$ as follows:

$$\mathcal{E}_y\Big|_{i_1,j_1+\frac{1}{2},k_1}^{n+1} = \bar{\alpha}_y^{(E,G)}\Big|_{i_1,j_1+\frac{1}{2},k_1} \mathcal{E}_y\Big|_{i_1,j_1+\frac{1}{2},k_1}^{n-1} + \bar{\beta}_y^{(E,G)}\Big|_{i_1,j_1+\frac{1}{2},k_1} \left[(\nabla \times \boldsymbol{\mathcal{H}})_y\Big|_{i_1,j_1+\frac{1}{2},k_1}^{n+\frac{1}{2}} + \mathcal{I}_G\Big|^{n+\frac{1}{2}} \right], \quad (20)$$

where:

$$\bar{\alpha}_y^{(E,G)}\Big|_{i_1,j_1+\frac{1}{2},k_1} = \frac{1 - \bar{Q}_y^{(E,G)}\Big|_{i_1,j_1+\frac{1}{2},k_1}}{1 + \bar{Q}_y^{(E,G)}\Big|_{i_1,j_1+\frac{1}{2},k_1}}, \quad (21)$$

$$\bar{\beta}_y^{(E,G)}\Big|_{i_1,j_1+\frac{1}{2},k_1} = \frac{\Delta t \Big/ \bar{\varepsilon}_y^{eff}\Big|_{i_1,j_1+\frac{1}{2},k_1}}{1 + \bar{Q}_y^{(E,G)}\Big|_{i_1,j_1+\frac{1}{2},k_1}}, \quad (22)$$

with:

$$\bar{Q}_y^{(E,G)}\Big|_{i_1,j_1+\frac{1}{2},k_1} = \frac{\Delta t}{2\bar{\varepsilon}_y^{eff}\Big|_{i_1,j_1+\frac{1}{2},k_1} R_G}, \quad (23)$$

i_1, j_1, k_1 being the spatial indices relevant to the source. In (20), the term:

$$\mathcal{I}_G\Big|^{n+\frac{1}{2}} = \frac{v_G(t)}{R_G}\Big|_{t=(n+1/2)\Delta t} \quad (24)$$

denotes the time-discretised nominal current delivered by the generator. Similarly, the electric field distribution within the feed point of dipole #2 is updated by using the finite-difference equation:

$$\mathcal{E}_y\Big|_{i_2,j_2+\frac{1}{2},k_2}^{n+1} = \bar{\alpha}_y^{(E,L)}\Big|_{i_2,j_2+\frac{1}{2},k_2} \mathcal{E}_y\Big|_{i_2,j_2+\frac{1}{2},k_2}^{n-1} + \bar{\beta}_y^{(E,L)}\Big|_{i_2,j_2+\frac{1}{2},k_2} (\nabla \times \boldsymbol{\mathcal{H}})_y\Big|_{i_2,j_2+\frac{1}{2},k_2}^{n+\frac{1}{2}}, \quad (25)$$

where:

$$\bar{\alpha}_y^{(E,L)}\Big|_{i_2,j_2+\frac{1}{2},k_2} = \frac{1 - \bar{Q}_y^{(E,L)}\Big|_{i_2,j_2+\frac{1}{2},k_2}}{1 + \bar{Q}_y^{(E,L)}\Big|_{i_2,j_2+\frac{1}{2},k_2}}, \quad (26)$$

$$\bar{\beta}_y^{(E,L)}\Big|_{i_2,j_2+\frac{1}{2},k_2} = \frac{\Delta t \Big/ \bar{\varepsilon}_y^{eff}\Big|_{i_2,j_2+\frac{1}{2},k_2}}{1 + \bar{Q}_y^{(E,L)}\Big|_{i_2,j_2+\frac{1}{2},k_2}}, \quad (27)$$

$$\bar{Q}_y^{(E,L)}\Big|_{i_2,j_2+\frac{1}{2},k_2} = \frac{\Delta t}{2\bar{\varepsilon}_y^{eff}\Big|_{i_2,j_2+\frac{1}{2},k_2} R_L}, \quad (28)$$

and with i_2, j_2, k_2 denoting the spatial indices relevant to the load. The total voltage and current signals excited at the input terminals of the antenna pair, regarded as a two-port microwave network, are readily computed as:

$$v_m(t) = -\int_{C_{V_m}} \boldsymbol{E}(\boldsymbol{r},t)\cdot d\boldsymbol{l} \simeq -\mathcal{E}_y(t)\Big|_{i_m, j_m+\frac{1}{2}, k_m}, \tag{29}$$

$$i_m(t) = \oint_{C_{I_m}} \boldsymbol{H}(\boldsymbol{r},t)\cdot d\boldsymbol{l} \simeq \mathcal{H}_x\Big|_{i_m, j_m+\frac{1}{2}, k_m+\frac{1}{2}}^{n+\frac{1}{2}} - \mathcal{H}_x\Big|_{i_m, j_m+\frac{1}{2}, k_m-\frac{1}{2}}^{n+\frac{1}{2}} - \mathcal{H}_z\Big|_{i_m+\frac{1}{2}, j_m+\frac{1}{2}, k_m}^{n+\frac{1}{2}} + \mathcal{H}_z\Big|_{i_m-\frac{1}{2}, j_m+\frac{1}{2}, k_m}^{n+\frac{1}{2}}, \tag{30}$$

where C_{V_m} is an open contour extending along the delta gap, and C_{I_m} a closed contour path wrapping around the driving point of dipole $\#m$ $(m=1,2)$. Under the mentioned assumptions, the normalized incident and reflected waves are evaluated as:

$$a_1(f) = \frac{1}{2}\left[\frac{V_1(f)}{\sqrt{R_0}} + I_1(f)\sqrt{R_0}\right], \tag{31}$$

$$b_1(f) = \frac{1}{2}\left[\frac{V_1(f)}{\sqrt{R_0}} - I_1(f)\sqrt{R_0}\right], \tag{32}$$

$$a_2(f) = 0, \tag{33}$$

$$b_2(f) = \frac{V_2(f)}{\sqrt{R_0}}, \tag{34}$$

where $R_0 = R_G$ denotes the reference resistance and $V_m(f) = \mathsf{F}[v_m(t)]$, $I_m(f) = \mathsf{F}[i_m(t)]$, $\mathsf{F}[\cdot]$ being the usual Fourier transform operator. Therefore, the scattering parameters of the structure can be easily determined as:

$$S_{11}(f) = \frac{b_1(f)}{a_1(f)}, \tag{35}$$

$$S_{21}(f) = \frac{b_2(f)}{a_1(f)}. \tag{36}$$

As it appears from Fig. 4a, the return-loss level is slightly affected by the Tx–Rx antenna separation that, on the other hand, is primarily responsible for the parasitic coupling level between the radiating elements. The impact of the antenna elevation above the ground has been also analyzed (see Fig. 4b). It is worth noting that, as h_d decreases, the fundamental resonant frequency of the dipole is shifted down because of the proximity effect of the soil. On the other hand, the ground influence on the S_{21} parameter is remarkable only at high frequencies, where the coupling level between the two radiating elements tends to decrease as the dipoles approach the air-ground interface.

In the performed numerical computations, a ten-cell uniaxial perfectly matched layer (UPML) absorbing boundary condition for lossy media [19] has been used at the outer FDTD mesh boundary to simulate the extension of the space lattice to infinity. As outlined in [19], the

UPML is indeed perfectly matched to the inhomogeneous medium formed by the upper air region and the lossy material modelling the soil. So, no spurious numerical reflections take place at the air-ground interface. In particular, a quartic polynomial grading of the UPML conductivity profile has been selected in order to have a nominal reflection error $R_{PML} \simeq e^{-16}$.

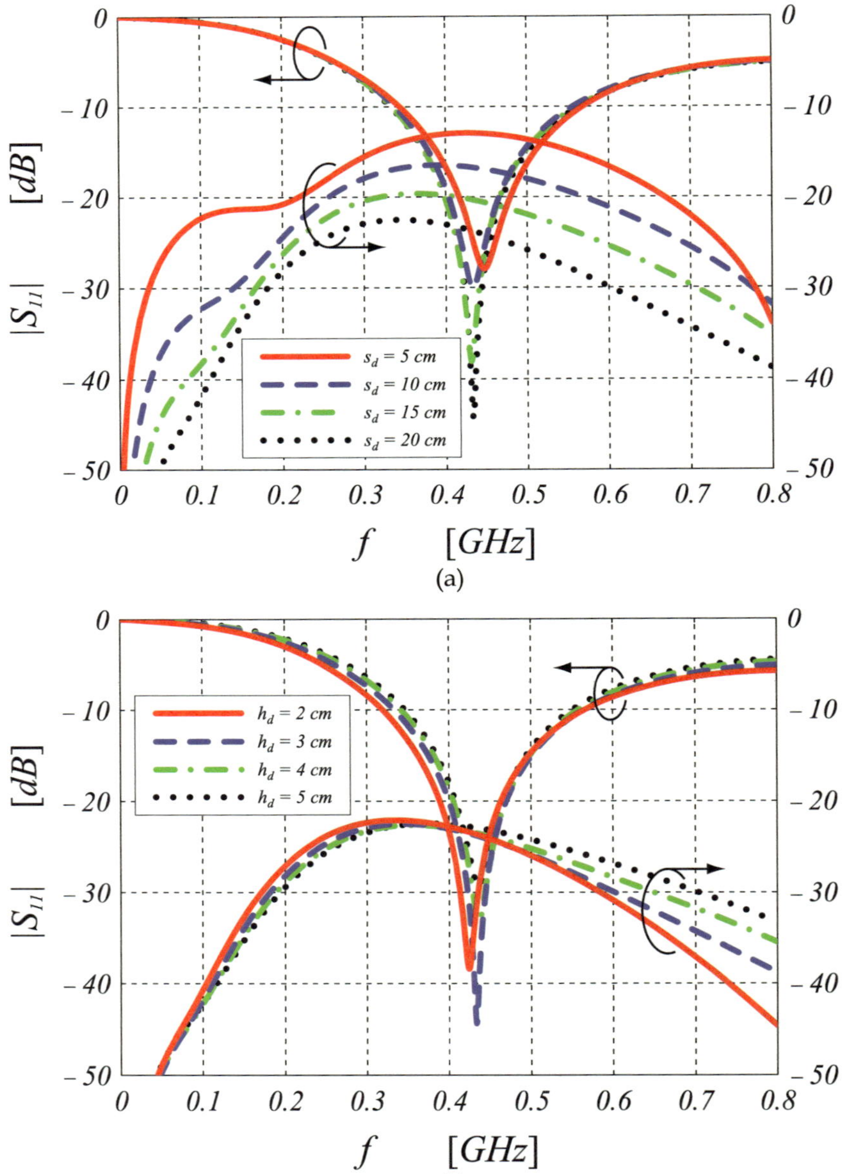

Fig. 4. Frequency behaviour of the scattering parameters of the dipole pair for different Tx-Rx separations (a) and antenna elevations (b) over the ground modelled as a lossy homogeneous half-space with electrical properties $\varepsilon_r = 6$ and $\sigma = 15 mS/m$.

4. The radar detection of buried pipes

In this section, emphasis is devoted to the analysis of the dipole antenna pair located above a lossy homogeneous/inhomogeneous material half space where an infinitely long pipe is buried (see Fig. 5). In such configuration, the transmit element of the radar unit emits an electromagnetic pulse that propagates into the ground, where it interacts with the target, modelled as a $y-$ directed circular cylinder having diameter $D_p = 30\,cm$, buried at a depth $h_p = 40\,cm$. This interaction results in a diffracted electromagnetic field which is measured by the receive element of the radar. By changing the location of the radar on the soil interface and recording the output of the receive antenna as function of time (or frequency) and radar location, one obtains the scattering data, which can be processed to get an image of the subsurface.

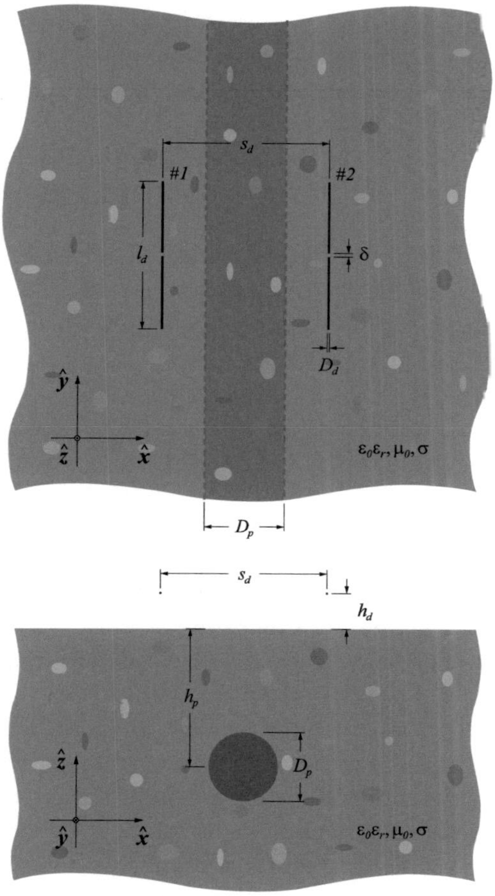

Fig. 5. Top and cross-sectional view of a resistively loaded dipole antenna pair located above a lossy inhomogeneous ground, where an infinitely long pipe is buried. Structure characteristics: $l_d = 40\,cm$, $D_d = 5\,mm$, $\delta = 2.5\,mm$, $s_d = 20\,cm$, $h_d = 3\,cm$, $D_p = 30\,cm$, $h_p = 40\,cm$.

The parasitic coupling level between transmit and receive antennas is a critical parameter in the design of ground-penetrating radars and satisfactory levels are usually achieved by empirical design methods. Anyway, the prediction of coupling levels already at the design stage enhances structure reliability, while also improving design cycle. To this end, the locally conformal FDTD model presented in Section 2 has been usefully adopted. In this way, as it can be noticed in Fig. 6, it has been found that the antenna return-loss parameter S_{11} is not strongly affected by the buried target which, conversely, has a significant impact on the frequency behaviour of the coupling level between the radiating elements, due to the electromagnetic field interference processes occurring at the receiver end.

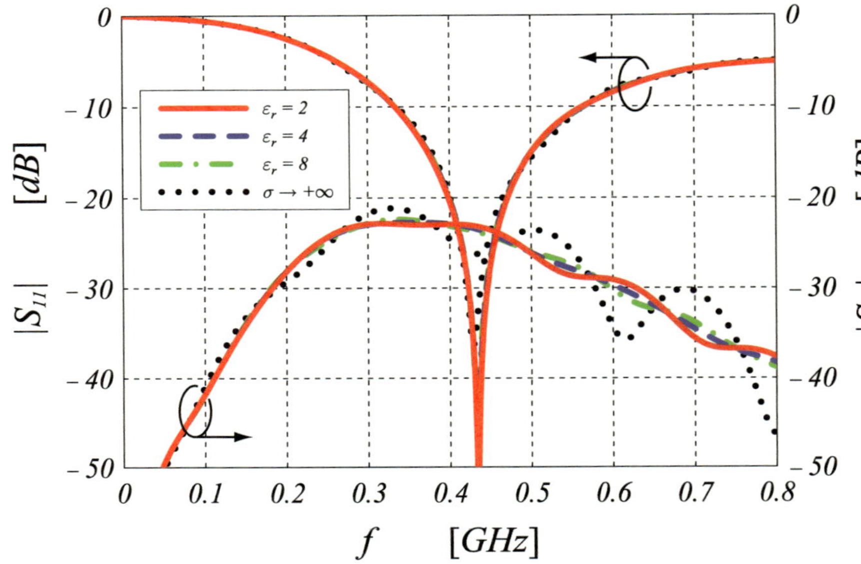

Fig. 6. Frequency behaviour of the scattering parameters of the dipole antenna pair for different electrical properties of the buried pipe. Structure characteristics: $l_d = 40cm$, $D_d = 5mm$, $\delta = 2.5mm$, $s_d = 20cm$, $h_d = 3cm$, $D_p = 30cm$, $h_p = 40cm$.

4.1 Analysis of sub-surface scattering processes

The considered antenna pair has been used to analyze the sub-surface scattering phenomena arising from the field interaction with a PVC pipe buried at a depth $h_p = 30cm$. As it can be easily inferred, the intensity of the radio signal diffracted by the pipe and measured at the input terminals of the receive antenna strongly depends on the electrical and geometrical properties of the target. In particular, the peak-to-peak level of the signal increases as the diameter, and hence the radar cross section of the pipe becomes larger (see Fig. 7). Another noticeable phenomenon is the sub-surface excitation of creeping waves. Such waves, propagating along the pipe surface, give rise to late-time pulse contributions in the received radio signal which can be clearly noticed in Fig. 7 as the radius of the object increases. Furthermore, it is worth noting that the strength of the creeping wave contribution tends to reduce with the pipe size because of the field attenuation relevant to the extra-path length.

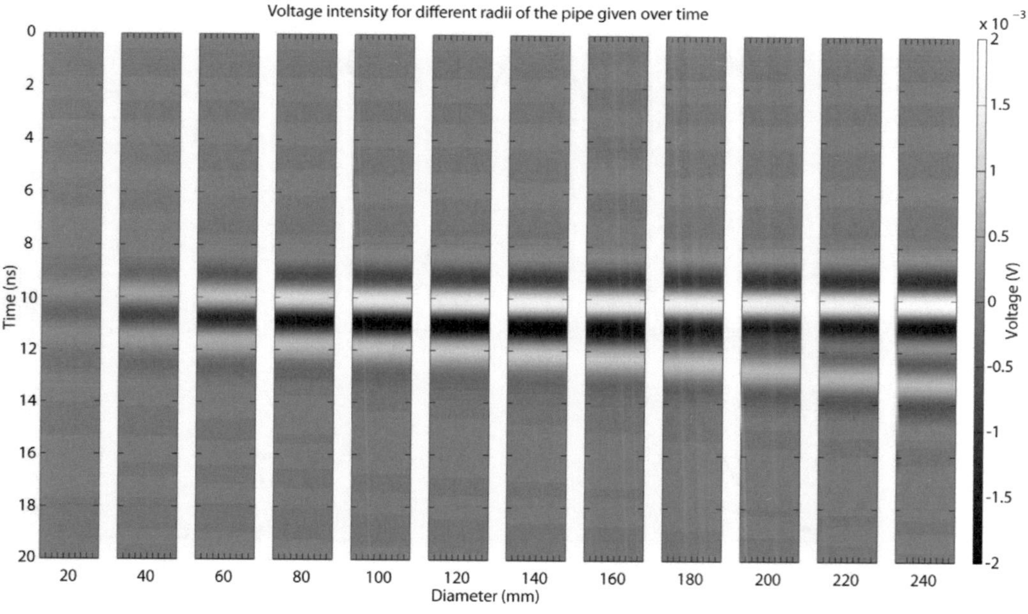

Fig. 7. Time-domain behaviour of the radio signal diffracted by a PVC pipe buried at a depth $h_p = 30cm$ in a lossy ground with electrical properties $\varepsilon_r = 6$ and $\sigma = 15 mS/m$. The effect of the sub-surface excitation of creeping waves can be noticed.

4.2 Impact of ground-embedded inhomogeneities

An invariable feature of real-life soils is heterogeneity. Without taking into account the inhomogeneities altering the idealized nature of the considered ground model, it becomes a futile effort to design a complex GPR system that will perform well over a real-life soil. To overcome this limitation, a realistic ground model has been developed by simulating small ellipsoidal scatterers embedded in the soil (see Fig. 5). The size, location and electrical properties of these inhomogeneities are determined randomly within preset limits. In particular, the maximum dimensions of the scatterers are $a_x = 10cm$, $a_y = 10cm$, and $a_z = 2cm$ in x, y, and z coordinate direction, respectively. In addition, all inhomogeneities have randomly selected values of the relative permittivity according to the following Gaussian probability distribution:

$$\varphi_{\mu_\varepsilon, \sigma_\varepsilon^2}(\varepsilon_r) = \frac{1}{\sqrt{2\pi}\sigma_\varepsilon} e^{-\frac{(\varepsilon_r - \mu_\varepsilon)^2}{2\sigma_\varepsilon^2}}, \tag{37}$$

with mean $\mu_\varepsilon = 13.7$, and standard deviation $\sigma_\varepsilon = 4.2$ (see Fig. 8). As it appears from Fig. 9, the ground-embedded inhomogeneities have a considerable impact on the coupling coefficient S_{21} of the dipole pair especially at high frequencies ($f > 0.6GHz$) where the radiated field assumes a quasi-optical behaviour and the diffraction processes arising from the interaction with the inhomogeneities tend to be significant, and could mask the detection of the buried target (see Fig. 10).

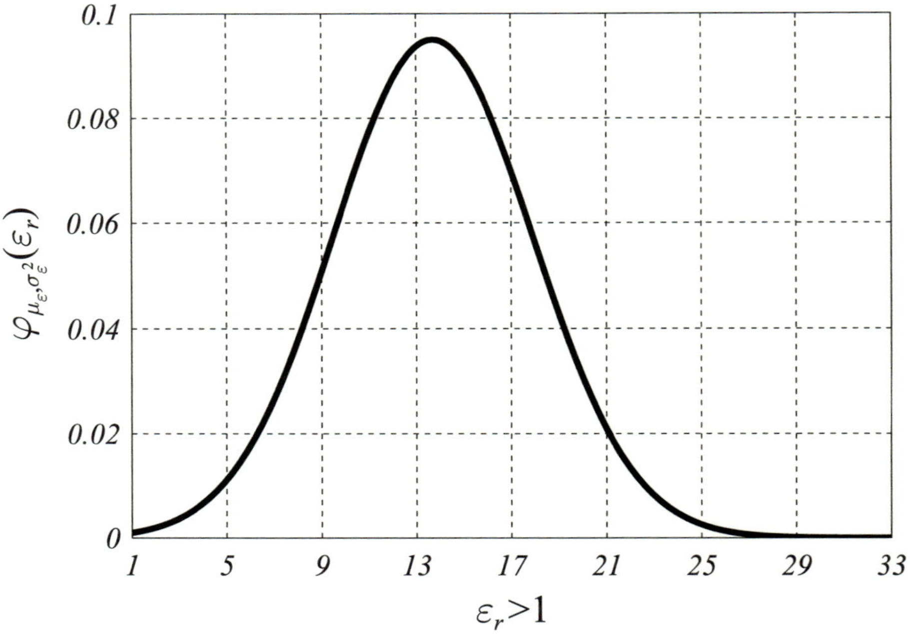

Fig. 8. Gaussian probability distribution of the relative permittivity of ground-embedded inhomogeneities.

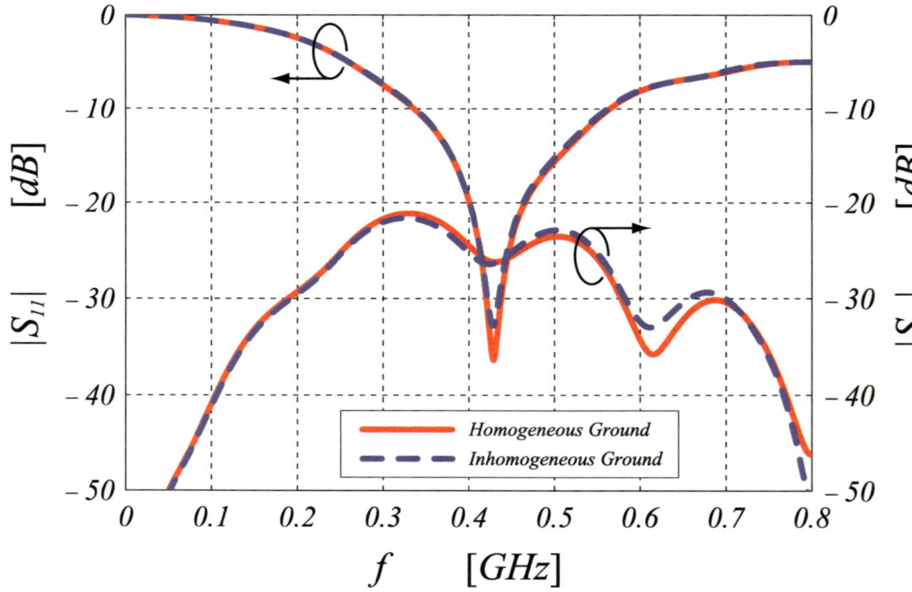

Fig. 9. Effect of the ground inhomogeneities on the frequency behaviour of the scattering parameters of the dipole pair. Structure characteristics: $l_d = 40cm$, $D_d = 5mm$, $\delta = 2.5mm$, $s_d = 20cm$, $h_d = 3cm$, $D_p = 30cm$, $h_p = 40cm$.

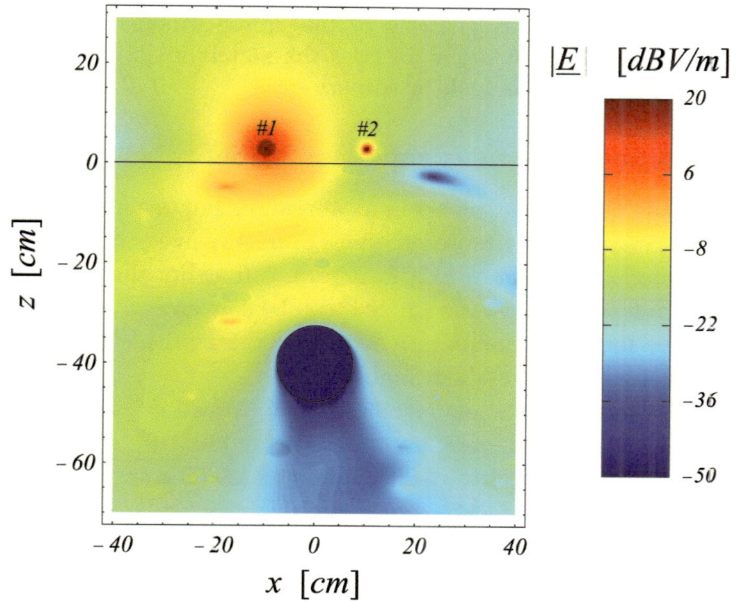

(a)

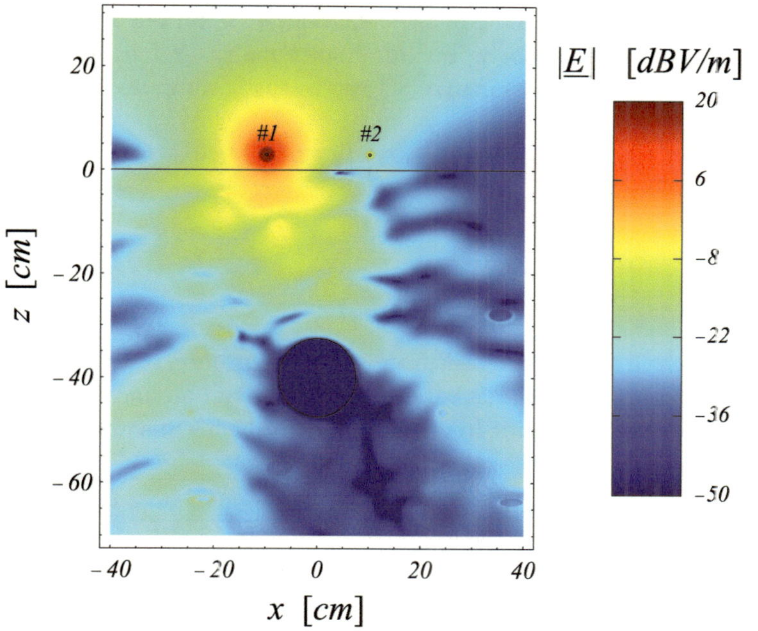

(b)

Fig. 10. Spatial distribution of the electric field excited along the vertical cut-plane of the antenna pair elevated over a lossy inhomogeneous ground, where an infinitely long metallic pipe is buried. Structure characteristics: $l_d = 40cm$, $D_d = 5mm$, $\delta = 2.5mm$, $s_d = 20cm$, $h_d = 3cm$, $D_p = 30cm$, $h_p = 40cm$. Working frequency: $f = 500MHz$ (a), $f = 1GHz$ (b).

As outlined in [11], the rigorous analysis of the aforementioned subsurface scattering phenomena is very important in order to asses the suitability of detection and imaging algorithms for GPR applications in realistic scenarios.

5. Frequency-independent equivalent circuit model

Typically, electromagnetic field solvers and measurement systems, such as network analyzers, generate $S-$ parameter representations of microwave components and antennas. However, circuit simulators, such as SPICE, require conventional equivalent circuits with lumped frequency-independent parameters that can be conveniently modelled within CAD software. In view of this, a technique for extracting the equivalent circuit from a given $S-$ parameter representation is highly desirable [20].

The proposed procedure for the equivalent circuit extraction is based on a heuristic modification of the Cauer's network synthesis technique [10]. Resistive elements are introduced to model metal, dielectric, and radiation losses. The scattering matrix $\underline{\mathbf{S}}$ of the antenna pair is modelled by means of an equivalent network composed of shunt $\pi-$ networks as shown in Fig. 11, where ideal transformers are to be included in order to assure the physical realizability of the network. In particular, RLC $\pi-$ subnetworks are required to properly model the high-frequency resonant phenomena taking place into the structure, whereas the capacitive $\pi-$ subnetwork is needed to describe the circuital behaviour of the antenna pair in the quasi-static regime. A simple analysis of the network shown in Fig. 11 allows determining the admittance matrix $\underline{\mathbf{Y}}^{eq}$ of the equivalent circuit. Then, the corresponding scattering matrix $\underline{\mathbf{S}}^{eq}$ can be easily computed as follows:

$$\underline{\mathbf{S}}^{eq} = \left(\underline{\mathbf{I}} - \hat{\underline{\mathbf{Y}}}^{eq}\right)\left(\underline{\mathbf{I}} + \hat{\underline{\mathbf{Y}}}^{eq}\right)^{-1}, \tag{38}$$

where $\underline{\mathbf{I}}$ is the identity matrix, and:

$$\hat{\underline{\mathbf{Y}}}^{eq} = R_0 \underline{\mathbf{Y}}^{eq}, \tag{39}$$

R_0 denoting the selected reference resistance at the input ports of the transmit and receive antennas. The synthesis procedure is based on an iterative non-linear fitting procedure [8] aimed to minimize, across a specific frequency band B, the mean-square error functional:

$$e_B\left(R_{ij}^{(p)}, L_{ij}^{(p)}, C_{ij}^{(p)}\right) = \sqrt{\int_B \left\| \underline{\mathbf{S}}^{FDTD}(f) - \underline{\mathbf{S}}^{eq}\left(f, R_{ij}^{(p)}, L_{ij}^{(p)}, C_{ij}^{(p)}\right) \right\|^2 df}, \tag{40}$$

$\underline{\mathbf{S}}^{FDTD}$ being the frequency-dependent scattering matrix of the structure computed by means of the locally conformal FDTD technique. In this way, it has been possible to evaluate numerically the circuital parameters, listed in Table 1, of the radar unit located above the lossy homogeneous material half space with relative permittivity $\varepsilon_r = 6$ and electrical conductivity $\sigma = 15 mS/m$. As shown in Fig. 12, a few resonant $\pi-$ subnetworks are necessary to describe adequately the antenna return-loss and mutual coupling coefficient in the frequency band of interest.

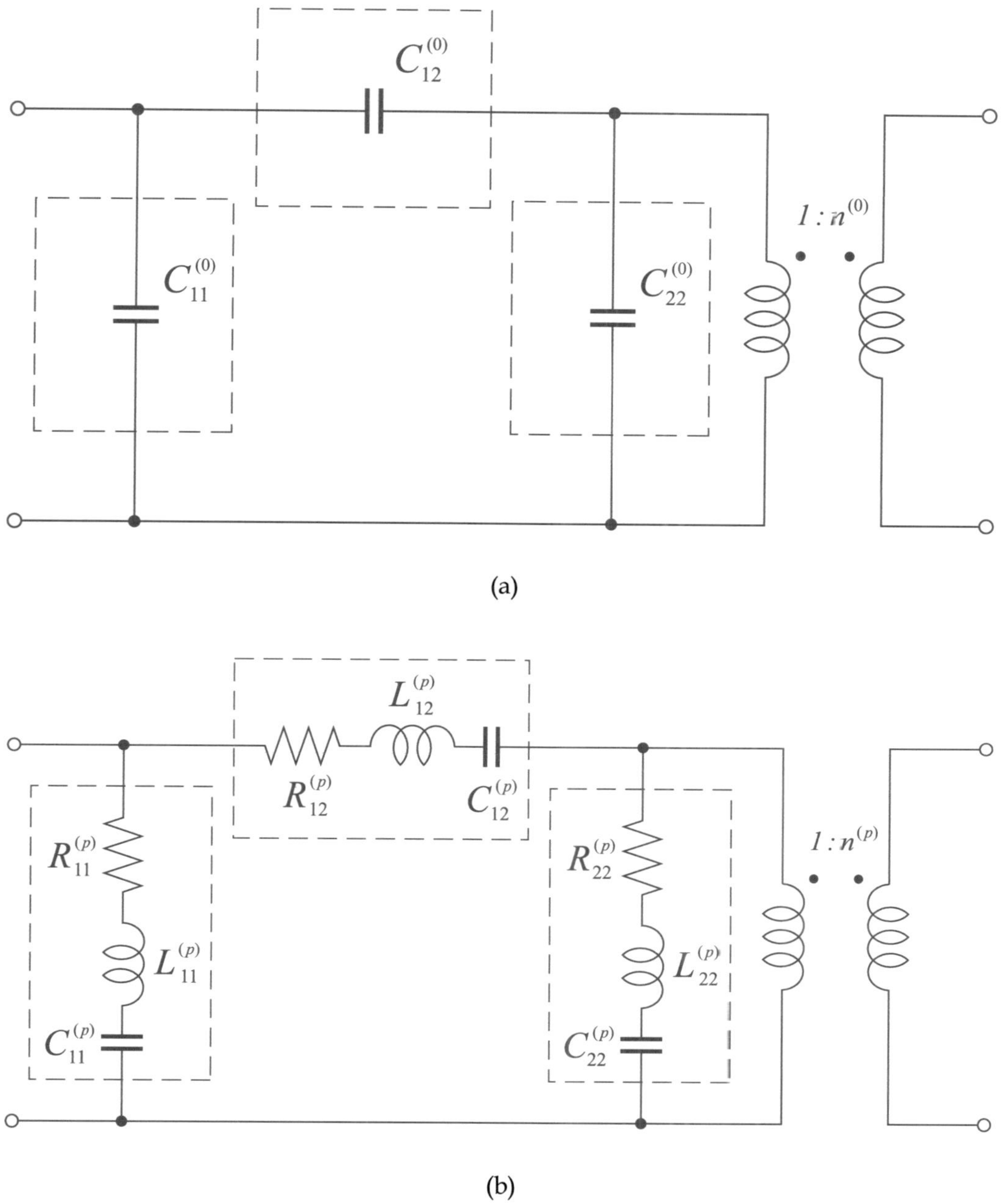

Fig. 11. Frequency-independent equivalent circuit model. Quasi-static (a), and $p-$ th high-frequency (b) $\pi-$ subnetwork. For symmetrical structures the transformation ratio $n^{(p)}$ is equal to ± 1 for each subnetwork.

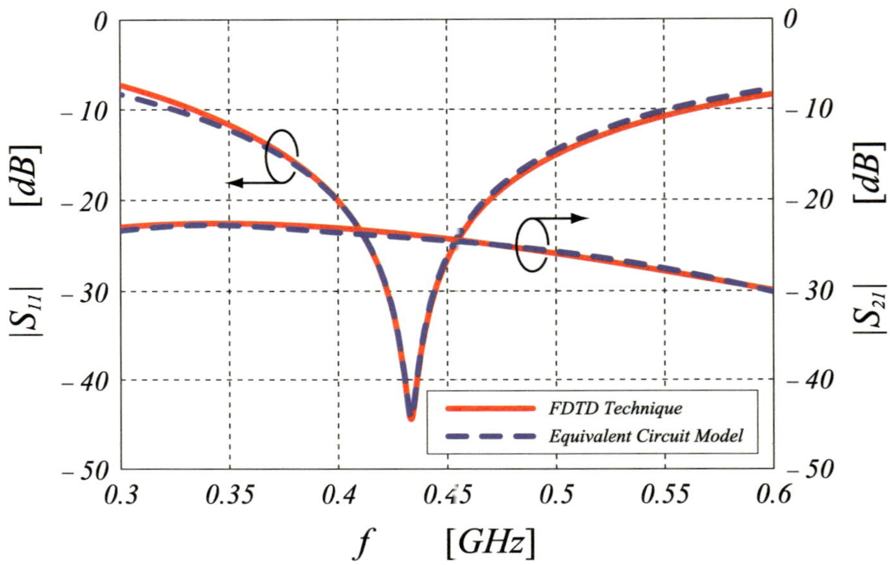

Fig. 12. Magnitude of the scattering parameters of the dipole pair versus frequency. A good agreement between the FDTD numerical results and those obtained by means of the equivalent circuit model is observed. Structure characteristics: $l_d = 40cm$, $D_d = 5mm$, $\delta = 2.5mm$, $s_d = 20cm$, $h_d = 3cm$.

$C_{11}^{(0)} = C_{22}^{(0)}$	$0.579\ pF$
$C_{21}^{(0)}$	$14.562\ fF$
$n^{(0)}$	+1
$R_{11}^{(1)} = R_{22}^{(1)}$	$208.073\ \Omega$
$L_{11}^{(1)} = L_{22}^{(1)}$	$0.141\ \mu H$
$C_{11}^{(1)} = C_{22}^{(1)}$	$1.565\ pF$
$R_{21}^{(1)}$	$1.307\ k\Omega$
$L_{21}^{(1)}$	$1.507\ \mu H$
$C_{21}^{(1)}$	$0.147\ pF$
$n^{(1)}$	+1
$R_{11}^{(2)} = R_{22}^{(2)}$	$322.581\ \Omega$
$L_{11}^{(2)} = L_{22}^{(2)}$	$0.092\ \mu H$
$C_{11}^{(2)} = C_{22}^{(2)}$	$0.235\ pF$
$R_{21}^{(2)}$	$1.838\ k\Omega$
$L_{21}^{(2)}$	$1.084\ \mu H$
$C_{21}^{(2)}$	$0.020\ pF$
$n^{(2)}$	+1

Table 1. Circuital parameters relevant to the equivalent circuit of the antenna pair. Structure characteristics: $l_d = 40cm$, $D_d = 5mm$, $\delta = 2.5mm$, $s_d = 20cm$, $h_d = 3cm$.

6. Conclusion

The full-wave analysis of electromagnetic sensing of buried pipes with GPR in realistic scenarios has been carried out. An enhanced locally conformal FDTD technique, useful to accurately model complex electromagnetic structures as well as ground-embedded inhomogeneities with arbitrary shape and material parameters, has been adopted. By using this scheme, an extensive parametric analysis of the antenna scattering parameters and radiated near-field spatial distribution has been performed for different Tx–Rx antenna separations and elevations over the ground, taking into account the presence of buried metallic and dielectric targets, as well as soil-embedded ellipsoidal inhomogeneities with arbitrary size, location and electrical properties. The obtained numerical results provide a physical insight into the underlying mechanisms of subsurface scattering and antenna mutual coupling processes. Finally, a frequency-independent equivalent circuit, useful to be employed in CAD tools, has been derived from the antenna scattering parameters, showing that including the effect of just a few resonant modes yields high numerical accuracy.

7. Appendix

In order to validate the accuracy of the proposed locally conformal FDTD scheme a number of test cases have been considered. Here the results obtained for the computation of the fundamental resonant frequency of a dielectric resonator enclosed in a metallic cavity are presented. The structure under consideration (see Fig. 13a) has been already analyzed in [5]. It consists of a perfectly conducting metallic cavity of dimensions $a = b = 50\,mm$ and $c = 30\,mm$, loaded with a cylindrical dielectric (ceramic) puck having diameter $D = 36\,mm$,

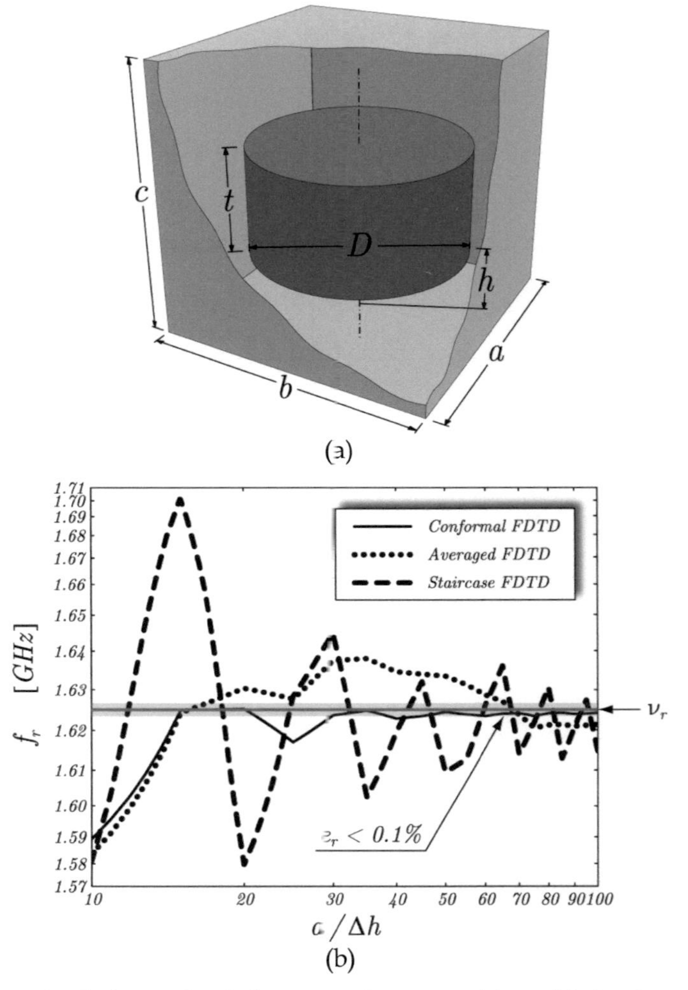

Fig. 13. Geometry of a dielectric loaded rectangular cavity (a), and behaviour of the relevant fundamental resonant frequency f_r as function of the FDTD mesh size Δh (b). Shown is the confidence region where the relative error e_r with respect to the reference resonant frequency $\nu_r = 1.625\,GHz$ [5] is smaller than 0.1%. Structure characteristics: $a = b = 50\,mm$, $c = 30\,mm$, $D = 36\,mm$, $t = 16\,mm$, $h = 7\,mm$. Relative permittivity of the dielectric puck: $\varepsilon_r = 37$.

height $t = 16\ mm$ and relative dielectric constant $\varepsilon_r = 37$. The puck is suspended at a distance of $h = 7\ mm$ from the bottom of the cavity. Since the dielectric permittivity of the resonator is rather high, the effect of the orthogonal Cartesian mesh being not conform to the resonator shape is expected to be noticeable. Here the structure is analyzed by means of a standard FDTD scheme featuring the traditional staircase approximation of the resonator's contour, and by means of the weighted averaging approach proposed in [7], and the locally conformal FDTD technique detailed in Section III. The numerical results obtained from these FDTD schemes are compared against the ones reported in [5] resulting from the use of a commercial Transmission Line Matrix (TLM) method-based solver. To this end, a cubic FDTD mesh having fixed spatial increment Δh has been adopted to analyze the structure. As it appears in Fig. 13b, this example clearly demonstrates the suitability of the proposed approach to efficiently handle complex metal-dielectric structures with curved boundaries. The proposed locally FDTD scheme introduces a significant improvement in accuracy over the stair-casing approximation, converging very quickly to the reference value. Such feature is thus of crucial importance to optimize the design of antennas for ground-penetrating radar applications.

8. References

[1] Caratelli D. & Cicchetti R., (2003). A full-wave analysis of interdigital capacitors for planar integrated circuits, *IEEE Trans. Magnetics*, Vol. 39(No. 3): 1598-1601.

[2] Caratelli D., Cicchetti R., Bit-Babik G., & Faraone A., (2006). A perturbed E-shaped patch antenna for wideband WLAN applications, *IEEE Trans. Antennas Propagat.*, Vol. 54(No. 6): 1871–1874.

[3] Caratelli D., Yarovoy A., & Ligthart L. P., (2007). Antennas for ground-penetrating radar applications, Delft University of Technology, *Tech. Rep. IRCTR–S–032–07*.

[4] Caratelli D., Yarovoy A., & Ligthart L. P., (2008). Full-wave analysis of cavity-backed resistively-loaded bow-tie antennas for GPR applications, *Proc. European Microwave Conference*, Amsterdam, the Netherlands, pp. 204-207.

[5] Chuma J., Sim C. W., & Mirshekar-Syahkal D., (1999). Computation of resonant frequencies of dielectric loaded rectangular cavity using TLM method, *IET Electron. Lett.*, Vol. 35(No. 20): 1712–1713.

[6] Daniels D., (2004). *Ground Penetrating Radar*, 2nd ed., IEE Press.

[7] Dey S. & Mittra R., (1999). A conformal finite-difference time-domain technique for modeling cylindrical dielectric resonators, *IEEE Trans. Microwave Theory Tech.*, Vol. 47(No. 9): 1737–1739.

[8] Fletcher R., (1980). *Practical methods of optimization*, John Wiley.

[9] Freundorfer A., Iizuka K., & Ramseier R., (1984). A method of determining electrical properties of geophysical media, *J. Appl. Phys.*, Vol. 55: 218–222.

[10] Guillemin E. A., (1965). *Synthesis of Passive Network: Theory and Methods Appropriate to the Realization and Approximation Problems*, John Wiley.

[11] Gürel L. & Oguz U., (2001). Simulations of ground-penetrating radars over lossy and heterogeneous grounds, IEEE Trans. Geosci. Remote Sensing, Vol. 39(No. 6): 1190–1197.

[12] Gürel L. & Oguz U., (2003). Optimization of the transmitter–receiver separation in the ground-penetrating radar, *IEEE Trans. Antennas Propagat.*, Vol. 51(No. 3): 362–370.

[13] Iizuka K., Freundorfer A. P., Wu K. H., Mori H., Ogura H., & Nguyen V., (1984). Step-frequency radar, *J. Appl. Phys.*, Vol. 56: 2572–2583.

[14] Kaneda N., Houshmand B., & Itoh T., (1997). FDTD analysis of dielectric resonators with curved surfaces, *IEEE Trans. Microwave Theory Tech.*, Vol. 45(No. 9): 1645–1649.

[15] Maloney J. G. & Smith G. S., (1993). A study of transient radiation from the Wu-King resistive monopole - FDTD analysis and experimental measurements, *IEEE Trans. Antennas Propagat.*, Vol. 41(No. 5): 668–676.

[16] Montoya T. P. & Smith G. S., (1996). A study of pulse radiation from several broad-band loaded monopoles, *IEEE Trans. Antennas Propagat.*, Vol. 44(No. 8): 1172–1182.

[17] Moray R. M., (1974). Continuous subsurface profiling by impulse radar, *Proc. Eng. Found. Conf. Amer. Soc. Civil Eng.*, pp. 213–232.

[18] Peter L. Jr., Young J. D., & Daniels J., (1994). Ground penetration radar as a subsurface environmental sensing tool, *Proc. IEEE*, Vol. 82: 1802–1822.

[19] Taflove A. & Hagness S. C., (2005) *Computational Electrodynamics: The Finite Difference Time Domain Method*, 3rd ed., Artech House.

[20] Timmins I. & Wu K., (2000). An efficient systematic approach to model extraction for passive microwave circuits, *IEEE Trans. Microwave Theory Tech.*, Vol. 48(No. 9): 1565–1573.

[21] Yee K. S., (1966). Numerical solution of initial boundary value problems involving Maxwell's equations, *IEEE Trans. Antennas Propagat.*, Vol. 14(No. 3): 302–307.

18

Impact of Ultra Wide Band Emission on Next Generation Weather RADAR and the Downlink of UMTS2600

Bazil Taha Ahmed[1] and Miguel Calvo Ramon[2]
[1]*Universidad Autonoma de Madrid,*
[2]*Universidad Politecnica de Madrid*
Spain

1. Introduction

The Federal Communications Commission (FCC) agreed in February 2002 to allocate 7.5 GHz of spectrum for unlicensed use of ultra-wideband (UWB) devices for communication applications in the 3.1–10.6 GHz frequency band, the move represented a victory in a long hard-fought battle that dated back decades. With its origins in the 1960s, when it was called time-domain electromagnetic, UWB came to be known for the operation of sending and receiving extremely short bursts of RF energy. With its outstanding ability for applications that require precision distance or positioning measurements, as well as high-speed wireless connectivity, the largest spectrum allocation ever granted by the FCC is unique because it overlaps other services in the same frequency of operation. Previous spectrum allocations for unlicensed use, such as the Unlicensed National Information Infrastructure (UNII) band have opened up bandwidth dedicated to unlicensed devices based on the assumption that "operation is subject to the following two conditions:

1. This device may not cause harmful interference. Harmful interference is defined as interference that seriously degrades, obstructs or repeatedly interrupts a radio communication service.
2. This device must accept any interference received, including interference that may cause undesired operation. This means that devices using unlicensed spectrum must be designed to coexist in an uncontrolled environment.

Devices utilizing UWB spectrum operate according to similar rules, but they are subject to more stringent requirements because UWB spectrum underlays other existing licensed and unlicensed spectrum allocations. In order to optimize spectrum use and reduce interference to existing services, the FCC's regulations are very conservative and require very low emitted power.

UWB has a number of advantages which make it attractive for consumer communications applications. In particular, UWB systems

- Have potentially low complexity and low cost;
- Have noise-like signal characteristics;
- Are resistant to severe multipath and jamming;
- Have very good time domain resolution.

In 1988, the NEXRAD Agencies established the WSR-88D (Weather Surveillance Radar 88 Doppler) Radar Operations Centre (ROC) in Norman, Oklahoma. The ROC employees come from the National Weather Service, Air Force, Navy, FAA, and support contractors. The ROC provides centralized meteorological, software, maintenance, and engineering support for all WSR-88D systems. WSR-88D systems will be modified and enhanced during their operational life to meet changing requirements, technology advances, and improved understanding of the application of these systems to real-time weather operations. The ROC also operates WSR-88D test systems for the development of hardware and software upgrades to enhance maintenance, operation, and provide new functionality.

NEXRAD is used to warn the people of the United States about dangerous weather and its location. Meteorologists can now warn the public to take shelter with more notice than any previous radar. There are 158 operational NEXRAD radar systems deployed throughout the United States and at selected overseas locations. The maximum range of the NEXRAD radar is 250 nautical miles. The NEXRAD network provides significant improvements in severe weather and flash flood warnings, air traffic safety, flow control for air traffic, resource protection at military bases, and management of water, agriculture, forest, and snow removal.

The spectrum for UMTS lies between 1900 MHz to 2025 MHz and 2110 MHz to 2200 MHz. For the satellite service an own sub-band in the UMTS spectrum is reserved (uplink 1980 MHz to 2010 MHz, downlink 2170 MHz to 2200 MHz). The remaining spectrum for terrestrial use is divided between two modes of operation. In the FDD (Frequency Division Duplex) mode there are two equal bands for the uplink (1920 MHz to 1980 MHz) and for the downlink (2110 MHz to 2170 MHz). In the operation mode TDD (Time division duplex) uplink and downlink are not divided by use of different frequency carriers but by using different timeslots on the same carrier. So there is no need for a symmetrical spectrum but the remaining unpaired spectrum can be used.

The European Conference of Postal and Telecommunications administrations (CEPT) have recommended that the 2500-2690 MHz band should be reserved for the use by licensed UMTS services. It has been recommended that the 2500-2570 and 2620-2690 MHz bands should be paired for UMTS FDD deployment with frequency blocks in multiples of 5 MHz. Here after we will dominate this system by UMTS2600.

In (Hamalainen et al., 2002) the coexistence of the UWB system with GSM900, UMTS/WCDMA, and GPS has been investigated. They have evaluated the level of the interference caused by different UWB signal to the three up mentioned systems. Also they have evaluated the performance degradation of UWB systems in the presence of narrow bandwidth interference and pulsed jamming. They have given the bit error rate (BER) of the above mentioned systems for different pulse length.

In (Hamalainen et al., 2004) the coexistence of the UWB system with IEEE802.11a and UMTS in Modified Saleh-Valenzuela Channel has been studied. The UWB system performance has been studied in the presence of multiband interference. The interference sources considered are IEEE802.11a and UMTS which are operating simultaneously with their maximum system bandwidths. The system under consideration is single band and single user UWB link operating at data rate of 100 Mbps without error correction coding. They have given the bit error rate (BER) of the UWB system for different types of modulation (Direct sequence and Time Hopping).

The interference between the UMTS and the UWB system has been studied in (Giuliano et al, 2003). The free space propagation model has been used to calculate the UWB signal propagation loss. It has been concluded that, a carrier frequency of 3.5 GHz is the minimum allowable value for UWB device transmitting at 100 Mbps in order to avoid harmful interference between UMTS and UWB. In (Hamalainen et al., 2001a), the effect of the in band interference power caused by different kinds of UWB signal at UMTS/WCDMA uplink and downlink frequency bands has been investigated. UWB frequency spectra have been produced by using several types of narrow pulse waveforms. They have concluded that one can reduce interfering UWB power by using different waveforms and pulse widths to avoid UMTS frequencies without any additional filtering. In (Hamalainen et al., 2001 b) the effect of the in band interference power caused by three different kinds of UWB signal on GPS L1 and GSM-900 uplink band has been studied. UWB frequency spectra have been generated by using several types of narrow pulse waveforms based on Gaussian pulse. In band interference power has been calculated over the IF bandwidth of the two victim receiver as a function of the UWB pulse width. Also the signal attenuation with distance has been presented.

In (Ahmed et al., 2004), the effect of the UWB on the DCS-1800 and GSM-900 macrocell downlink absolute range using the (Line of Sight) propagation model between the UWB transmitter and the mobile receiver has been studied without taking into account the shadowing factor within the propagation loss model.

In (ITU, 2003), the effect of UWB system on fixed service system (point to point and Fixed Wireless Access (FWA) systems in bands from 1 to 6 GHz has been investigated. It has been concluded that, when the UWB transmitter is in LOS with the two systems antennas, the effect is very high when the UWB power density is -41.3 dBm/MHz.

2. UWB effect on the NEXRAD

RADAR systems performance (detection) is almost the optimum when the Signal to Noise Ratio (SNR) is 16 dB or more. Any extra interference due to communications systems degrades the performance (probability of detection with constant range or the range with constant probability of detection) of the radar system. Thus, the extra interference should not exceed a given value. In practice, extra interference should be within the following range:

$$I_{extra} \leq 0.1\, P_{n-RADAR} \tag{1}$$

where

I_{extra} is the extra interference due to other communications systems,

$P_{n\text{-}RADAR}$ is the RADAR receiver noise calculated as:

$$P_{n-RADAR}(dB) = -114 + 10\log_{10}(BW_{MHZ}) + NF(dB) \tag{2}$$

where

- BW_{MHz} is the radar system IF bandwidth measured by MHz .
- NF(dB) is the RADAR receiver noise figure measured in dB.

The UWB interference power I_{UWB} is calculated by:

$$I_{UWB} = P_{UWB} - L_{UWB}(d) + L_{exra} + G_{Ant} \tag{3}$$

where

- P_{UWB} is the UWB EIRP in dBm in the radar band,

- $L_{UWB}(d)$ is the propagation loss between the UWB device and the RADAR system as a function of the distance between the UWB source and the radar,
- L_{extra} is the extra propagation loss due to tree or building insertion loss when is applicable,
- G_{Ant} is the RADAR antenna gain in the direction of the UWB transmitter.
- α is the second propagation exponent of the UWB signal with a typical value of 4 to 5 depending on the surrounding environment.

Using the Two-Slope Propagation Model, the UWB signal propagation loss L_{UWB} in dB at a distance d can be given as (Ciccoganini et al., 2005):

$$L_{UWB}(d) \approx \begin{cases} 20\log_{10}\left(\frac{4\pi d}{\lambda}\right) & d \le R_b \\ 20\log_{10}\left(\frac{4\pi R_b}{\lambda}\right) + 10\alpha\log_{10}\left(\frac{d}{R_b}\right) & R_b < d \end{cases} \tag{4}$$

where λ is the wavelength and R_b is the break point distance given by (Ahmed et al., 2002):

$$R_b = \frac{4\, h_{UWB}\, h_{RADAR}}{\lambda} \tag{5}$$

where

- h_{UWB} is the UWB antenna height,
- h_{RADAR} is the RADAR antenna height.

3. UWB effect on the UMTS2600 downlink performance

To account for UWB interference, an extra source of interference is added linearly to the UMTS2600 intracellular interference I_{UMTS}. The interference power is calculated by assuming the UWB source to be at different distances from the UMTS2600 receiver (the mobile station). Therefore, the interference power generated by a device UWB, I_{UWB}, is given by (in dBm) (Ahmed, Ramon, 2008):

$$I_{UWB} = P_{UWB} - L_{UWB}(d) + G_{UMTS} \tag{6}$$

where

- P_{UWB} is the UWB EIRP in dBm in the UMTS2600 band.
- $L_{UWB}(d)$ is the path-loss between the UWB device and the UMTS2600 receiver which varies with the separation distance, d in m, and
- G_{UMTS} is the UMTS2600 antenna gain.

Given that UWB devices are typically low power, short range devices, then the line-of-sight path-loss model is often most appropriate for distances less than 5m. Thus the UWB signal propagation loss in dB is calculated as (Ahmed, Ramon, 2008):

$$L_{UWB}(d) \approx 20\,\log_{10}\left(\frac{4\pi}{\lambda}\right) + 20\,\log_{10}(d) = 40.92 + 20\log_{10}(d) \tag{7}$$

where λ is the wavelength at an operating frequency of 2.655 GHz.

The effect of the UWB interference is to reduce the UMTS2600 macrocell range or/and the macrocell capacity.

The macrocell downlink range R_{UMTS} with the existence of the UWB interference is given as (Ahmed, Ramon, 2008):

$$R_{UMTS} = R_{UMTS,o} \sqrt[s]{\frac{I_{UMTS}}{(I_{UMTS} + I_{UWB})}} \tag{8}$$

where RUMTS,o is the UMTS2600 downlink range without UWB interference.

The UMTS2600 normalized macrocell range R_n is given as:

$$R_n = \frac{R_{UMTS}}{R_{UMYTS,o}} = \sqrt[s]{\frac{I_{UMTS}}{(I_{UMTS} + I_{UWB})}} \tag{9}$$

where s is the UMTS2600 outdoor signal propagation exponent (3.5 to 4.5).

The UMTS2600 normalized downlink capacity C_n is given as (Ahmed, Ramon, 2008):

$$C_n = \frac{C_{UMTS}}{C_{UMTS,o}} = \left(\frac{I_{UMTS}}{I_{UMTS} + I_{UWB}}\right) \tag{10}$$

where C_{UMTS} is the UMTS2600 downlink capacity with the UWB interference, and $C_{UMTS,o}$ is the UMTS2600 downlink capacity without the UWB interference.

4. Results

For an outdoor environment (UWB transmitter out side of any building), the FCC maximum permitted UWB EIRP power density for the frequency range 2.7 to 3.0 GHz is -61.3 dBm/MHz while it is -51.3 dBm/MHz for indoor environment (UWB transmitter is within a given building).

We study the effect of the UWB system on the NEXRAD system assuming that the RADAR receiver noise is – 114 dBm, its operating frequency is 2.9 GHz, the second propagation exponent α is 4 and that its antenna height is 30 m. Here we have assumed that, the UWB maximum allowed interference is -124 dBm (10 dB protection) which give a rise to about 2.5% reduction of the NEXRAD range. Fig. 1 shows the NEXRAD vertical pattern.

Fig. 2 shows the acceptable UWB power density for three different UWB antenna heights. It can be noticed that the coordinate distance (minimum distance between the UWB transmitter and the Radar) is almost 0 km when the UWB antenna height is 3 m. The coordinate distance will be 1.12 and 1.50 km when the UWB antenna height is 15 and 30 m respectively. Second and third cases (UWB antenna height of 15 to 30 m) should be avoided as far as possible. At an UWB antenna height of 30m, the UWB interference will be injected to the NEXRAD receiver through the NEXRAD antenna main-lobe. Thus, the UWB effect will be the maximum.

Fig. 3 shows the acceptable UWB power density for three different UWB antenna heights assuming that some trees are between the UWB antenna and the RADAR antenna and that the tree absorption loss is 10 dB. It can be noticed that the coordinate distance is 0 km when the UWB antenna height is 3 m. The coordinate distance will be 0 and 0.48 km when the UWB antenna height is 15 and 30 m respectively.

Fig. 4 shows the acceptable UWB power density for three different UWB antenna heights assuming that the UWB transmitter is within a high building and that the wall absorption loss is 10 dB. It can be noticed that the coordinate distance is 0 km when the UWB antenna height is 3 m. The coordinate distance will be 1.12 and 1.50 km when the UWB antenna height is 15 and 30 m respectively.

For the above three mentioned cases, it has been assumed that, the RADAR main beam is in the direction of the UWB transmitter and that the RADAR antenna has a tilt of 0.0°.

Fig. 5 shows the acceptable UWB power density for three different UWB antennas tilting assuming that the UWB antenna height is 3 m. It can be noticed that the coordinate distance is 0 km when the UWB antenna tilt is 0°. Also, the coordinate distance will be 0 km when the UWB antenna tilt is 3° or 6°. Thus, the effect of the UWB is null with any positive RADAR antenna tilt of 3 degrees or more assuming that the UWB antenna height is 3 m.

Fig. 6 shows the acceptable UWB power density for three different UWB antennas tilting assuming that the UWB antenna height is 30 m. It can be noticed that the coordinate distance is 1.5 km when the UWB antenna tilt is 0°. The coordinate distance will be 0 km for UWB antenna tilt angle of 3° and 6°.

The same results are applicable for an operating frequency of 3 GHz. For a distance of 100m between the UWB transmitter and the Radar, the UWB EIRP power density at 3 GHz should be -84 dBm/MHz or lower.

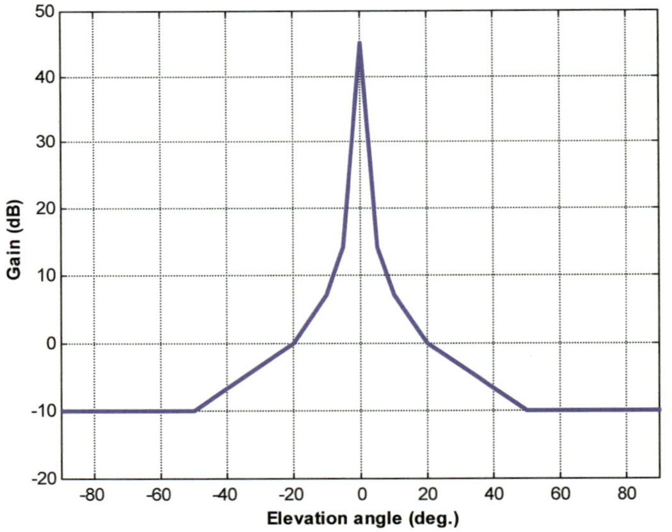

Fig. 1. NEXRAD Vertical Antenna Pattern.

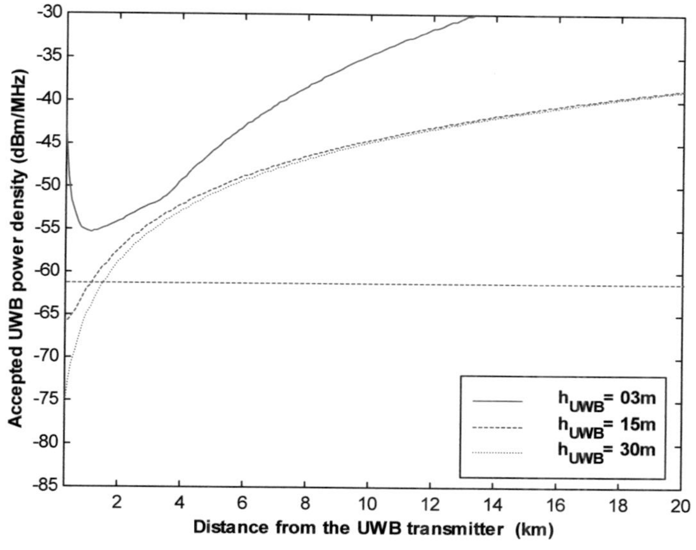

Fig. 2. Maximum permitted UWB EIRP for an outdoor environment for three different UWB antenna height (RADAR antenna height = 30m).

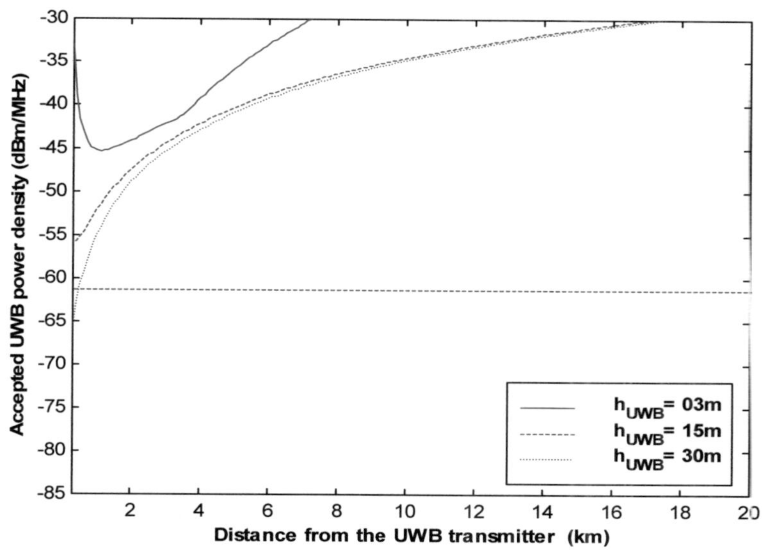

Fig. 3. Maximum permitted UWB EIRP for an outdoor environment for three different UWB antenna height (RADAR antenna height = 30m and 10 dB tree absorption loss).

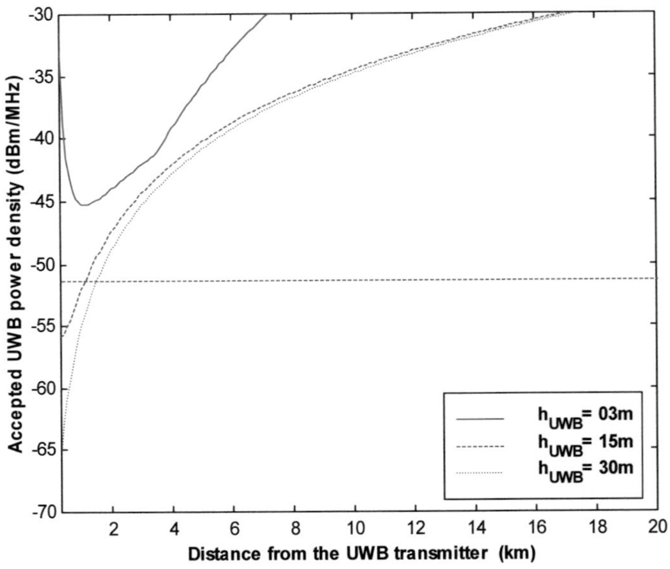

Fig. 4. Maximum permitted UWB EIRP for an indoor environment for three different UWB antenna height (RADAR antenna height = 30m and 10 dB wall absorption loss).

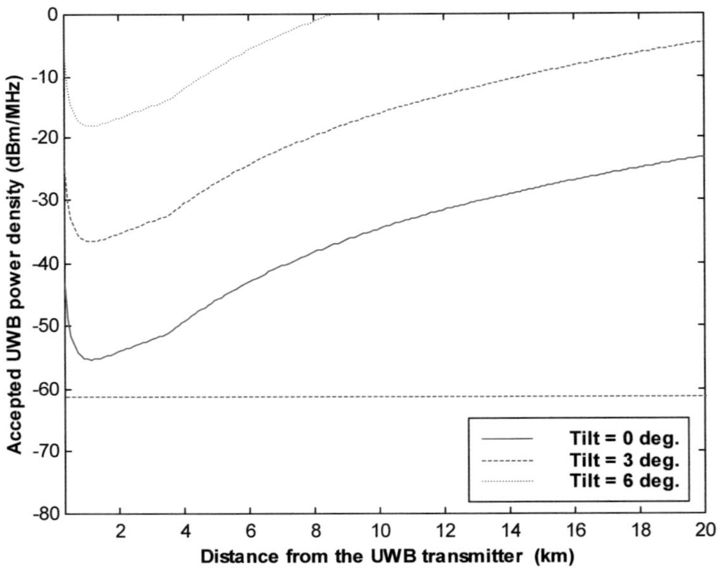

Fig. 5. Maximum permitted UWB EIRP for an outdoor environment for three RADAR antenna tilt (UWB antenna height = 3m and RADAR antenna height = 30m).

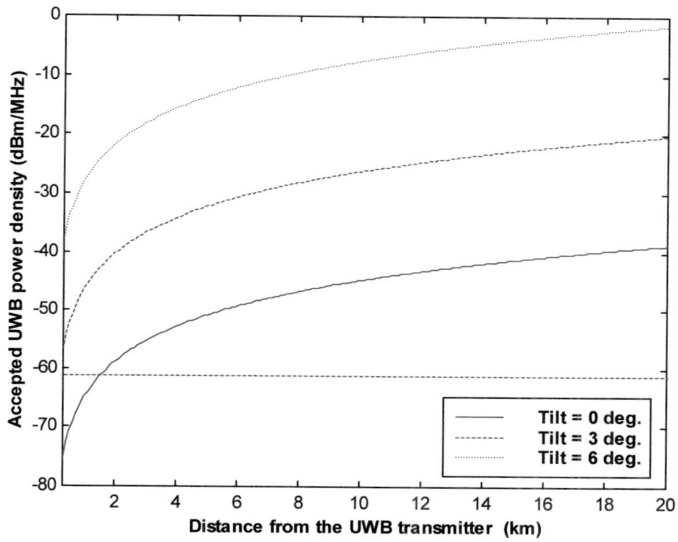

Fig. 6. Maximum permitted UWB EIRP for an outdoor environment for three RADAR antenna tilt (UWB antenna height = 30m and RADAR antenna height = 30m).

Here we address the effect of the UWB system on the downlink of the UMTS2600 system. In the analysis we assume that the UWB data rate is higher than the UMTS2600 chip rate, i.e., the UWB bit rate is higher than 4 Mbps. In Fig. 7, the UWB interference power on the UMTS2600 downlink (i.e. interference as seen at the mobile) is plotted assuming an average P_{UWB} of -51.3 dBm/MHz within the UMTS2600 bandwidth.

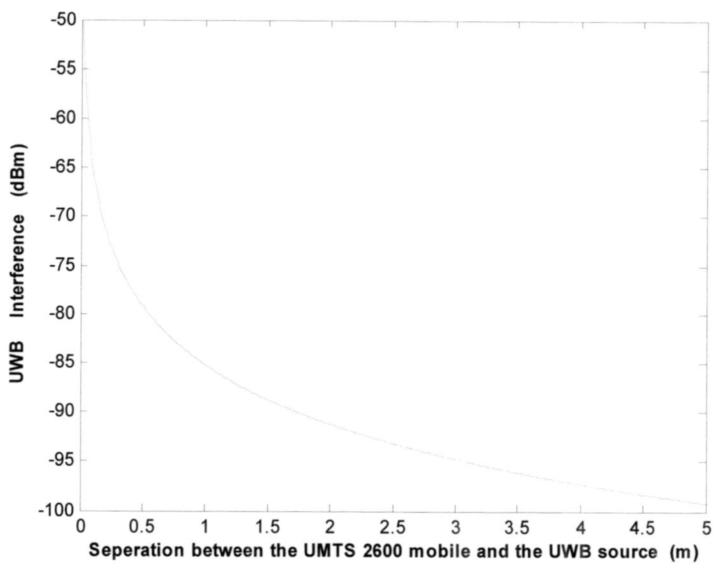

Fig. 7. UWB interference as a function of the separation between the UWB transmitter and the UMTS2600 mobile (P_{UWB} = -51.3 dBm/MHz).

We study the case of voice service (G_p = 25 dB and $(E_b/N_o)_{req}$ = 6 dB) (Ahmed, Ramon, 2008) assuming an UMTS2600 interference of -88 dBm (14 dB noise rise) and UWB power density of -51.3 dBm/MHz. Fig. 8 shows the downlink macrocell normalized range as a function of the separation between the UMTS2600 mobile and the UWB transmitter for three different values of the propagation exponent s. It can be noticed that the UWB signal creates a high interference which reflects a macrocell normalized range reduction of 26% when the separation is 1m. For larger separation, the interference is lower and thus the range reduction is also lower.

Fig. 9 shows the downlink macrocell normalized capacity as a function of the separation between the UMTS2600 mobile and the UWB transmitter for the same UWB power density. It can be noticed that the UWB signal creates a high interference which gives arise a macrocell normalized capacity reduction of 66 % when the separation is 1m. For larger separation, the interference is lower and thus the normalized capacity reduction is also lower.

Thus, it can concluded that, the UWB recommended power density of -51.3 dBm recommended by FCC is very high and its effect on the UMTS2600 system is dramatic i.e., a reduction of 26% of the macrocell range or a reduction of 66% of the cell capacity. For this reason lower UWB power density should be studied.

Let us now study the case data service (G_p = 14.25 dB and $(E_b/N_o)_{req}$ = 4.25 dB) assuming an UMTS2600 total interference of -92.0 dBm (10 dB noise rise and thus highly loaded macrocell). Fig. 10 shows the downlink macrocell normalized range as a function of the UWB power density. It can be noticed that for a distance of 1m, the macrocell normalized range increases with the reduction of the UWB power density. If we consider that the UWB system is un harmful when the UMTS range reduction is 1% or less then, the recommended UWB power density should be -74 dBm/MHz or lower. This power density is well below the FCC and the ETSI recommendations.

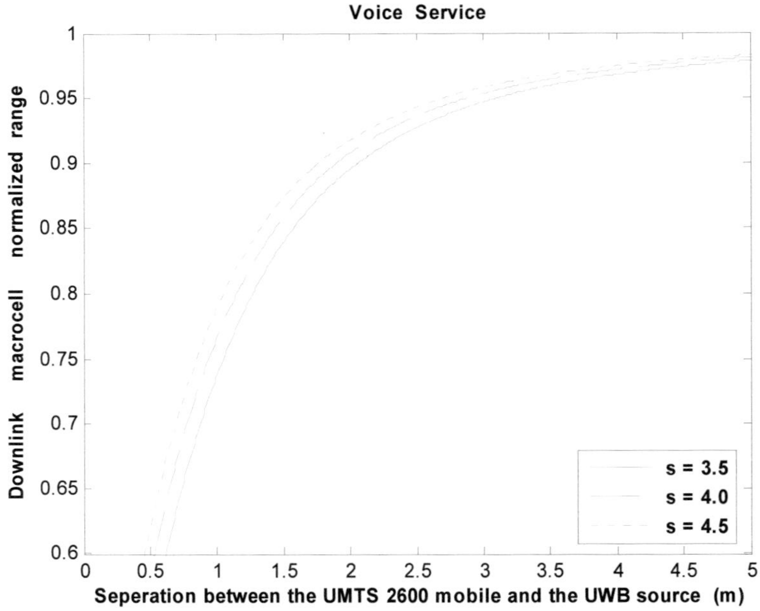

Fig. 8. Effect of the UWB interference on the macrocell range as a function of the separation between the UWB transmitter and the UMTS2600 mobile (P_{UWB} = -51.3 dBm/MHz).

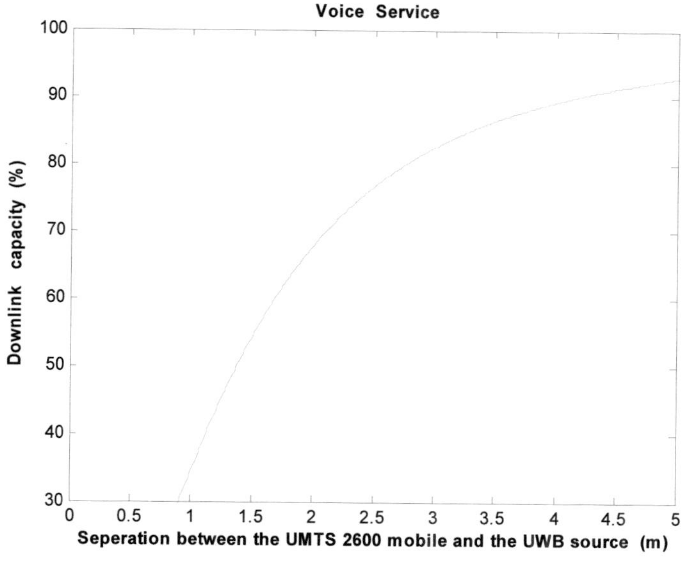

Fig. 9. Effect of the UWB interference on the macrocell normalized capacity as a function of the separation between the UWB transmitter and the UMTS2600 mobile (P_{UWB} = -51.3 dBm/MHz).

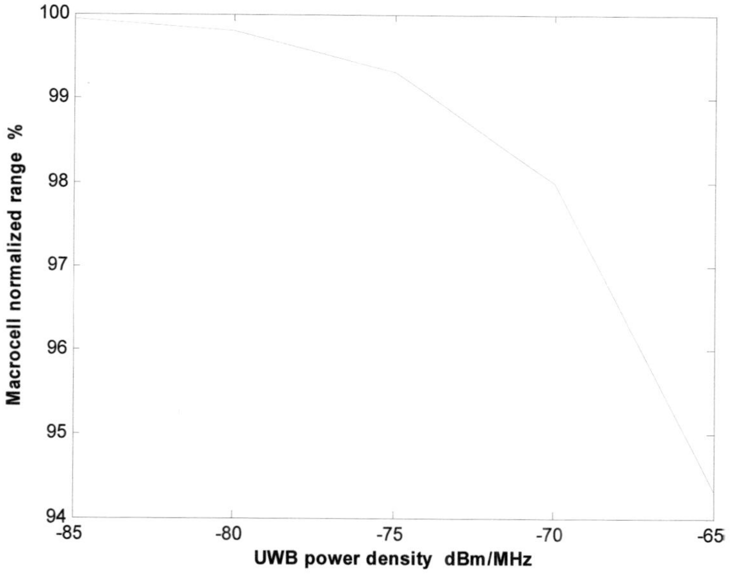

Fig. 10. Effect of the UWB interference on the macrocell normalized range as a function of the UWB power density.

Fig. 11 shows the downlink macrocell normalized capacity as a function of the UWB power density. It can be noticed that for a distance of 1m, the macrocell normalized capacity increases with the reduction of the UWB power density. If we consider that the UWB system is un harmful when the UMTS capacity reduction is 1% or less then, the recommended UWB power density should be -79 dBm/MHz or lower. Also, this power density is well below the FCC and the ETSI recommendations.

Then we study the case of multiple UWB transmitters with one UWB transmitter at each 4×4 m^2 area of the indoor environment assuming P_{UWB} of -82 dBm/MHz, 18 UWB transmitters and noise rise of 10 dB. Fig. 12 shows the downlink macrocell normalized range as a function of the UMTS2600 mobile location for three different values of s. It can be noticed that the UWB signal creates a high interference which will drastically reduce the macrocell normalized range when the UMTS receiver is located at 0 to 0.15m. At a distance higher than 0.4 m from the nearest UWB transmitter, the macrocell range reduction is less than 1%.

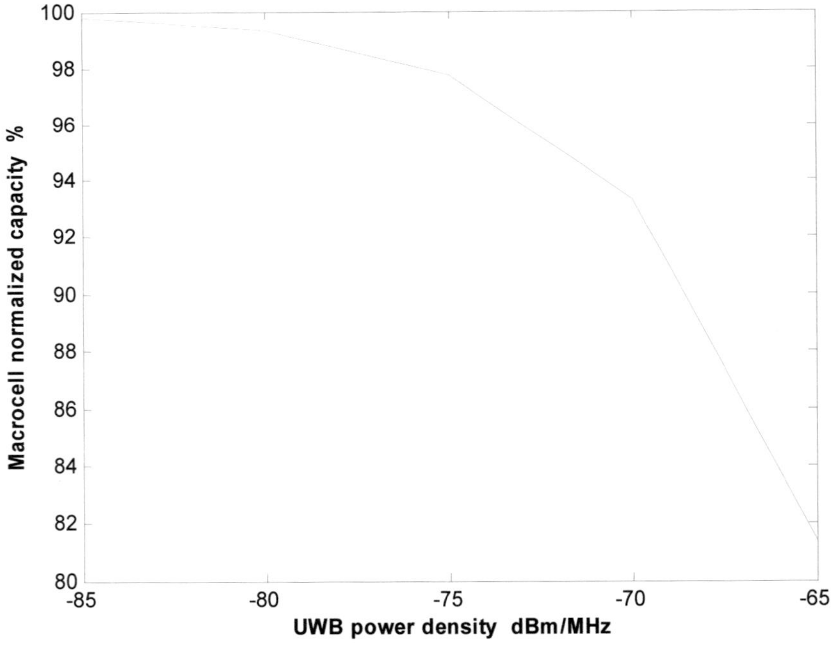

Fig. 11. Effect of the UWB interference on the macrocell normalized capacity as a function the UWB power density.

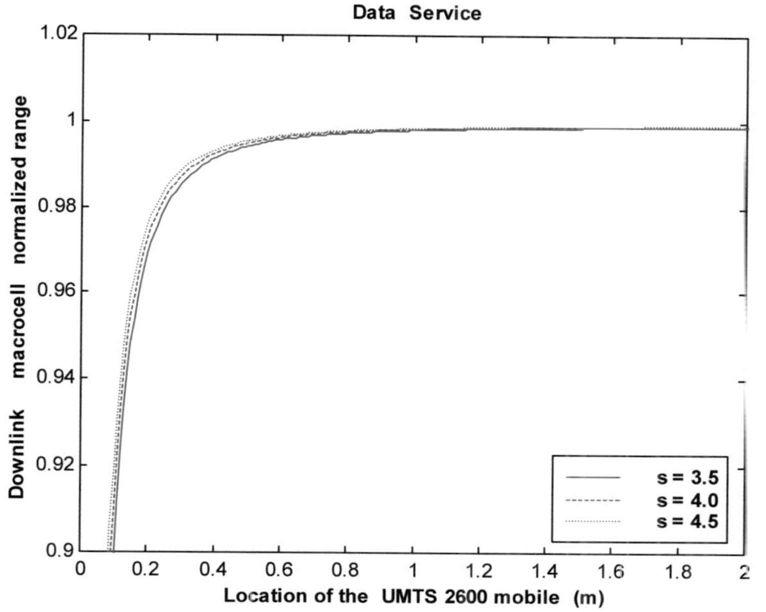

Fig. 12. Effect of the UWB interference on the macrocell range as a function of the UMTS2600 mobile location (P_{UWB} = -82 dBm/MHz) for multi UWB transmitters and 9.5 dB noise rise.

Fig. 13 shows the downlink macrocell capacity as a function of the UMTS2600 mobile location. It can be noticed that the UWB signal creates a high interference which produces a high macrocell normalized capacity reduction when the UMTS2600 receiver is located at a distance less than 0.2m from the nearest UWB transmitter. For a distance greater than 1m from the nearest UWB transmitter, the effect of the UWB transmitters is almost null (less than 1% capacity reduction).

It can be concluded that, for the case of single UWB transmitter, the UMTS2600 can easily tolerate the UWB interference when the UWB EIRP is -79 dBm/MHz for a distance between the UWB transmitter and the UMTS mobile of 1m or higher. For the case of multi UWB transmitter, the UMTS can easily tolerate the UWB interference when the UWB EIRP is -82 dBm/MHz. The above mentioned numbers are valid for highly loaded macrocells, i.e., 90% loaded macrocell. For lower loaded macrocells (50-70)%, the UMTS2600 system can tolerate (1.2-3) dB lower UWB power density. If we reduce the critical distance to 0.5m, we have to lower the maximum accepted UWB power density by 6 dB.

For the common UMTS systems that functions within the band (1.92-2.17) GHz, the tolerable UWB power density is 2 dB lower than the tolerable UWB power density of UMTS2600, i.e., for single UWB transmitter, the common UMTS system can tolerate UWB power density of -81 dBm/MHz. For multi UWB transmitters, the common UMTS can tolerate UWB power density of -84 dBm/MHz.

Fig. 14 shows the FCC, ETSI and our recommended UWB power density masks. It can be noticed that, for a frequency lower than 3.1 GHz, our recommended mask has always lower accepted UWB power density than the FCC mask. For a frequency of 950 MHz to 3.1 GHz, our mask has lower accepted UWB power density than the ETSI mask.

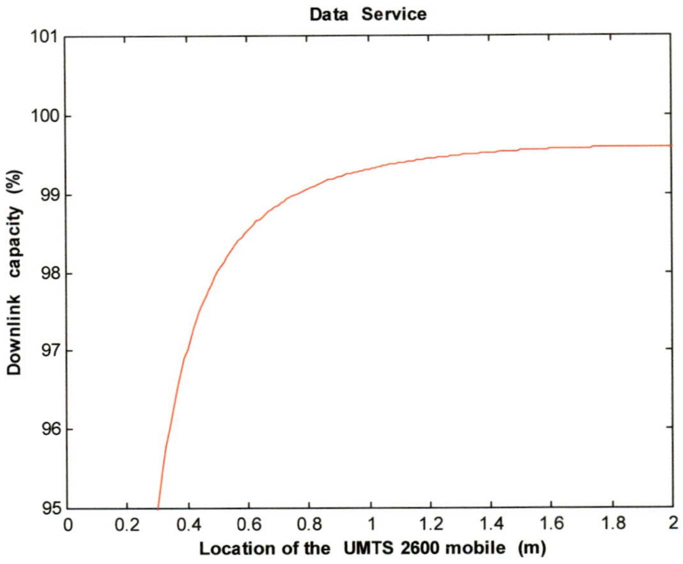

Fig. 13. Effect of the UWB interference on the macrocell normalized capacity as a function of the UMTS2600 mobile location (P_{UWB} = -82 dBm/MHz) for multi UWB transmitters and 9.5 dB noise rise.

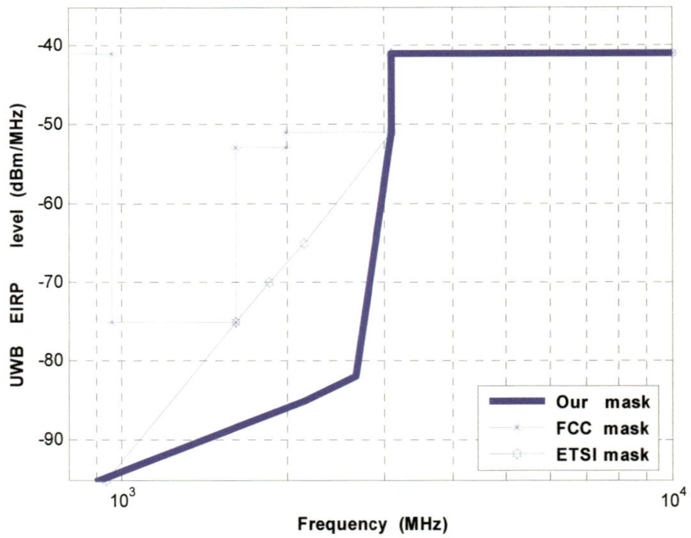

Fig. 14. FCC, ETSI and our recommended accepted UWB power density.

5. Conclusions

The effect of the UWB interference on the on Next Generation Weather RADAR (NEXRAD) has been presented. The coordination distance has been given for different scenarios, i.e., for different UWB antenna heights, different NEXRAD antenna tilt and when a group of trees exist between the UWB antenna and the NEXRAD antenna. Also the case of UWB transmitter in indoor environment has been studied. It has been noticed that, for high antenna tilt of NEXRAD antenna, the effect of the UWB system is null. When the antenna heights of both UWB and NEXRAD are the same the effect of the UWB will be the maximum. Also, it has been noticed that, when a group of trees or a given obstacle exist between the UWB and NEXRAD antennas, the effect of the UWB system will be lower than the case of clear path between the antennas of both systems.

The effect of the UWB transmitters on the UMTS2600 downlink for different configuration and environments has been studied. It has been noticed that, for the case of low UWB power density (-79 dBm/MHz), the effect of the UWB signals is low when the distance between the UWB transmitter and the UMTS2600 receiver is greater than 1m. For the case of multi UWB transmitters, the accepted UWB power density is 1 to 5 dB lower than the accepted UWB power density of the UWB single transmitter case. The UWB power density reduction depends on the number of the UWB transmitters and their spatial density, i.e., the higher is the number of UWB transmitters and their spatial density, higher should be the reduction of the UWB power density. If we reduce the critical distance to 0.5m, we have to lower the maximum accepted UWB power density by 6 dB. It has been noticed that, the effect of the UMTS2600 signal propagation exponent (s) is very little when it has a value of 3.5 to 4.5.

6. References

Ahmed B. T., Ramon M. C. & Ariet L. H., 2002," Capacity and Interference Statistics of Highways W-CDMA Cigar-Shaped Microcells (Uplink Analysis), IEEE Communications Letters, Vol. 6, No. 5, pp. 172-174.

Ahmed B. T., Ramon M. C. & Ariet L. H., 2004, "Impact of Ultra Band (UWB) on Macrocell Downlink of DCS-1800 and GSM-900 Systems", Radioenginnering, Vol. 14, No.1, pp. 51-55.

Ahmed B. T., Ramón M. C.,2008, "On the Impact of Ultra-Wideband (UWB) on Macrocell Downlink of UMTS and CDMA-450 Systems", IEEE Electromagnetic Compatibility, Vol. 5, No. 2, pp. 406-412.

Ciccoganini W., Durantini A., and Cassioli D., 2005, "Time domain propagation measurements of the UWB Indoor Channel Using PN-Sequence in the FCC-Compliant Band 3.6-6 GHz", IEEE trans. Antennas and Propagation, Vol. 53, No. 4, pp. 1542-1549.

Giuliano R., Mazzenga F., Vatalaro F., "On the interference between UMTS and UWB systems", pp: 339 - 343, IEEE Conference on Ultra Wideband Systems and Technologies, 2003 , 16-19 Nov. 2003.

Hamalinen M., Hovinen V., Iinatti J., Latva-aho M., 2001: "In band Interference Power Caused by Different Kinds of UWB Signals at UMTS/WCDMA Frequency Bands", , the 2001 IEEE Radio and Wireless Conference, RAWCON 2001, pp. 97-100, Waltham-Boston, Massachusetts, USA, Aug. , 2001.

Hamalinen M., Iinatti J., Hovinen V., Latva-aho M., 2001," In band Interference of Three Kind of UWB Signals in GPS L1 Band and GSM900 Uplink Band", the 12th International Symposium on Personal, Indoor and Mobile Radio Communications, PIMRC2001, pp. D 76-80, USA, Sep - Oct , 2001.

Hamalainen M. , Hovinrn V. , Tesi R., Iinatti J. & Latava-aho M., 2002, " On the UWB System Coexistance with GSM900, UMTS/WCDMA, and GPS", IEEE Journal on Selected Areas in Communications, Vol. 20, No. 9, pp. 1712-1721.

Hamalinen M., Tesi R., Iinatti J.," UWB co-existence with IEEE802.11a and UMTS in modified Saleh-Valenzuela channel", Ultra Wideband Systems, 2004, pp. 45 - 49, May 2004.

Holma H. , Toskala A., 2000, "WCDMA for UMTS", John Wiley & Sons.

ITU Document 1-8/29-E, 2003, "Updating of Preleminary Study on Coexistance Betwwen UWB and the Fixed Service in Band from 1 to 6 GHz".

High-Precision Time-of-Arrival Estimation for UWB Localizers in Indoor Multipath Channels

Marzieh Dashti[1], Mir Ghoraishi[1], Katsuyuki Haneda[2] and Jun-ichi Takada[1]

[1]*Tokyo Institute of Technology*
[2]*Aalto University School of Science and Technology*
[1]*Japan*
[2]*Finland*

1. Introduction

The global positioning system (GPS) has found application in many different fields, in areas where there is a good line-of-sight (LoS) to GPS satellites, this technique provides a good estimate of the location of user terminal (UT). However, in indoor and dense urban environments, localization has always been a more challenging problem for several reasons. Typically the GPS signal is not strong enough to penetrate through most materials. As soon as an object obscures the GPS satellite from the UT's view, the signal is corrupted. This constrains the usefulness of GPS to open environments, and limits its performance in forests or in dense urban environments, as retaining a lock on the GPS signals becomes more difficult. GPS typically becomes completely useless inside buildings. However there is an increasing need for accurate localization in cluttered environments, in addition to open spaces. In commercial applications for example, the tracking of inventory in warehouses or cargo ships is an emerging need. In military applications the problem of blue force tracking, i.e., knowing where friendly forces are, is of vital importance. This is not a problem in open environments where systems can rely on GPS, but in dense urban or indoor environment, no satisfactory solution exists. Navigation in GPS-denied environment is also a pressing military need. For example untethered robots operating in enclosed environments such as urban canyons or inside buildings need accurate positioning to safely navigate. Indoor localization is of great importance for the applications that a person or a vehicle enter a building and accurately tracking its position over time is needed and the position estimate should have a precision of under one meter, i.e. on the order of some of the building feature dimensions, such as hallway width.

1.1 Indoor localization

To address the problem of localization in cluttered environments, deploying a wireless sensor network (WSN) composed of fixed sensors emitting radio signals is considered in (Jourdan, 2006). Once sensors are deployed, the location of them is known. For example the sensors can be placed outside and rely on GPS, or UTs can place them inside and determine their locations by survey or other means (e.g. an accurate map). The UT can then extract range estimates to the sensors from the received signals (for instance by time-of-arrival (ToA) estimation), and then use the range estimates in a triangulation technique to determine its own

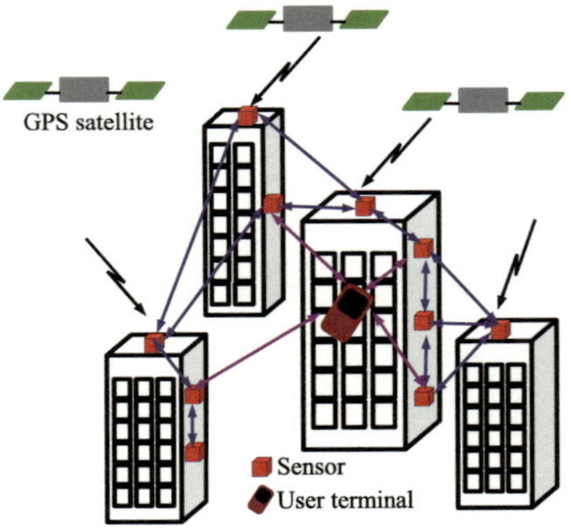

Fig. 1. Indoor localization architecture

position. Among various approaches (Falsi et al., 2006; Gezici et al., 2005; Lee & Scholtz, 2002; Low et al., 2005), use of ultra-wideband (UWB) signal is a promising technology in particular for indoor application. The sensors transmit UWB signals in the case, which own inherent delay resolution and ability to penetrate obstacles (Jourdan, 2006; Win & Scholtz, 1998; 2002). Further information on the fundamentals of UWB can be found in (Jourdan, 2006) and the references therein. The indoor localization architecture is illustrated on Fig. 1. Assume an emergency scenario as an example application of localization, in response to an emergency in a high rise building, a network of UWB sensors is deployed. These sensors provide a localization network to responders (or UTs) moving inside the building. The role of the sensors is to provide the UT with ranges. The UT would broadcast a UWB signal, so that the sensors each measure the range to the UT, share the information and infer the location of the UT. The quality of range measurements will degrade with distance, so that the distance between sensors is of concern. The ranging algorithm can also easily accommodate range constraints between sensors and UTs (Jourdan, 2006). The described localization architecture is called user terminal based localization technique, i.e. the localization task is performed by the UT itself and the sensors do not need to interact with one another to perform the localization task. The UT-based technology requires the installation of client software on the UT to determine its location. On the other hand, network-based techniques utilize the service provider's network infrastructure to identify the location of the UT. The advantage of network-based techniques is that they can be implemented non-intrusively, without affecting the UTs. One of the key challenges of network-based techniques is the requirement to work closely with the service provider, as it entails the installation of hardware and software within the operator's infrastructure. Often, a legislative framework, such as E911, would need to be in place to compel the cooperation of the service provider as well as to safeguard the privacy of the information. The focus of this chapter is exclusively on the range estimation between UT and AP sensors, and the network architecture is not discussed in remainder of this chapter. This implies that issues related to the communication connectivity between sensors, etc., are not presented. Without lost of generality of ranging analysis, any of UT-based or network-based localization systems can be assumed.

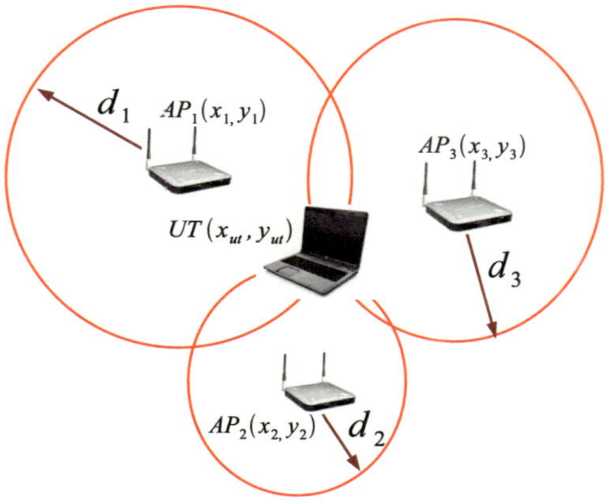

Fig. 2. Time-of-arrival triangulation of ranges to determine location

1.2 ToA-based ranging

An appropriate method for the indoor ranging problem is based on timing. Assuming the sensors and the UT are synchronized in time, the UT can calculate the time of arrival (time-of-flight) of a signal by comparing its time stamp at transmission to its ToA. This can then be converted to a distance by multiplying the time-of-flight by the speed of light. Since the accuracy of ToA estimation increases with the signal-to-noise ratio (SNR) and the bandwidth (BW) (Gezici et al., 2005), UWB (for a given SNR) will typically achieve great ToA accuracy compared to narrower band signals. Note that in general the sensors and the UT will not have a common time reference, in which case variants of this method must be used. In the round-trip method, the UT transmits a UWB signal to a certain sensor (IEEE Std, 2007). Once it is received at the sensor, it is retransmitted and in turn received by the UT. By comparing the time of original transmission to the ToA (and accounting for the processing time required for the sensor to retransmit the signal), a time-of-flight can be determined (Lee & Scholtz, 2002). In 2D, three sensors are sufficient to generate a position estimate. Fig. 2 shows the ToA triangulation of ranges to determine location. The technique assumes that three (or more) ranges $c \times t_1$, $c \times t_2$ and $c \times t_3$ (where c is speed of light) are gathered from APs $i = 1, 2,$ and 3, respectively, with known locations (x_i, y_i), as in Fig. 2. Solving the set of equations of

$$\begin{cases} c \times t_1 = \sqrt{(x_1 - x_{ut})^2 + (y_1 - y_{ut})^2} \\ c \times t_2 = \sqrt{(x_2 - x_{ut})^2 + (y_2 - y_{ut})^2} \\ c \times t_3 = \sqrt{(x_3 - x_{ut})^2 + (y_3 - y_{ut})^2} \end{cases} \tag{1}$$

translates to finding the intersection of the three circles (or spheres in three dimensions), yielding the unknown coordinates of the UT. Time of arrival is appealing due to its application to the general network architecture, however, the associate asynchronous ranging requires two or more messages per ranging session. This requirement may potentially increase the network traffic considerably (Jourdan, 2006). If the AP sensors are synchronized, but not the UT, then the time-of-difference-of-arrival (TDoA) method can be used. In this case the UT transmits a UWB signal, and the TDoA is computed at two sensors. The UT is then located on a hyperbola with foci at the sensors.

In the remainder of the chapter, for simplicity and without loss of generality, we assume that the sensors and the UT have a common time-reference.

1.3 Challenges to UWB ranging

Let's refer to a range measurement between a transmitter (Tx) and a receiver (Rx) as a direct path (DP) measurement if the range is obtained from the signal traveling along a straight line between the two points. Range measurements are typically corrupted by multipath fading, thermal noise, DP blockage, and DP excess delay. Multipath fading is due to destructive and constructive interference of signals at the receiver arriving via different propagation paths. This makes the detection of DP , if present, challenging. When the received signals are from reflections, resulting in measured ranges larger than the true distances. The difficulty is due to DP excess delay incurred by propagation of the partially obstructed DP through different materials, such as walls. When such a partially obstructed DP signal is observed as first arrival, the propagation time depends not only upon the traveled distance, but also upon the materials it encountered. Because the propagation of electro-magnetic signals is slower in some materials than in the air, the signal arrives with excess delay, again yielding a range estimate larger than the true one. The effect of DP blockage and DP excess delay is the same: they both add a positive bias to the true range between UT and sensor, so that the measured range is larger than the true value. This positive error has been identified as a limiting factor in UWB ranging performance (Falsi et al., 2006; Lee & Scholtz, 2002), so it must be accounted.

1.4 Contribution of the chapter

This chapter reviews the ToA estimation algorithms and then employs a threshold-based ToA estimation algorithm to calculate the range between Tx and Rx nodes in an indoor multipath environment. A practical threshold setting technique is introduced. For the purpose of this study, a set of empirical data obtained to create a baseline for comparative performance evaluation of ranging algorithms. The measured ranging error is used as a criteria to evaluate the ToA estimation algorithm.

2. UWB ToA estimation

As described in previous section, ToA estimation technique used with UWB transmission can be used for accurate indoor ranging. The transmitter sends out a UWB ranging signal $\sqrt{E_{\mathrm{tx}}}p_{\mathrm{tx}}(t)$ where p_{tx} is the monocycle pulse waveform with normalized energy after passing through a root-raised cosine bandpass filter with bandwidth, BW, adopted from (IEEE Std, 2007), and E_{tx} is the pulse energy. Standard UWB pulse with BW=0.5GHz is shown in Fig. 3. The received UWB signal in multipath channel is represented as

$$r(t) = \sum_{i=0}^{I} \sqrt{E_{\mathrm{tx}}}\alpha_i p_i(t) + w(t) \quad (2)$$

where α_i is the complex path-gain of the ith path assuming $I + 1'$ effective multipaths in the channel, and $w(t)$ is zero mean Gaussian random noise with variance σ_w^2. In the LoS scenario the first arriving multipath is the direct path and the remaining I multipaths arrive to the receiver after one or more interactions (scattering, reflection, diffraction) in the channel. It is known that the UWB waveform is distorted during interactions to the wireless channel. A simplifying assumption is to consider this distortion negligible, i.e. $p_i(t) = p_{\mathrm{tx}}(t - \tau_i)$ with $\tau_i = l_i/c$ representing the delay of the ith multipath, c is the

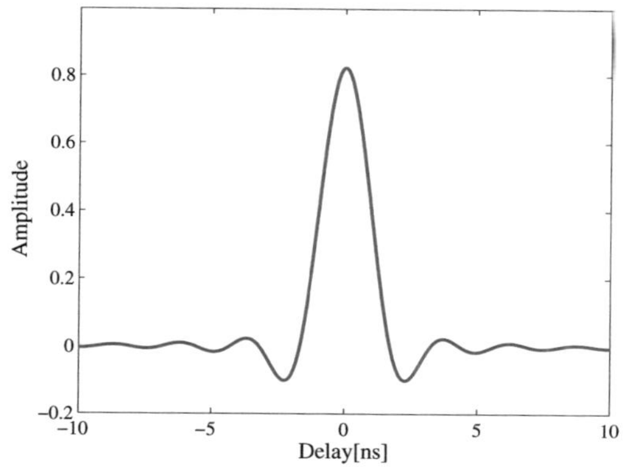

Fig. 3. Standard UWB pulse, BW=0.5GHz.

propagation velocity of electromagnetic wave and l_i is the path length, i.e. the distance ith path travels from transmitter to receiver. A comparative study of different UWB transceivers employed for ToA estimation is presented in (Guvenc et al., 2006). Matched-filter (MF) receiver is known to obtain better decision for ToA estimation than the energy-detector (ED) however it requires the knowledge of the received pulse shape which may not be available in practice (Sahinoglu et al., 2003). Another disadvantage of the MF is the requirement of Nyquist rate sampling for accurate ranging, hence complex analogue-to-digital converters are required. On the other hand ED based estimator is preferable due to its low complexity, e.g. implementation at sub-Nyquist sampling rates (Sahinoglu et al., 2003). A disadvantage for the ED is the squared noise component at its output which makes a poor performance in low SNR scenarios. The ED output samples can be represented as

$$z(n) = \int_{nT}^{(n+1)T} |r(t)|^2 dt \tag{3}$$

where T is sampling period, $n \in 1, ..., N$ is sample index and N represents the total number of received samples. The integrator performs the sampling operation by successively integrating the squared received signal. First arriving sample is presented by n_0 which in the LoS scenario corresponds to the direct path with the propagation delay τ_0. If the transmitter and receiver are by some means synchronized their range d_0 can then be obtained from the relation $d_0 = c\tau_0$.

2.1 Review of UWB ToA estimation algorithms

Several methods for ToA estimation of UWB signals can be found in (Falsi et al., 2006; Gezici et al., 2005; Low et al., 2005), i.e., examples of maximum likelihood based ranging, low-complexity estimators include a maximum peak detection method and a threshold detection method. In the latter a threshold is chosen a priori, and the ToA is defined as the instant when the received amplitude goes above this threshold (Falsi et al., 2006). In spite of its simplicity, this method works quite well for UWB signals, especially in high SNR environments. In this chapter the performance of method is further analyzed by details.

2.1.1 Maximum likelihood based ranging

Range estimation based on Maximum likelihood (ML) techniques is explained (Sahinoglu et al., 2003). Starting from the assumption of full a-priori information, various degrees of assumptions about the unknown parameters are considered (Gezici et al., 2005), (Falsi et al., 2006), (Win & Scholtz, 2002). In addition, the generalized ML ratio test approach and the sub-Nyquist sampling ML approaches are discussed (Win & Scholtz, 1998),(Guvenc & Sahiinoglu, 2005). ML ToA estimation (equivalently, ranging) can be performed by means of a correlator with a template signal that is perfectly matched to the received multipath signal, and by determining the time delay of the template for which the correlator output is maximized (Gezici et al., 2005). That optimal template can be expressed as

$$p_{\mathrm{tmp}}(t) = \sum_{i=0}^{I} \alpha_i p_i(t - \tau_i) \tag{4}$$

However, this ML receiver cannot be implemented in practice, since the received waveform has unknown parameters to be estimated, i.e., delays, amplitudes, and pulse shapes of different multipath components (MPCs)(Guvenc et al., 2008). A ML Estimation with no prior information is described in (Guvenc et al., 2008), which in this method, the ML estimation of the delays and the channel coefficients has high computational complexity as it requires a search over different values of delays. As explained, the ML estimation based on the channel estimation problem provides an optimal ToA estimate. However, it is quite impractical for UWB channels with a large number of MPCs as the estimation of all the channel parameters is computationally very intensive. Therefore, a simplified version of the ML approach, called generalized maximum likelihood (GML) technique, is studied in (Win & Scholtz, 1998). Assuming that the strongest MPC has been detected, the GML technique estimates the ToA by performing a search over a smaller search space. Specifically, the paths prior to the strongest MPC are investigated. In (Win & Scholtz, 1998), an iterative technique with lower complexity is proposed which uses certain prior statistics of the UWB channel to define a new search space and to set a stopping rule. The ML techniques can be impractical in many cases due to their computational complexity and high sampling rate requirements. In (Guvenc et al., 2008), ToA estimators that operate at low sampling rates and with various levels of a-priori information are described. These methods require prior information related to the statistics of individual ED samples, which, such prior information may not be available in practice.

2.1.2 Low complexity ranging techniques

In this section, peak detection algorithms (Falsi et al., 2006), two step TOA estimation approaches, ranging with dirty templates (Yaung & Giannakis, 2004), (Yaung & Giannakis, 2005), and threshold based ranging algorithms (Dardari & Win, 2006)-(Scholtz & Lee, 2002) are discussed.

- Peak detection techniques The maximum energy selection (MES) approach is a simple technique that estimates the ToA according to the strongest sample. It can be formulated as

$$\begin{aligned} \tau_{\mathrm{D}} &= n_{\mathrm{D}} T, \\ n_{\mathrm{D}} &= \underset{n}{\operatorname{argmax}} \left(z[n] \right), \end{aligned} \tag{5}$$

  where τ_{D} is detected sample as direct path. In typical UWB channels, the number of samples which include both desired signal and noise can be much larger than 1 when

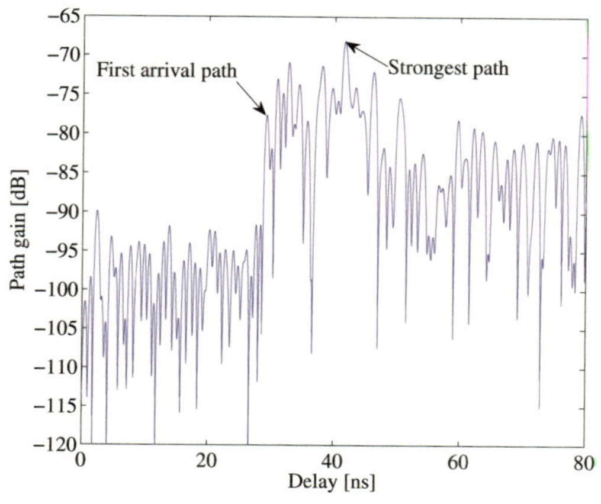

Fig. 4. An example of CIR where the FAP is not the same as SP.

the sampling interval is on the order of a pulse duration. Therefore, there can be a large delay between the first path and the strongest path, and the MES technique can yield large estimation errors. In the other hand, the strongest path (SP) in many cases even in LoS scenario is not the first arrival path (FAP) as shown in figure 4. In ranging analysis, FAP should be detected rather than SP. The MES technique in (5) may not yield an accurate ToA (range) estimate in the presence of MPCs prior to the strongest one. In order to improve the accuracy of range estimation, (Falsi et al., 2006) proposes to consider the largest N correlation peaks, and to select the time delay corresponding to the peak with the smallest time index. In (Falsi et al., 2006), the authors also propose two improvements for this technique in non-resolvable channels.

- Ranging with Two-Step ToA Estimators

  It is of significant importance to perform accurate range estimation without employing high sampling rates in UWB ranging systems. In order to have low-power and low-complexity implementations, symbol-rate or frame-rate samples should be considered (Gezici et al., 2005), which can, however, increase the time to perform ToA estimation considerably. One technique to perform reasonably accurate range estimation based on low rate samples is to employ two-step ToA estimators (Gezici et al., 2008). In order to perform ToA estimation in short time intervals, the first step of the two-step ToA estimator in (Gezici et al., 2008) performs a coarse timing estimate using energy detection (similarly, the dirty template approach can be used). Then, the second step refines the ToA estimate based on a statistical change detection algorithm (Guvenc et al., 2008).

- Ranging with Dirty Templates

  Another low complexity ToA estimator is the dirty-template technique introduced in (Yaung & Giannakis, 2004), (Yaung & Giannakis, 2005), which operates on symbol rate samples. The main idea behind this technique is to use the received signal itself as a dirty template and then to perform cross-correlations with the symbol-length portions of the received signal. The main advantage of the dirty-template technique is its low complexity, however, the estimates obtained by this technique can have an ambiguity equal to the extent of the noise-only region between consecutive symbols.

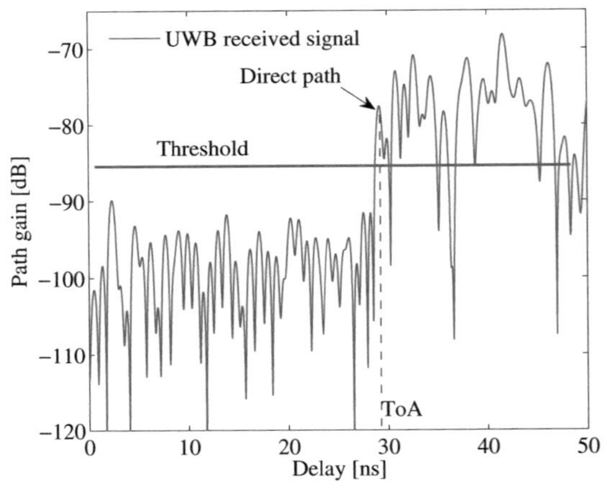

Fig. 5. Threshold-based ToA estimation on received UWB signal

- Threshold-based Ranging

  Threshold based ranging algorithms compare the received signal samples against a threshold in order to identify the first arriving MPC (Dardari & Win, 2006). In (Scholtz & Lee, 2002), a simple threshold based algorithm is introduced, which compares absolute values of the direct samples of the received signal (or, samples of the correlator/energy detector output) against a threshold. The ToA of the signal is determined by the index of the first sample that exceeds the threshold. Threshold-based ToA estimation algorithm, in general, is illustrated on Fig. 5. Since the signal energy and the channel impulse response are unknown, the threshold can be determined based on the noise floor. In next section, a detailed survey on threshold setting is presented.

2.2 Survey on threshold-based ranging techniques

In previous sections, it was explained that the signal peak detection for the estimation of the n_0, discussed in (Haneda et al., 2009), (Sahinoglu et al., 2003), is not reliable in multipath environments even in LoS scenario due to multipath interference to the first arriving signal. The simplest approach then is comparing the received samples to a presumed threshold level. In the threshold-based ranging systems after acquiring the received signal sample energy from (3), the decision is made on the first threshold-exceeding sample n_{D}. The corresponding delay can be obtained as

$$\tau_{\mathrm{D}} = n_{\mathrm{D}} T,$$
$$n_{\mathrm{D}} = \arg\min_{n} \left(z(n) > \xi \right), \quad (6)$$

where ξ is a presumed threshold. Threshold-based ToA estimation is specifically attractive due to its low complexity and computational burden which are crucial for many low cost battery powered devices, e.g. wireless sensor networks and radio frequency identification (RFID). These search algorithms compare individual signal samples with a certain threshold in order identify the first arriving signal and obtain the range information (Dardari & Win, 2006). The important challenge in the threshold-based approach is how to pick up a threshold level to have least ToA estimation error which is equivalent to the minimum ranging error. In next section, a review on the existed threshold-setting techniques is given.

2.2.1 Threshold setting techniques

Few threshold selection methods have been proposed in the literature, in fact in most researches a known optimum threshold, usually derived by trial and error, is presumed. In (Dardari & Win, 2006) it was observed that the performance of the ToA estimation depends on the threshold value but no comment was delivered on how to set it. It is reported that the threshold-based estimator is more convenient in high signal-to-noise ratio (SNR). Moreover the sensitivity of the system to the threshold value was discussed without proposing how to set it. It was argued that the threshold level can be selected based on the noise level or signal peak (Dizdarevic & Witrisal, 2007; Haneda et al., 2009). In (Guvenc & Sahinoglu, 2005), two different threshold based ToA estimation algorithms are proposed. The thresholds of those algorithms can be set based on the noise level, or, using a normalized threshold as in (Guvenc & Sahinoglu, 2005). In either case, certain statistics of the UWB channels should be used to select parameters for the search space and stopping rules accurately. The problem in those approaches is that the derivation of SNR without any knowledge of the ToA in practice is a challenge itself. The threshold has been selected as a normalized value between maximum and minimum energy samples to improve the performance in (Guvenc & Sahinoglu, 2005), although the dependency to the SNR and signal level is preserved. In (Guvenc & Sahinoglu, 2005) the Kurtosis of the signal samples is used as a metric for threshold selection. Unlike the SNR of the received signal, Kurtosis captures both the statistics of individual channel realizations, and the relative energy of the signal to noise. It is shown that using the Kurtosis metric, estimation error can be significantly decreased compared to fixed threshold. In the so far discussed methods the threshold is always assumed a fixed value. The common disadvantage for all of these methods is that they use a presumed threshold value, which is system and environment dependent. In (Xu & Law, 2008) and (Dashti et al., 2008) a threshold based on the delay of first arriving signal was proposed. Proposing the delay-based threshold selection technique is postulated on the fact that the first arriving signal power decreases as wireless nodes separate from each other due to the propagation loss. Consequently the threshold for detecting the signal is a function of delay. In (Xu & Law, 2008) the threshold value is an exponential function whose parameters are obtained by solving a numerical equation with a knowledge of the noise level and through a further optimization process. The design parameters of the threshold are sensitive to the channel condition and still it is necessary to predetermine the noise level. The common problem of the discussed threshold setting methods is that none of them was optimized based on the design parameters such as the ranging error. The problem of threshold-setting is experimentally addressed in next section.

3. Ranging based on measured propagation channels

To examine the performance of ToA-based ranging in the real scenario a measurement campaign was conducted in an office room. The threshold-based ranging algorithm was applied to get the range of the measured data d_{m}. Since the real separation of the transmitter and receiver d_0 is well known from the coordinates of the transmitter and receiver nodes, the experimental ranging error is obtained as

$$e_{\mathrm{m}} = d_{\mathrm{m}} - d_0 \tag{7}$$

The ranging error is used as criteria to evaluate the ranging algorithms. The details of the measurement and ranging results are given in next sections.

Fig. 6. The view of the measurement environment

3.1 Experiment scenario and channel measurement results

The database is obtained in the typical office room in National Institute of Information and Communications Technology (NICT), Japan. The view and layout of the measurement environment are shown in Fig. 6 and 7. The description of the experiment is as follows. The transmitter antenna could be positioned at almost any place in the room by the aid of a large spatial precision scanner covering the whole areas of the room, whereas the receiver antenna was fixed at the corner of the room. The transmitter to receiver distance varied from 0.6 to 9.3 m. In total, 4200 spatial samples of transfer function were measured on the transmitter side by a vector network analyzer (VNA). The maximum detectable ToA was 200 ns, sufficient to capture dominant propagation paths. LoS was assured in most Tx positions except for two arrays where LOS was obstructed due to television displays. The whole measurement took about 6 hours, but there was no moving object during the measurement so that the time-invariance of the angular-delay channel characteristics was ensured. Phase drift of the VNA was carefully compensated during the measurement by performing an internal calibration procedure of the VNA once in two hours. Details of the measurement specifications are summarized in Table 1, for more details please refer to (Dashti et al., 2008). Since the measurement data and all the analysis in this study were done in allocated sub-bands in IEEE 802.15.4a standard, the specifications of this standard are briefly described. The IEEE802.15.4a standard defines several frequency bands for the realization of piconets as summarized in Table 39i-UWB PHY band allocation in (IEEE Std, 2007). The band plan defines three band groups: sub-gigahertz, low, and high bands. The frequency band of interests is selected from the band plan of IEEE802.15.4a-2007 standard. The band plan defines low and high bands spanning from 3.1 to 4.8 GHz and from 6.2 to 10.3 GHz, respectively. Bandwidth of each channel is either 0.5 or more than 1.0 GHz. Transfer functions measured within the band of interests are extracted for data processing. We will not consider sub-gigahertz band, since the measurement data do not cover that frequency. In the low and high bands, there are mandatory bands where the ranging operation should be assured most (channels 3 and 9), and others are optional. Bandwidth of each mandatory channel is 499.2 MHz with various center frequencies. There are also optional channels with larger bandwidth (channels 4, 7, 11 and 15) with bandwidth of 1331.2 MHz. Ranging accuracy were assessed in all low-band and high-band UWB channels.

The transfer function is converted to a channel impulse response (CIR) by the inverse Fourier transform. Frequency spectrum of the root-raised cosine pulse is applied as a windowing

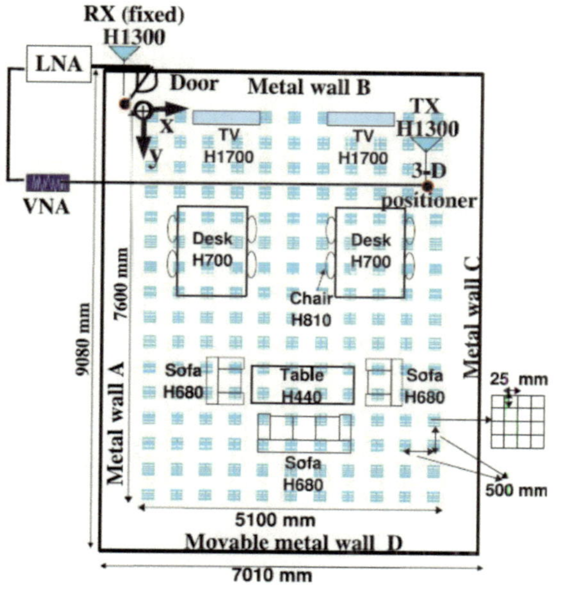

Fig. 7. Layout of the measurement environment

Frequency band	3.1-10.6 GHz
Band-width (B)	500 MHz
Measurement equipment	Vector network analyzer Room-wide spatial scanner Low-noise amplifier (30 dB)
No. of frequency sweeps	1501
Antenna	UWB monopole
Transmitted power	−17 dBm
Coverage area dimensions	5.1 × 7.6 m^2
Wireless nodes range	0.6m to 9.3 m
Wireless nodes height	1.3 m above the floor

Table 1. Experiment parameters

function. The root-raised cosine pulse is denoted in the time domain as

$$r(t) = \frac{4\beta}{\pi\sqrt{T_p}} \frac{\cos\left(\frac{(1+\beta)\pi t}{T_p}\right) + \frac{T_p}{4\beta t}\sin\left(\frac{(1-\beta)\pi t}{T_p}\right)}{1 - \left(\frac{4\beta t}{T_p}\right)^2} \tag{8}$$

where β and T_p is a roll-off factor and pulse length specified in the standard (see (Molisch et al, 2004), pp.82-83). CIR is calculated for all the Tx locations.

Power of the direct and strongest paths is shown in Fig. 8 against Tx-Rx distances. Results from channels 2 and 4, which are in the low band, and 11, which is in the high band, are shown. The figures revealed the following findings. Channels with wider bandwidth show less gain variation of the direct and strongest paths. Comparison of results from channels 2 and 4 revealed that the variation of path gain values is less in channel 4. The two channels have

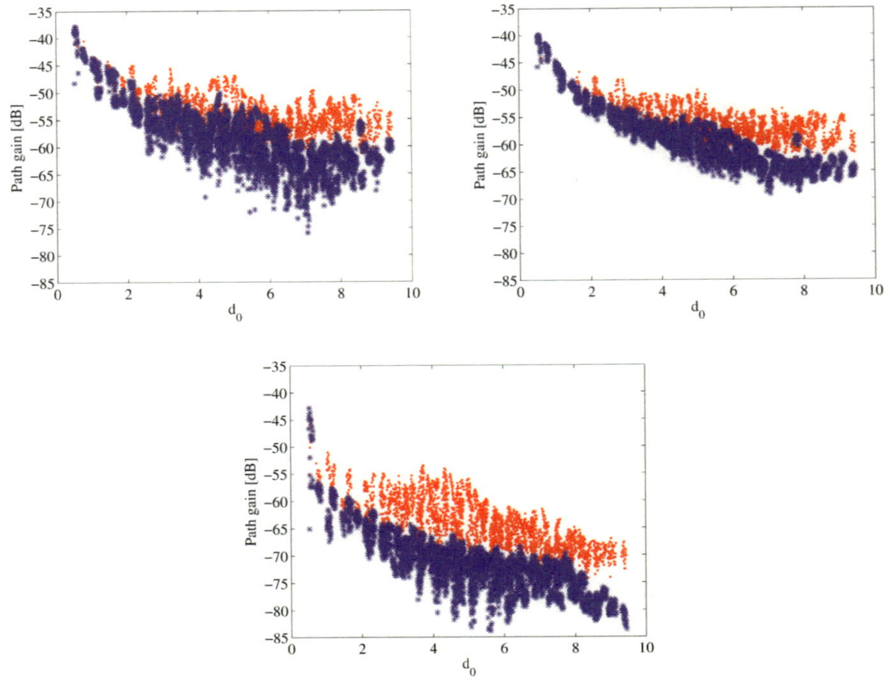

Fig. 8. distribution of measured direct path gain (blue stars) and strongest path (red dots) in channels (a) 2 (b) 4 (c) 11

the same center frequency, but channel 4 has about three times larger bandwidth than channel 2. The narrower bandwidth leads to poorer delay resolution, which causes the fluctuation of power in direct and strongest paths due to the fading with non-resolvable signal components around the paths. As a result, the gain of the first and strongest paths is slightly higher in those channels. The same trend was observed in other channels with the same center frequency and different bandwidths, such as channels 5 and 7, 9 and 11, and 13 and 15. This is the same observation as reported in the work of Alsindi et al. (Alsindi et al., 2007).

Difference of the path gain between the high and low bands are 5 to 15 dB. The path gain in the high band was smaller value than the low band as expected. The largest and smallest gain was observed in channels 1 and 11, respectively. The level of path gain is almost the same in the low band, while 5 dB gain difference was observed within the high band. Channels 5 and 11 showed the largest and smallest gain in the high band, respectively. The channel with the highest frequency did not show the smallest path gain, probably because of the frequency characteristics of antenna gain.

Fig. 9 shows the example of a measured received signal. It depicted that due to the effect of multipath interference the strongest path is not necessarily the direct path even under the LoS condition. Multipath interference leads to fading and causes the strongest path spread over the delay axis. In ranging analysis, direct path should be detected rather than strongest path. In this example the ToA of direct path is estimated wrongly from expected ToA. The ranging error is modeled in (Dashti et al., 2010).

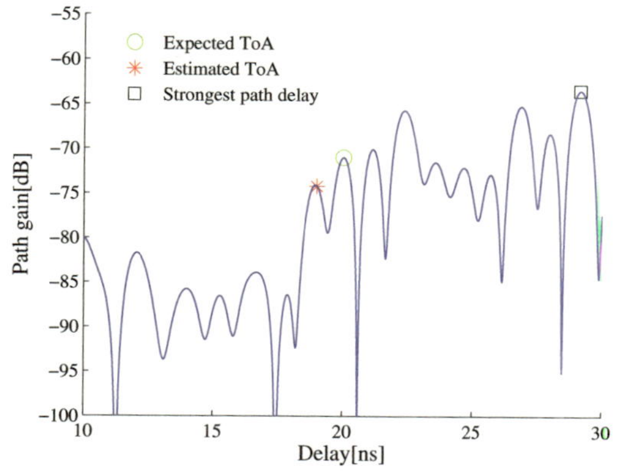

Fig. 9. An example of received signal where the strongest path is in a delay from LoS as a result of destructive multipath interference (Dashti et al., 2010).

3.2 Ranging with fixed threshold value

The fixed threshold value can be optimized based on noise level or peak signal level. Two threshold-based methods are introduced to detect the signal component corresponding to the first path: the leading edge detection, which set the threshold based on noise level, and the search back method, which the detection threshold level is given by the power of strongest path (SP). Coherent detection is assumed in both ranging methods. Schematic representation of these two methods is shown in Fig. 10.

3.2.1 Search-back method

Search-back method utilizes the strongest path of CIRs to detect the direct path. It has been commonly reported that the first path is not always the strongest path, particularly in NLoS scenarios due to LoS blockage. As it was discussed earlier, this could happen even in LoS situations due to multipath propagation. Specially, power of delayed paths could be greater than the first path because of overlapping multipaths arriving at the same delay time. In other cases, the first path suffers from destructive fading due to surrounding non-resolvable multipaths. The search back method first finds the strongest path, and then looks for a peak arriving before the strongest path which has greater power than a detection threshold level.
We proposed an iterative search-back algorithm to calculate the noise floor (NF) to be used in the detection of first path. In the first iteration, the algorithm detects the strongest path, and then calculates the noise floor by averaging over the interval $[0, (t_{sp} - t_c)]$, where t_c is delay resolution. The interval is t_c less than the SP delay to exclude the effect of SP signal. To remove the effect of side lobes, t_c was chosen 1 ns. In next iterations this process is repeated for new time interval $[0, (t_i - t_c)]$, and it will continue to find the new peak value and the new NF. Here t_i is the time delay of the peak detected in the $i-$ th iteration. The algorithm will be continue until finding the first peak higher than the NF by predefined search-back threshold value, γ_S, which is dependent on system bandwidth. Fig. 11 shows the flowchart of the proposed iterative algorithm. P_i and NF_i in the flowchart are peak value and NF in the $i-$ th iteration. Obviously the value of NF is erroneous in the first iteration but it will give the real NF and first detected path after enough iterations. γ_S level which the algorithm used for detecting of first path is chosen different for each subband. To obtain the optimum γ_S which gives lowest error, we calculated the ranging error using several γ_S, such as 5, 10,15 and 20 dB. Concerning

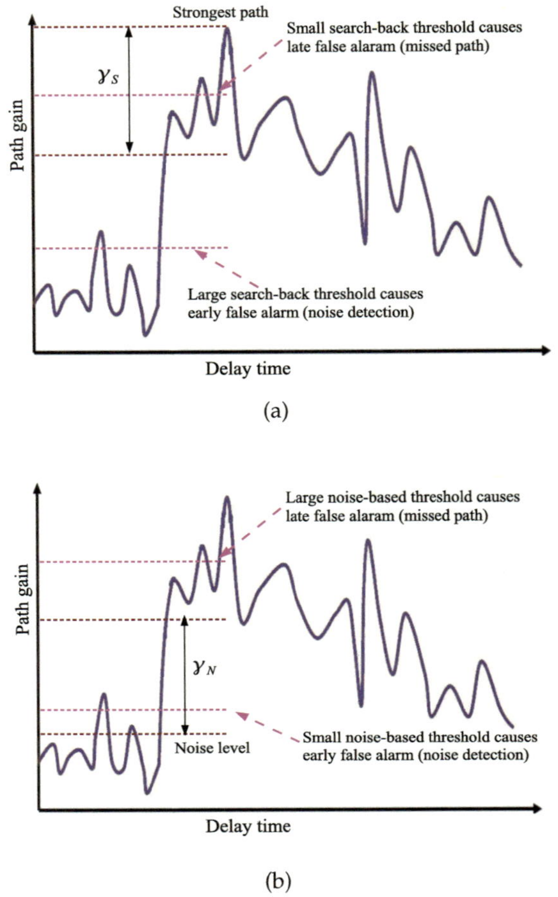

Fig. 10. (a) Search-back detection method vs. (b) leading edge detection method

the difference of SP signal level in different sub bands, these optimum γ_S values are different for channels with different bandwidth. For instance, γ_S was chosen 15 dB for Channel 4. Same analysis was done for the other subbands, however we hesitate to show the ranging results of all of them for the sake of conciseness. For higher BW the algorithm search for the first peak above 15dB from NF, γ_S is chosen 10dB for channels with lower BW. It is observed that the NF decreases for higher bands and also decrease by increasing the bandwidth. The peak value decreases in higher bands and also decreases by increasing the bandwidth. Since path loss increases as the frequency increases. This algorithm has the advantage of obtaining the result after a few number of iterations for the far points. Also for the close points (Tx and Rx close together) in the lower frequency bands, the averaging over longer intervals in the first iteration seems to be reliable by using this algorithm. For instance for an arbitrary position in the room in channel 3, by applying the mentioned iterative algorithm, after only 2 iterations, we could detect the correct first path. The ranging error for this position is 0.2 m, which is a relatively small error while the real distance between Tx and Rx is 4.6 m. However the required ranging accuracy depends on the application. The calculated NF for this position

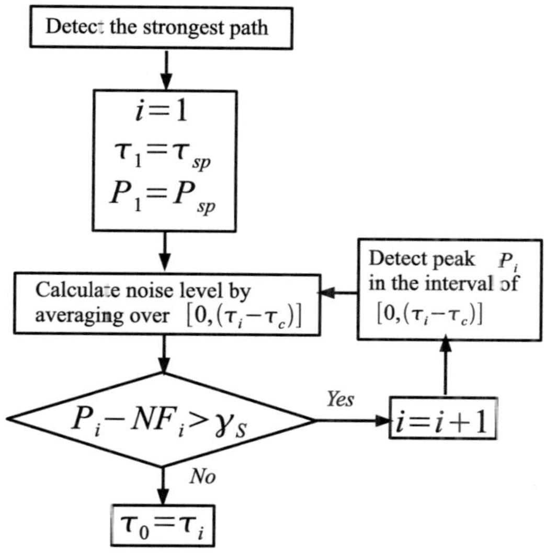

Fig. 11. Flowchart of search-back algorithm

is -72 dB. The power level of first path is 14.2 dB more than the calculated NF. Evaluation of ranging accuracy were assessed in all channels. The ranging result shows the algorithm works well for almost all of the positions, however ranging errors are observed in some cases. We categorized the ranging errors to two main categories, relatively small positive/minus errors and large positive/minus errors. When the peak of channel response gets a little shifted from the expected ToA to shorter/longer ToA, resulting small errors in ToA estimations. In some far positions from the Tx antenna large ranging errors are observed. These large errors may be produced by the occurrences of undetected path conditions, or false estimation of NF by proposed algorithm. For instance in an arbitrary position where large minus error happened, the calculated NF for that point is -104 dB, and the first detected path level is 14dB higher than this NF, however this peak is not the real first arrival path, so causes relatively large minus ranging error. In the proposed first path detection, the detecting of first peak started from SP , going to the origin, and it continues till finding the first peak higher than calculated NF by γ_S value. This algorithm has the advantage of detecting the peak after a few iteration numbers in many cases. However for some cases the algorithm cannot detect the first path, and SP is detected as first path. Detection algorithm started from origin and going to SP may eliminate the error of such these cases. In following leading edge algorithm is described.

3.2.2 Leading edge method

In leading edge method, the fixed threshold value can be optimized based on noise level. We refer this method as noise level based threshold. Leading edge detection is the most primitive method to detect the first path. The device monitors a time series of correlator outputs in a coherent detector. Provided that the power monitor, like a received signal strength indicator in a general receiver, knows the noise level of the receiver in advance, it can detect the first path when a signal level exceeds a certain level. The first output sample exceeding noise level by a predefined threshold value will be detected as ToA, i.e. ToA is the delay time of the

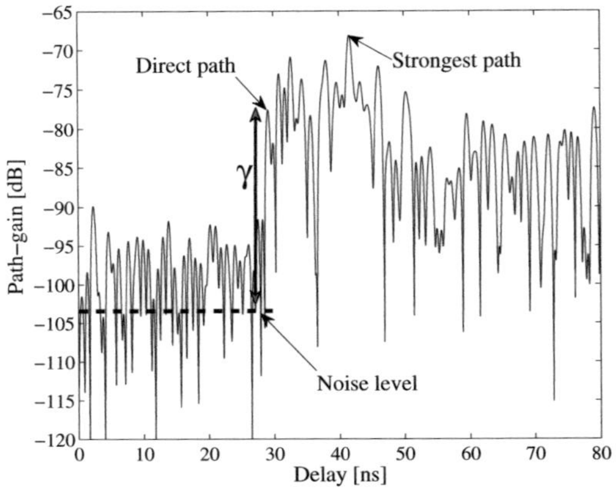

Fig. 12. Noise level based threshold for ToA estimation

earliest received sample that fulfills the condition of:

$$\tau_{\mathrm{D}} = n_{\mathrm{D}} T \quad (9)$$

$$n_{\mathrm{D}} = \underset{n}{\operatorname{argmin}} \left(z[n] > l_{\mathrm{N}} + \gamma \right) \quad (10)$$

where γ is the presumed fixed threshold value and l_{N} is the noise level. γ can be optimized for individual UWB subbands in order to have the minimal ranging errors. The principle of noise level based ToA estimation algorithms is summarized in Figure 12. However, there are two cases the method fails to detect the first path: miss and noise detection. The miss detection (late false alarm) occurs if the level of the detection threshold is greater than the power of the fist path, while the noise detection refers to the case where a noise peak is wrongly detected as the first path. The noise detection is regarded as a early false alarm.

The Fig. 13 shows the superior performance of leading edge against search-back method for channel 3. The ranging results in all channels revealed that the leading edge detection always outperforms the search back method. This is because the search back method uses strongest path. As reported in the channel modeling result, strongest paths fluctuate in power, resulting in larger fluctuation of the level difference between the first and strongest paths. Therefore, the search back method needs to increase the search back level in order to capture the first path perfectly. The larger search back level, however, results in increasing probability of noise detection, resulting in the degradation of the mean detection probability. On the other hand, the leading edge detection suffers from the power fluctuation less. According to the channel modeling result, smaller power fluctuation was observed in channels with wider bandwidth. In such channels, the first path detection probability of the search back method is comparable with that of the leading edge method. The search back method achieves perfect detection probability on the diagonal line of the room, but miss and noise detection starts to occur once the Tx location is getting off from the diagonal line. This means that the performance is largely dependent on spatial multipath characteristics. While it was not found in the leading edge detection because of its robustness to the varying multipath structure. The miss detection is most visible in near-wall Tx locations. It is generally seen that in leading edge method, smaller path gain leads to lower threshold values in order to capture first paths correctly. Hence the threshold value indicates larger values when it is optimized in the limited areas to rule out

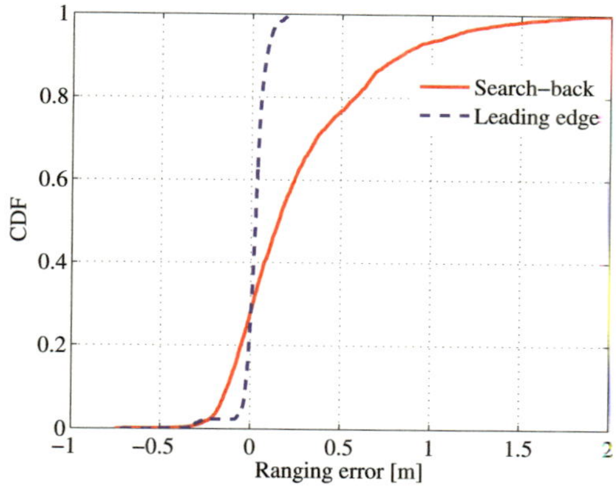

Fig. 13. Comparison of leading edge and search-back methods

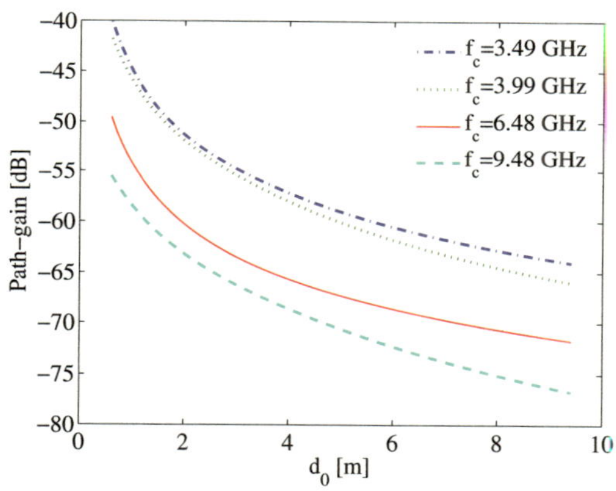

Fig. 14. Direct path-gain in different subbands with different center frequencies

Tx locations with low signal levels. The same trend is observed in the search back level, but the fluctuation of the value is small over different center frequencies and bandwidths In the leading edge method named noise level based threshold approach, noise level can be assumed initially as a fixed value or can be calculated based on initial part of the signal. We categorized noise level based threshold ToA estimation concerning presumption or estimation of noise level. In following section more description is given.

- Presumed noise level A prior knowledge about the noise can be assumed to set the l_N as a single value, i.e. in equation (10), l_N is presumed single noise level. We assumed thermal noise level given by $l_N = k_B T_k B$ where k_B is the Boltzmann constant, B is the system bandwidth and T_k is the absolute temperature in kelvin.

  Fig. 14 shows the best fit for the measured FAP path gain as a function of Tx-Rx distances for different channels. It is observed that the FAP path gain decreases in higher subbands since the path loss increases, Hence γ in equation (10) was optimized for each channel individually in order to have minimal ranging errors. Fig. 15 shows the optimum value of threshold for all different channels. γ_{opt} varies from 30 dB for channel No.1 with lowest

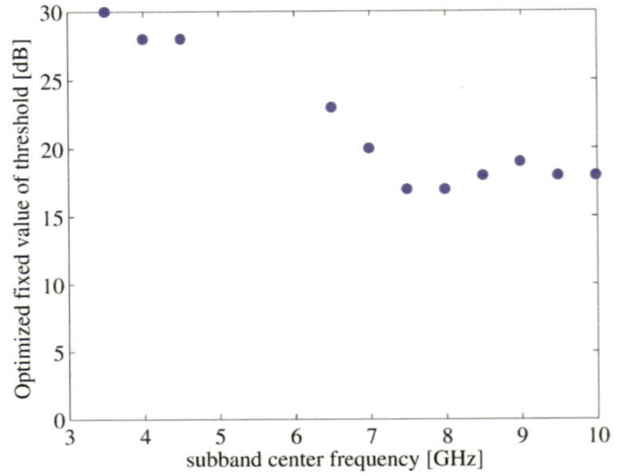

Fig. 15. Optimized fixed value of threshold for different subbands with different center frequencies

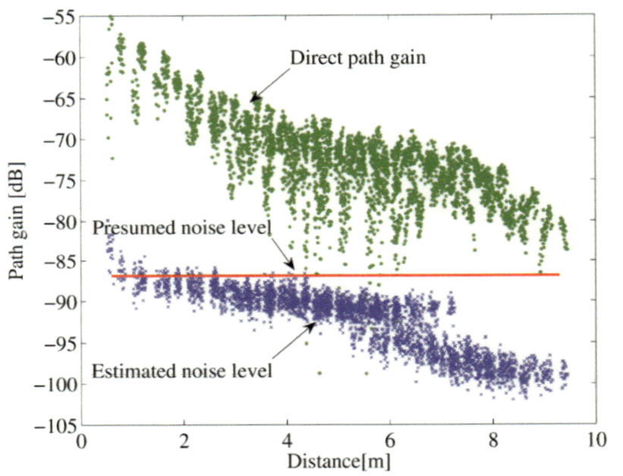

Fig. 16. Measured direct path gain against presumed and estimated noise level

center frequency, $f_c = 3.49$ GHz, to 15 dB for channel No.15 with highest center frequency, $f_c = 9.98$ GHz,. Fig. 15 shows direct path gain compared with presumed fixed l_N . The direct path gain decreases with longer Tx-Rx distance while noise level is a single value, therefore the differences of direct path gain and noise level are not a single value for all Tx-Rx distances. As Fig. 16 shows, the difference between direct path gain and l_N, vary in a wide range. Due to this wide variation, presetting a single value for γ, which gives minimal ranging errors for all possible Tx-Rx distances, is a challenge.

- Estimated noise level In this approach instead of presuming a single noise level, we estimate the noise level based on the initial part of the received signal, i.e. in equation (10), l_N is not a single value but it is calculated for each channel realization. Fig. 15 shows direct path gain compared with estimated l_N for different Tx-Rx distances. In (Dashti et al., 2008) the variance of ranging error of estimated noise level approach with those obtained by presuming the l_N are compared. It was shown that by estimating l_N, variance of ranging error dramatically decreases in all channels, However still the algorithm fails in some cases. Setting a fixed threshold value is not reliable due to variation of direct path gain on different Tx-Rx distances. Since direct path gain decreases with longer Tx-Rx distance, threshold value also can be set to decrease with Tx-Rx distance. We proposed a delay-dependent threshold selection method in next section.

3.3 Ranging with delay-dependent threshold setting

In previous section two fixed threshold based methods (leading edge vs. search-back) are introduced and the ranging performance of them are compared. The performance degradation in the search back method is due to the gain fluctuation of the first and strongest paths, which is most remarkable in the high band. The selection of the optimum threshold level for these two ranging methods still remains an important issue.

As it was described in previous section, we introduce a technique to set the threshold as a function of Tx-Rx distance instead of a fixed value as in conventional noise level based threshold methods. In this method, we preset a delay-dependent threshold function $\xi(n)$. The received samples are then compared to the respective threshold values, $\xi(n)$. The arrival time of the first sample crossing the respective threshold value within time interval $[0, t_{\mathrm{SP}}]$ is estimated as ToA, where t_{SP} is the delay time of the SP. Fig. 17 (a) shows the basic of the proposed method. In this method estimation or assumption of noise level is not needed. As described, algorithm searches for a first received sample crossing its respective threshold. In some cases there is no peak located in the detected sample, n_{D} th sample, as shown in Fig. 17 (b), due to resolution of system and algorithm. The algorithm then search for a nearest peak value in the interval of $[n_{\mathrm{D}}T - t_c, n_{\mathrm{D}}T + t_c]$, where t_c is set according to the resolution of system.

As a reliable delay-dependent threshold the standard path-gain model is employed, which is to predict the expectation of E_{n_0} at any indoor position, according to the IEEE802.15.4a standard channel model (Molisch et al, 2004). This model is generic and widely used for the indoor UWB channel modeling applications. In following IEEE802.15.4a standard path gain model is briefly explained. The parameters of the model are also extracted by fitting measurement data to the described path gain model.

In the IEEE802.15.4a standard, path gain in a UWB channel is defined as:

$$G(f,d) = G(f)G(d) \tag{11}$$

Path gain is a function of the distance and frequency. In this model, it is assumed that the distance and frequency dependent effects are spreadable. The separation reduces the complex two-dimensional path gain modeling to one-dimensional problem. The frequency dependency of the channel path gain is modeled as:

$$G(f) \propto \sqrt{f^{-k}} \tag{12}$$

In IEEE802.15.4a model the distance dependence of the path gain is described by the conventional power law for simplicity as:

$$G(d) = G_{\mathrm{R}} \times \left(\frac{d}{d_{\mathrm{R}}}\right)^{q} \tag{13}$$

Combining (11), (12) and (13) yields the following equation in dB for total path gain.

$$G(d) = G_{\mathrm{R}} - 20klog\left(\frac{f}{f_{\mathrm{R}}}\right) - 10qlog\left(\frac{d}{d_{\mathrm{R}}}\right) \tag{14}$$

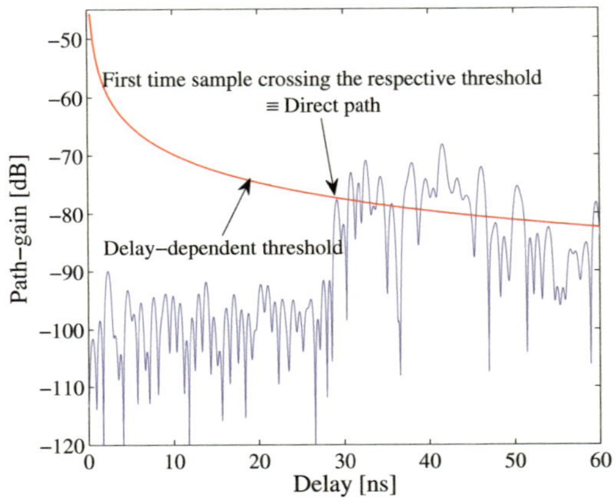

(a)

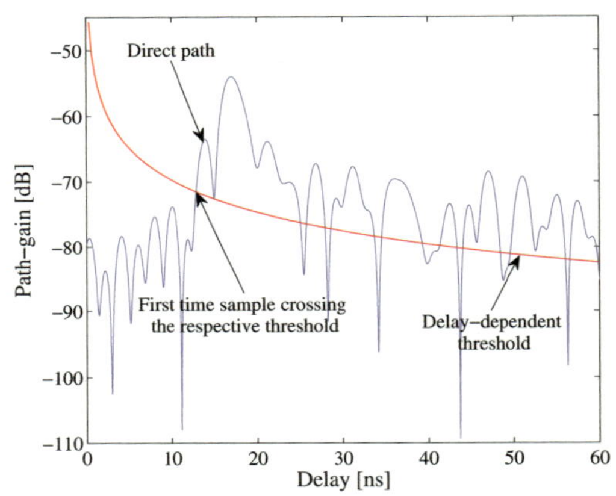

(b)

Fig. 17. An example of Delay-dependent threshold against a measured channel impulse response a) a peak located in n_{D} sample b) n_{D} sample is not a peak hence algorithm search for nearest peak in the interval of $[n_{\mathrm{D}}T - t_c, n_{\mathrm{D}}T + t_c]$

which in essence states that the path-gain is influenced by attenuation due to the frequency f and the transmitter to receiver separation d. The decaying exponent due the frequency and the distance are expressed as k and q, respectively while G_{R}, f_{R} and d_{R} are the reference path-gain, frequency and distance respectively (Molisch et al, 2004).

Fig. 18 shows the distribution of measured path gain within the scanned area in the room. X-axix and Y-axix represent the coordinate of the transmitter in X-Y plane in the area covered. Measured direct path gain distribution for lowest and highest subbands , which are channel 2 and 14 respectively, are shown in the Fig. 18 (a) and (b) . Figure depicts the dependency of the path gain to the distance and frequency. The parameters of the model were extracted by fitting measurement data to the described path loss model. Following procedure was performed for determination of model parameters similar to method presented in (Haneda et al., 2007):

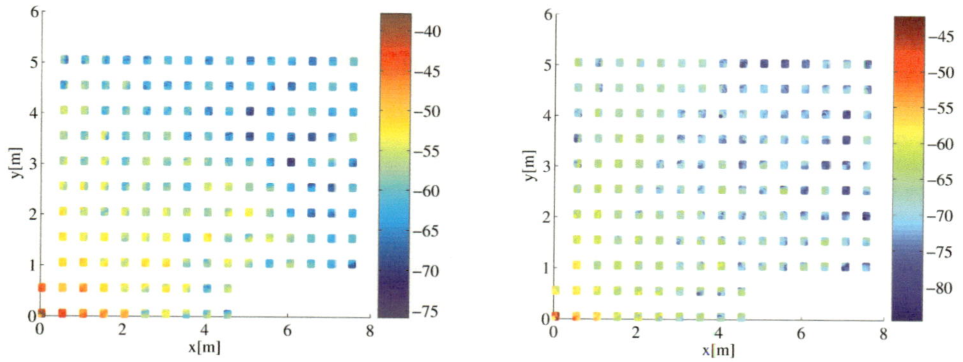

Fig. 18. distribution of measured direct path gain within the scanned area in the room in channels (a) 2 (b) 14

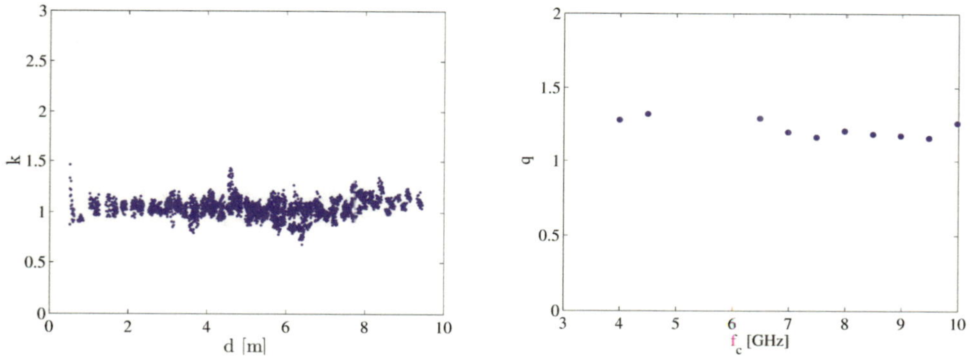

Fig. 19. Path loss model parameters (a) Dependency of k to Rx-Tx distance (b) Dependency of q to frequency

- Frequency decaying factor determination: Frequency decaying factor, k, was derived using equation (12). To observe the variation of k on the Tx-Rx distance, derivation was done for all the possible Tx-Rx distances. It was assured that the frequency decaying factor is almost constant for all possible distances. The variation range was between 0.96 to 1.22 and the mean value is 1.12. Dependency of k to Rx-Tx distance is shown in Fig. 19(a).
- Distance decaying factor determination: the distance decaying factor, q, was derived using (13). To observe the variation of the distance decaying factor, derivation was done for all possible frequency samples. It was assured that the variation of n is negligible for different frequency samples. The variation of q was between 1.15 to 1.32. The mean value of all samples, 1.22, could be represented the distance decaying factor. Dependency of q to frequency is shown in Fig. 19(b).
- Initial path gain determination: initial path loss value, G_R, was calculated using (14) by replacing the obtained frequency and distance decaying factor from the above two steps. Obtained G_R from our measured data was 39.07 [dB].

The specific values for these parameters for the indoor LoS scenario are reported as f_R = 5 GHz, d_R = 1 m, G_R= -35.4 dB, k = 0.03 and q = 1.6 for the indoor office and G_R = -43.9 dB, k = 1.12 and q = 1.8 for the residential environment (Molisch et al, 2004). Following the same approach, corresponding parameters for the measured values were derived as G_R = -39 dB, k = 1.12 and q = 1.22. These parameters are slightly different from those proposed by the standard model due to specific environment. Good fit, typical for all subchannels, is observed which indicates the appropriateness of the model to be used for the threshold setting.
The standard path-gain formula was applied as the proposed delay-dependent threshold to get the range of the measured data d_m. Standard deviation of the ranging error σ_e obtained from the path-gain threshold and the best fixed threshold are presented in (Dashti et al., 2009). It is observed that the path-gain threshold gives a lower ranging error in all subchannels with a stable performance over all frequency bands. The performance of the fixed threshold ranging however is frequency dependent due to different path-loss and interference (Dashti et al., 2008).

3.4 Effect of center frequency and band width

An important finding from Fig. 8 is that given the wide dynamic range of signal levels over varying distance, it is hard to find one optimum threshold which achieves the perfect direct path detection everywhere in one office room. The inherent problem here is that the limited transmit power hinders the signals from reaching more than several meters away.
Another finding is that channel 4 is able to provide reliable ranging in almost all the locations of the room. The result from channel 7 indicated that the noise detection is the main source of error in many Tx locations. In wall-side Tx locations, however, the miss detection becomes a dominant source of error. The miss detection is attributed to the weak direct paths close to the noise level, making its detection difficult. The results of channel 11, which showed the smallest path gain among the channels, is dominated both by the noise and miss detection. In that channel, even the strongest paths are as weak as, or weaker than the noise level. Systems operated in the high band often faces this issue. It is therefore very essential to introduce a technique to improve the signal to noise ratio, such as channel averaging functionality for noise reduction and beam forming for increased signal level, in the receiver. Accurate ranging in the low band is promising even under the transmit power restrictions, while the use of high band necessitates a fundamental countermeasure against the low signal level at the receiver.
It turned out that the gain of direct and strongest paths quickly decreases with increasing frequency. The restriction of the transmit spectral density further limits the service coverage. Still, ranging in the low band reveals promising performance, while accurate ranging can only be performed in a very limited areas in the high band. For example the ranging method in the highest frequency band allows accurate ranging only within 1 m range relative to the device. This fact implies that accurate ranging in NLoS scenario is even less promising due to excess path loss due to whatever path obstruction. It is important to note that the most influential factor in the accurate ranging in NLoS scenarios would be the limited transmit power, rather than the LoS blockage and multipath propagation.
It is also found that the detection probability has obvious dependency on the bandwidth. There are four combinations of bands with the same center frequencies and different bandwidths. It was found that channels with wider bandwidths give rise to lower detection probability. The trend becomes remarkable as the frequency increases. This is a natural consequence of the observation in the channel modeling that the wider bandwidth gives the lower power of the direct and strongest path, which resulted in increased probability

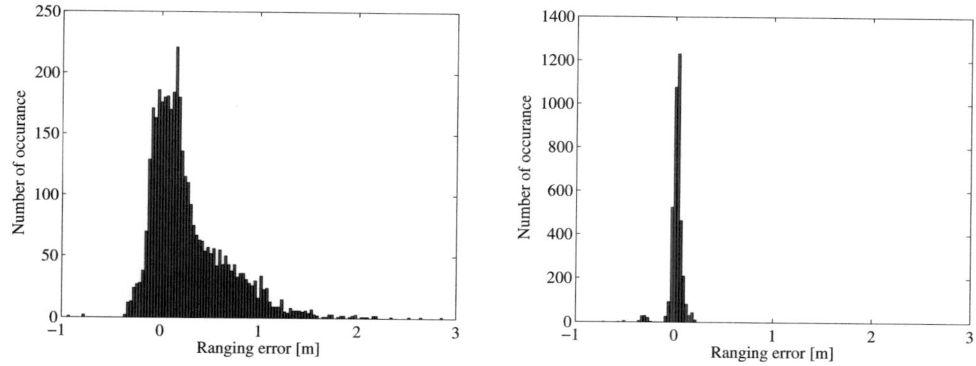

Fig. 20. Histogram of the ranging error (a) channel 5 with bandwidth of 500 MHz (b) channel 7 with bandwidth of 1.0 GHz

of noise detection. From the obtained ranging results in different subbands, it is observed that wider bandwidth provides better estimation accuracy of the distance because of the finer delay resolution, as is commonly reported. In Fig. 20 the histogram of ranging errors in channel 5 and 7 with the bandwidth of 500 MHz and 1.0 GHz is shown. It can be concluded that wider bandwidth always gives better detection and estimation accuracy of the direct paths in the low band, while the superiority of the wider bandwidth is becoming less visible as the center frequency goes higher. Use of wider bandwidth does not always provide better ranging performance, particularly in the high band. In contrast to the well-known observation that wider system bandwidth gives rise to better accuracy of range estimation, performance of the range detection revealed the opposite trend, particularly in the high band. This is because wider bandwidth leads to lower gain of direct and strongest paths. Systems with wider bandwidth clearly outperforms those with narrower bandwidth in the low band, but that would not be necessarily the case in the high band.

4. Summary and future trends

In this chapter the motivations for research on indoor ranging/localization using ultra-wideband systems is described and a literature review is given. UWB time-based ranging and ToA estimation algorithms are reviewed and threshold-based ToA estimation algorithm is provided. A measurement campaign for the indoor ranging is introduced and the obtained results are inspected. A practical method is proposed for setting the threshold value. This method is based on the path-loss of the signal which can be predicted by the standard channel model. The applicability is checked experimentally. The effect of bandwidth on distribution of the ranging error is discussed. There are a few directions that one might take to extend this research:

- A practical threshold setting technique is introduced based on the standard channel model for the indoor environments (Dashti et al., 2011). Proposed threshold setting technique is validated using a set of channel measurement data acquired in a typical office room. More channel measurement should be performed in different indoor environments in order to validate the applicability of the proposed threshold-setting technique in different environments to evaluate the generality of the method.

- Some practical issues remain unresolved. In particular perfect clock synchronization between transmitter and receiver is assumed. This assumption is unlikely in practice. Solutions to this problem like round-trip measurement have been mentioned, but they need to be implemented and validated in practice. At a deeper level, understanding and quantifying how the synchronization error impacts the accuracy will help in designing a practical system.
- In the system model explained, it is assumed that the transmitter sends out a UWB waveform. It is known that the UWB waveform is distorted during interactions to the wireless channel. For the simplicity of the simulation it is assumed that this distortion is negligible, one might take a more practical received signal to extend this research.
- More practical scenario should be considered, the case that UT antenna pattern is distorted by near objects and the UT orientation is random. Ranging results with the antenna proximity to the human head are presented in (Dashti et al., 2010). It should be noted that the human body is just one of the sources of distortion. Even it is quite possible that the antenna pattern is distorted by the antenna itself and the chassis of UT. Deep understanding of antenna pattern distortion and its effect on ToA estimation can be considered.
- Since this research area is fairly new, there are many different and important ways to contribute to indoor localization technology. There is a need for comprehensive measurements and modeling for indoor localization specific applications. As such the emerging UWB technology promises a solution for combating the indoor multipath condition. As a result the implementation of UWB measurement system and indoor channel modeling for localization is an important area for further research. In addition, analyzing the effect of bandwidth on the ranging error could be accomplished by examining bandwidths in excess of 60 GHz. The following can also be conducted as a continuation of the research work, namely, comparing the performance of super resolution algorithms to the UWB system for indoor localization.

5. References

Alsindi, N., Alavi, B. & Pahlavan, K. (2007). Measurement and modeling of ultra wideband TOA- based ranging in indoor multipath environments, In: *IEEE Trans. Veh. Tech.*

Dardari, D. & Win. M. (2006). Threshold-based time-of-arrival estimators in UWB dense multipath channels, In: *Proc. IEEE Int. Conf. Commun. (ICC)*, pp. (4723-4728), vol. 10, Istanbul, Turkey

Dashti, M., Ghoraishi, M., Haneda, K., Takizawa, K. & Takada, J. (2008). Distance dependent threshold ToA estimation, In: *Proceedings of IEICE WBS Technical Meeting, WBS2008-53*

Dashti, M., Ghoraishi & Takada, J.(2009). Optimum Threshold for Ranging Based on ToA Estimation Error Analysis, In: *20th Personal, Indoor and Mobile Radio Communications Symposium 2009(PIMRC09)*

Dashti, M., Khatun, A., Laitinen,T., Al-Hadi, A.A., Haneda, K., Ghoraishi, M. & Takada, J. (2010). UWB Ranging with Antenna Proximity to the Human Head, In: *APMC 2010*

Dashti, M., Ghoraishi, M., Haneda, K., Takizawa, K. & Takada, J. (2010). Sources of ToA Estimation Error in LoS Scenario, In: *ICUWB*

Dashti, M., Ghoraishi, M., Haneda, K., Takizawa, K. & Takada, J. (2011). Statistical Analysis of Ranging Error and Optimum Threshold for Indoor UWB Positioning, Submitted to: *EURASIP*

Dizdarevic, V. & Witrisal, K. (2007). Statistical UWB Range Error Model for the Threshold Leading Edge Detector, In: *International Conference on Information, Communications and Signal Processing, ICICS*

Ellis, J. & Rouzet, P. (2004). P802.15.4a Alt PHY selection criteria, In: *doc. IEEE802.15-04-0232-16-004a.*

Falsi, C., Dardari,D., Mucchi,L. & Win,M. Z. (2006). Time of arrival estimation for UWB localizers in realistic environments, In: *EURASIP J. Appl. Signal Processing (Special Issue on Wireless Location Technologies and Applications)*

Gezici,S., Sahinoglu, Z., Molisch, A., Kobayashi, h. & Poor, H. (2008). Two-step time of arrival estimation for pulse based ultra-wideband systems, In: *EURASIP Journal on Advances in Signal Processing*, vol. 2008, Article ID 529134, 11 pages

Gezici, S., Tian, Z, Giannakis, G., Kobayashi, H., Molisch, A.F., Poor, H. & Sahinoglu, Z. (2005). localization via ultra-wideband radios: a look at positioning aspects for future sensor networks, In: *IEEE signal processing Magazine*, pp. (22:70-84)

Guvenc, I. & Sahinoglu, Z. (2005). TOA estimation with different IR-UWB transceiver types, In: *Proc. IEEE Int. Conf. UWB*, pp. (426-431), Zurich, Switzerland

Guvenc,I.; Sahinoglu, Z.; Molisch,A. & Orlik, P. (2005). Non-coherent TOA estimation in IR-UWB systems with different signal waveforms, In: *Proc. IEEE Int. Workshop on Ultrawideband Networks (UWBNETS)*, pp. (245-251)

Guvenc, I. & Sahingolu, Z. (2005). Threshold-based TOA estimation for impulse radio UWB systems, In: *Proc. IEEE Int. Conf. UWB*, pp. (420-425), Zurich, Switzerland

Guvenc, I. & Sahinoglu, Z. (2005). Threshold selection for UWB ToA estimation based on kurtosis analysis, In: *IEEE Commun. Lett.*, pp. (1025-1027), Vol. 9, No. 12

Guvenc, I., Shahinoglu, Z. & Orlik, P. (2006). TOA Estimation for IR-UWB Systems with Difference transceiver Types, In: *IEEE Trans. on Microwave Theory and Techniques*, Vol. 54, No. 4

Guvenc, I., Gezici, S. & Sahinoglu, Z. (2008). Ultra-wideband range estimation: Theoretical limits and practical algorithms, In: *Proc. IEEE International Conference on Ultra-Wideband (ICUWB 2008)*, pp. (93-96), Hannover, Germany

Haneda, H., Takizawa, K., Takada, J., Dashti, M. & Vainikainen, P. (2009). Performance Evaluation of Threshold-Based UWB Ranging Methods-Leading Edge vs. Search Back-, In: *3rd European Conference on Antennas and Propagation*, pp. (3673-3677)

Haneda, K., Takada, J., Takizawa K. (2007). Ultra Wideband Path Loss Modelling in a Line-of-Sight Office Environment. *2nd European Conference on Antennas and Propagation (EuCAP 2007)*, Nov. 2007 (Edinburgh, UK).

IEEE Std. (2007). Wireless Medium Access Control (MAC) and Physical Layer (PHY) Specifications for Low-Rate Wireless Personal Area Networks(WPANs), In: *IEEE Std 802.15.4a-2007*, pp. (81-83)

Jourdan, D. (2006). Wireless Sensor Network Planning with Application to UWB Localization in GPS-Denied Environments, In: *Doctoral Thesis at Massachusetts Institute of Technology*

Lee, J. & Scholtz, R. (2002). Ranging in a dense multipath environment using an UWB radio link, In: *IEEE Journal on selected Areas in Communications*, pp. (20(9):1677-1683)

Low, Z., Cheong, J., Law, C., Ng, W., & Lee, Y. (2005). Pulse detection algorithm for Line-of-Sight (LoS) UWB ranging application, In: *IEEE Antennas and Wireless Propagation Letters*, pp. (4:63-67)

Molisch, A.F., Balakrishnan, v, Cassioli, D., Chong, C.C., Emami, S., Fort, A., Karedal, J., Kunisch, J., Schantz, H., Schuster, U. & Siwiak, K. (2004). IEEE 802.15.4a channel model - final report, IEEE 802.15 WPAN Low Rate Alternative PHY Task Group 4a(TG4a), In: *Tech. Rep.*

Sahinoglu, Z., Gezici, S. & Guvenc, I. (2003). Ultra-wideband Positioning Systems: Theoretical Limits, Ranging Algorithms, and Protocols, In: *Cambridge university press*

Scholtz , R. & Lee, J. (2002). Problems in modeling UWB channels, In: *Proc. IEEE Asilomar Conf. Signals, Syst. Computers*, pp. (706-711), , vol. 1, Pacific Grove, CA

Win, M. & Scholtz, R. (1998). On the robustness of ultra-wide bandwidth signals in dense multipath environments, In: *IEEE Commun. Lett.*, pp. (2(2):51-53)

Win, M. & Scholtz, R. (2002). Characterization of ultra-wide bandwidth wireless indoor communications channel: A communication theoretic view, In: *IEEE J. Select. Areas Commun.*, pp. (20(9):1613-1627)

Xu, C. & Law, C. (2008). Delay-dependent threshold selection for UWB ToA estimation, In: *IEEE Communication letters*, pp. (380-382), Vol. 12, No.5

Yang, L. & Giannakis, G. (2004). Ultra-wideband communications: An idea whose time has come, In: *IEEE Sig. Processing Mag.*, pp. (26-54), vol. 21, no. 6

Yang, L. & Giannakis, G. (2005). Timing ultra-wideband signals with dirty templates, In: *IEEE Trans. Commun.*, pp. (1952-1963), vol. 53, no. 11

20

Novel Mechanisms for Location-Tracking Systems

Giuseppe Destino and Giuseppe Abreu
University of Oulu, Centre for Wireless Communications
Finland

1. Introduction

The need of location information is rapidly emerging in many wireless application scenarios Hightower & Borriello (2001); Poslad (2009); Vossiek et al. (2003). For instance, in home and office environments, location-based services are developed to improve the efficiency of the working environment, to localize printers, mobile-phones, people, etc. In warehouse, industrial and hospital application scenarios, location information can be used to track assets and persons. In military and rescuing applications, positioning technologies can be utilized for real-time monitoring of soldiers in the troop, track machines and cars Destino & Abreu (2009a); Destino et al. (2007).

Location-information, however, is also emerging as a requirement for the next generation of wireless communication technologies. For instance, for mobile networks, the 23rd of September 2010, the Federal Communications Commission (FCC) unanimously approved new rules for the use of unlicensed TV white space spectrum. It was stated that devices will be able to access to the TV white space spectrum if they will able to determine their locations and to identify the unused channels at that location. Yet another emerging area where positioning will play a major role is the Internet-of-things (IoT) Scott & Benlamri (2010). In this case, context and location-awareness will be fundamental for the development of smart technologies that will allow "Things" (computer, mobile-phones, objects, sensors, actuators, etc.) to be autonomous and energy-efficient.

Motivated from all the above, a lot of researches are devoted to the development of accurate positioning technologies based on satellite radios like the Global Positioning System (GPS), or short- and medium-range radio technologies such as Wi-Fi, Bluetooth and Ultra-wide band (UWB). In particular, UWB technology has seen a strong surge of interests because of its high accurate ranging capabilities and energy efficiency Dardari et al. (2008.); Gezici et al. (2005); Yihong et al. (2004).

This chapter is intended as a survey on current state-of-the-art localization techniques for large-scale and single-hop networks, and for the latter case, a dedicated section will be also devoted for Non-Line-of-Sight (NLOS) mitigation mechanisms. Finally, considering a low-data-rate impulse radio (LDR-IR) UWB ranging model Denis et al. (2007), the performance of the described algorithms will be shown for Line-of-Sight (LOS) and mixed LOS/NLOS channel conditions in both single-hop and multi-hop network topologies.

2. Modeling of the localization problem

Consider a network of N nodes deployed in the η-dimensional space. We shall assume that N_{A} nodes are anchors and N_{T} nodes are targets, where an anchor is a node whose location is known *a priori*, while a *target* is a node whose position is yet to be determined.
Denote by $\mathbf{p}_i \in \mathbb{R}^{\eta}$ the position (Euclidean coordinates) of the i-th node such that $\mathbf{p}_i \triangleq \mathbf{a}_i$ and $\mathbf{p}_i \triangleq \mathbf{z}_j$ for $1 \leq i \leq N_A$ and $N_A + 1 \leq i \leq N$, respectively.
The Euclidean distance between the i-th and the j-th node is defined as

$$d_{ij} \triangleq \|\mathbf{p}_i - \mathbf{p}_j\|_{\mathrm{F}}, \tag{1}$$

where $\|\cdot\|_{\mathrm{F}}$ is the Frobenius norm, while a measurement (ranging) of d_{ij} is given by

$$\tilde{d}_{ij} = \begin{cases} d_{ij} + b_{ij} + n_{ij}, & \text{if either } \mathbf{p}_i = \mathbf{z}_i \text{ or } \mathbf{p}_j = \mathbf{z}_j, \\ d_{ij}, & \text{if both } \mathbf{p}_i = \mathbf{a}_i \text{ and } \mathbf{p}_j = \mathbf{a}_j \end{cases} \tag{2}$$

where n_{ij} and b_{ij} indicates small(noise) and large(bias) errors.
Extensive measurement campaigns can be found in the literature in order to characterize the statistics of n_{ij} and b_{ij} for different radio-technologies Gentile & Kik (2006); Joon-Yong & Scholtz (2002); Mao et al. (2007); Patwari et al. (2003). In the case of Low-Data-Rate Ultra-Wideband (LDR-UWB) we adopt the model proposed in Denis et al. (2007), which summarizes as follows.
Define the biased distance d'_{ij} as $d'_{ij} \triangleq d_{ij} + b_{ij}$ and consider such a variable as a random variate conditioned upon the true Euclidean distance d_{ij} and governed by the probability density functions p_C

$$p_C(d'_{ij}|d_{ij}, C) = \frac{G_C}{d_{ij}\sqrt{2\pi}\sigma_C} \exp\left(\frac{\left(\frac{d'_{ij}}{d_{ij}} - 1\right)^2}{2\sigma_C^2}\right) + \lambda_C E_C \frac{1_{d'_{ij} > d_{ij}}}{d_{ij}} \exp\left(\frac{-\lambda_C(d'_{ij} - d_{ij})}{d_{ij}}\right), \tag{3}$$

where $1_{d'_{ij} > d_{ij}} = 1$ if $d'_{ij} > d_{ij}$ and 0 otherwise, $\{G_C, \sigma_C\}$ and $\{E_C, \lambda_C\}$ are the weights and parameters of Gaussian and Exponential mixture components and $C \triangleq \{\mathrm{LOS}, \mathrm{NLOS}, \mathrm{NLOS}^2\}$ refers to a ranging error model without bias (LOS), with small bias (NLOS) and large bias (NLOS^2). Furthermore, consider that the channel C is also a function of the distance d_{ij}, and the probability of LOS, NLOS or NLOS^2 can be computed as

$$W_C(d_{ij}) = \frac{\xi}{\sqrt{2\pi}\varsigma_C} \exp\left(\frac{-\left(d_{ij} - d_0\right)^2}{2\varsigma_0^2}\right), \tag{4}$$

where d_0 and ς_0 are reference values (typical $d_0 = 10$ and $\varsigma_0 = 4.6$) and ξ ensures that $W_{LOS}(d_{ij}) + W_{NLOS}(d_{ij}) + W_{NLOS^2}(d_{ij}) = 1$ (for instance ξ is 10 when $d_0 = 10$ and $\varsigma_0 = 4.6$).
Once, the biased distance in computed, then the distance measurement $\tilde{d}_{ij}$ is obtained as in equation (2), where n_{ij} is a zero-mean Gaussian random variable with variance σ_{ij}^2.

In figure 1 we exemplify the LDR-UWB ranging model and we show the histograms and pdfs of $\tilde{d}_{ij}$ obtained for $d_{ij} = 10$, $\sigma_{ij} = 0.7$ and bias-distance parameters $\{G_C, \sigma_C\}$ and $\{E_C, \lambda_C\}$ given by

	G_C	σ_C	E_C	λ_C
LOS	0	0.0068	0	0
NLOS	0.31	0.0102	0.69	47.013
NLOS^2	0.26	0.0129	0.74	8.4331

Table 1. Setting of the parameters for the UWB-LDR ranging model given in equation (3)

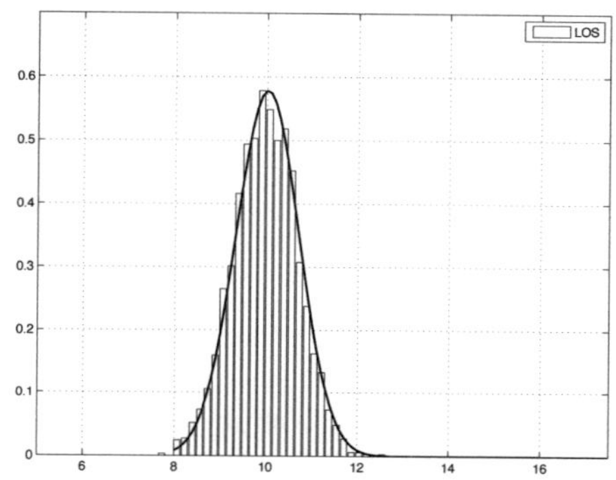

(a) LDR-UWB LOS model

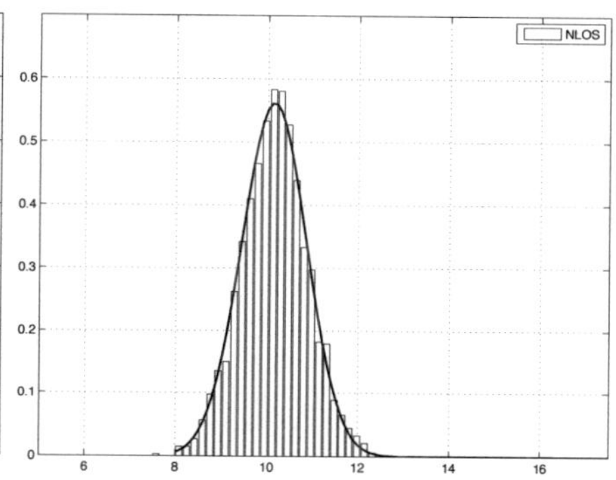

(b) LDR-UWB NLOS model

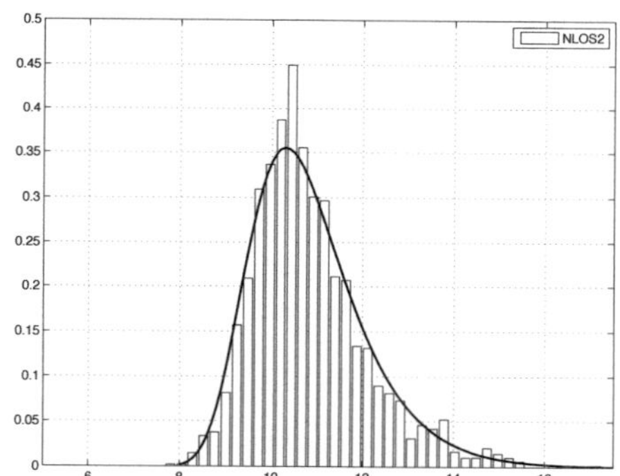

(c) LDR-UWB strong NLOS model

Fig. 1. Example of the biased distance d'_{ij} in different channel conditions.

In many application scenarios, however, it is assumed that the ranging model is unknown and it cannot be accurately estimated because of scarcity of information. Therefore, we consider non-parametric localization methods such as the minimization of a Weighted Least Square (WLS) objective function,

$$\min_{\hat{\mathbf{Z}} \in \mathbb{R}^{N_T \times \eta}} f_R(\hat{\mathbf{Z}}), \tag{5}$$

with

$$f_{\mathrm{R}}(\hat{\mathbf{Z}}) \triangleq \sum_{ij\in\mathcal{H}} w_{ij}\left(\tilde{d}_{ij}-\hat{d}_{ij}\right)^2 = \sum_{ij\in\mathcal{H}} w_{ij}\left(\tilde{d}_{ij}-\|\mathbf{a}_i-\hat{\mathbf{z}}_j\|_{\mathrm{F}}\right)^2+\sum_{ij\in\mathcal{H}} w_{ij}\left(\tilde{d}_{ij}-\|\hat{\mathbf{z}}_i-\hat{\mathbf{z}}_j\|_{\mathrm{F}}\right)^2, \tag{6}$$

where $\mathcal{H}$ is the set of indexes related to connected links, $\hat{d}_{ij} \triangleq \|\hat{\mathbf{p}}_i - \hat{\mathbf{p}}_j\|_{\mathrm{F}}$ is the distance obtained from the estimates of the i-th and j-th nodes, and w_{ij} is a weight Costa et al. (2006); Destino & G. (2009) related to the "concern" Boyd & Vandenberghe (2004) over the term $(\tilde{d}_{ij} - \hat{d}_{ij})$.

In the localization problem posed as in equation (5), several challenges are met and the one that has attracted a large research community is the design of efficient minimization techniques Costa et al. (2006),Biswas, Liang, Toh & Wang (2006),Ding et al. (2008),Destino & Abreu (2009c),Wymeersch et al. (2009). In the sequel, this issue will be addressed and the most effective state-of-the-art solutions will be described in details.

2.1 WLS localization methods in large scale networks

Rewrite the objective function given in equation (6) as

$$f_{\mathrm{R}}(\hat{\mathbf{Z}}) = \left\|\mathbf{W} \circ \left(\tilde{\mathbf{D}} - \mathcal{D}(\hat{\mathbf{P}})\right)\right\|_{\mathrm{F}}^2, \tag{7}$$

where the ij-th element of $\mathbf{W}$ is the weight w_{ij}, $\circ$ is the Hadamard product and

$$\hat{\mathbf{D}} = \mathcal{D}(\hat{\mathbf{P}}) \triangleq \sqrt{\mathbf{1}_N \cdot \mathrm{diag}\left(\hat{\mathbf{P}}\cdot\hat{\mathbf{P}}^{\mathrm{T}}\right)^{\mathrm{T}} + \mathrm{diag}\left(\hat{\mathbf{P}}\cdot\hat{\mathbf{P}}^{\mathrm{T}}\right)\cdot\mathbf{1}_N^{\mathrm{T}} - 2\cdot\hat{\mathbf{P}}\cdot\hat{\mathbf{P}}^{\mathrm{T}}}, \tag{8}$$

where ^{T} indicates transpose, $\mathbf{1}_N$ is a column vector of N elements equal to 1, and diag$(\cdot)$ indicates a column vector containing the diagonal elements of its argument Dattorro (2005).

The localization problem given in equation (5) can then be approached in two different manners Dattorro (2005); Destino & Abreu (2009c); So & Ye (2005). The first one, which is the basis for the later described Classical Multidimensional Scaling (CMDS) Cox & Cox (2000) and Semidefinite Programming (SDP) methods, is to consider $\tilde{\mathbf{D}}$ as the observation of a multidimensional variable $\hat{\mathbf{D}}$. Therefore, the optimization problem can be formulated as matrix proximity optimization problem, in which the objective is to estimate the closest Euclidean Distance Matrix (EDM) $\hat{\mathbf{D}}$ to the observed EDM-sample $\tilde{\mathbf{D}}$. In so doing, the optimization problem benefits from the fact that the space of the EDM, denoted by $\mathbb{EDM}^N$, is related to the space of symmetric positive semidefinite matrixes, denoted by $\mathbb{S}_+^N$ with the linear relationship

$$\mathbf{K} \triangleq \mathcal{K}(\mathbf{D}) = -\frac{1}{2}\mathbf{J}\cdot(\mathbf{D})^{\circ 2}\cdot\mathbf{J}^{\mathrm{T}}, \tag{9}$$

where $^{\circ 2}$ indicates the element-wise square and

$$\mathbf{J} \triangleq \mathbf{I}_N - (\mathbf{1}_N \cdot \mathbf{1}_N^{\mathrm{T}})/N. \tag{10}$$

The search of the optimum matrix can therefore be constrained either to $\mathbb{S}_+^N$ or to $\mathbb{EDM}^N$, such that two different methods can be formulated. The first method is to solve the optimization problem as

$$\begin{aligned} \min_{\hat{\mathbf{K}}} \; & \left\| \mathcal{K}(\mathbf{W} \circ (\tilde{\mathbf{D}} - \hat{\mathbf{D}})) \right\|_F^2, \\ \text{s.t.} \; & \mathcal{K}(\hat{\mathbf{D}}) \in \mathbb{S}_+^N, \end{aligned} \tag{11}$$

and the second method is to formulate the problem as

$$\begin{aligned} \min_{\hat{\mathbf{D}}} \; & \left\| \mathbf{W} \circ (\tilde{\mathbf{D}} - \hat{\mathbf{D}}) \right\|_F^2. \\ \text{s.t.} \; & \hat{\mathbf{D}}^2 \in \mathbb{EDM} \end{aligned} \tag{12}$$

For the sake of illustration, in figure 2 we show the logic of the two approaches with an Euler diagram. The black and red arrows indicate the linear mapping from $\mathbb{S}_+^N$ to $\mathbb{EDM}^N$ given by equations (8) and (9) and viceversa, respectively. The yellow cicle describes the method 1 (optimization in $\mathbb{S}_+^N$) and the blue arrow method 2 (optimization in $\mathbb{EDM}^N$). In the following subsections we describe two state-of-the-art solutions based on method 1 and 2, namely the algebraic Classical Multidimensional Scaling (CMDS) technique and the Semi Definite Programming (SDP) method.

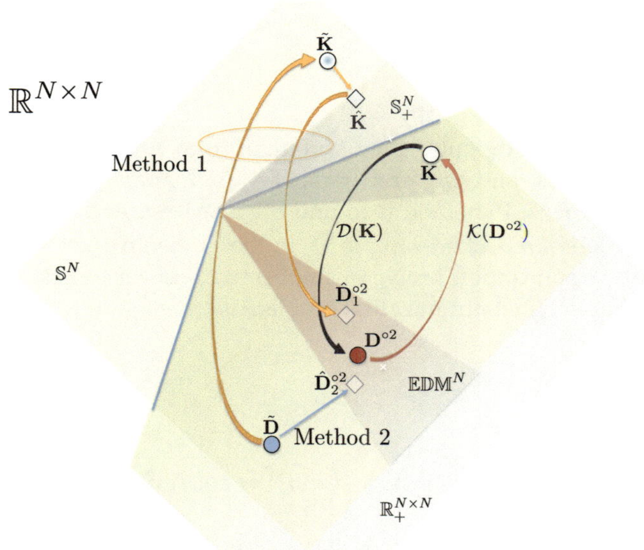

Fig. 2. Illustration of the matrix proximity problem with and Euler diagram. The notations $\mathbb{R}^{N\times N}$, $\mathbb{R}_+^{N\times N}\mathbb{S}^N$, $\mathbb{S}_+^N$ and $\mathbb{EDM}^N$ indicate the real, real-positive, real-symmetric, real-positive semidefinite and Euclidean Distance Matrix spaces.

As mentioned above, however, the WLS-based localization problem can also be approached in a different manner than a matrix proximity problem. The alternative indeed is to solve equation (5) directly over the unknown variables $\mathbf{z}_i$'s Gezici (2008). In this approach, the major difficulty is to handle the multiple minima with robust optimization methods. To this end, indeed, several techniques can be found in the literature which are proposed either as distributed or centralized algorithms. Amongst all, we will describe two algorithms that can benefit of a very low-computational cost, namely the Stress-of-a-MAjorizing-Complex-Objective-Function (SMACOF) Cox & Cox (2000) and the Range-Global Distance Continuation (R-GDC) Destino & Abreu (2009c); More & Wu (1997).

2.2 Classical Multidimensional Scaling (CMDS)

The CMDS is an algebraic technique to solve the localization problem posed as in equation (11). Specifically, the CMDS algorithm relies on the $\mathbb{EDM}^N - \mathbb{S}_+^N$ relationship given in equation (9) Schoenberg (1935) and it can be concisely summarized as

$$\hat{\mathbf{P}}_o = \left([\mathbf{U}]_{\text{UL}:N\times\eta} \cdot [(\mathbf{\Lambda})^{\frac{1}{2}}]_{\text{UL}:\eta\times\eta} \right)^{\text{T}}, \tag{13}$$

where $\hat{\mathbf{P}}_o$ is a representation of the desired estimates coordinates $\hat{\mathbf{P}}$ up to rigid motions (rotation, mirroring and shifting) and scaling, $[\cdot]_{\text{UL}:n\times q}$ denotes the n-by-q upper-left partition and the matrices $\mathbf{U}$ and $\mathbf{\Lambda}$ are the eigenvector and eigenvalue matrices (both in decreasing order) of $\tilde{\mathbf{K}} \triangleq \mathcal{K}(\widetilde{\mathbf{D}})$.

Notice however, that the CMDS performs optimally only if all pairwise links are observed and all weights are unitary. In the other cases, the accuracy of the solution computed via CMDS can be very poor and not sufficient for any location-based application. Furthermore, it is imperative to remind that the real nodes' location estimates $\hat{\mathbf{P}}$ are computed from $\hat{\mathbf{P}}_o$ applying a procrustes operation, which calculates the scaling, rotation, mirroring and shifting factors based on the location of the anchors.

2.3 Semi-definite Programming (SDP)

The SDP method is one of the most powerful algorithms for network localization and it is able to handle incomplete and imperfect data Biswas, Liang, Toh, Wang & Ye (2006). The fundamental idea of the SDP method is to find the EDM-estimate $\hat{\mathbf{D}} \triangleq [\hat{d}_{ij}]$ of rank at most $\eta + 2$ closest to the observed EDM-sample $\tilde{\mathbf{D}}$, in the Frobenius norm sense. Because of the rank-constraint, the optimization problem is not convex, nevertheless, a rank-relaxation can be adopted such that the final optimization problem is

$$\begin{aligned}
&\min_{\hat{\mathbf{K}},\{\hat{\mathbf{B}}_{ij}\}} \sum_{ij\in\mathcal{H}} w_{ij}\,\varepsilon_{ij} \\
&\text{s.t.}\quad [-\tilde{d}_{ij}\quad 1]\,\hat{\mathbf{B}}_{ij}\,[-\tilde{d}_{ij}\quad 1]^{\text{T}} = \varepsilon_{ij},\ \forall ij \\
&\qquad [\mathbf{0}_\eta\quad \mathbf{e}_i - \mathbf{e}_j]\,\hat{\mathbf{K}}\,[\mathbf{0}_\eta\quad \mathbf{e}_i - \mathbf{e}_j]^{\text{T}} = \nu_{ij},\ i,j \geq N_A \\
&\qquad [\mathbf{a}_i\quad -\mathbf{e}_j]\,\hat{\mathbf{K}}\,[\mathbf{a}_i\quad -\mathbf{e}_j]^{\text{T}} = \nu_{ij},\ i \leq N_A,\ \forall j \\
&\qquad \hat{\mathbf{B}}_{ij} \triangleq \begin{bmatrix} 1 & b_{ij} \\ b_{ij} & \nu_{ij} \end{bmatrix} \succeq 0 \\
&\qquad \hat{\mathbf{K}} \triangleq \begin{bmatrix} \mathbf{I}_\eta & \hat{\mathbf{Z}}^{\text{T}} \\ \hat{\mathbf{Z}} & \hat{\mathbf{Y}} \end{bmatrix} \succeq 0
\end{aligned} \tag{14}$$

where $\mathbf{0}_\eta$ is a vector of zeros and $\mathbf{e}_i \in \mathbb{R}^{N_T}$ the only non-zero element is a 1 at the i-th element. The SDP formulation can be optimally solved using standard convex SDP optimization software, such as SDPA, CSDP, SDPT3, SeDuMi[1], however, the computational complexity grows quickly with the number of variables and constraints.

2.3.1 SMACOF

The SMACOF technique is another optimization method, that in contrast to the SDP and C-MDS algorithm, operates on the space of the variables $\hat{\mathbf{z}}_i$'s. The fundamental idea in SMACOF is to find the minimum of a non-convex function by tracking the global minima of the so-called majored convex functions $\mathcal{T}(\hat{\mathbf{P}}, \mathbf{Y})$. As illustrate in figure 3 the majorinzing function is computed from from the original objective and a given point $\hat{\mathbf{P}} = \hat{\mathbf{X}}$. Mathematically, such a function is given by

$$\mathcal{T}(\hat{\mathbf{P}}, \mathbf{Y}) = \sum w_{ij}^2 \cdot \tilde{d}_{ij}^2 + \text{tr}\left(\hat{\mathbf{P}}^{\text{T}} \cdot \mathbf{H} \cdot \hat{\mathbf{P}}\right) - 2 \cdot \text{tr}\left(\hat{\mathbf{P}}^{\text{T}} \cdot \mathbf{A}(\mathbf{Y}) \cdot \mathbf{Y}\right), \tag{15}$$

where $\text{tr}(\cdot)$ denotes the trace, $\mathbf{Y} \in \mathbb{R}^{N \times \eta}$ is an auxiliary variable and the entries of $\mathbf{H}$ and $\mathbf{A}(\mathbf{Y})$ are given by

$$h_{ij} = \begin{cases} \sum\limits_{\substack{i=1 \\ i \neq j}}^{N} h_{ij}, \; i = j, \\ -w_{ij}^2, \; i \neq j, \end{cases} \tag{16a}$$

$$a_{ij} = \begin{cases} \sum\limits_{\substack{i=1 \\ i \neq j}}^{N} a_{ij}, \; i = j, \\ w_{ij}^2 \cdot \dfrac{\tilde{d}_{ij}}{\|\mathbf{y}_i - \mathbf{y}_j\|_2}, \; i \neq j. \end{cases} \tag{16b}$$

The SMACOF algorithm, therefore, consists of an iterative method that converges to a solution $\hat{\mathbf{P}}$ that depends on the initial estimate $\hat{\mathbf{P}}^{(0)}$. The main advantage is that at the n-th iteration the global minimum $\hat{\mathbf{P}}_{\min}^{(n)}$ of the majored function $\mathcal{T}(\hat{\mathbf{P}}, \mathbf{Y})$ with $\mathbf{Y} = \hat{\mathbf{P}}_{\min}^{(n-1)}$, can be computed in closed form via the Guttman transform,

$$\hat{\mathbf{P}}_{\min}^{(n)} = \mathbf{H}^{\dagger} \cdot \mathbf{A}\left(\hat{\mathbf{P}}_{\min}^{(n-1)}\right) \cdot \hat{\mathbf{P}}_{\min}^{(n-1)}, \tag{17}$$

where † denotes the pseudoinverse and $\mathbf{A}\left(\hat{\mathbf{P}}_{\min}^{(n-1)}\right)$ is the matrix with elements a_{ij}.

2.3.2 Nearly optimum WLS minimization

Recently, in Destino & Abreu (2009c) a novel low-complexity algorithm was proposed to solve the WLS optimization problem with nearly optimal performance. The minimization method, hereafter referred to as the R-GDC algorithm, is based on the global continuation method proposed in More & Wu (1997), which can be summarized as the iteration of three fundamental steps: *smoothing*, *minimization* and *continuation*. In the smoothing step the entire

[1] SeDuMi runs in Matlab©and uses the Self-Dual method for solving general convex optimization problems, etc.

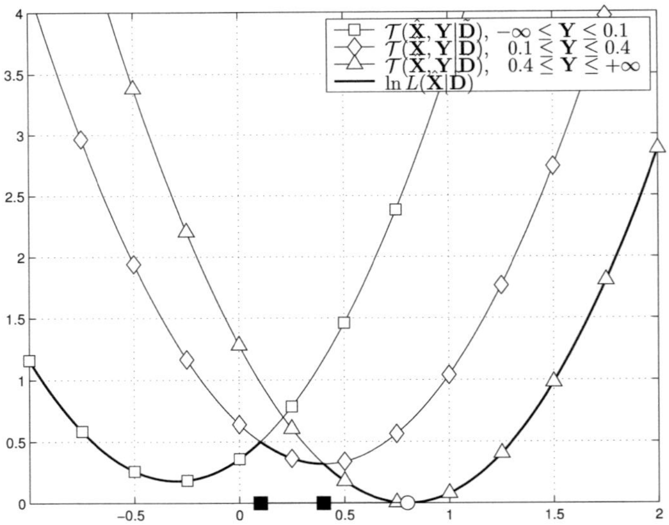

Fig. 3. Illustration of the majorizing functions $\mathcal{T}(\hat{\mathbf{P}} = \hat{\mathbf{X}}|\mathbf{Y}, \mathbf{D})$ in the optimization of WLS-objective function related to a source-localization problem in $\eta = 1$ dimension. The function $\ln L(\hat{\mathbf{X}}|\mathbf{D})$ is the WLS-objective with $\tilde{d} = d$. On the x-axis, we have plotted the network, where the anchors and the target are indicated with a black square and a white circle, respectively.

objective is approximated by function with a higher degree of differentiability (smoothed), obtained by means of a convolution of the original function with a Gaussian kernel $g(x;\lambda)$

$$g(x;\lambda) = \exp\left(-\frac{x^2}{\lambda^2}\right), \tag{18}$$

where the parameter λ controls the smoothing degree.
In the minimization step each of these smoothed functions is minimized using a conventional Newtonian algorithm Nocedal & Wright (2006). Finally, the continuation refers to the process of tracing the global minimum, which in practice is typically performed by initializing the minimization of the next smoothed objective with the latest solution.
In figure 4, for instance, an illustrative example of the GDC method is shown, where the non-convex objective function $s(x)$ is given by the sum of Gaussian functions. The dark and the thin lines indicate the original and the smoothed objective functions, respectively. The smoothed functions are obtained via the convolution of the original objective $s(x)$ with the Gaussian kernel $g(x;\lambda)$ given in equation (18). The algorithm starts with the minimization of the most smoothed function (largest λ), from which a new iteration will be initiated. This process is then repeated until $\lambda = 0$, from which the solution of the optimization problem is obtained.
In the context of network localization, this technique consists of

$$\hat{\mathbf{Z}}^{(k)} = \min_{\hat{\mathbf{Z}} \in \mathbb{R}^{N_T \times \eta}} \langle f_R \rangle_{\lambda^{(k)}} (\hat{\mathbf{Z}}), \; 1 \leq k \leq K, \tag{19}$$

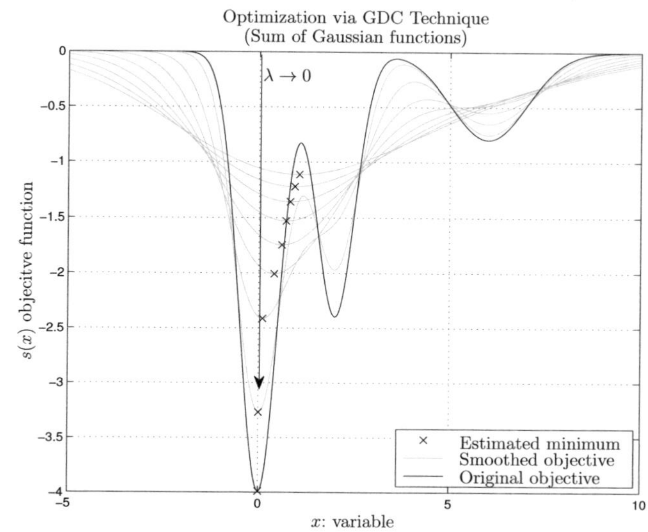

Fig. 4. Illustration of the GDC method. Starting from the original objective (dark line) and give a set of smoothing parameters λ, smoothed versions (thin line) of the original objective are computed. Iterating the process smooth-minimize-continue, the global optimum of the original objective can be found with high probability when the last minimization with $\lambda = 0$ is performed.

where $\langle f_R\rangle_{\lambda(k)}(\hat{\mathbf{Z}})$ is the smoothed variation of $f_R(\hat{\mathbf{Z}})$ and it is given by

$$\langle f_R\rangle_\lambda(\hat{\mathbf{Z}}) = \frac{1}{\pi}\int_{\mathbb{R}^\eta} \sum_{ij\in\mathcal{H}} w_{ij}\left(\tilde{d}_{ij} - \|\hat{\mathbf{p}}_i-\hat{\mathbf{p}}_j+\lambda\mathbf{u}\|_F\right)^2 \exp(-\|\mathbf{u}\|_F^2)\,d\mathbf{u} \tag{20}$$

$$= \sum_{ij\in\mathcal{H}} w_{ij}\cdot\left(\lambda^2+\tilde{d}_{ij}^2+\hat{d}_{ij}^2 - \lambda\sqrt{\pi}\tilde{d}_{ij}\,{}_1F_1\left(\frac{3}{2};1;\frac{\hat{d}_{ij}^2}{\lambda^2}\right)\exp\left(\frac{-\hat{d}_{ij}^2}{\lambda^2}\right)\right), \tag{21}$$

where $\Gamma(a)$ is the gamma function and ${}_1F_1(a;b;c)$ is the confluent hypergeometric function Abramowitz & Stegun (1965)., which can be efficiently evaluated as

$${}_1F_1\left(\frac{3}{2};1;s\right) = 1+\sum_{m=1}^{+\infty}\left(s^m\cdot\prod_{t=1}^{m}\left(\frac{1}{2t^2}+\frac{1}{t}\right)\right), \tag{22}$$

and (Abramowitz & Stegun, 1965, Eq. 13.5.1, pp. 508)

$${}_1F_1\left(\tfrac{3}{2};1;s\right)=\frac{2e^s}{\sqrt{\pi}}\sum_{p=0}^{P-1}\frac{s^{\frac{1}{2}-p}}{p!}\prod_{t=0}^{p-1}\left(t-\tfrac{1}{2}\right)^2-\frac{s^{-3/2}}{2\sqrt{\pi}}\sum_{m=0}^{M-1}\frac{(-s)^{-m}}{m!}\prod_{t=0}^{m-1}\left(\tfrac{3}{2}+t\right)^2+\mathcal{O}(|s|^{-M})+\mathcal{O}(|s|^{-P}) \tag{23}$$

The minimization step is then performed with a very low-complexity mechanism, namely the Broyden-Fletcher-Goldfarb-Shanno (BFGS), in which the gradient of the smoothed objective can be computed as

$$\nabla_{\hat{\mathbf{Z}}}\langle f_{\mathrm{R}}\rangle_{\lambda}(\hat{\mathbf{Z}}) \triangleq \sum_{ij\in\mathcal{H}} w_{ij}\left(2-\frac{\sqrt{\pi}\tilde{d}_{ij}}{\lambda}\exp\left(\frac{-\hat{d}^2_{ij}}{\lambda^2}\right){}_1F_1\left(\frac{3}{2};2;\frac{\hat{d}^2_{ij}}{\lambda^2}\right)\right)\times\left(\mathbf{e}_{ij}\otimes(\hat{\mathbf{z}}_j-\hat{\mathbf{z}}_i)\right),$$

where $\otimes$ indicates the Kronecker product and $\mathbf{e}_{ij} \in \mathbb{R}^{N_{\mathrm{T}}}$ are row-vectors with the i-th and the j-th element equal to 1 and -1 respectively.
Although any decreasing sequence of λ can be selected, the general rule-of-thumb is to compute $\lambda^{(0)}$ such that $\langle f_{\mathrm{R}}\rangle_{\lambda^{(0)}}$ is convex and then decreases λ linearly for K steps until $\lambda^{(K)} = 0$. In the specific case of $\langle f_{\mathrm{R}}\rangle_{\lambda}$ given by equation (21), the initial value $\lambda^{(0)}$ can be selected as

$$\lambda^{(0)} = \frac{\sqrt{\pi}}{2}\max_{ij\in\mathcal{H}} \tilde{d}_{ij}. \tag{24}$$

2.4 Enhanced LS-based localization methods for NLOS conditions

The NLOS problem, that is the presence of large errors in the ranging measurements, is yet a remaining challenge in the context of network localization, especially when multiple targets are localized simultaneously like in the algorithms described above. In this regard, only little articles tackle this problem and few algorithms are proposed to solve effectively this challenge Denis & Daniele (2004); Destino & G. (2010); Guvenc et al. (2007); Venkatesh & Buehrer (2007); Yu & Jay Guo (2008). For instance in Destino & G. (2009) a weighing strategy was derived in order to associate a lower weight to those measurements affected by bias and more weight to those ranging that are considered very reliable. To the best of our knowledge, however, the most effective localization techniques that compensate for the bias in a non parametric manner are the Sequential-Quadratic-Programming (SQP) and the Distance Contraction (DC) described in Yu & Jay Guo (2008) and Destino & Abreu (2009b); Destino & G. (2010), respectively.
In the remaining subsections, these algorithms are discussed in details. For the sake of convenience, hereafter, we simplify the notation such that the j-th index will be omitted in any subscript pair $_{ij}$. For instance, the symbol d_{iN_T} that refers to the Euclidean distance between the i-th anchor and the target will be simply denoted by d_i.

2.5 Sequential-quadratic-programming method

The SQP algorithm was proposed in Yu & Jay Guo (2008), and it consists of a constrained variation of the minimization problem given in equation (5). Specifically, the authors consider the bias errors as variables to be estimated. In essence, the SQP formulation of the source localization problem in NLOS conditions is given by

$$\begin{aligned}\hat{\mathbf{z}} = & \min_{\hat{\mathbf{z}}\in\mathbb{R}^{\eta},\hat{b}_i\in\mathbb{R}^+} \sum_{i=1}^{N_{\mathrm{A}}}\left(\tilde{d}_i-\hat{d}_i-\bar{h}_i\hat{b}_i\right)^2, \\ & s.t.\ \hat{d}_i \le \tilde{d}_i,\ \forall i. \\ & \hat{b}_i \le \min_{j=1,\ldots,N_{\mathrm{A}}}\{\tilde{d}_i+\tilde{d}_j-d_{i,j}\}, \\ & \hat{b}_i \ge 0\end{aligned} \tag{25}$$

where $d_{i,j}$ is the distance between the i-th and the j-th anchor and $\bar{h}_i$ is equal to 1 when it is assumed the presence of a bias and $\bar{h}_i = 0$ otherwise.
The effectiveness and the accuracy of this method depends on: a) the size of the feasibility region $\mathcal{I}$, which is obtained from the set of constraints $\hat{d}_i \leq \tilde{d}_i, \forall i$; b) the tightness of the bias upper bounds $\{u_i = \min_{j=1,\ldots,N_A} \{\tilde{d}_i + \tilde{d}_j - d_{i,j}\}\}$; c) the exactness of assumptions on the presence/absence of bias. Specifically, a) implies that the smaller is the size of the feasibility region, the more restricted is the domain of $\hat{\mathbf{z}}$ and therefore the more accurate is the solution; b) implies that the tighter are the bias upper-bounds, the more accurate are the estimations of the variables $\hat{b}_i$; c) implies that the larger is the number of exact h_i, the more appropriate are the bias corrections.
Because of these conditions, the SQP approach is somewhat heuristic and it leaves room to further improvements which can be achieved with the modification of the constraints in the optimization problem.

2.6 Least-square with distance contraction method

In this subsection, a completely novel technique for bias mitigation will be described. The method is based on "distance contractions" Destino & Abreu (2009b); Destino & G. (2010), which in plain words is the mechanism to contract (correct) the distance measurements such that $\tilde{d}_i < d_i$. This mechanism has the following advantages

a) it can improve the convexity of the objective function,

b) it relaxes the requirements on the ranging precision,

c) it can provide accurate location estimates,

d) it is low-complexity.

It was also shown that corrections can be made such that the WLS-objective function always convex and with minimum in the real target location. Such corrections relate to the geometry of the network and, it can be studied by observing the null space of the angle kernel matrix

$$\boldsymbol{\Omega} \triangleq \begin{bmatrix} 1 & \cos(\theta_{1T2}) & \ldots & \cos(\theta_{1TN_A}) \\ \cos(\theta_{1T2}) & 1 & \ddots & \cos(\theta_{2TN_A}) \\ \vdots & \ddots & \ldots & \vdots \\ \cos(\theta_{1TN_A}) & \cos(\theta_{2TN_A}) & \ldots & 1 \end{bmatrix}, \tag{26}$$

where θ_{iTj} is the angle between the (i,j) pair of anchors seen by the target. The example proposed in figure 5 illustrates the aforementioned concepts. Specifically, in the subfigure 5(a), the LS objective function is studied under the assumption of exact distance measurements, i.e. Visualizing the contour levels (lines) together with the convex area (dots) of the function, we can observe that, for the specific example, only one minimum exists. In the subfigure 5(b), the same type of study is carried out, but in contrast, the measurements are now assumed with positive bias (typical for NLOS channel conditions). In this case, the WLS objective function results with two minima, and both correspond to very inaccurate node location estimate. In addition, it is observed that the convex area is drastically decreased such that it is no longer convex where the real target location is. In the subfigure 5(c), the contraction mechanism is applied such that the ranging measurements are shorter than the true distances. It is

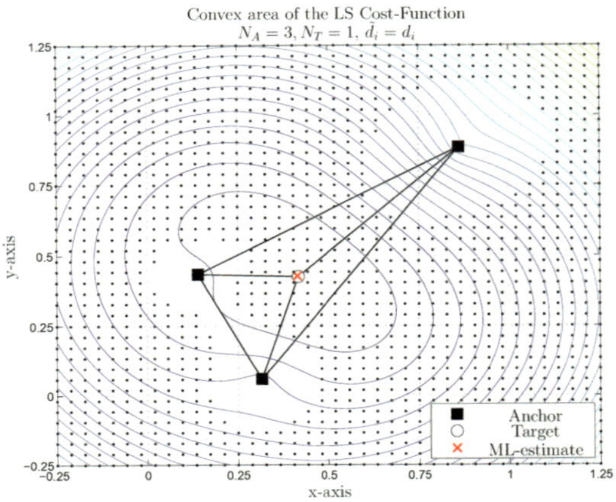

(a) Exact distances

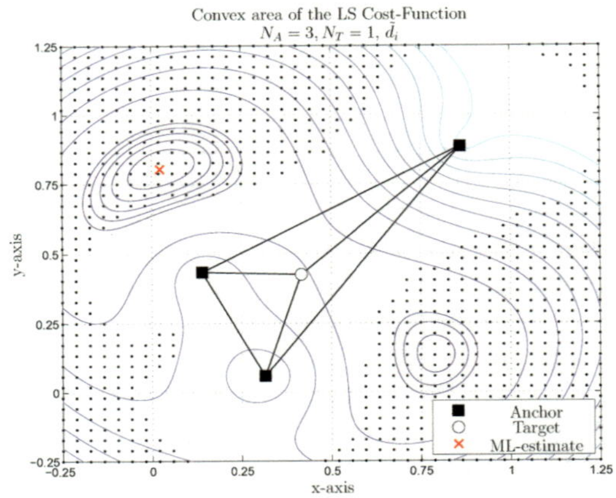

(b) Noisy measurements

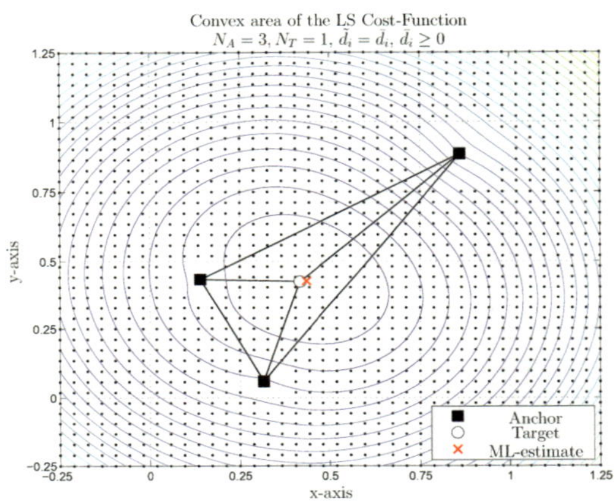

(c) Contracted distances

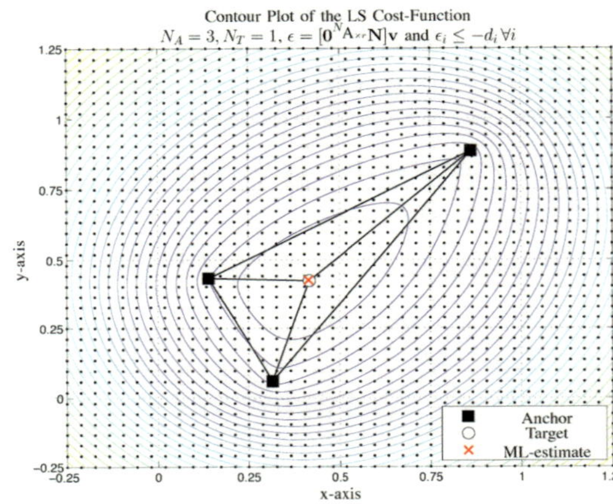

(d) Negative contracted distances

Fig. 5. Illustrative example of the distance contraction theory.

observed, that with these simple contractions the convex area is expanded around the true target location and the minimum is very close to the true location. Finally, in the subfigure 5(d), we show that with negative measurements and structured errors the objective function is convex everywhere and the minimum is exactly the true target's location.

In practice, however, there are several difficulties to deal with while utilizing the aforementioned distance contraction method, and perhaps, the largest one is the estimation of the angle kernel matrix $\mathbf{\Omega}$ especially, if only distance measurements are available.

Nevertheless, suboptimal and yet practical solutions can be considered via simply exploiting the fact that $\tilde{d}_i < d_i$ can improve on the optimization of the WLS-objective function. In Destino & G. (2010), for instance, a distance contraction based method was proposed, and thereby, the idea was to utilize the knowledge of the feasibility region in order to make the contraction. Specifically, the contracted distance of $\tilde{d}_i$, hereafter denoted by $\bar{d}_i$, is found as the distance between the nearest point $\bar{\mathbf{x}}_i \in \mathcal{I}$ to $\mathbf{a}_i$, which can be computed as

$$\bar{\mathbf{x}}_i = \max_{\hat{\mathbf{x}} \in \mathcal{I}} \left(\tilde{d}_i - \|\mathbf{a}_i - \hat{\mathbf{x}}\|_{\mathrm{F}}\right)^2, \tag{27}$$

$$\text{s.t.} \qquad \tilde{d}_i - \hat{d} \geq 0, \forall i.$$

Repeating the optimization above for all anchor-target distance measurements, a set of contracted distances $\bar{d}_i$'s is obtained, which will be utilized in the WLS optimization problem as

$$\hat{\mathbf{z}} = \min_{\hat{\mathbf{z}} \in \mathbb{R}^{\eta}} \sum_{i=1}^{N_{\mathrm{A}}} \left(\bar{d}_i - \hat{d}_i\right)^2. \tag{28}$$

3. Results

In this section we compare the performance of the nearly optimum ML (R-GDC), the SQP and DC algorithms in LOS and mixed LOS/NLOS channel conditions utilizing the LDR-IR UWB ranging model described in section 2. The results will be shown for both cases of a single-hop and a multi-hop network topology.
In particular, we shall measure the localization error

$$\epsilon \triangleq \|\mathbf{P} - \hat{\mathbf{P}}\|_{\mathrm{F}}^2, \tag{29}$$

and study its statistics via the location error probability

$$\Pi \triangleq \Pr\{\epsilon \leq \xi\}. \tag{30}$$

In figures 6 and 7 the results related to the single-hop scenarios are shown. We considered a network with 4 anchor nodes, which are placed at the corner of a square with edge 20 meters. The target is then randomly located within the convex-hull of the anchors.
The results obtained in LOS conditions show that all algorithms have similar performance. To mention some relevant figures, for instance, the localization error is below 40 cm with probability 0.5, and below 1 meter the 90% of the cases.
The performance degrades, although not significantly, when the LOS/NLOS channel conditions are considered. In this case, indeed, the SQP and DC are the best performing methods, which can provide a localization error of 1 and 3 meters with a probability of 0.5 and 0.9, respectively.
The next type of study is the evaluation of the proposed algorithms in a mesh(multi-hop) network. In this regard two approaches are considered, namely a full centralized (cooperative) and a multi-hop (target-centric) methods which are sketched in figures 8(a) and 8(b), respectively. The full centralized approach consists of collecting all information at the anchors and process this information jointly in order to estimate the location of the target. In the target-centric approach, instead, the objective is to exploit only the paths that link each anchor to the target and use those distance measurements to estimate the target's location.

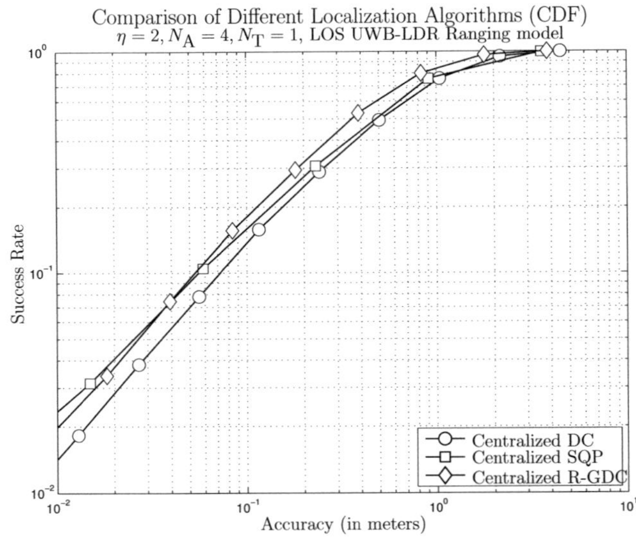

Fig. 6. Comparison of the localization accuracy achieved by different algorithms for the case of a single-hop scenario in LOS conditions.

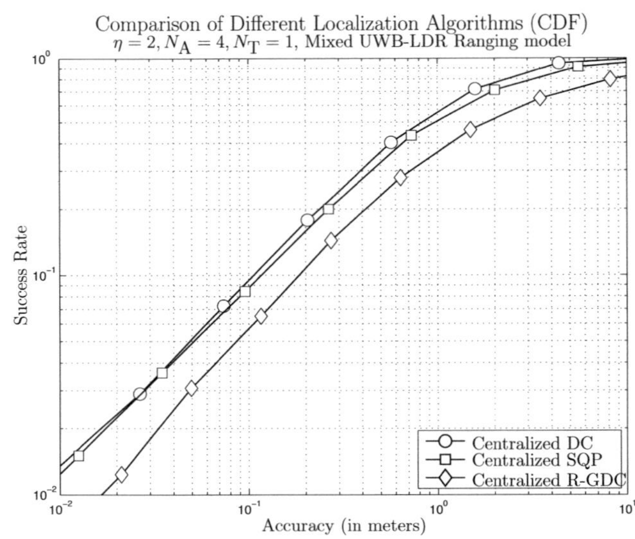

Fig. 7. Comparison of the localization accuracy achieved by different algorithms for the case of a multi-hop scenario in mixed LOS/NLOS conditions.

In figures 9 and 10, the results related to considered studies are shown. In the LOS conditions, the full centralized approach is the best performing method with an average localization accuracy per node of 50cm and 1m with probabilities 0.5 and 0.9, respectively.

In the mixed LOS/NLOS conditions, where distances are affected by bias errors, the SQP and DC are the best performing methods and amongst the latter, the DC can achieve the

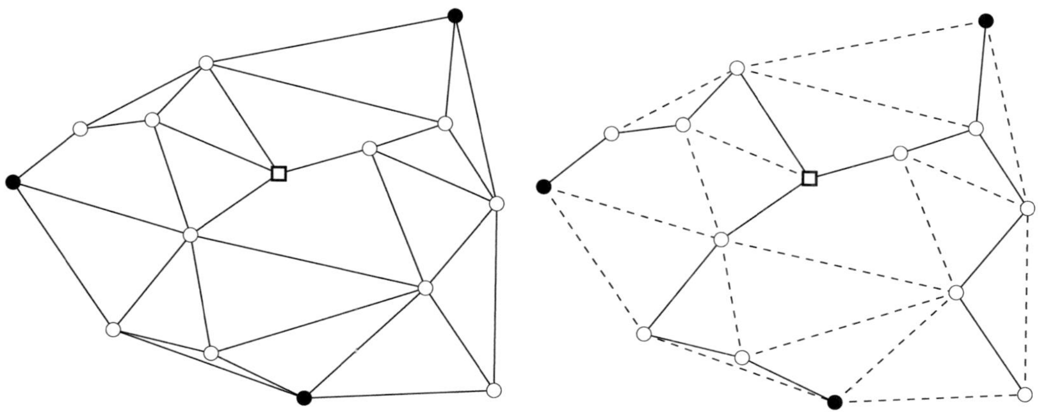

(a) Full Centralized (b) Target Centric

Fig. 8. Illustration of two different approaches for network localization.

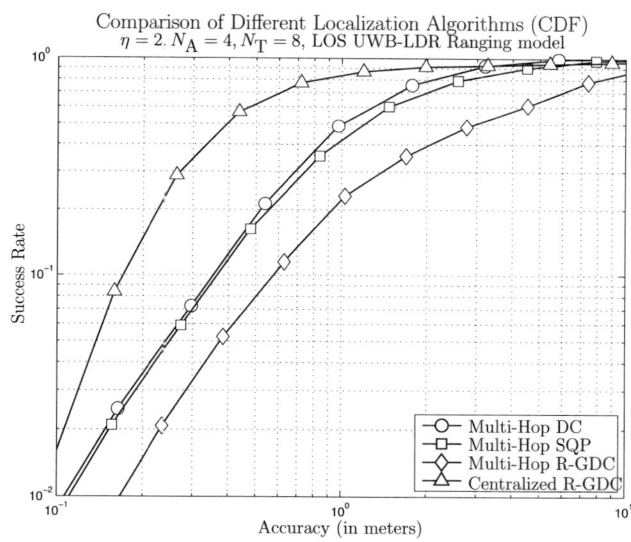

Fig. 9. Comparison of the localization accuracy achieved by different algorithms for the case of a multi-hop scenario in LOS conditions.

highest accuracy. Notice, moreover, that in this simulation set up, the target-centric approach can generally achieve a better accuracy than the centralized one. The reason is that in the target-centric approach minimizes the impact of wrong measurements and poor connectivity onto the localization error since, the problem to be solved is always a "single-hop" type positioning.

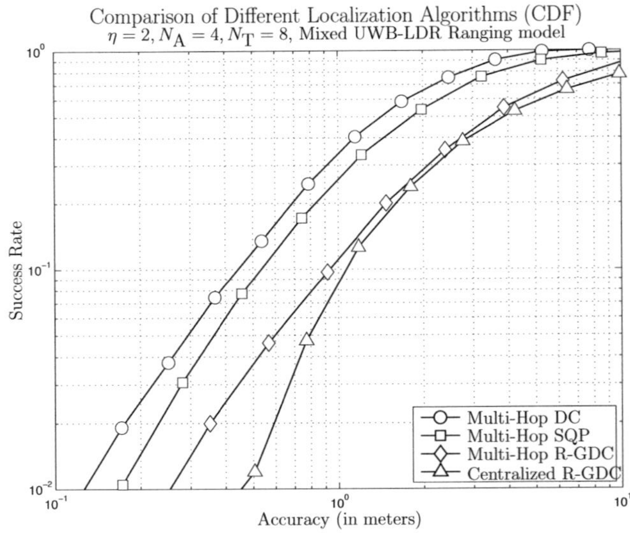

Fig. 10. Comparison of the localization accuracy achieved by different algorithms for the case of a multi-hop scenario in mixed LOS/NLOS conditions.

4. Conclusions

In this chapter, we have seen the most effective optimization-based localization methods described in the literature. We distinguished them in methods for large-scale and single-hop networks. We also addressed the NLOS problem and, we provided effective solutions for the single-hop scenario. In the simulation section, we also described a novel approach for network localization in NLOS conditions, which basically relies on a combination of a multi-hop routing with a single-hop localization method.

It was observed that such a technique can provide accurate location estimates, especially in the case of mixed LOS/NLOS conditions.

5. References

Abramowitz, M. & Stegun, I. A. (1965). *Handbook of Mathematical Functions with Formulas, Graphs, and Mathematical Tables*, 10 edn, Dover Publications.

Biswas, P., Liang, T.-C., Toh, K.-C. & Wang, T.-C. (2006). Semidefinite programming based algorithms for sensor network localization with noisy distance measurements, *ACM Transactions on Sensor Networks* 2(2): 188–220.

Biswas, P., Liang, T.-C., Toh, K.-C., Wang, T.-C. & Ye, Y. (2006). Semidefinite programming approaches for sensor network localization with noisy distance measurements, *IEEE Tansactions on Automation Science and Engineering* 3(4): 360–371.

Boyd, S. & Vandenberghe, L. (2004). *Convex Optimization*, Cambridge University Press.

Costa, J. A., Patwari, N. & Hero, A. O. (2006). Distributed multidimensional scaling with adaptive weighting for node localization in sensor networks, *ACM Journal on Sensor Networks* 2(1): 39–64.

Cox, T. F. & Cox, M. A. A. (2000). *Multidimensional Scaling*, 2 edn, Chapman & Hall/CRC.

Dardari, D., Chong, C. & Win, M. Z. (2008.). Threshold-based time-of-arrival estimators in UWB dense multipath channels, *IEEE Transactions on Communications* 56(8): 1366–1378.

Dattorro, J. (2005). *Convex Optimization and Euclidean Distance Geometry*, Meboo Publishing.

Denis, B. & Daniele, N. (2004). NLOS ranging error mitigation in a distributed positioning algorithm for indoor UWB ad-hoc networks, *Proc. IEEE Intern. Workshop on Wireless Ad-Hoc Netw.*, pp. 356–360.

Denis, B., He, L. & Ouvry, L. (2007). A flexible distributed maximum log-likelihood scheme for UWB indoor positioning, *Proc. IEEE 4th Workshop on Positioning, Navigation and Communication*, pp. 77–86.

Destino, G. & Abreu, G. (2009a). Advanced location-tracking systems in home, automotive and public transportation environments, *IEEE Personal Indoor Mobile Radio Communication*, pp. 1908 – 1912.

Destino, G. & Abreu, G. (2009b). Reformulating the least-square source localization problem with contracted distances, *Proc. IEEE 43th Asilomar Conference on Signals, Systems and Computers*.

Destino, G. & Abreu, G. (2009c). Solving the source localization prolem via global distance continuation, *Proc. IEEE International Conference on Communcations*.

Destino, G. & G., A. (2009). Weighing strategy for network localization under scarce ranging information, *IEEE Transactions on Wireless Communications* 8(7): 3668 – 3678.

Destino, G. & G., A. (2010). Improving source localization in NLOS conditions via ranging contraction, *IEEE Workshop on Positioning, Navigation and Communication*, pp. 56 – 61.

Destino, G., Macagnano, D., de Abreu, G. T. F., Denis, B. & Ouvry, L. (2007). Localization and tracking for LDR-UWB systems, *Proc. IST Mobile & Wireless Communications Summit*.

Ding, Y., Krislock, N., Qian, J. & Wolkowicz, H. (2008). Sensor network localization, euclidean distance matrix completions, and graph realization, *Proc. ACM 1st International workshop on Mobile entity localization and tracking in GPS-less environments*.

Gentile, C. & Kik, A. (2006). An evaluation of ultra wideband technology for indoor ranging, *Proc. IEEE Global Telecommunications Conference (GLOBECOM)*, pp. 1–6.

Gezici, S. (2008). A survey on wireless position estimation, *Wireless Personal Communications* 44(3): 263–282.

Gezici, S., Tian, Z., Giannakis, G., Kobayashi, H., Molisch, A., Poor, H. & Sahinoglu, Z. (2005). Localization via ultra-wideband radios: a look at positioning aspects for future sensor networks, *IEEE Signal Processing Magazine* 22(4): 70–84.

Guvenc, I., Chia-Chin, C. & Watanabe, F. (2007). NLOS identification and mitigation for UWB localization systems, *Proc. IEEE Wireless Comm. and Netw. Conf. (WCNC)*, pp. 1571 – 1576.

Hightower, J. & Borriello, G. (2001). Location systems for ubiquitous computing, *IEEE Computer* 34(8): 57 – 66.

Joon-Yong, L. & Scholtz, R. (2002). Ranging in a dense multipath environment using an UWB radio link., *IEEE Journal on Selected Areas in Communications* 20: 1667–1683.

Mao, G., Fidan, B. & Anderson, B. D. O. (2007). Wireless sensor network localization techniques, *Computer Networks: The Intern. J. of Comp. and Telecomm. Networking* 51(10): 2529–2553.

More, J. & Wu, Z. (1997). Global continuation for distance geometry problems, *SIAM Journal on Optimization* 7: 814–836.

Nocedal, J. & Wright, S. (2006). *Numerical Optimization*, Springer.

Patwari, N., Dea, R. J. O. & Wang, Y. (2003). Relative location estimation in wireless sensor networks, *IEEE Transactions on Signal Processing* 51(8): 2137–2148.

Poslad, S. (2009). *Ubiquitous Computing: Smart Devices, Environments and Interactions*, 978-0-470-03560-3, Wiley.

Schoenberg, I. J. (1935). Remarks to Maurice Frechet's article "Sur la definition axiomatique d'une classe d'espace distances vectoriellement applicable sur l'espace de Hilbert, *The Annals of Mathematics* 36(3): 724–732.

Scott, K. & Benlamri, R. (2010). Context-aware services for smart learning spaces, *IEEE Transactions on Learning Technologies*, 3(3): 214–227.

So, A. M.-C. & Ye, Y. (2005). Theory of semidefinite programming for sensor network localization, *Proc. ACM-SIAM 16th Annual Symposium on Discrete Algorithms*, pp. 405 – 414.

Venkatesh, S. & Buehrer, R. M. (2007). NLOS mitigation using linear programming in ultrawideband location-aware networks, *IEEE Trans. Veh. Technol.* 56(5, Part 2): 3182 – 3198.

Vossiek, M., Wiebking, L., Gulden, P., Wieghardt, J., Hoffmann, C. & Heide, P. (2003). Wireless local positioning, *IEEE Microwave Magazine* 4(4): 77 – 86.

Wymeersch, H., Lien, J. & Win, M. (2009). Cooperative localization in wireless networks, *IEEE Proceedings* 97(2): 427–450.

Yihong, Q., Suda, H. & Kobayashi, H. (2004). On time-of-arrival positioning in a multipath environment, *Proc. IEEE 60th Vehicular Technology Conference (VTC'04 Fall)*, Vol. 5, pp. 3540– 3544.

Yu, K. & Jay Guo, Y. (2008). Improved positioning algorithms for nonline-of-sight environments, *IEEE Trans. Veh. Technol.* 57(4): 2342–2353.